W9-CAL-382

Mercury Hazards to Living Organisms

Dedication

This book is dedicated to my family:

Jeannette, Renée, David, Charles, Julie, and Eb.

Preface

To appreciate the complexities of the element mercury, it is enlightening to review the role of Mercury — the god — in Greek, Roman, and Celtic mythology. Both the element and the mythological god assume complex and ever-changing forms and functions.

In Greek mythology, for example, at the height of Aegean civilization around 1600 BCE, Mercury (or Hermes as the Greeks called him) was known as the god of shepherds and animal fertility (Aldington and Ames, 1968). Later, he was known as the god of travelers, whom he guided on their perilous ways. Mercury's images were placed at crossroads and at forks in the road. A natural extension of this role is that Mercury was also charged with conducting the souls of the dead to the underworld. In primitive times, voyages were usually made for commercial purposes; Mercury was consequently the god of profit — both lawful and unlawful — and the god of games of chance. Because traders required subtle and persuasive words, Mercury also became the god of eloquence. To these various functions, Mercury added that of messenger to Zeus. Not surprisingly, this runner to the gods was honored by athletes and to him was attributed the invention of boxing and racing. The classic aspect of Mercury (Hermes) is that of an athlete god. The early Greeks portrayed him as a mature man with a thick, long beard, his hair bound with a ribbon and falling in curls to his shoulders. Later, he was represented as a young gymnast, with a lithe, graceful body. His hair at that time was short and crisp, and his features fine. He often wore a round, winged hat — a *petasus* — and on his feet were winged sandals. In his hand he holds a winged staff — a *caduceus* — around which serpents are entwined. In Roman mythology, the name Mercury is connected with the root *merx* (merchandise) and *mercari* (to deal, trade). The Roman Mercury appeared about 500 BCE and was exclusively the god of merchants. To portray him, Roman artists drew upon representations of Hermes, portraying him as beardless with a *caduceus*, a purse in his hand, and *petasus* on his head. In Celtic mythology, circa 200 BCE to 800 CE, Mercury was the most popular of all Roman gods to be adopted by the Celts, his role being that of a chieftain-type god, the inventor of all arts, the presiding deity of commerce, and the guardian of travelers (Aldington and Ames, 1968).

The element mercury follows a similar convoluted course. In coastal environments, for example, where toxicological aspects of elemental mercury and mercury compounds have been extensively reviewed and summarized (Eisler, 1981, 2000), most authorities agree on five points. First, forms of mercury with relatively low toxicity can be transformed into forms with very high toxicity through biological and other processes. Second, uptake of mercury directly from seawater or through biomagnification in marine food chains returns mercury directly to humans in concentrated forms. Third, mercury uptake may result in genetic changes. Fourth, naturally elevated levels of mercury in some marine fishes — such as tuna, swordfish, and marlin — emphasize the complexity of both natural mercury cycles and anthropogenic impact on those cycles. Finally, human use of mercury needs to be curtailed because — in contrast to many other contaminants — the difference between tolerable natural background levels of mercury and harmful effects in the environment is exceptionally small.

REFERENCES

Aldington, R. and D. Ames (translators). 1968. *New Larousse Encyclopedia of Mythology.* Hamlyn Publishers, New York, 500 pp.

Eisler, R. 1981. *Trace Metal Concentrations in Marine Organisms.* Pergamon, Elmsford, NY. 687 pp.

Eisler, R. 2000. Mercury. *Handbook of Chemical Risk Assessment: Health Hazards to Humans, Plants, and Animals. Vol. 1, Metals*, p. 313–409. Lewis Publishers, Boca Raton, FL.

Acknowledgments

For expediting access to the technical literature, I thank the librarians and staff of the National Library of Medicine in Bethesda, Maryland, and the National Agricultural Library in Beltsville, Maryland. I am also indebted to numerous colleagues who furnished me with current reprints and recommendations for additional technical sources. The editorial input of Gail Renard and other Taylor & Francis staff is gratefully acknowledged.

About the Author

Ronald Eisler received the B.A. degree from New York University in biology and chemistry, and M.S. and Ph.D. degrees from the University of Washington. On his retirement from federal service as a senior research biologist in 2004, he was a recognized national and international authority on chemical risk assessment to wildlife and other natural resources. In his 45-year federal career, he served with the U.S. Department of the Interior in New Jersey, Maryland, and Washington D.C., the U.S. Environmental Protection Agency in Rhode Island, and the U.S. Army Medical Service Corps in Colorado. He has conducted research and monitoring studies on pollution issues in the Pacific Northwest, the Territory of Alaska, Colorado, the Marshall and Marianas Islands, all along the eastern seaboard of the U.S. Atlantic coast, the Adirondacks Region of New York, the Red Sea, and the Gulf of Mexico. Since 1955 he has authored more than 145 technical articles — including five books and 16 book chapters — on contaminant hazards to plants, animals, and human health. He has held several adjunct professor appointments and taught for extended periods at the Graduate School of Oceanography of the University of Rhode Island, and the Department of Biology of American University in Washington, D.C. He also served as Visiting Professor and Resident Director of Hebrew University's Marine Biology Laboratory located in Eilat, Israel. In retirement, he actively consults and writes on chemical risk assessment.

Eisler resides in Potomac, Maryland, with his wife Jeannette, a teacher of French and Spanish.

BOOKS BY RONALD EISLER

2004. *Biogeochemical, Health, and Ecotoxicological Perspectives on Gold and Gold Mining*. CRC Press, Boca Raton, FL. 356 pp.

2000. *Handbook of Chemical Risk Assessment: Health Hazards to Humans, Plants, and Animals. Vol. 1, Metals; Vol. 2, Organics; Vol. 3, Metalloids, Radiation, Cumulative Index to Chemicals and Species*. Lewis Publishers, Boca Raton, FL. 1903 pp.

1981. *Trace Metal Concentrations in Marine Organisms*. Pergamon Press, Elmsford, NY. 687 pp.

List of Tables

List of Figures

Contents

Part 1 Mercury Sources, Uses, Properties, Transport, Poisoning Mechanisms, and Treatment

Part 2 Mercury Concentrations in Field Collections of Abiotic Materials, Plants, and Animals

Part 3 Lethal and Sublethal Effects of Mercury under Controlled Conditions

Part 4 Case Histories

Part 5 Proposed Mercury Criteria, Concluding Remarks

PART 1

Mercury Sources, Uses, Properties, Transport, Poisoning Mechanisms, and Treatment

Chapter 1

Introduction

The element mercury, also known as quicksilver (symbol Hg for *hydrargyrum*), and its compounds have no known normal metabolic function. Their presence in the cells of living organisms represents contamination from natural and anthropogenic sources; all such contamination must be regarded as undesirable and potentially hazardous (U.S. National Academy of Sciences [USNAS] 1978). Accumulation of mercury in tissues is reportedly associated with an excess risk of myocardial infarction (Salonen et al., 1995; Gualler et al., 2002), increased risk of death from coronary heart disease and cardiovascular disease (Salonen et al., 1995), and accelerated progression of carotid atherosclerosis (Salonen et al., 2000).

The most important ore of mercury, cinnabar (mercuric sulfide), has been mined continuously since 415 BCE (Clarkson and Marsh, 1982). In the period before the industrial revolution, mercury was used extensively in gold extraction and in the manufacture of felt hats and mirrors; in the 1800s, it was used in the chloralkali industry, in the manufacture of electrical instruments, and as a medical antiseptic; and since 1900, it has been used in pharmaceuticals, in agricultural fungicides, in the pulp and paper industry as a slimicide, and in the production of plastics (Clarkson and Marsh, 1982). World use of mercury is estimated at 10,000 to 15,000 metric tons annually (Boudou and Ribeyre, 1983), of which the United States accounts for about 18.0% (Clarkson and Marsh, 1982). Today, mercury is a leading public health concern, as judged by the increases in regulations governing mercury emissions, in mercury fish advisories, in clinical studies, and in media attention (Kales and Goldman, 2002; Hightower and Moore, 2003; Weil et al., 2005).

The first cases of fatal inorganic mercury poisoning in humans were reported for two men in a European chemical laboratory in 1865, and the first documented human poisoning from agricultural exposure to an organomercury compound was in 1940 (Das et al., 1982). Human exposure to mercury compounds via dermal, dietary, and respiratory routes has severe consequences. For example, 1600 infants in Argentina showed symptoms of mercury poisoning after a laundry treated their diapers with a mercury disinfectant; numerous poisonings resulted from ingestion of mercury-contaminated fish, pork, seafoods, and grains; and from occupational exposure in Nicaraguan mercury fungicide applicators via respiration (Elhassani, 1983). Sporadic incidences of human poisonings have occurred in the United States, the Soviet Union, and Canada; while major epidemics have been reported in Japan, Pakistan, Guatemala, Ghana, Yugoslavia, and Iraq (Bakir et al., 1973; Clarkson and Marsh, 1982; Das et al., 1982; Elhassani, 1983; Greener and Kochen, 1983). In Iraq, for example, a major mercury poisoning occurred in the early 1970s, with 6530 hospital admissions and 459 hospital deaths (Bakir et al., 1973; Amin-Zaki et al., 1974; Marsh et al., 1977, 1987; Das et al., 1982; Elhassani, 1983; Choi et al., 1978; Inouye, 1989). In 1970, the Iraqi government decided to import seed grain from Mexico, owing to a serious wheat shortage throughout Iraq. About 73,200 metric tons of seed wheat were imported and distributed between September and November of 1971. The seed had been treated with a methylmercury fungicide and colored with a red dye; it was delivered in 50-kg sacks with a warning written in Spanish. A skull and crossbones was painted on the outside of each sack to indicate that the seed had been treated with poison. However, both Spanish and the skull and crossbones insignia were totally unfamiliar to the Iraqi farmers and the people decided to make home-made bread instead of using the grain for seed; about half the wheat was consumed for this purpose. The water-soluble red dye was washed off the wheat, with the assumption that the mercury would be equally soluble (it was not). Before human consumption of the wheat, the farmers fed the wheat for a few days to chickens and other livestock without apparent effect; it was not realized that a lengthy latency period was involved. The wheat flour contained 4.8 to 14.6 mg methylmercury/kg, far in excess of the 0.5 mg Hg/kg regulatory limit. Consumption of the bread lasted 1 to 2 months, and the first cases of poisoning appeared at the end of December 1971. In 1972, there were 6530 reported cases within 18 months, including 459 deaths, among Iraqi farmers who ate bread made from seed wheat treated with a methylmercury fungicide. Babies congenitally affected by methylmercury born in Iraq during this outbreak showed clinical features of mental retardation, motor disorders resembling cerebral palsy, and other physical and mental disturbances. Histopathological examination of the brain of victims in cases of death showed degenerative neuronal damage and abnormal development of the cerebral cortex. There was no effective antidote to counteract the effects of methylmercury on the central nervous system.

At Oak Ridge, Tennessee, liquid metallic mercury was used during the 1950s and 1960s for uranium processing in a nuclear weapons plant (Silver et al., 1994). Accidental releases of mercury of about 100 tons — equivalent to that released at Minamata Bay, Japan, in which hundreds died — resulted in heavy accumulations of mercury in the waste pipes and as a "lake of Hg^o" of about 7000 L under the plant sites. However, unlike Minamata, Japan (which is discussed extensively in Chapter 10), there have been no documented human health problems at the Oak Ridge site (Silver et al., 1994). The mercury at Oak Ridge is released into local waters at about 75 kg annually, at which rate it will take more than 1000 years to disappear, although microbial activities transform the mercury found in local waters within hours (Barkay et al., 1991).

Mercury contamination is the most serious environmental threat to fishery and wildlife resources in the southeastern United States, with fish consumption advisories issued in the ten states comprising this region (Facemire et al., 1995). In March 1989, the Florida Department of Health and Rehabilitative Services issued a health advisory prohibiting consumption of predatory fishes, such as largemouth bass (*Micropterus salmoides*), bowfin (*Amia calva*), and gar (*Lepisosteus* spp.) in southern Florida, and the entire Everglades watershed has been closed to hunting of alligators (*Alligator mississippiensis*) due to excessive mercury in edible tissues of alligators (Roelke et al., 1991; Sundolf et al., 1994; Kannan et al., 1998). Methylmercury concentrations of 2.0 to 3.0 mg/kg fresh weight (FW) muscle were documented in largemouth bass and other fish-eating species (Fleming et al., 1995), or 4 to 10 times the allowable limit in foods for human consumption (Eisler, 2000, 2004a). About 0.91 million ha of the South Florida Everglades ecosystem are currently under fish consumption advisories because of mercury contamination (Atkeson et al., 2003).

The number of mercury-contaminated fish and wildlife habitats has, in general, progressively increased worldwide, almost all as a direct result of anthropogenic activities (Boudou and Ribeyre, 1983). Long-range atmospheric transport of mercury since 1940 has resulted in elevated mercury loadings in remote Canadian lakes up to 1400 km from the closest industrial centers (Lucotte et al., 1995). Since 1985, annual mercury accumulation rates in flooded Florida Everglades soils averaged 53.0 $\mu g/m^2$, which is about 4.9 times greater than rates observed around the turn of the century; the increase is attributed to increased global and regional deposition and is similar to increases reported for lakes in Sweden and the northern United States (Rood et al., 1995). In 1967, the Swedish medical board banned the sale of fish that contained high concentrations of organomercury salts, originating from about 40 lakes and rivers (Das et al., 1982). More than 10,000 Swedish lakes have been closed to fishing because of excessive mercury loadings (Lindqvist et al., 1991). In 1970, after the discovery of high levels of mercury in fish from Lake St. Clair, Canada, restrictions on fishing and the sale of fish were imposed in many areas of the United States and Canada (Das et al., 1982). Since 1970, a total of 26 of the 48 states in the conterminous United States have reported mercury pollution in their waters as a direct result of human activities. These states have banned sport or commercial fishing in mercury-contaminated waters, or have issued health warnings about the consequences of eating mercury-contaminated fish or seafood from selected water courses, or have placed restrictions on fish consumption from certain streams, lakes, or rivers polluted with mercury (USNAS, 1978; Becker and Bigham, 1995). At present, more than 40 states have issued health advisories restricting fish consumption based on unacceptable mercury contents; similar advisories have been issued in North America, Europe, and Asia (Atkeson et al., 2003). Fish consumption advisories on mercury issued by the U.S. Food and Drug Administration (FDA) are effective in reducing fish consumption; in Massachusetts; for example, there was an average monthly decline in fish consumption of 1.4 servings among pregnant women following an FDA advisory on the health risks of mercury (Oken et al., 2003).

Poisoning of game birds and other wildlife in Sweden, apparently by seeds treated with organomercurials, was first noted in 1960 (Das et al., 1982). Massive kills of the grey heron (*Ardea cinerea*) in the Netherlands during 1976 were attributed to a combination of low temperatures, undernourishment, and high body burdens of mercury (Van der Molen et al., 1982). Mercury contamination has resulted in the closure of pheasant and partridge hunting areas in Alberta, Canada (Mullins et al., 1977). Declining numbers of wading birds in southern Florida are attributed to mercury contamination of their food supply (Sundolf et al., 1994).

Most authorities on mercury hazards to living organisms now agree on six points. First, mercury and its compounds have no known biological function, and its presence in living organisms is undesirable and potentially hazardous. Second, forms of mercury with relatively low toxicity can be transformed into forms with very high toxicity through biological and other processes. Third, methylmercury can be bioconcentrated in organisms and biomagnified through food chains, returning mercury directly to human and other upper trophic level consumers in concentrated form. Fourth, mercury is a mutagen, teratogen, and carcinogen, and causes embryocidal, cytochemical, and histopathological effects. Fifth, high body burdens of mercury normally encountered in some species of fish and wildlife from remote locations emphasize the complexity of natural mercury cycles and human impacts on these cycles. And finally, the anthropogenic use of mercury should be curtailed because the difference between tolerable natural background levels of mercury and harmful effects in the environment is exceptionally small. These, and other aspects of mercury and its compounds in the environment as a result of anthropogenic or natural processes, are the subject of many reviews.*

The current book builds extensively on these reviews, emphasizing recent and credible information on mercury uses and sources to the environment; biological, physical, and chemical properties of mercury and its compounds; documentation and significance of mercury concentrations in abiotic materials, plants, invertebrates, amphibians, reptiles, elasmobranchs, fishes, birds, and humans, and other mammals; lethal and sublethal effects of inorganic and organic mercury compounds to plants, animals, and humans; biological, chemical, and physical variables known to

* Montague and Montague (1971), D'Itri (1972), Friberg and Vostal (1972), Hartung and Dinman (1972), Jernelov et al. (1972, 1975), Keckes and Miettinen (1972), Buhler (1973), Eisler (1973, 1978, 1981, 1987, 2000, 2004a, 2004b), Holden (1973), Eisler and Wapner (1975), Smith and Smith (1975), World Health Organization [WHO] (1976, 1989, 1990, 1991), D'Itri and D'Itri (1977), U.S. National Academy of Sciences [USNAS] (1978), Eisler et al. (1978, 1979), Beijer and Jernelov (1979), Birge et al. (1979), Magos and Webb (1979), Nriagu (1979), Reuhl and Chang (1979), Takizawa (1979), U.S. Environmental Protection Agency [USEPA] (1980, 1985, 1997), Jenkins (1980), Clarkson and Marsh (1982), Das et al. (1982), Boudou and Ribeyre (1983), Elhassani (1983), Clarkson et al. (1984), Robinson and Touvinen (1984), Wren (1986), Scheuhammer (1987), Clarkson (1990), Lindqvist (1991), Lindqvist et al. (1991), Zillioux et al. (1993), U.S. Public Health Service [USPHS] (1994), Watras and Huckabee (1994), Hamasaki et al. (1995), Porcella et al. (1995), Heinz (1996), Thompson (1996), Wiener and Spry (1996), Sigel and Sigel (1997), De Lacerda and Salomons (1998), Wolfe et al. (1998), Ebinghaus et al. (1999), Boening (2000), National Research Council [NRC] (2000), Takizawa and Osame (2001), and Wiener et al. (2003).

modify toxicity and accumulation; in-depth analysis of case histories of mercury poisoning in Minamata, Japan, and of current (Brazil) and historical (United States) use of mercury in gold extraction; listing of proposed mercury criteria for the protection of natural resources and human health, together with a critical risk assessment of these criteria; recommendations for additional research; and concluding remarks. A general index and a biological species index are also included.

REFERENCES

Amin-Zaki, L., S. Elhassani, M.A. Majeed, T.W. Clarkson, R.A. Doherty, and M. Greenwood. 1974. Intra-uterine methylmercury poisoning in Iraq, *Pediatrics*, 54, 587–595.

Atkeson, T., D. Axelrad, C. Pollman, and G. Keeler. 2003. *Integrating Atmospheric Mercury Deposition and Aquatic Cycling in the Florida Everglades: An Approach for Conducting a Total Maximum Daily Load Analysis for an Atmospherically Derived Pollutant. Integrated Summary. Final Report.* 272 pp. Available from Mercury and Applied Science MS6540, Florida Dept. Environmental Conservation, 2600 Blair Stone Road, Tallahassee, FL 32399-2400. See also <http://www.floridadep.org/labs/mercury/index.htm>

Bakir, F., S.F. Damluji, L. Amin-Zaki, M. Murthada, A. Khalidi, N.Y. Al-Rawi, S. Tikriti, H.I. Dahir, T.W. Clarkson, J.C. Smith, and R.A. Doherty. 1973. Methylmercury poisoning in Iraq, *Science*, 181, 230–241.

Barkay, T., R. Turner, A. VandenBrook, and C. Liebert. 1991. The relationships of Hg (II) volatilization from a freshwater pond to the abundance of *mer* genes in the gene pool of the indigenous microbial community, *Microbial Ecology*, 21, 151–161.

Becker, D.S. and G.N. Bigham. 1995. Distribution of mercury in the aquatic food web of Onondaga Lake, New York, *Water Air Soil Pollut.*, 80, 563–571.

Beijer, K. and A. Jernelov. 1979. Methylation of mercury in natural waters. I J.0. Nriagu (Ed.), *The Biogeochemistry of Mercury in the Environment*, p. 201–210. Elsevier/North-Holland Biomedical Press, New York.

Birge, W.J., J.A. Black, A.G. Westerman, and J.E. Hudson. 1979. The effect of mercury on reproduction of fish and amphibians. In J.0. Nriagu (Ed.), *The Biogeochemistry of Mercury in the Environment*, p. 629–855t. Elsevier/North-Holland Biomedical Press, New York.

Boening, D.W. 2000. Ecological effects, transport, and fate of mercury: a general review, *Chemosphere*, 40, 1335–1351.

Boudou, A. and F. Ribeyre. 1983. Contamination of aquatic biocenoses by mercury compounds: an experimental toxicological approach. Pages 73–116 in J.0. Nriagu (ed.). *Aquatic Toxicology*. John Wiley, NY.

Buhler, D.R. (Ed.). 1973. *Mercury in the western environment. Proceedings of a workshop, Portland, Oregon, February 25–26, 1971.* Oregon State University, Corvallis. 360 pp.

Choi, B.H., L.W. Lapham, L. Amin-Zaki, and T. Saleem. 1978. Abnormal neuronal migration, deranged cerebral cortical organization, and diffuse white matter astrocytosis of human fetal brain: a major effect of methylmercury poisoning *in utero, J. Neuropathol. Exp. Neurol.*, 37, 719–733.

Clarkson, T.W. 1990. Human health risks from methylmercury in fish, *Environ. Toxicol. Chem.*, 9, 957–961.

Clarkson, T.W., R. Hamada, and L. Amin-Zaki. 1984. Mercury. In J.O. Nriagu (Ed.), *Changing Metal Cycles and Human Health*, p. 285–309. Springer-Verlag, Berlin.

Clarkson, T.W. and D.O. Marsh. 1982. Mercury toxicity in man. In A.S. Prasad (Ed.), *Clinical, Biochemical, and Nutritional Aspects of Trace Elements,* Vol. 6, p. 549–568. Alan R. Liss, Inc., New York.

Das, S.K., A. Sharma, and G. Talukder. 1982. Effects of mercury on cellular systems in mammals — a review, *Nucleus (Calcutta)*, 25, 193–230.

De Lacerda, L.D. and W. Salomons. 1998. *Mercury from Gold and Silver Mining: A Chemical Time Bomb?.* Springer, Berlin, 146 pp.

D'Itri, F.M. 1972. *Mercury in the aquatic ecosystem.* Tech. Rep. 23, Inst. Water Res., Michigan State University, Lansing. 101 pp.

D'Itri, P. and F.M. D'Itri. 1977. *Mercury Contamination: A Human Tragedy.* John Wiley, New York. 311 pp.

Ebinghaus, R., R.R. Turner, S.D. Lacerda, O. Vasiliev, and W. Salomons (Eds.). 1999. *Mercury Contaminated Sites*. Springer, Berlin.

Eisler, R. 1973. Annotated bibliography on biological effects of metals in aquatic environments (No. 1-567), *U.S. Environ. Protect. Agen., Ecol. Res. Ser.*, Rept. EPA-R3-73-007, 1–287.

Eisler, R. 1978. Mercury contamination standards for marine environments. In J.H. Thorp and J.W. Gibbons (Eds.), *Energy and Environmental Stress in Aquatic Systems*, p. 241–272. U.S. Dept. Energy Symp. Ser. 48, CONF-771114. Available from Natl. Tech. Infor. Serv., U.S. Dept. Commerce, Springfield, VA 22161.

Eisler, R. 1981. *Trace Metal Concentrations in Marine Organisms.* Pergamon, Elmsford, NY. 687 pp.

Eisler, R. 1987. Mercury hazards to fish, wildlife, and invertebrates: a synoptic review, *U.S. Fish Wildl. Serv. Biol. Rep.*, 85(1.10), 1–90.

Eisler, R. 2000. Mercury. In *Handbook of Chemical Risk Assessment: Health Hazards to Humans, Plants, and Animals,* Vol. 1, Metals, p. 313–409. Lewis Publishers, Boca Raton, FL.

Eisler, R. 2004a. Mercury hazards from gold mining to humans, plants, and animals, *Rev. Environ. Contam. Toxicol.*, 181, 139–198.

Eisler, R. 2004b. *Biogeochemical, Health, and Ecotoxicological Perspectives on Gold and Gold Mining*. CRC Press, Boca Raton, FL. 355 pp.

Eisler, R., D.J. O'Neill Jr., and G.W. Thompson. 1978. Third annotated bibliography on biological effects of metals in aquatic environments (No. 1293-2246), *U.S. Environ. Protect. Agen., Ecol. Res. Ser.*, Rept. 600/3-78-005, 1–487.

Eisler, R., R.M. Rossoll, and G.A. Gaboury. 1979. Fourth annotated bibliography on biological effects of metals in aquatic environments (No. 2247-3132), *U.S. Environ. Protect. Agen., Ecol. Res. Ser.*, Rept. 600/3-79-085, 1–592.

Eisler, R. and M. Wapner. 1975. Second annotated bibliography on biological effects of metals in aquatic environments (No. 568-1292), *U.S. Environ. Protect. Agen., Ecol. Res. Ser.*, Rept. EPA-600/3-75-008, 1–400.

Elhassani, S.B. 1983. The many faces of methylmercury poisoning, *J. Toxicol.*, 19, 875–906.

Facemire, C., T. Augspurger, D. Bateman, M. Brim, P. Conzelmann, S. Delchamps, E. Douglas, L. Inmon, K. Looney, F. Lopez, G. Masson, D. Morrison, N. Morse, and A. Robison. 1995. Impacts of mercury contamination in the southeastern United States, *Water Air Soil Pollut.*, 80, 923–926.

Fleming, L.E., S. Watkins, R. Kaderman, B. Levin, D.R. Ayyar, M. Bizzio, D. Stephens, and J.A. Bean. 1995. Mercury exposure to humans through food consumption from the Everglades of Florida, *Water Air Soil Pollut.,* 80, 41–48.

Friberg, L. and J. Vostal (Eds.). 1972. *Mercury in the Environment.* CRC Press, Boca Raton, FL. 215 pp.

Greener, Y. and J.A. Kochen. 1983. Methyl mercury toxicity in the chick embryo, *Teratology*, 28, 23–28.

Guallar, E., M.I. Sanz-Gallardo, P.V. Veer, P. Bode, A. Aro, J. Gomez-Arcena, J.D. Kark, R.A. Riemersma, J.M. Martin-Moreno, and F.J. Kok. 2002. Mercury, fish oils, and the risk of myocardial infarction, *New Engl. J. Med.*, 347, 1747–1754.

Hamasaki, T., H. Nagawe, Y. Yoshioka, and T. Sato. 1995. Formation, distribution, and ecotoxicity of methylmetals of tin, mercury, and arsenic in the environment, *Crit. Rev. Environ. Sci. Technol.*, 25, 45–91.

Hartung, R. and E.D. Dinman (Eds.). 1972. *Environmental Mercury Contamination.* Ann Arbor Press, Ann Arbor, MI.

Heinz, G.H. 1996. Mercury poisoning in wildlife. In A. Fairbrother, L.N. Locke, and G.L. Hoff (Eds.), *Noninfectious Diseases of Wildlife, 2nd edition*, p. 118–127. Iowa State Univ. Press, Ames.

Hightower, J.M. and D. Moore. 2003. Mercury levels in high-end consumers of fish, *Environ. Health Perspect.*, 111, 604–608.

Holden, A.V. 1973. Mercury in fish and shellfish, a review, *Jour. Food Technol.*, 8, 1–25.

Inouye, M. 1989. Teratology of heavy metals: mercury and other contaminants, *Cong. Anom.*, 29, 333–344.

Jenkins, D.W. 1980. In *Biological Monitoring of Toxic Trace Metals, Vol. 2, Toxic Trace Metals in Plants and Animals of the World, Part II, Mercury*, p. 779–982. U.S. Environ. Protection Agency, Rep. 600/3-80-091.

Jernelov, A., R. Hartung, P.B. Trost, and R.E. Bisque. 1972. Environmental dynamics of mercury. Pages 167-201 In R. Hartung and B.D. Dinman (Eds.), *Environmental Mercury Contamination,* p. 167–201. Ann Arbor Science Publ., Ann Arbor, MI.

Jernelov, A., L. Landner, and T. Larsson. 1975. Swedish perspectives on mercury pollution, *J. Water Pollut. Control Fed.*, 47, 810–822.

Kales, S.N. and R.H. Goldman. 2002. Mercury exposure: current concepts, controversies, and a clinic's experience, *J. Occup. Environ. Med.*, 44, 143–154.

Kannan, K., R.G. Smith Jr., R.F. Lee, H.L. Windom, P.T. Heitmuller, J.M. Macauley, and J.K. Summers. 1998. Distribution of total mercury and methyl mercury in water, sediment, and fish from south Florida estuaries, *Arch. Environ. Contam. Toxicol.*, 34, 109–118.

Keckes, S. and J.J. Miettinen. 1972. Mercury as a marine pollutant. In M. Ruivo (Ed.), *Marine Pollution and Sea Life*, p. 276–289. Fishing Trading News (books) Ltd., London.

Lindqvist, O. (Ed.). 1991. Mercury as an environmental pollutant. *Water Air Soil Pollut.,* 56, 1–847.

Lindqvist, O., K. Johansson, M. Aastrup, A. Andersson, L. Bringmark, G. Hovsenius, L. Hakanson, A. Iverfeldt, M. Meili, and B. Timm. 1991. Mercury in the Swedish environment — recent research on causes, consequences and corrective methods. *Water Air Soil Pollut.*, 55, 1–261.

Lucotte, M., A. Mucci, C. Hillaire-Marcel, P. Pichet, and A. Grondin. 1995. Anthropogenic mercury enrichment in remote lakes of northern Quebec (Canada), *Water Air Soil Pollut.*, 80, 467–476.

Magos, L. and M. Webb. 1979. Synergism and antagonism in the toxicology of mercury. In J.O. Nriagu (Ed.), *The Biogeochemistry of Mercury in the Environment*, p. 581–599. Elsevier/North-Holland Biomedical Press, New York.

Marsh, D.O., T.W. Clarkson, C. Cox, G.J. Myers, L. Amin-Zaki, and S. Al-Tikriti. 1987. Fetal methylmercury toxicity: relationship between concentration in single strands of maternal hair and child effects, *Arch. Neurol.*, 44, 1017–1022.

Marsh, D.O., G.J. Myers, T.W. Clarkson, L. Amin-Zaki, and S. Al-Tikriti. 1977. Fetal methylmercury toxicity: new data on clinical and toxicological aspects. *Trans. Ar. Neurol. Assoc.*, 102, 1–3.

Montague, K. and P. Montague. 1971. *Mercury.* Sierra Club, NY. 158 pp.

Mullins, W.H., E.G. Bizeau, and W.W. Benson. 1977. Effects of phenyl mercury on captive game farm pheasants, *J. Wildl. Manage.*, 41, 302–308.

National Research Council (NRC). 2000. *Toxicological Effects of Methylmercury.* National Academy of Sciences, National Academy Press, 368 pp.

Nriagu, J.O. (Ed.). 1979. *The Biogeochemistry of Mercury in the Environment.* Elsevier/North-Holland Biomedical Press, New York. 696 pp.

Oken, E., K.P. Kleinman, W.E. Berland, S.R. Simon, J.W. Rich-Edwards, and M.W. Sillman. 2003. Decline in fish consumption among pregnant women after a national mercury advisory, *Obstet. Gynecol.*, 102, 346–351.

Porcella, D., J. Huckabee, and B. Wheatley (Eds.). 1995. Mercury as a global pollutant. *Water Air Soil Pollut.,* 80, 1–1336.

Reuhl, K.R. and L.W. Chang. 1979. Effects of methylmercury on the development of the nervous system: a review, *NeuroToxicology*, 1, 21–55.

Robinson, J.B. and O.H. Touvinen. 1984. Mechanisms of microbial resistance and detoxification of mercury and organomercury compounds: physiological, biochemical and genetic analysis, *Microbiol. Rev.*, 48, 95–124.

Roelke, M.E., D.P. Schultz, C.F. Facemire, and S.F. Sundlof. 1991. Mercury contamination in the free-ranging endangered Florida panther (*Felis concolor coryi*), *Proc. Am. Assoc. Zoo Veterin.*, 1991, 277–283.

Rood, B.E., J.F. Gottgens, J.J. Delfino, C.D. Earle, and T.L. Crisman. 1995. Mercury accumulation trends in Florida Everglades and Savannas marsh flooded soils, *Water Air Soil Pollut.*, 80, 981–990.

Salonen, J.T., K. Seppanen, T.a. Lakka, R. Salonen, and G.A. Kaplan. 2000. Mercury accumulation and accelerated progression of carotid atherosclerosis: a population-based prospective 4-year follow-up study in men in eastern Finland, *Atherosclerosis*, 148, 265–273.

Salonen, J.T., K. Seppanen, K. Nyyssonen, H. Korpela, J. Kauhanen, M. Kantola, J. Tuomilehto, H. Esterbauer, F. Tatzber, and R. Salonen. 1995. Intake of mercury from fish, lipid peroxidation, and the risk of myocardial infarction and coronary, cardiovascular, and any death in eastern Finnish men, *Circulation*, 91, 645–655.

Scheuhammer, A.M. 1987. The chronic toxicity of aluminium, cadmium, mercury, and lead in birds: a review, *Environ. Pollut.*, 46, 263-295.

Sigel, A. and H. Sigel (Eds.). 1997. *Metal Ions in Biological Systems, Vol. 34, Mercury and Its Effects on Environment and Biology.* Marcel Dekker, New York.

Silver, S., G. Endo, and K. Nakamura. 1994. Mercury in the environment and the laboratory, *J. Japan. Soc. Water Environ.*, 17, 235–243.

Smith, W.E. and A.M. Smith. 1975. *Minamata.* Holt, Rinehart, and Winston, New York. 192 pp.

Sundolf, S.F., M.G. Spalding, J.D. Wentworth, and C.K. Steible. 1994. Mercury in livers of wading birds (Ciconiiformes) in southern Florida, *Arch. Environ. Contam. Toxicol.*, 27, 299–305.

Takizawa, Y. 1979. Epidemiology of mercury poisoning. In J.O. Nriagu (Ed.), *The Biogeochemistry of Mercury in the Environment*, p. 325–365. Elsevier/North Holland Biomedical Press, Amsterdam.

Takizawa, Y. and M. Osame (Eds.). 2001. *Understanding of Minamata Disease: Methylmercury Poisoning in Minamata and Niigata, Japan.* Japan Public Health Assoc., Tokyo.

Thompson, D.R. 1996. Mercury in birds and terrestrial mammals. In W.N. Beyer, G.H. Heinz, and A.W. Redmon-Norwood (Eds.), *Environmental Contaminants in Wildlife: Interpreting Tissue Concentrations*, p. 341–356. CRC Press, Boca Raton, FL.

U.S. Environmental Protection Agency (USEPA). 1980. Ambient water quality criteria for mercury, *U.S. Environ. Protection Agen. Rep. 440/5-80-058.* Available from Natl. Tech. Infor. Serv., 5285 Port Royal Road, Springfield, VA 22161.

U.S. Environmental Protection Agency (USEPA). 1985. Ambient water quality criteria for mercury — 1984. *U.S. Environ. Protection Agen. Rep. 440/5-84-026.* 136 pp. Available from Natl. Tech. Infor. Serv., 5285 Port Royal Road, Springfield, VA 22161.

U.S. Environmental Protection Agency (USEPA). 1997. Mercury study report to Congress, *U.S. Environ. Protection Agen. Publ. 452R-97-004.* Washington, D.C.

U.S. National Academy of Sciences (USNAS). 1978. *An Assessment of Mercury in the Environment.* Natl. Acad. Sci., Washington, D.C., 185 pp.

U.S. Public Health Service (USPHS). 1994. *Toxicological profile for mercury (update), TP-93/10.* U.S. PHS, Agen. Toxic Substances Dis. Registry, Atlanta, GA. 366 pp.

Van der Molen, E.J., A.A. Blok, and G.J. De Graaf. 1982. Winter starvation and mercury intoxication in grey herons (*Ardea cinerea*) in the Netherlands, *Ardea*, 70, 173–184.

Watras, C.J. and J.W. Huckabee (Eds.). 1994. *Mercury Pollution: Integration and Synthesis.* Lewis Publishers, Boca Raton, FL, 727 pp.

Weil, M., J. Bressler, P. Parsons, K. Bolla, T. Glass, and B. Schwartz. 2005. Blood mercury levels and neurobehavioral function, *J. Am. Med. Assoc.*, 293, 1875–1882.

Wiener, J.G., D.P. Krabbenhoft, G.H. Heinz, and A.M. Scheuhammer. 2003. Ecotoxicology of mercury. Pages 409–463. In D.J.Hoffman, B.A. Rattner, G.A. Burton Jr., and J. Cairns Jr. (Eds.), *Handbook of Ecotoxicology, 2nd edition*, p. 409–463. Lewis Publishers, Boca Raton, FL.

Wiener, J.G., and D.J. Spry. 1996. Toxicological significance of mercury in freshwater fish. In W.N. Beyer, G.H. Heinz, and A.W. Redmon-Norwood (Eds.). *Environmental Contaminants in Wildlife: Interpreting Tissue Concentrations*, p. 297–339. CRC Press, Boca Raton, FL.

Wolfe, M.E., S. Schwarzbach, and R.A. Sulaiman. 1998. Effects of mercury on wildlife: a comprehensive review, *Environ. Toxicol. Chem.*, 17, 146–160.

World Health Organization (WHO). 1976. *Mercury. Environmental Health Criteria 1.* WHO, Geneva, 131 pp.

World Health Organization (WHO). 1989. *Mercury — Environmental Aspects. Environmental Health Criteria 86.* WHO, Geneva. 115 pp.

World Health Organization (WHO). 1990. *Methylmercury. Environmental Health Criteria 101.* WHO, Geneva. 144 pp.

World Health Organization (WHO). 1991. *Inorganic Mercury. Environmental Health Criteria 118.* WHO, Geneva. 168 pp.

Wren, C.D. 1986. A review of metal accumulation and toxicity in wild mammals. I. Mercury, *Environ. Res.*, 40, 210–244.

Zillioux, E.J., D.B. Porcella, and J.M. Benoit. 1993. Mercury cycling and effects in freshwater wetland ecosystems, *Environ. Toxicol. Chem.*, 12, 2245–2264.

CHAPTER 2

Mercury Uses and Sources

Mercury is a metallic element easily distinguished from all others by its liquidity at even the lowest temperatures occurring in moderate climates (Anon., 1948a). Mercury was unknown to the ancient Hebrews and early Greeks; and the first mention of mercury was by Theophrastus, writing around 300 BCE, stating that mercury was extracted from cinnabar by treatment with copper and vinegar (Anon., 1948a).

The most important ore of mercury, cinnabar (mercuric sulfide), has been mined continuously since 415 BCE (Clarkson and Marsh, 1982). Historically, the five primary mining areas for mercury were the Almeden district in Spain, the Idrija district in Slovenia, the Monte Amiata district in Italy, and various locales in Peru, the United States (Ferrara, 1999), especially California and Texas, as well as sites in Russia, Hungary, Mexico, and Austria (Anon., 1948a). The Almeden mines over the past 2500 years were the most important, having produced about 280,000 metric tons of mercury, or about 35.0% of the estimated total global production of about 800,000 tons (Ferrara, 1999). Major producers of mercury now include the former Soviet Union, Spain, Yugoslavia, and Italy (USPHS, 1994). In the United States, mercury consumption rose from 1305 metric tons in 1959 to 2359 tons in 1969 (Montague and Montague, 1971). Mining of mercury in the United States has decreased in recent years owing to decreasing demand and falling prices; the last operating mine in the United States that produced mercury as its main product closed in 1990 (Jasinski, 1995). In the late 1970s, world use of mercury was estimated at 10,000 to 15,000 metric tons annually (Boudou and Ribeyre, 1983), of which the United States used about 18.0% (Clarkson and Marsh, 1982). Accurate data on recent mercury consumption in the United States was difficult to obtain. In 1987, the United States imported 636 metric tons; this fell to 329 tons in 1988 and 131 tons in 1989 (USPHS, 1994). Domestic production of mercury produced as a by-product was 207 metric tons in 1990, 180 tons in 1991, and 160 tons in 1992; during this same period, the United States imported 15 tons in 1990, 56 tons in 1991, and 100 tons in 1992 (USPHS, 1994).

Atmospheric input of mercury has tripled over the past 150 years (Morel et al., 1998). The atmosphere plays an important role in the mobilization of mercury with 25.0 to 30.0% of the total atmospheric mercury burden of anthropogenic origin (USNAS, 1978), although more recent estimates of 67.0% are significantly higher (Morel et al., 1998). As a direct result of human activities, mercury levels in river sediments have increased fourfold since precultural times, and two- to fivefold in sediment cores from lakes and estuaries (Das et al., 1982). Analyses of sediment cores of North American mid-continental lakes show that mercury deposition rates increased by a factor of 3.7 since 1850, at a rate of about 2.0% annually (Rolfhus and Fitzgerald, 1995). During the past 100 years, more than 500,000 metric tons of mercury has entered the atmosphere, hydrosphere, and surface soils, with eventual deposition in subsurface soils and sediments (Das et al., 1982).

2.1 USES

Cinnabar (HgS), the main mercury ore, was used as a red pigment long before refining processes for elemental mercury were implemented (Wiener et al., 2003). In the 16th century, elemental mercury in combination with other compounds was considered a powerful medicinal agent (Anon., 1948a). Mercuro zinc cyanide [$Zn_3Hg(CN)_8$], also known as Lister's antiseptic, was used in the early days of antiseptics, circa 1880s, in the form of "cyanide gavage" or "cyanide wool" in general surgery. The stronger mercurial ointments were used to kill cutaneous parasites and to control itching. Until the advent of antibiotics, mercury salts were a key treatment for syphilis and other venereal diseases. Mercuric salts, especially the chloride and iodide, are powerful antiseptics and until the 1940s were routinely used in surgery to kill bacteria; however, their use was contraindicated in patients with actual or potential renal inflammation, patients with scarlet fever (risk to throat), and eclampsia (uterus sensitivity). Mercurochrome, di-sodium hydroxy-mercuro dibromo fluorescein, has been widely used in the United States and the United Kingdom as an antiseptic (Anon., 1948a), and is still sold in drugstores. The volatility of mercury and many of its compounds causes their absorption by the lungs and is an unintended consequence of external application; this could account for the occurrence of chronic mercurial poisoning in certain trades. Mercury is largely used in affectations of the alimentary canal and allegedly has value in many cases of heart disease and arterial degeneration. In cases of intestinal obstruction, elemental mercury has been administered — up to 454.0 grams — without ill effects, the weight of the mercury being sufficient to remove the obstruction (Anon., 1948a).

In the period prior to the industrial revolution, mercury was used extensively in gold extraction and the manufacture of felt hats and mirrors; in the 1800s, it was used in the chloralkali industry, the manufacture of electrical instruments, and as a medical antiseptic; and since 1900, it has been used in pharmaceuticals, agricultural fungicides, the pulp and paper industry as a slimicide, and the production of plastics (Clarkson and Marsh, 1982). In 1892, the process of producing chlorine and caustic soda from brine (sodium chloride) was developed (Paine, 1994). Electrolysis of brine using a liquid mercury cathode to produce chlorine at the anode and a sodium–mercury amalgam at the cathode is still used worldwide, with significant mercury contamination of the biosphere; however, the process is increasingly under replacement using mercury-free components (Paine, 1994). Until the late 1940s, mercury was used in the manufacture of fur felt hats from furs of rabbit, hare, muskrat, nutria, and beaver (Anon., 1948). Mercuric nitrate was used to remove fur from the hides. The fur fiber is harvested and the hide, now denuded of fur and useless for hat manufacturing purposes, discarded. A "mad hatter" syndrome was sometimes reported among workers in the manufacture of fur felt hats, with symptoms consistent with that of inorganic mercury poisoning, namely, tremors, excessive salivation, irritability, and excitement (Anon., 1948; Norton, 1986).

Mercury catalysts were used in the production of acetaldehyde, acetic acid, and vinyl chloride (Silver et al., 1994). The major use of mercury in the decade between 1959 and 1969 was as a cathode in the electrolytic preparation of chlorine and caustic (Nriagu, 1979; Paine, 1994; Table 2.1). In 1968, this use accounted for about 33.0% of the total U.S. demand for mercury (USEPA, 1980). During that same period, electrical apparatus accounted for about 27.0% of U.S. mercury consumption; industrial and control instruments, such as switches, thermometers, barometers, and general laboratory appliances, 14.0%; antifouling and mildew-proofing paints, 12.0%; mercury formulations to control fungal diseases of seeds, bulbs, and vegetables, 5.0%; and dental amalgams, pulp and paper manufacturers, pharmaceuticals, metallurgy and mining, and catalysts, 9.0% (Table 2.1; USEPA, 1980). Mercury, however, is no longer registered for use in antifouling paints, or for the control of fungal diseases of bulbs (USEPA, 1980). In India and other developing countries, however, $HgCl_2$ is commonly used for the preservation of seeds and by farmers in fruit preservatives after harvest and to inhibit growth of microorganisms (Ghosh et al., 2004). In Kenya, certain soaps containing skin-lightening agents also contained high concentrations of inorganic

Table 2.1 Industrial and Other Uses of Mercury in the United States in 1959 and 1969

Use	1959		1969	
	Tons	Percent	Tons	Percent
Chloralkali process	201	15.5	716	31.4
Electrical apparatus	308	23.6	644	27.3
Antifouling and mildew paints	121	9.3	336	14.2
Control devices	213	16.3	241	10.2
Dental preparations	63	4.8	105	4.5
Catalysts	33	2.5	102	4.3
Agriculture	110	8.4	93	3.9
Laboratory	38	2.9	71	3.0
Pharmaceuticals	59	4.5	25	1.1
Pulp and paper mill	150	11.5	19	0.8
Metal amalgamation	9	0.7	7	0.3
Total	**1305**		**2359**	

[a] *Source:* Modified from Montague K. and P. Montague. 1971. *Mercury.* Sierra Club, New York. 158 pp.

mercury (Ohno et al., 2004). And in Mexico and developing countries, merthiolate (an ethylmercury thiosalicylate compound) is currently used as a preservative in medical vaccines and as a skin antiseptic (Ohno et al., 2004).

Mercury in its various forms is still widely available in thermometers, fungicides, hearing aid and watch batteries, paints, mercurial drugs, antiquated cathartics, and ointments (Rumack and Lovejoy, 1986). The most recent uses of mercury and its compounds are in the manufacture of lighting fixtures such as fluorescent, metal halide, and mercury vapor lamps; dental amalgams; mining and reprocessing of gold; batteries; and paint manufacture (USNAS, 1978; Gonzalez, 1991; Sundolf et al., 1994; Barbossa et al., 1995; Gustafson, 1995; Lacerda et al., 2004; Liang et al., 2004).

2.2 SOURCES

Major inputs of mercury to the environment are mainly from natural sources, with significant and increasing amounts contributed from human activities. The atmosphere plays an important role in the mobilization of mercury, with an estimated 25.0 to 30.0% of the total atmospheric burden of anthropogenic origin (USNAS, 1978). The global anthropogenic atmospheric emission of mercury is estimated at 900 to 6200 tons annually, of which the United States contributed 300 metric tons in 1990 with 31.0% of the total from combustion of fossil fuels by power plants (Chu and Porcella, 1995). Atmospheric deposition is generally acknowledged as the major source of mercury to watersheds. In northern Minnesota watersheds, for example, atmospheric deposition was the primary source of mercury. Geologic and point source contributions were not significant. Transport from soils and organic materials may also be important, but the mercury from these sources probably originate from precipitation and direct atmospheric sorption by watershed components (Swain and Helwig, 1989; Sorensen et al., 1990). In Sweden, increased mercury concentrations in lakes are attributed to increased atmospheric emissions and deposition of mercury, and to acid rain (Hakanson et al., 1990). Airborne particulates may contribute to the high mercury levels found in some marine dolphins and whales (Rawson et al., 1995). A total of 60 to 80 metric tons of mercury is deposited from the atmosphere into the Arctic each year; the main sources of mercury to the Arctic are Eurasia and North America from combustion of fossil fuels to produce electricity and heat (Pacyna and Keeler, 1995). However, elevated mercury concentrations in fish muscle (0.5 to 2.5 mg/kg fresh weight [FW]) from remote Arctic lakes over extended periods (1975 to 1993) are sometimes due to natural sources of mercury (Stephens, 1995). Atmospheric deposition of mercury into the Great

Lakes from sources up to 2500 km distant are documented at annual deposition rates of 15.0 μg Hg/m² (Glass et al., 1991). In South Florida, 80.0 to 90.0% of the annual mercury deposition occurs during the summertime wet season (Guentzel et al., 1995). Dry deposition processes are important for the flux of inorganic mercury and methylmercury to Swedish forested ecosystems; for methylmercury, the most important deposition route is from the air to a relatively stable form in litter (Munthe et al., 1995).

2.2.1 Natural Sources

The total amount of mercury in various global reservoirs is estimated at 334.17 billion metric tons; almost all of this amount is in oceanic sediments (98.75%) and oceanic waters (1.24%), and most of the rest is in soils (Table 2.2). Living aquatic organisms are estimated to contain only 7.0 metric tons of mercury (Clarkson et al., 1984; Table 2.2).

The largest pool of methylmercury in freshwater biota is found in fish tissues (Porcella, 1994; Watras et al., 1994), and fly larvae are alleged to play an important role in mercury cycling from feeding on beached fish carcasses, as judged by observations on blowfly (*Calliphora* sp.) adult egg layers, eggs, larvae, pupae, and emerging adults feeding on carcasses of brook trout, *Salvelinus fontinalis* (Sarica et al., 2005). Specifically, methylmercury that accumulated in blowfly larvae is retained in pupae but eliminated by adults following emergence (Sarica et al., 2005).

Mercury from natural sources enters the biosphere directly as a gas, in lava (from terrestrial and oceanic volcanic activity), in solution, or in particulate form; cinnabar (HgS), for example, is a common mineral in hot spring deposits and a major natural source of mercury (Das et al., 1982). The global cycle of mercury involves degassing of the element from the Earth's crust and evaporation from natural bodies of water, atmospheric transport — mainly in the form of mercury vapor — and deposition of mercury back onto land and water. Oceanic effluxes of mercury are tied to equatorial upwelling and phytoplankton activity and may significantly affect the global cycling of this metal. If volatilization of mercury is proportional to primary production in the world's oceans, oceanic phytoplankton activity represents about 36.0% of the yearly mercury flow to the atmosphere, or about 2400 tons per year (Kim and Fitzgerald, 1986). Riverine input — with sediments containing up to 0.5 mg Hg/kg DW — of mercury influences mercury content of outer shelf areas in southeastern Brazil where most of the offshore oil fields are located (Lacerda et al., 2004). Mercury finds its way into sediments, particularly oceanic sediments, where the retention time can be lengthy (Table 2.2), and where it may continue to contaminate aquatic organisms (Lindsay and Dimmick, 1983). Estimates of the quantities of mercury entering the atmosphere from degassing of the surface of the planet vary widely, but a commonly quoted figure is 30,000 tons annually (Clarkson et al., 1984).

Table 2.2 Amount of Mercury in Some Global Reservoirs and Residence Time

Reservoir	Mercury Content[a] (metric tons)	Residence Time[b]
Atmosphere	850	6–90 days
Soils	21,000,000	1000 years
Freshwater	200	—
Freshwater biota (living)	4	—
Ocean water	4,150,000,000	2000 years
Oceanic biota (living)	3	—
Ocean sediments	330,000,000,000	> 1 million years

[a] From USNAS, 1978.
[b] Modified from Clarkson, T.W., R. Hamada, and L. Amin-Zaki. 1984. Mercury. In J.O. Nriagu (Ed.), *Changing Metal Cycles and Human Health*, p. 285–309. Springer-Verlag, Berlin.

Mercury is emitted from volcanoes into the atmosphere, along with large quantities of lead, cadmium, and bismuth (Hinkley et al., 1999). About 6000 tons of mercury are discharged into the atmosphere every year from all sources (Fitzgerald, 1986); and from all volcanoes, about 60 tons or about 1.0% of the total (Varekamp and Buseck, 1986).

Virtually all mercury in the Florida Everglades from natural sources (39.0% of the total mercury deposited) is attributed to release from the soil through natural processes, including microbial transformations of inorganic and organic mercury to methylmercury (Sundolf et al., 1994). Major sources of mercury in humans — other than residing in mercury-contaminated environments — include consumption of large predatory fish, such as tuna and swordfish, caught hundreds of kilometers offshore in clean ocean waters (Fitzgerald, 1986; WHO, 1990, 1991). The naturally elevated concentrations of mercury in these species is discussed in greater detail later.

Terrestrial vegetation functions as a conduit for the transport of elemental mercury from the geosphere to the atmosphere (Leonard et al., 1998a). Estimated mercury emissions from plants in the Carson River Drainage Basin of Nevada — an area heavily contaminated with mercury from historical gold mining activities — over the growing season (0.5 mg Hg/m^2) add to the soil mercury emissions (8.5 mg Hg/m^2) for a total landscape emission in that area of 9.0 mg Hg/m^2. In one species (tall whitetop, *Lepidium latifolium*), as much as 70.0% of the mercury taken up by the roots during the growing season was emitted to the atmosphere (Leonard et al., 1998a). Factors known to increase the flux of elemental mercury from terrestrial plants growing in soils with high (34.0 to 54.0 mg Hg/kg soil DW) levels of mercury include increasing air temperature in the range 20.0 to 40.0°C, increasing irradiance, increasing soil mercury concentrations, and increasing leaf area (Leonard et al., 1998b).

2.2.2 Anthropogenic Sources

In 1995, approximately 1900 metric tons of anthropogenic mercury entered the atmosphere, mostly (75.0%) from the combustion of fossil fuels (Pacyna and Pacyna, 2002). About 56.0% of global mercury atmospheric emissions came from Asian countries, with Europe and North America combined contributing less than 25.0%; gaseous elemental mercury (Hg^o) comprised 53.0% of total atmospheric emissions, gaseous Hg^+ 37.0%, and particle-associated mercury the remainder (Semkin et al., 2005).

Large-scale mining of mercury in North America ceased around 1990 because of low prices and stringent environmental regulations (Rytuba, 2003). In the United States, mercury is now produced only as a by-product from presently operating gold mines where environmental regulations require its recovery, and from the reprocessing of precious metal mine tailings and gold-placer sediments. Many of the mercury mines in the California Coast Ranges contain waste rock that contributes mercury-rich sediments to nearby watersheds. At some mines, the release of mercury in acidic drainage is a significant source of mercury to watersheds, where it is taken up by fish and other organisms (Rytuba, 2003).

Atmospheric transport of anthropogenic mercury may contaminate remote ecosystems (Watras et al., 1994; Lin et al., 2001). In one case, a remote lake in northern Wisconsin with no surface inflow and negligible groundwater inflow received about 100.0 mg Hg/ha during 1988 to 1990, an input that could account for the elevated mercury burdens found in water, sediments, and fish (Wiener et al., 1990, 2003; Watras et al., 1994).

Several human activities that contribute significantly to the global input of mercury include the combustion of fossil fuels; mining and reprocessing of gold, copper, and lead; operation of chlor-alkali plants; runoff from abandoned cinnabar mines; wastes from nuclear reactors, pharmaceutical plants, oil refining plants, and military ordnance facilities; incineration of municipal solid wastes and medical wastes; offshore oil exploration and production; disposal of batteries and fluorescent lamps; and the mining, smelting, use, and disposal of mercury (USNAS, 1978; Das et al., 1982; Gonzalez, 1991; Lodenius, 1991; Facemire et al., 1995; Gustafson, 1995; Atkeson et al., 2003;

Lacerda et al., 2004; Liang et al., 2004). In one case, more than 5.5 million kg of elemental mercury were released into the Carson River Drainage Basin in Nevada during historic mining operations — now closed — in which mercury was used to amalgamate gold and silver ore (Gustin et al., 1995). Mercury concentrations in sample tailings were as high as 1570.0 mg/kg. The air directly over the site contained 1.0 to 7.1 ng Hg/m^3, and was as high as 240.0 ng/m^3 in October 1993. The estimated range of mercury flux to the atmosphere from the site was 37.0 to 500.0 ng/m^2 hourly (Gustin et al., 1995).

Mercury emissions from electric utilities constitute the largest uncontrolled source of mercury to the atmosphere (USEPA, 1997), and globally it accounts for up to 59.0% of the total annual atmospheric loading of mercury from both natural and anthropogenic sources (Fitzgerald, 1986; Fitzgerald and Clarkson, 1991; WHO, 1976, 1991; Mason et al., 1994; USEPA, 2000; Lamborg et al., 2002). Coal-fired power plants are now considered the greatest source of environmental mercury in the United States, and the only significant source that remains unregulated (Maas et al., 2004). In 1994, about 50 metric tons of mercury were emitted into the biosphere from coal-burning power plants in the United States, with lesser amounts from oil- and gas-combustion units (Finkelman, 2003). Available technologies now installed in waste combustion and medical incinerators are recommended for installation in coal-fired plants and may reduce mercury emissions by as much as 90.0% (Maas et al., 2004). In fact, the U.S. Environmental Protection Agency made steady progress throughout the 1990s in reducing mercury emissions from power plants, although this effort abated in recent years owing to the high costs of pollution abatement (Trasande et al., 2005). The economic costs of methylmercury neurotoxicity from these plants was estimated using a hypothetical model. In that scenario, Trasande et al. (2005) aver that methylmercury-induced loss of intelligence would affect between 316,000 and 637,000 children each year in the United States, as judged by umbilical cord blood concentrations greater than 5.8 μg methylmercury/L, a level associated with IQ loss. This intelligence loss is translated into diminished economic activity equivalent to at least U.S. $8.7 billion annually (range $2.2 to $43.8 billion). The model proposed by Trasande et al. (2005) requires verification.

Logging and forest fires can contribute to the bioavailability of mercury (Garcia and Carignan, 2005). Watersheds impacted by clear-cut logging, or burnt forest ecosystems, release mercury into the biosphere with significant increases in the flesh of predatory fish from impacted drainage lakes when compared to reference watersheds (Garcia and Carignan, 2005).

Most of the daily intake of mercury compounds is in the form of methylmercury derived from dietary sources, primarily fish, and to a lesser extent elemental mercury from mercury vapor in dental amalgams, and ethylmercury added as an antiseptic to vaccines (Mottet et al., 1985; USNAS, 2000; Clarkson et al., 2003; Dye et al., 2005). Dental amalgams, which may contain up to 50.0% by weight of metallic mercury, may also constitute a significant source of mercury in some cases (Summers et al., 1993). Amalgam mercury is imperfectly stable, slowly leaching from the mercury-silver or mercury-gold amalgam through the action of oral bacteria and exacerbated by chewing. Following placement or removal of fillings, up to 200.0 mg mercury is eliminated in the feces, with subsequent selection of mercury-resistant bacteria for degradation. Normal mastication may result in body accumulations of 10.0 μg daily through either intestinal uptake or respiratory intake of mercury vapor released during chewing (Summers et al., 1993).

Consumption of fish contaminated with mercury was associated with elevated mercury concentrations in the blood and hair of about 1200 Amerindians in northwestern Ontario, Canada (Phelps and Clarkson, 1980). The source was inorganic mercury catalysts from a chloralkali plant. In this case, about 9 tons of mercury were discharged into wastewaters between 1962 and 1979, at which time mercury usage for the production of chlorine and caustic was drastically reduced (Phelps and Clarkson, 1980). In aquatic ecosystems, removal of the source of anthropogenic mercury — such as chloralkali plants — results in a slow decrease in the mercury content of sediments and biota (USNAS, 1978). The rate of loss depends, in part, on the initial degree of contamination, the chemical form of mercury, physical and chemical conditions of the system, and the hydraulic

Table 2.3 Mercury Loss (in kg) from Five Operating Chloralkali Plants in Canada between 1986 and 1989 via Wastewater, Air Emissions, Products, and Solid Wastes

	1986	1987	1988	1989	Total
Wastewater	88	55	52	46	241
Air emissions	680	831	572	547	2630
Products	71	79	81	72	303
Solid wastes	449	662	1010	1196	3317
Total	1288	1627	1715	1861	6491[a]

[a] Individual plant losses of total mercury from the five chloralkali plants during the 4-year period 1986 to 1989 were 290, 1089, 1247, 1721, and 2144 kg, respectively.

Source: Paine, P.J. 1994. Compliance with chlor-alkali mercury regulations, 1986 to 1989: status report, *Report EPS 1/HA/2, 1-43.* Available from Environmental Protection Publications, Environment Canada, Ottawa, Ontario K1A OH3.

turnover time (USNAS, 1978). Between 1986 and 1989, throughout all Canada, a total of five mercury-cell chloralkali plants were still operating (Paine, 1994). By comparison, during the period 1935 through the mid-1970s, about 15 chloralkali plants were operating; the development of alternative technologies to replace the mercury cell was largely responsible for the plant closures. Total mercury losses to the environment from these five operating plants during the 4-year period from 1986 to 1989 were about 6.5 metric tons (Table 2.3).

Mercury contamination of more than 500,000 miners, adjacent Indian populations, and numerous populations of fish and wildlife is one consequence of the gold rush that took place in the early 1980s in the Amazon region of Brazil. Metallic mercury was used to agglutinate the fine gold particles through amalgamation. During this process, large amounts of mercury were lost to the river and soil; additional mercury was lost as vapor to the atmosphere during combustion of the amalgamated gold to release the gold (Barbossa et al., 1995). Elemental mercury used in seals of three trickling filters in municipal wastewater treatment plants — each seal contained several hundred kilograms of mercury — leaked repeatedly, discharging 157.0 grams of mercury and 0.4 grams of methylmercury daily; the use of mercury seals for this purpose should be discontinued (Gilmour and Bloom, 1995). Also discontinued, in 1967, by Finland is the use of phenylmercury compounds as slimicides in the pulp industry (Lodenius, 1991). In Canada and the United States, mercurial compounds were used to control fungi in pulp paper processing plants, with subsequent mercury contamination of edible fish muscle (Silver et al., 1994).

Abandoned mercury mines may contribute excess mercury loadings and other contaminants to the environment. For example, mercury mines in western Turkey that were gradually abandoned owing to low demand, low prices, and increasing environmental concern over mercury adversely affected adjacent water resources (Gemici, 2004). One abandoned mine located 5 km west of Beydag, Turkey, that operated from 1958 through 1986 with a total production of 2045 metric tons of mercury during this period released metal-rich, acidic drainage affecting groundwater and adjacent stream water quality through decreasing pH; elevated levels of silicon, aluminum, magnesium, calcium, and potassium; increasing precipitation of iron oxides; and increasing sulfates, manganese, iron, and arsenic. Most of the mine water and groundwater samples exceeded drinking water standards for aluminum, iron, manganese, arsenic, nickel, and cadmium. Mercury concentrations in all samples were below the Turkish drinking water standard of 1.0 μg/L for human health; however, two samples contained 0.3 and 0.5 μg Hg/L and were above the USEPA mercury criterion for aquatic life protection of < 0.012 μg/L (Gemici, 2004).

In the Florida Everglades, 61.0% of the mercury is due to atmospheric deposition from anthropogenic sources, especially municipal solid waste combustion facilities (15.0%), medical waste incinerators (14.0%), paint manufacturing and application (11.0%), electric utility industries (11.0%) and private residences (2.0%) through combustion of fossil fuels, and electrical apparatus, including fluorescent, metal halide, and mercury vapor lights (6.0%). All other anthropogenic sources combined — including sugarcane processing, the dental industry, open burning, and sewage sludge disposal — accounted for about 3.0% of the total mercury emitted to the environment (Sundolf et al., 1994). Recent and more extensive studies (Dvonch et al., 1999; Atkeson et al., 2003) on mercury monitoring in the Florida Everglades ecosystem show that more than 95.0% of the mercury loading is from atmospheric deposition; that 92.0% of the total deposition came from local sources, such as municipal waste combustion, medical waste incinerators, electric utility boilers (coal, oil, gas), commercial and industrial boilers, and hazardous waste incinerators; and that long-distance transport of mercury wastes from certain sources are becoming increasingly important in Florida, especially mercury from waste incinerators and other atmospheric emitters, increased release of mercury from drainage and soil disturbance, and from hydrologic changes. Atkeson et al. (2003) believe that more data is needed on the sources of mercury deposition in the Florida Everglades, especially atmospheric mercury loadings from nonlocal sources.

Reduction in mercury emissions and other contaminants from municipal waste incinerators using the best available technology is not always completely successful. For example, the Angers, France, solid waste incinerator plant in operation since 1974 was upgraded in 2000 to comply with the new European standards (Glorennec et al., 2005). Mean mercury emissions were reduced about 13.0%, from 26.8 (18.6 to 69.4) kg/year to 23.3 (< 1.0 to 72.5) kg/year; and mean ambient air concentrations by about 62.0%, from 0.0004 $\mu g/m^3$ (max. 0.001 $\mu g/m^3$) to 0.00015 $\mu g/m^3$ (max. 0.0028 $\mu g/m^3$); however, the maximum values in both comparisons were larger in 2000 to 2001 than in 1975 to 1979 (Glorennec et al., 2005).

2.3 SUMMARY

Mercury has been mined continuously for at least 2400 years for use in gold recovery (until the present time), in the manufacture of felt hats and mirrors (1700s), in the chloralkali industry to manufacture chlorine and caustic (since the 1800s), as a fungicide in agriculture and paper production (1900s), and currently in lighting fixtures, batteries, paints, dentistry, and in medicine to kill bacteria and cutaneous parasites. Major anthropogenic sources of mercury to the biosphere include combustion of fossil fuels from power plants, municipal solid waste combustion facilities, medical waste incinerators, paint manufacturing, and gold extraction. Combustion of mercury-containing fossil fuels may account for up to 60.0% of the global mercury burden from human activities. Use of mercury to amalgamate gold in Amazonia has resulted in mercury contamination of at least 500,000 miners and numerous fish and wildlife populations. And use of inorganic mercury catalysts in chloralkali plants has caused contamination of fish in waterways and elevated blood and mercury levels in Canadian Amerindians who consumed these fish.

World production of mercury in recent years is estimated at 10,000 to 15,000 metric tons annually; major producers of mercury now include the former Soviet Union, Spain, the former Yugoslavia, and Italy. The total amount of mercury in various global reservoirs is estimated at 334 billion metric tons, mostly in ocean sediments (98.75%) and ocean waters (1.24%). Only 7 metric tons of mercury are believed to be present in living aquatic organisms. During the past 100 years, an estimated 500,000 tons of mercury entered the biosphere, with eventual deposition in subsurface sediments. Mercury inputs to the biosphere are mainly from natural sources, but with significant and increasing amounts contributed from human activities. Soil bacteria and terrestrial plants expedite flux rates of elemental mercury from the geosphere to the atmosphere. The atmosphere plays an important role in the mobilization of mercury, with about half the total atmospheric mercury burden of anthropogenic origin. Atmospheric transport of mercury may contaminate remote ecosystems hundreds of kilometers distant.

REFERENCES

Anon. 1948. Hat manufacture, *Encyclop. Britannica*, 11, 249–250.

Anon. 1948a. Mercury, *Encylop. Britannica*, 15, 269–272.

Atkeson, T., D.Axelrad, C. Pollman, and G. Keeler. 2003. *Integrating Atmospheric Mercury Deposition and Aquatic Cycling in the Florida Everglades: An Approach for Conducting a Total Maximum Daily Load Analysis for an Atmospherically Derived Pollutant. Integrated Summary. Final Report.* 272 pp. Available from Mercury and Applied Science MS6540, Florida Dept. Environmental Conservation, 2600 Blair Stone Road, Tallahassee, FL 32399-2400. See also <http://www.floridadep.org/labs/mercury/index.htm>

Barbossa, A.C., A.A. Boischio, G.A. East, I. Ferrari, A. Goncalves, P.R.M. Silva, and T.M.E. da Cruz. 1995. Mercury contamination in the Brazilian Amazon. Environmental and occupational aspects, *Water Air Soil Pollut.*, 80, 109–121.

Boudou, A. and F. Ribeyre. 1983. Contamination of aquatic biocenoses by mercury compounds: an experimental toxicological approach. In J.0. Nriagu (Ed.). *Aquatic Toxicology*, p. 73–116. John Wiley, NY.

Chu, P. and D.B. Porcella. 1995. Mercury stack emissions from U.S. electric utility power plants, *Water Air Soil Pollut.*, 80, 135–144.

Clarkson, T.W., R. Hamada, and L. Amin-Zaki. 1984. Mercury. Pages 285–309 In J.O. Nriagu (Ed.). *Changing Metal Cyclesand Human Health*, p. 285–309. Springer-Verlag, Berlin.

Clarkson, T.W., L. Magos, and G.J. Myers. 2003. The toxicology of mercury — current exposures and clinical manifestations, *New Engl. J. Med.*, 349, 1731–1737.

Clarkson, T.W. and D.O. Marsh. 1982. Mercury toxicity in man. In A.S. Prasad (Ed.). *Clinical, Biochemical, and Nutritional Aspects of Trace Elements*, Vol. 6, p. 549–568. Alan R. Liss, Inc., New York.

Das, S.K., A. Sharma, and G. Talukder. 1982. Effects of mercury on cellular systems in mammals — a review, *Nucleus (Calcutta)*, 25, 193–230.

Dvonch, J.T., J.R. Graney, G.J. Keeler, and R.K. Stevens. 1999. Utilization of elemental tracers to source apportion mercury in south Florida precipitation, *Environ. Sci. Technol.*, 33, 4522–4527.

Dye, B.A., S.E. Schober, C.F. Dillon, R.L. Jones, C. Fryar, M. McDowell, and T.H. Sinks. 2005. Urinary mercury concentrations associated with dental restorations in adult women aged 16–49 years; United States, 1999–2000; *Occup. Environ. Medic.*, 62, 368–375.

Facemire, C., T. Augspurger, D. Bateman, M. Brim, P. Conzelmann, S. Delchamps, E. Douglas, L. Inmon, K. Looney, F. Lopez, G. Masson, D. Morrison, N. Morse, and A. Robison. 1995. Impacts of mercury contamination in the southeastern United States, *Water Air Soil Pollut.*, 80, 923–926.

Ferrara, R. 1999. Mercury mines in Europe: assessment of emissions and environmental contamination. Pages 51–72 In R. Ebinghaus, R.R. Turner, L.D. Lacerda, O. Vasiliev, and W. Salomons (Eds.). *Mercury Contaminated Sites*, p. 51–72. Springer, Berlin.

Finkelman, R.B. 2003. Mercury in coal and mercury emissions from coal combustion. In J.E. Gray (Ed.). *Geologic Studies of Mercury by the U.S. Geological Survey*, p. 10–13. USGS circular 1248. Available from USGS Information Services, Box 25286, Denver, CO 80225.

Fitzgerald, W.F. 1986. Cycling of mercury between the oceans and the atmosphere. In P.Buat-Menard (Ed.). *The Role of Air-Sea Exchange in Geochemical Cycling*, p. 363–408. D. Reidel Publ., Boston.

Fitzgerald, W.F. and T.W. Clarkson. 1991. Mercury and monomethylmercury: present and future concerns, *Environ. Health Perspect.*, 96, 159–166.

Garcia, E. and R. Carignan. 2005. Mercury concentrations in fish from forest harvesting and fire-impacted Canadian boreal lakes compared using stable isotopes of nitrogen, *Environ. Toxicol. Chem.*, 24, 685–693.

Gemici, U. 2004. Impact of acid mine drainage from the abandoned Halikoy mercury mine (western Turkey) on surface and groundwater, *Bull. Environ. Contam. Toxicol.*, 72, 482–489.

Ghosh, S.K., S. Ghosh, J. Chaudhuri, R. Gachhui, and A. Mandal. 2004. Studies on mercury resistance in yeasts isolated from natural sources, *Bull. Environ. Contam. Toxicol.*, 72, 21–28.

Gilmour, C.C. and N.S. Bloom. 1995. A case study of mercury and methylmercury dynamics in a Hg-contaminated municipal wastewater treatment plant, *Water Air Soil Pollut.*, 80, 799–803.Glass, G.E., J.A. Sorensen, K.W. Schmidt, G.R. Rapp, Jr., D. Yap, and D. Fraser. 1991. Mercury deposition and sources for the upper Great Lakes region, *Water Air Soil Pollut.*, 56, 235–249.

Glorennec, P., D. Zmirou, and D. Bard. 2005. Public health benefits of compliance with current E.U. emissions standards for municipal waste incinerators: a health risk assessment with the CalTox multimedia exposure model, *Environ. Int.*, 31, 693–701.

Gonzalez, H. 1991. Mercury pollution caused by a chlor-alkali plant, *Water Air Soil Pollut.*, 56, 83–93.

Guentzel, J.L., W.M. Landing, G.A. Gill, and C.D. Pollman. 1995. Atmospheric deposition of mercury in Florida: the FAMS project (1992–1994), *Water Air Soil Pollut.*, 80, 393–402.

Gustafsson, E. 1995. Swedish experiences of the ban on products containing mercury, *Water Air Soil Pollut.*, 80, 99–102.

Gustin, M.S., G.E. Taylor, Jr., and T.L. Leonard. 1995. Atmospheric mercury concentrations above mercury contaminated mill tailings in the Carson River drainage basin, NV, *Water Air Soil Pollut.*, 80, 217–220.

Hakanson, L., T. Andersson, and A. Nilsson. 1990. Mercury in fish in Swedish lakes — linkages to domestic and European sources of emission, *Water Air Soil Pollut.*, 50, 171–191.

Hinkley, T.K, P.J. Lamothe, S.A. Wilson, D.L. Finnegan, and T.M. Gerlach. 1999. Metal emissions from Kilauea, and a suggested revision of the estimated worldwide output by quiescent degassing of volcanoes, *Earth Planet. Sci. Lett.*, 170(3), 315–325.

Jasinski, S.M. 1995. The materials flow of mercury in the United States, *Resour. Conserva. Recycling*, 15, 145–179.

Kim, J.P. and W.F. Fitzgerald. 1986. Sea-air partitioning of mercury in the equatorial Pacific Ocean, *Science*, 231, 1131–1133.

Lacerda, L.D., C.E. Rezende, A.R.C. Ovalle, and C.E.V. Carvalho. 2004. Mercury distribution in continental shelf sediments from two offshore oil fields in southeastern Brazil, *Bull. Environ. Contam. Toxicol.*, 72, 178–185.

Lamborg, C.H., W.F. Fitzgerald, J. O'Donnell, and T. Torgersen. 2002. A non-steady-state compartmental model of global scale mercury biogeochemistry with interhemispheric gradients, *Geochim. Cosmochim. Acta*, 66, 1105–1118.

Leonard, T.L., G.E. Taylor Jr., M.S. Gustin, and G.C.J. Fernandez. 1998a. Mercury and plants in contaminated soils. 1. Uptake, partitioning, and emission to the atmosphere, *Environ. Toxicol. Chem.*, 17, 2063–2071.

Leonard, T.L., G.E. Taylor Jr., M.S. Gustin, and C.J. Fernandez. 1998b. Mercury and plants in contaminated soils. 2. Environmental and physiological factors governing mercury flux to the atmosphere, *Environ. Toxicol. Chem.*, 17, 2072–2079.

Liang, L.N., J.T. Hu, D.Y. Chen, Q.F. Zhou, B. He, and G.B. Jiang. 2004. Primary investigation of heavy metal contamination status in molluscs collected from Chinese coastal sites, *Bull. Environ. Contam. Toxicol.*, 72, 937–944.

Lin, C.J., M.D. Cheng, and W.H. Schroeder. 2001. Transport patterns and potential sources of total gaseous mercury measured in Canadian high Arctic in 1995, *Atmospher. Environ.*, 35, 1141–1154.

Lindsay, R.C. and R.W. Dimmick. 1983. Mercury residues in wood ducks and wood duck foods in eastern Tennessee, *Jour. Wildl. Dis.*, 19, 114–117.

Lodenius, M. 1991. Mercury concentrations in an aquatic ecosystem during twenty years following abatement of the pollution source, *Water Air Soil Pollut.*, 56, 323–332.

Maas, R.P., S.C. Patch, and K.R. Sergent. 2004. A Statistical Analysis of Factors Associated with Elevated Hair Mercury Levels in the U.S. Population: An Interim Progress Report. Tech. Rep. 04-136, Environ. Qual. Inst., Univ. North Carolina, Ashville. 13 pp.

Mason, R.P., W.F. Fitzgerald, and F.M.M. Morel. 1994. The biogeochemistry of elemental mercury — anthropogenic influences, *Geochim. Cosmochim. Acta*, 58, 3191–3198.

Montague, K. and P. Montague. 1971. *Mercury.* Sierra Club, New York. 158 pp.

Morel, F.M., A.M.L. Kreapiel, and M. Amyot. 1998. The chemical cycle and bioaccumulation of mercury, *Annu. Rev. Ecol. Syst.*, 29, 543–566.

Mottet, N.K., C.M. Shaw, and T.M. Burbacher. 1985. Health risks from increases in methylmercury exposure, *Environ. Health Perspect.*, 63, 133–140.

Munthe, J., H. Hultberg, and A. Ivefeldt. 1995. Mechanisms of deposition of methylmercury and mercury to coniferous forests, *Water Air Soil Pollut.*, 80, 363–371.

Norton, S. 1986. Toxic responses of the central nervous system. In C.D. Klaassen, M.O. Amdur, and J. Doull (Eds.). *Casarett and Doull's Toxicology, third edition*, p. 359–386. Macmillan, New York.

Nriagu, J.O. (Ed.). 1979. *The Biogeochemistry of Mercury in the Environment.* Elsevier/North-Holland Biomedical Press, New York. 696 pp.

Ohno, H., R. Doi, Y. Kashima, S. Murae, T. Kizaki, Y. Hitomi, N. Nakano, and M. Harada. 2004. Wide use of merthiolate may cause mercury poisoning in Mexico, *Bull. Environ. Contam. Toxicol.*, 73, 777–780.

Pacyna, J.M. and G.J. Keeler. 1995. Sources of mercury in the Arctic, *Water Air Soil Pollut.*, 80, 621–632.

Pacyna, E.G. and J.M Pacyna. 2002. Global emission of mercury from anthropogenic sources in 1995, *Water Air Soil Pollut.*, 137, 149–165.

Paine, P.J. 1994. Compliance with chlor-alkali mercury regulations, 1986–1989: status report, *Report EPS 1/HA/2*, 1–43. Avail. from Environmental Protection Publications, Environment Canada, Ottawa, Ontario K1A OH3.

Phelps, R.W. and T.W. Clarkson. 1980. Interrelationships of blood and hair mercury concentrations in a North American population exposed to methylmercury, *Arch. Environ. Health*, 35, 161–168.

Porcella, D.B. 1994. In C.J. Watras and J.W. Huckabee (Eds.), *Mercury Pollution — Integration and Synthesis*, p. 3–19. CRC Press, Boca Raton, FL.

Rawson, A.J., J.P. Bradley, A. Teetsov, S.B. Rice, E.M. Haller, and G.W. Patton. 1995. A role for airborne particulates in high mercury levels of some cetaceans, *Ecotoxicol. Environ. Safety*, 30, 309–314.

Rolfhus, K.R. and W.F. Fitzgerald. 1995. Linkages between atmospheric mercury deposition and the methylmercury content of marine fish, *Water Air Soil Pollut.*, 80, 291–297.

Rumack, B.H. and F.H. Lovejoy Jr. 1986. Clinical toxicology. In C.D. Klaassen, M.O. Amdur, and J. Doull (Eds.). *Casarett and Doull's Toxicology, third edition*, p. 879–901. Macmillan, New York.

Rytuba, J.J. 2003. Mercury from mineral deposits and potential environmental impact, *Environ. Geol.*, 43, 226–238.

Sarica, J., M. Amyot, J. Bey, and L. Hare. 2005. Fate of mercury accumulated by blowflies feeding on fish carcasses, *Environ. Toxicol. Chem.*, 24, 526–529.

Semkin, R.G., G. Mierle, and R.J. Neureuther. 2005. Hydrochemistry and mercury cycling in a high Arctic watershed, *Sci. Total Environ.*, 342, 199–221.

Silver, S., G. Endo, and K. Nakamura. 1994. Mercury in the environment and the laboratory, *J. Japan. Soc. Water Environ.*, 17, 235–243.

Sorensen, J.A., G.E. Glass, K.W. Schmidt, J.K. Huber, and G. R. Rapp, Jr. 1990. Airborne mercury deposition and watershed characteristics in relation to mercury concentrations in water, sediments, plankton, and fish of eighty northern Minnesota lakes, *Environ. Sci. Technol.*, 24, 1716–1727.

Stephens, G.R. 1995. Mercury concentrations in fish in a remote Canadian arctic lake, *Water Air Soil Pollut.* 80, 633–636.

Summers, A.O J. Wireman, M.J. Vimy, F.L. Lorscheider, B. Marshall, S.B. Levy, S. Bennett, and L. Billard. 1993. Mercury released from "silver" fillings provokes an increase in mercury- and antibiotic-resistant bacteria in oral and intestinal flora of primates, *Antimicrob. Agents Chemother.*, 37, 825–834.

Sundolf, S.F., M.G. Spalding, J.D. Wentworth, and C.K. Steible. 1994. Mercury in livers of wading birds (Ciconiiformes) in southern Florida, *Arch. Environ. Contam. Toxicol.*, 27, 299–305.

Swain, E.B. and D.D. Helwig. 1989. Mercury in fish from northeastern Minnesota lakes: historical trends, environmental correlates, and potential sources, *J. Minnesota Acad. Sci.*, 55, 103–109.

Trasande, L., P.J. Landrigan, and C. Schechter. 2005. Public health and economic consequences of methyl mercury toxicity to the developing brain, *Environ. Health Perspect.*, 113, 590–596.

U.S. Environmental Protection Agency (USEPA). 1980. Ambient water quality criteria for mercury, *U.S. Environ. Protection Agen. Rep. 440/5-80-058.* Available from Natl. Tech. Infor. Serv., 5285 Port Royal Road, Springfield, VA 22161.

U.S. Environmental Protection Agency (USEPA) 1997. *Mercury Study Report to Congress, Volumes I-VIII.* USEPA Rept. EPA-452/R-97-003.

U.S. Environmental Protection Agency (USEPA). 2000. Emissions and generation resource integrated database (E-GRID), <*http://www.epa.gov/airmarkets/egrid*>.

U.S. National Academy of Sciences (USNAS). 1978. *An Assessment of Mercury in the Environment.* Natl. Acad. Sci., Washington, D.C., 185 pp.

U.S. National Academy of Sciences (USNAS). 2000. *Toxicological Effects of Methyl Mercury.* National Research Council, National Academy Press, Washington, D.C. 368 pp.

U.S. Public Health Service (USPHS). 1994. *Toxicological profile for mercury (update), TP-93/10.* U.S. PHS, Agen. Toxic Substances Dis. Registry, Atlanta, GA. 366 pp.

Varekamp, J.C. and P.R. Buseck. 1986. Global mercury flux from volcanic and geothermal sources, *Appl. Geochem.*, 1, 65–73.

Watras, C.J., N.S. Bloom, R.J.M. Hudson, S. Gherini, R. Munson, S.A. Claas, K.A. Morrison, J. Hurley, J.G. Wiener, W.F. Fitzgerald, R. Mason, G. Vandal, D. Powell, R. Rada, L. Rislove, M. Winfrey, J. Elder, D. Krabbenhoft, A. W. Andren, C. Babiarz, D.B. Porcella, and J.W. Huckabee. 1994. Sources and fates of mercury and methylmercury in Wisconsin lakes. Pages 153–177 In C.J. Watras and J.W. Huckabee (Eds.). *Mercury Pollution: Integration and Synthesis*, p. 153–177. Lewis Publ., Boca Raton, FL.

Wiener, J.G., W.F. Fitzgerald, C.J. Watras, and R.G. Rada. 1990. Partitioning and bioavailability of mercury in an experimentally acidified Wisconsin lake, *Environ. Toxicol. Chem.*, 9, 909–918.

Wiener, J.G., D.P. Krabbenhoft, G.H. Heinz, and A.M. Scheuhammer. 2003. Ecotoxicology of mercury. In D.J.Hoffman, B.A. Rattner, G.A. Burton Jr., and J. Cairns Jr. (Eds.)., *Handbook of Ecotoxicology, 2nd edition*, p. 409–463. Lewis Publ., Boca Raton, FL.

World Health Organization (WHO). 1976. *Mercury. Environmental Health Criteria 1*. WHO, Geneva, 131 pp.

World Health Organization (WHO). 1990. *Methylmercury. Environmental Health Criteria 101*. WHO, Geneva. 144 pp.

World Health Organization (WHO). 1991. *Inorganic Mercury. Environmental Health Criteria 118*. WHO, Geneva. 168 pp.

CHAPTER 3

Properties

Mercury, a silver-white metal that is liquid at room temperature and highly volatile, can exist in three oxidation states: elemental mercury (Hg^o), mercurous ion (Hg_2^{2+}), and mercuric ion (Hg^{2+}). It can be part of both inorganic and organic compounds (USEPA, 1980; Clarkson et al., 1984). All mercury compounds interfere with thiol metabolism, causing inhibition or inactivation of proteins containing thiol ligands and ultimately to mitotic disturbances (Das et al., 1982; Elhassani, 1983; Hook and Hewitt, 1986). The mercuric species is the most toxic inorganic chemical form, but all three forms of inorganic mercury may have a common molecular mechanism of damage in which Hg^{2+} is the toxic species (Clarkson and Marsh, 1982). Mercury poisoning and treatment is discussed in more detail in Chapter 4.

3.1 PHYSICAL

Pure mercury is a coherent, silvery-white mobile liquid with a metallic luster (Anon., 1948). In thin layers it transmits a bluish-violet light. It freezes at about minus 39°C with contraction, forming a white, ductile, malleable mass easily cut with a knife, and with cubic crystals. When heated, the metal expands uniformly, boiling at 357.01°C at 760 mm, and vaporizing at about 360.0°C. The vapor is colorless. Mercury forms two well-defined series of salts: the mercurous salts derived from the oxide Hg_2O, and the mercuric salts from the oxide HgO. Mercuric oxide occurs in two forms: a bright red crystalline powder and as an orange-yellow powder. The yellow form is the most reactive and is transformed into the red when heated at 400.0°C. Heating the red form results in a black compound, which regains its color on cooling; on further heating to 630.0°C, it decomposes to mercury and oxygen (Anon., 1948).

Mercurous and mercuric chloride, known respectively as calomel and corrosive sublimate, are two of the most important salts of mercury (Table 3.1). Other halogenated mercury salts include mercurous bromide, Hg_2Br_2, a yellowish-white powder insoluble in water; mercuric bromide, $HgBr_2$, comprised of white crystals sparingly soluble in cold water but readily soluble in hot water; mercurous iodide, Hg_2I_2, a yellowish-green powder; and mercuric iodide, which exists in two crystalline forms. Other mercuric and mercurous compounds include nitrates, nitrites, sulfides, sulfates, phosphides, phosphates, and ammonium salts (Anon., 1948).

3.2 CHEMICAL

Elemental mercury is relatively inert in dry air, oxygen, nitrous oxide, carbon dioxide, ammonia, and some other gases at room temperatures (Anon., 1948). In damp air, it slowly becomes coated

Table 3.1 Some Properties of Mercury and Its Compounds[a]

Property	Elemental Mercury	Mercurous Chloride	Mercuric Chloride	Methylmercury Chloride
Empirical formula	Hg	Hg_2Cl_2	$HgCl_2$	CH_3HgCl
Molecular weight	200.59	472.09	271.52	251.09[c]
Chlorine, %	0	15.02	26.12	14.12[c]
Mercury, %	100	84.98	73.88	79.89[c]
Melting point, °C	–38.87	Sublimes at 400–500	277	170[c]
Density	13.534	7.15	5.4	4.063[c]
Solubility, mg/L (ppm)				
in water	0.056	2.0	74,070	1,016[d]
in benzene	2.387[b]	Insoluble	5,000	6,535[e]

[a] All data from *Merck Index* (1976), except where indicated.
[b] Spencer and Voigt (1961).
[c] Weast and Astle (1982).
[d] Eisler (unpublished), 72-h equilibrium value.
[e] Eisler (unpublished), 24-h equilibrium value.

with a film of mercurous oxide. When heated in air or oxygen, it is transformed into the red mercuric oxide, which decomposes into mercury and oxygen on continued heating at higher temperatures. Mercury dissolves many metals to form compounds called amalgams (Anon., 1948).

Chemical speciation is probably the most important variable influencing mercury toxicity, but mercury speciation is difficult to quantify, especially in natural environments (Boudou and Ribeyre, 1983). Mercury compounds in an aqueous solution are chemically complex. Depending on pH, alkalinity, redox, and other variables, a wide variety of chemical species are liable to be formed, having different electrical charges and solubilities. For example, $HgCl_2$ in solution can speciate into $Hg(OH)_2$, Hg^{2+}, $HgCl^+$, $Hg(OH)^-$, $HgCl_3^-$, and $HgCl_4^{2-}$; anionic forms predominate in saline environments (Boudou and Ribeyre, 1983). In the aquatic environment, under naturally occurring conditions of pH and temperature, mercury may also become methylated by biological or chemical processes, or both (Beijer and Jernelov, 1979; USEPA, 1980; Ramamoorthy and Blumhagen, 1984; Zillioux et al., 1993; Figure 3.1) — although biological methylation is limited (Callister and Winfrey, 1986). Methylmercury is the most hazardous mercury species due to its high stability, its lipid solubility, and its possession of ionic properties that lead to a high ability to penetrate membranes in living organisms (Beijer and Jernelov, 1979; Hamasaki et al., 1995). In general, essentially all mercury in freshwater fish tissues is in the form of methylmercury; however, methylmercury accounts for less than 1.0% of the total mercury pool in a lake (Regnell, 1990).

All mercury discharged into rivers, bays, or estuaries as elemental (metallic) mercury, inorganic divalent mercury, or phenylmercury or alkoxyalkyl mercury can be converted into methylmercury compounds by natural processes (Jernelov, 1969; Figure 3.1). Mercury methylation in ecosystems depends on mercury loadings, microbial activity, nutrient content, pH and redox conditions, suspended sediment load, sedimentation rates, and other variables (USNAS, 1978; Compeau and Bartha, 1984; Berman and Bartha, 1986; Callister and Winfrey, 1986; Jackson, 1986). Net methylmercury production was about 10 times higher in reduced sediments than in oxidized sediments (Regnell, 1990). The finding that certain microorganisms are able to convert inorganic and organic forms of mercury into the highly toxic methylmercury or dimethylmercury has made it clear that any form of mercury is highly hazardous to the environment (USEPA, 1980, 1985). The synthesis of methylmercury by bacteria from inorganic mercury compounds present in the water or in the sediments is the major source of this molecule in aquatic environments (Boudou and Ribeyre, 1983; Nakamura, 1994). This process can occur under both aerobic and anaerobic conditions (Beijer and Jernelov, 1979; Clarkson et al., 1984), but seems to favor anaerobic conditions (Olson and Cooper, 1976; Callister and Winfrey, 1986). Transformation of inorganic mercury to an organic form by

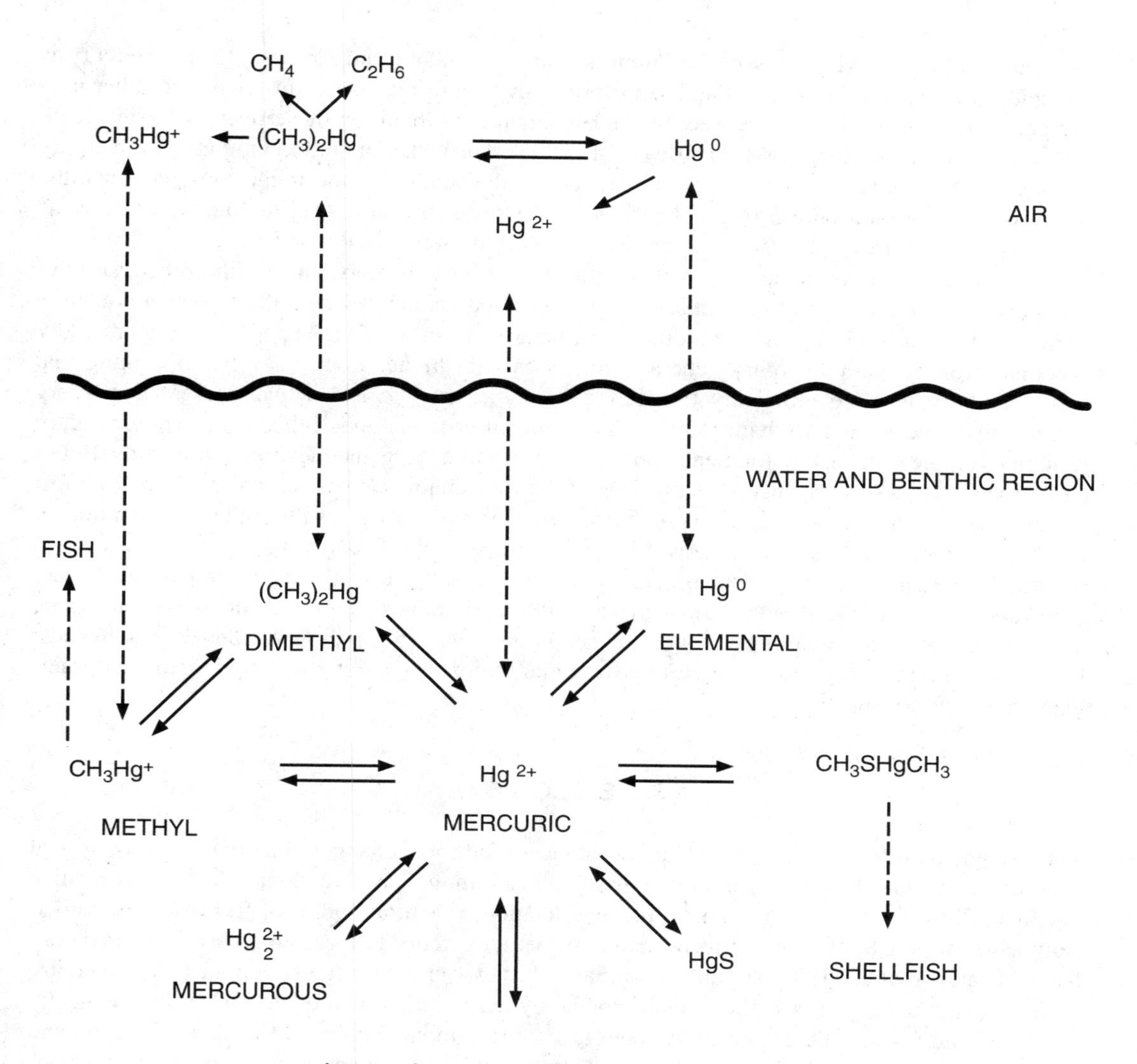

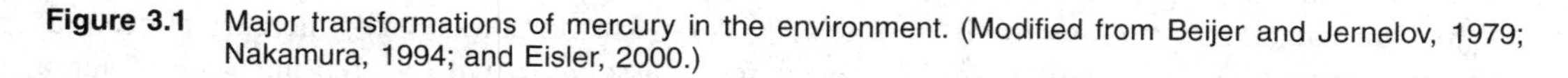

Figure 3.1 Major transformations of mercury in the environment. (Modified from Beijer and Jernelov, 1979; Nakamura, 1994; and Eisler, 2000.)

bacteria alters its biochemical reactivity and hence its fate (Windom and Kendall, 1979; Figure 3.1). Methylmercury is decomposed by bacteria in two phases. First, hydrolytic enzymes cleave the C–Hg bond, releasing the methyl group. Second, a reductase enzyme converts the ionic mercury to the elemental form, which is then free to diffuse from the aquatic environment into the vapor phase. These demethylating microbes appear to be widespread in the environment; they have been isolated from water, sediments, and soils and from the gastrointestinal tract of mammals, including humans (Clarkson et al., 1984). Some strains of microorganisms contain mercuric reductase, which transforms inorganic mercury to elemental mercury, and organomercurial lyase, which degrades organomercurials to elemental mercury (Baldi et al., 1991).

Humic substances can reduce inorganic divalent mercury (Hg^{2+}) to elemental mercury (Hg^{o}). In aquatic environments, Hg^{o} was highest under anoxic conditions, in the absence of chloride, and at pH 4.5 (Allard and Arsenie, 1991). Under these conditions, about 25.0% of 400.0 μg Hg^{2+}/L was reduced to Hg^{o} in 50 h. Production of Hg^{o} was reduced in the presence of europium ions and by methylated carboxyl groups in the humic substances (Allard and Arsenie, 1991). Mercury is efficiently transferred through wetlands and forests in a more reactive form relative to other land use patterns, resulting in an increased uptake by organisms inhabiting these rivers or downstream impoundments and drainage lakes (Hurley et al., 1995). The behavior and accumulation of mercury in forest soils of Guyana, South America, is related to the penetration of humic substances and the progressive adsorption onto iron oxy-hydroxides in the mineral horizons; flooding of these soils may lead to a release of 20.0% of the mercury initially present (Roulet and Lucotte, 1995).

Methylmercury is produced by methylation of inorganic mercury present in both freshwater and saltwater sediments, and accumulates in aquatic food chains in which the top-level predators usually contain the highest concentrations (Clarkson and Marsh, 1982). The percent of total mercury accounted for by methylmercury generally increases with higher trophic levels, confirming that methylmercury is more efficiently transferred to higher trophic levels than inorganic mercury compounds (Becker and Bigham, 1995). Organomercury compounds other than methylmercury decompose rapidly in the environment, and behave much like inorganic mercury compounds (Beijer and Jernelov, 1979). In organisms near the top of the food chain, such as carnivorous fishes, almost all mercury accumulated is in the methylated form, primarily as a result of the consumption of prey containing methylmercury; methylation also occurs at the organism level by way of mucus, intestinal bacteria, and enzymatic processes, but these pathways are not as important as diet (Huckabee et al., 1979; Boudou and Ribeyre, 1983). In tissues of marine flounders, inorganic mercury compounds are strongly bound to metallothioneins and high-molecular-weight ligands; however, methylmercury has a low affinity for metallothioneins and is strongly lipophilic (Barghigiani et al., 1989).

3.3 BIOLOGICAL

The biological cycle of mercury is delicately balanced, and small changes in input rates, geophysical conditions, and the chemical form of mercury can result in increased methylation rates in sensitive systems (USNAS, 1978). For example, the acidification of natural bodies of freshwater is statistically associated with elevated concentrations of methylmercury in the edible tissues of predatory fishes (Clarkson et al., 1984). Acidification has a stronger effect on the supply of methylmercury to the ecosystem than on specific rates of uptake by the biota (Bloom et al., 1991). In chemically sensitive waterways, such as poorly buffered lakes, the combined effects of acid precipitation and increased emissions of mercury to the atmosphere (with subsequent deposition) pose a serious threat to the biota if optimal biomethylation conditions are met (USNAS, 1978). In remote lakes of the Adirondack mountain region in upstate New York, fish contain elevated mercury concentrations in muscle; mercury loadings in fish were directly associated with decreasing water column pH and increasing concentrations of dissolved organic carbon (DOC), although high DOC concentrations may complex methylmercury, thus diminishing its bioavailability (Driscoll et al., 1995). At high concentrations of monomeric aluminum, the complexation of methylmercury with DOC decreases, enhancing the bioavailability of methylmercury (Driscoll et al., 1995).

Mercury excretion in mammals is through the urine and feces, and depends on the form of mercury, dose, and time postexposure (Goyer, 1986). With mercury vapor, there is minor initial loss from exhalation and major loss via fecal excretion. With inorganic mercury, fecal loss is predominant after initial exposure, and renal excretion increases with time. With methylmercury, about 90.0% is excreted in feces after acute or chronic exposure and does not change over time (Goyer, 1986).

Based on animal studies, all forms of mercury can cross the placenta to the fetus (Goyer, 1986). Fetal uptake of elemental mercury in rats — possibly due to its high solubility in lipids — is 10 to 40 times higher than uptake after exposure to inorganic mercury salts. Concentrations of mercury in the fetus after exposure to alkylmercuric compounds, when compared to elemental mercury, are twice those found in maternal tissues, and methylmercury levels in fetal red blood cells are 30.0% higher than in maternal cells. The positive fetal maternal gradient and increased mercury concentration in fetal erythrocytes enhance fetal toxicity to mercury, especially after exposure to alkylmercury. Maternal milk contains about 5.0% of the mercury concentration of maternal blood; however, neonatal exposure to mercury may be greatly augmented by nursing (Goyer, 1986).

Elemental or metallic mercury is oxidized, probably via catalases, to divalent mercury after absorption into body tissues (Goyer, 1986). Most inhaled mercury vapor absorbed into erythrocytes is transformed into divalent mercury, but some is also transported as metallic mercury to the brain, where biotransformation may occur. Some metallic mercury may cross the placenta into the fetus. Oxidized metallic mercury is then accumulated by brain and fetus. Organomercurials also undergo biotransformation to divalent mercury compounds in tissues by cleavage of the carbon–mercury bond, with no evidence of organomercury formation by mammalian tissues. Phenylmercurials are converted to inorganic mercury more rapidly than the shorter-chain methylmercurials. Phenyl and methoxyethyl mercurials are excreted at about the same rate as inorganic mercury, whereas methylmercury excretion is slower (Goyer, 1986). The half-time persistence of methylmercurials in mammalian tissues is about 70 days and seems to follow a linear pattern. For inorganic mercury salts, the biological half-time is about 40 days. And for elemental mercury or mercury vapor, the half-time persistence in tissues ranges between 35 and 90 days and also appears to be linear (Goyer, 1986).

3.4 BIOCHEMICAL

Mercury binds strongly with sulfhydryl groups, and has many potential target sites during embryogenesis; phenylmercury and methylmercury compounds are among the strongest known inhibitors of cell division (Birge et al., 1979). In mammalian hepatocytes, the L-alanine carrier contains a sulfhydryl group that is essential for its activity and is inhibited by mercurials (Sellinger et al., 1991). In the little skate (*Raja erinacea*), $HgCl_2$ inhibits Na^+-dependent alanine uptake and Na^+/K^+-ATPase activity, and increases K^+ permeability. Inhibition of Na^+-dependent alanine in skate hepatocytes by $HgCl_2$ is attributed to three different concentration-dependent mechanisms: (1) direct interaction with the transporters; (2) dissipation of the Na^+ gradient; and (3) loss of membrane integrity (Sellinger et al., 1991). Organomercury compounds, especially methylmercury, cross placental barriers and can enter mammals by way of the respiratory tract, gastrointestinal (GI) tract, skin, or mucus membranes (Elhassani, 1983). When compared with inorganic mercury compounds, organomercurials are more completely absorbed, are more soluble in organic solvents and lipids, pass more readily through biological membranes, and are slower to be excreted (Clarkson and Marsh, 1982; Elhassani, 1983; Greener and Kochen, 1983). Biological membranes, including those at the blood–brain interface and placenta, tend to discriminate against ionic and inorganic mercury, but allow relatively easy passage of methylmercury and dissolved mercury vapor (Greener and Kochen, 1983). As judged by membrane model studies, it appears that electrically neutral mercurials are responsible for most of the diffusion transport of mercury, although this movement is modified significantly by pH and mercury speciation. It appears, however, that the liposolubility of methylmercury is not the entire reason for its toxicity and does not play a major role in its transport. This hypothesis should be examined further in studies with living membranes (Boudou et al., 1983).

In liver cells, methylmercury forms soluble complexes with cysteine and glutathione, which are secreted in bile and reabsorbed from the GI tract. In general, however, organomercurials undergo cleavage of the carbon–mercury bond, releasing ionic inorganic mercury (Goyer, 1986). Mercuric mercury induces synthesis of metallothioneins, mainly in kidney cells. Mercury within renal cells becomes localized in lysosomes (Goyer, 1986).

The finding of naturally elevated mercury concentrations in marine products of commerce is worrisome to regulatory agencies charged with protection of human health. This topic is discussed later in Chapter 6 and Chapter 12. At this time, however, many authorities argue that mercury–selenium interactions are the key to methylmercury bioavailability in human diets. In the case of marine mammals and seabirds, for example, mercury is accumulated from the diet mainly as methylmercury and then transformed into the less toxic inorganic mercury. Accordingly, most of the tissue mercury found in high concentrations of these two marine groups is inorganic mercury (Itano et al., 1984; Thompson, 1990; Law, 1996; Yang et al., 2002). Many authorities aver that selenium detoxifies inorganic mercury by forming complexes in a 1:1 molar ratio (Koeman et al., 1973; Ping et al., 1986; Palmisano et al., 1995). Field studies on marine mammal livers corroborate the equimolar ratio of mercury and selenium (Koeman et al., 1973; Pelletier, 1985; Cuvin-Aralar and Furness, 1991; Kim et al., 1996); moreover, mercuric selenide (HgSe) — an inert end product of mercury detoxification in marine mammals — was found in the livers of marine mammals and birds (Martoja and Berry, 1980; Rawson et al., 1995; Nigro and Leonzio, 1996). A proposed mercury detoxification model in higher-trophic marine animals involves transformation of dietary methylmercury to inorganic mercury by reactive oxygen species (Suda et al., 1992; Hirayama and Yasutake, 2001; Yasutake and Hirayama, 2001); gut microflora (Rowland et al., 1984; Rowland, 1988); and selenium (Iwata et al., 1982). Inorganic mercury binds to metallothioneins or forms an equimolar Hg–Se complex, and subsequently combines with high molecular weight compounds in liver (Ikemoto et al., 2004). Glutathione molecules solubilize HgSe; the complex then binds to a specific protein (Gailer et al., 2000). The protein-bound Hg–Se complex is thought to be the precursor of mercuric selenide, which occurs in macrophages of marine mammals (Nigro, 1994; Rawson et al., 1995; Nigro and Leonzio, 1996).

3.5 MERCURY TRANSPORT AND SPECIATION

The global mercury cycle involves mercury release from geological and industrial processes into water and the atmosphere, followed by sedimentation via rainfall and by microbial metabolism that releases mercury from soil and sediments and transforms mercury from one chemical form to another (WHO, 1976, 1989, 1990, 1991; Fitzgerald, 1986; Fitzgerald and Clarkson, 1991; Barkay, 1992).

Atmospheric-borne mercury, including anthropogenic mercury, is deposited everywhere, including remote areas of the globe, hundreds of kilometers from the nearest mercury source, as evidenced by its presence in ancient lake sediments (Fitzgerald et al., 1998) and glacial ice (Schuster et al., 2002). In Amituk Lake in the Canadian Arctic, recent annual deposition of mercury was estimated at 15.1 kg, about 56.0% from snowpack and the rest from precipitation (Semkin et al., 2005). This represents a dramatic increase from historic annual burdens of 6.0 kg mercury annually in this remote area (Semkin et al., 2005); the effects of this increase on Arctic watersheds is unknown.

Many important sources affecting global mercury cycles emit elemental metallic mercury (Hg^o) in gaseous form, and to a lesser extent gaseous and particulate species of Hg^{2+} (USEPA, 1997; Pacyna et al., 2001). Gaseous and particulate Hg^{2+} are removed from the atmosphere through rainfall and dry deposition, thus limiting long-range transport (Lindberg and Stratton, 1998; Ebinghaus et al., 2002). Inorganic Hg^{2+} can be readily reduced to Hg^o by natural processes in terrestrial and aquatic ecosystems (Nakamura, 1994; Zhang and Lindberg, 2001). Elemental mercury can be oxidized in the atmosphere to Hg^{2+}, which is removed through wet and dry deposition (Lindberg et al., 2002). About 67.0% of the mercury in global fluxes is a result of human activities and the rest from natural emissions (Mason et al., 1994). Soils and sediments are the primary sinks for atmospherically derived mercury; however, these enriched pools can be remobilized through volatilization, leaching, and erosion (Wiener et al., 2003).

Mercury speciation varies in atmospheric, aquatic, and terrestrial environments. In the atmosphere, mercury is in the form of gaseous elemental Hg^o (95.0%), Hg^{2+} (called reactive gaseous mercury), and trace amounts of methylmercury (Lindberg and Stratton, 1998; Schroeder and Munthe, 1998). Particulate and reactive mercury in the atmosphere travels short distances, usually less than 50 km, and has a residence time of about 1 year (Mason et al., 1994). Reactive gaseous mercury is assumed to be $HgCl_2$, with some $Hg(NO_3)_2 \bullet H_2O$ in the gas phase (Stratton et al., 2001). Reactive gaseous mercury may comprise a majority of atmospheric gaseous mercury at some locations (e.g., springtime in Alaska) and this component is rapidly removed from the atmosphere by wet and dry deposition (Lindberg et al., 2002; Munthe et al., 2001) and available for methylation once deposited (Lindberg et al., 2002). Mercury point sources and rates of particle scavenging are key factors in atmospheric transport rates to sites of methylation and subsequent entry into the marine food chain (Rolfhus and Fitzgerald, 1995). Airborne soot particles transport mercury into the marine environment either as nuclei for raindrop formation or by direct deposition on water (Rawson et al., 1995). In early 1990, both dimethylmercury and monomethylmercury species were found in the subthermocline waters of the equatorial Pacific Ocean; the formation of these alkyl-mercury species in the low oxygen zone suggests that Hg^{2+} is the most likely substrate (Mason and Fitzgerald, 1991).

In aquatic environments, Hg^o and methylmercury species are the most common, with concentrations low, usually in the picogram-per-liter to microgram-per-liter range, except in the vicinity of anthropogenic or natural mercury sources (Wiener et al., 2003). The speciation of mercury in water is influenced by redox, pH, and ligands (Gill and Bruland, 1990). In most aerated surface waters near pH 7.0, ion-pair formation for CH_3^+ and Hg^{2+} is dominated by dissolved organic matter and chloride (Ravichandran et al., 1999). Under anoxic conditions, Hg^{2+} and CH_3^+ are present mainly as sulfide and sulfhydryl ion pairs (Benoit et al., 1999a; Gill et al., 1999). Complexed Hg^{2+} sulfides are less available for methylation (Benoit et al., 1999b). Concentrations of total mercury in uncontaminated, unfiltered freshwater samples range from 0.3 to 8.0 ng/L, but range from 10.0 to 40.0 ng/L near mercury sources (Wiener et al., 2003), and up to 1000.0 ng/L in waters contaminated by mercury tailings from gold mines (Eisler, 2004).

Sulfate-reducing bacteria are the most important mercury-methylating agents in aquatic environments (Gilmour et al., 1992, 1998), with the most important site of methylation at the oxic–anoxic interface in sediments (Pak and Bartha, 1998); a similar pattern is documented for wetlands (Krabbenhoft et al., 1995; Branfireum et al., 1996; Gilmour et al., 1998). In sediments, microbial methylation of mercury is fastest in the upper profiles where the rate of sulfate reduction is greatest (Hines et al., 2000; King et al., 2001). Methylcobalamin, produced by bacteria, is the active methyl donor to the Hg^{2+} ion; methylcobalamin reacts with Hg^{2+} to form CH_3Hg^+ (Choi et al., 1994). Methylation also occurs, but to a lesser degree, in aerobic freshwater and seawater, in aquatic plants, and in mucosal slime and intestines of fish (Wiener et al., 2003). Abiotic formation of CH_3Hg^+ compounds in sediments is documented; however, amounts formed are small when compared to biotic processes (Nagase et al., 1988; Falter and Wilken, 1998). Demethylation occurs via abiotic and biotic processes in the near-surface sediments and in the water column (Winfrey and Rudd, 1990; Oremland et al., 1991; Sellers et al., 1996). An oxidative demethylation pathway similar to that of the degradation of methanol or monomethylamine by methanogens has been proposed for methylmercury degradation (Marvin-DiPasquale et al., 1998, 2000). Photodegradation of methylmercury in surface waters of freshwater lakes is documented at rates up to 18.0% daily and is quantitatively important in mercury budgets of that ecosystem (Sellers et al., 1996, 2001; Branfireun et al., 1998; Krabbenhoft et al., 2002); the end products of mercury photodemethylation are not known with certainty (Krabbenhoft et al., 1998). Frequently, however, methylmercury concentrations in aerobic lakewater surfaces increase during sunlight hours or remain unchanged (Sicialiano et al., 2005). This phenomenon is linked to dissolved organic matter (DOM) and solar radiation — specifically to certain fractions of DOM that generates methylmercury when exposed to sunlight, especially in water from lakes with logged watersheds. The mechanism to account for methylmercury production is not clear at present; however, it corrects the conventional wisdom that methylmercury is rapidly photodegraded (Sicialiano et al., 2005).

Terrestrial soils are a significant contributor of mercury to surface waters (Mason et al., 1994; Cooper and Gillespie, 2001; Grigal, 2002; Gabriel and Williamson, 2004). Moreover, up to 60.0% of the atmospherically deposited mercury that reaches lakes originates from the associated terrestrial watershed (Krabbenhoft and Babiarz, 1992; Lorey and Driscoll, 1999; Grigal, 2002). The main biogeochemical reactions affecting the transport and speciation of mercury in the terrestrial watershed include formation of mercury ligands, mercury adsorption and desorption, and elemental mercury reduction and volatilization (Gabriel and Williamson, 2004). In terrestrial environments, OH^-, Cl^-, and S^- ions have the greatest impact on inorganic mercury–ligand formation. Under oxidized surface soil conditions, $Hg(OH)_2$, $HgCl_2$, $HgOH^+$, HgS (cinnabar), and Hg^0 are the dominant inorganic mercury forms and usually bind to organic and mineral ions. Under reducing conditions, common mercury forms are $HgSH^+$, HgOHSH, and HgClSH, and many are further bound to both inorganic and organic ligands. The following organomercurials predominated in terrestrial soils: CH_3HgCl > CH_3HgOH > free CH_3Hg^+ (Gabriel and Williamson, 2004).

In upland soils, mercury is mostly in the form of Hg^{2+} species sorbed to organic matter in the humus layer and a lesser extent to soil minerals (Kim et al., 1997). The overall adsorption of mercury to mineral and organic particles is positively correlated, in order of importance, with surface area, organic content, cation exchange capacity, and grain size (Gabriel and Williamson, 2004). Mercury adsorption and desorption to mineral and organic surfaces is strongly influenced by pH and dissolved ions; for example, increased Cl^- and decreased pH — alone or together — can decrease mercury adsorption, and clays and organic soils have the highest capability of adsorbing mercury (Gabriel and Williamson, 2004). Common forms of methylated mercury in soils depend on pH (Reimers and Krenkel, 1974; Jackson, 1998). At pH 2 to about 4.7, the most common forms were CH_3HgCl > free CH_3Hg^+ > CH_3HgOH; at pH 4.7 to about 7.5, the most common forms were CH_3HgCl > CH_3HgOH > free CH_3Hg^+; and at pH 7.5 to 10, this order was CH_3HgOH > CH_3HgCl >> free CH_3Hg^+ (Reimers and Krenkel, 1974; Jackson, 1998). Reduction of abiotic inorganic mercury increases with increasing electron donors, low redox potential, and sunlight intensity (Gabriel and Williamson 2004). Factors that increase mercury volatilization from soils include increased soil permeability, higher temperatures, and increased sunlight intensity; therefore, increased volatilization is expected in tropical climates. A decrease in mercury adsorption and an increase in soil moisture can also increase volatilization (Gabriel and Williamson, 2004).Gabriel and Williamson (2004) recommend that additional research be conducted on inorganic mercury–ligand formation in water and runoff and its effects on methylmercury formation in soils, and on quantification of the sources and transport characteristics of methylmercury in terrestrial environments. The mercury–ligand form exiting the terrestrial watershed will strongly influence the mercury/methylmercury bioaccumulation potential in surface waters. Accordingly, more analyses are needed to determine the mercury forms in terrestrial watershed runoff in dissolved and particulate fractions (Gabriel and Williamson, 2004).

3.6 MERCURY MEASUREMENT

Techniques for analysis of different mercury species in biological samples and abiotic materials include atomic absorption, cold vapor atomic fluorescence spectrometry, gas-liquid chromatography with electron capture detection, neutron activation, and inductively coupled plasma mass spectrometry (Takizawa, 1975; Lansens et al., 1991; Schintu et al., 1992; Porcella et al., 1995). Methylmercury concentrations in marine biological tissues are detected at concentrations as low as 10.0 μg Hg/kg tissue using graphite furnace sample preparation techniques and atomic absorption spectrometry (Schintu et al., 1992).

3.7 SUMMARY

Physical, chemical, biological, and biochemical properties of mercury are briefly reviewed; mercury transport and speciation processes are summarized; and analytical techniques listed for mercury measurement.

In mammals, all forms of mercury can cross the placenta to the fetus and interfere with thiol metabolism. Chemical speciation is probably the most important variable influencing mercury toxicity; however, speciation is difficult to quantify. In freshwater lakes, for example, mercury speciation depends, in part, on pH, alkalinity, redox, and microbial activity. Most authorities agree that all mercury species discharged into natural bodies of water can be converted into methylmercurials — the most toxic form — at rates influenced, in part, by mercury loadings, nutrient content, sedimentation rates, and suspended sediment loadings. Bioavailability of methylmercury to aquatic biota is highly dependent on lake chemistry, mercury deposition rates, dissolved organic carbon, and other variables. Methylmercury, in turn, is decomposed abiotically and by bacteria containing mercuric reductase and organomercurial lyase; these demethylating strains of bacteria are common in the environment and have been isolated from water, sediments, soils, and the GI tract of humans and other mammals. Other mechanisms can reduce inorganic Hg^{2+} to Hg^{o} in freshwater, with Hg^{o} production higher under anoxic conditions, in a chloride-free environment, and at pH 4.5.

Methylmercury concentrations in tissues of marine fishes can now be detected at levels greater than 10.0 μg/kg tissue using graphite furnace sample preparation techniques and atomic absorption spectrometry.

REFERENCES

Allard, B. and I. Arsenie. 1991. Abiotic reduction of mercury by humic substances in aquatic system — an important process for the mercury cycle, *Water Air Soil Pollut.*, 56, 457–464.

Anon. 1948. Mercury, *Encylop. Brittanica*, 15, 269–272.

Baldi, F., F. Semplici, and M. Filippelli. 1991. Environmental applications of mercury resistant bacteria, *Water Air Soil Pollut.*, 56, 465–475.

Barghigiani, C., D. Pellegrini, and E. Carpene. 1989. Mercury binding proteins in liver and muscle of flat fish from the northern Tyrrhenian Sea, *Comp. Biochem. Physiol.*, 94C, 309–312.

Barkay, T. 1992. Mercury cycle. In *Encyclopedia of Microbiology*, Vol. 3, p. 65–74. Academic Press, San Diego.

Becker, D.S. and G.N. Bigham. 1995. Distribution of mercury in the aquatic food web of Onondaga Lake, New York, *Water Air Soil Pollut.*, 80, 563–571.

Beijer, K. and A. Jernelov. 1979. Methylation of mercury in natural waters. In J.0. Nriagu (Ed.). *The Biogeochemistry of Mercury in the Environment, p. 201–210.* Elsevier/North-Holland Biomedical Press, New York.

Benoit, J.M., C.C. Gilmour, R.P. Mason, and A. Heyes. 1999a. Sulfide controls on mercury speciation and bioavailability to methylating bacteria in sediment and pore waters, *Environ. Sci. Technol.*, 33, 951–957.

Benoit, J.M., R.P. Mason, and C.C. Gilmour. 1999b. Estimation of mercury-sulfide speciation and bioavailability in sediment pore waters using octanol-water partitioning, *Environ. Toxicol. Chem.*, 18, 2138–2141.

Berman, M. and R. Bartha. 1986. Levels of chemical versus biological methylation of mercury in sediments, *Bull. Environ. Contam. Toxicol.*, 36, 401–404.

Birge, W.J., J.A. Black, A.G. Westerman, and J.E. Hudson. 1979. The effect of mercury on reproduction of fish and amphibians. In J.0. Nriagu (Ed.), *The Biogeochemistry of Mercury in the Environment,* p. 629–655. Elsevier/North-Holland Biomedical Press, New York.

Bloom, N.S., C.J. Watras, and J.P. Hurley. 1991. Impact of acidification on the methylmercury cycle of remote seepage lakes, *Water Air Soil Pollut.*, 56, 477–491.

Boudou, A., D. Georgescauld, and J.P. Desmazes. 1983. Ecotoxicological role of the membrane barriers in transport and bioaccumulation of mercury compounds. In J.0. Nriagu (Ed.), *Aquatic Toxicology,* p. 117–136. John Wiley, New York.

Boudou, A. and F. Ribeyre. 1983. Contamination of aquatic biocenoses by mercury compounds: an experimental toxicological approach. In J.O. Nriagu (Ed.), *Aquatic Toxicology*, p. 73–116. John Wiley, New York.

Branfireum, B.A., A. Heyes, and N.T. Roulet. 1996. The hydrology and methylmercury dynamics of a Precambrian shield headwater peatland, *Water Resour. Res.*, 32, 1785–1794.

Branfireun, B.A., D. Hilbert, and N.T. Roulet. 1998. Sinks and sources of methylmercury in a boreal catchment, *Biogeochemistry*, 41, 277–291.

Callister, S.M. and H.R. Winfrey. 1986. Microbial methylation of mercury in upper Wisconsin River sediments, *Water Air Soil Pollut.*, 29, 453–465.

Choi, S.C., T. Chase, and R. Bartha. 1994. Enzymatic catalysis of mercury methylation by *Desulfovibrio desulfuricans LS*, *Appl. Environ. Microbiol.*, 60, 1342–1346.

Clarkson, T.W., R. Hamada, and L. Amin–Zaki. 1984. Mercury. In J.O. Nriagu (Ed.), *Changing Metal Cycles and Human Health*, p. 285–309. Springer-Verlag, Berlin.

Clarkson, T.W. and D.O. Marsh. 1982. Mercury toxicity in man. In A.S. Prasad (Ed.), *Clinical, Biochemical, and Nutritional Aspects of Trace Elements,* Vol. 6, p. 549–568. Alan R. Liss, Inc., New York.

Compeau, G. and R. Bartha. 1984. Methylation and demethylation of mercury under controlled redox, pH, and salinity conditions, *Appl. Environ. Microbiol.*, 48, 1202–1207.

Cooper, C.M. and W.B. Gillespie. 2001. Arsenic and mercury concentrations in major landscape components of an intensively cultivated watershed, *Environ. Pollut.*, 111, 67–74.

Cuvin-Aralar, M.L.A. and R.W. Furness. 1991. Mercury and selenium interaction: a review, *Ecotoxicol. Environ. Safety*, 21, 348–364.

Das, S.K., A. Sharma, and G. Talukder. 1982. Effects of mercury on cellular systems in mammals — a review, *Nucleus (Calcutta)*, 25, 193–230.

Driscoll, C.T., V. Blette, C. Yan, C.L. Schofield, R. Munson, and J. Holsapple. 1995. The role of dissolved organic carbon in the chemistry and bioavailability of mercury in remote Adirondack lakes, *Water Air Soil Pollut.*, 80, 499–508.

Ebinghaus, R. H.H. Kock, C. Temme, J.W. Einax, A.G. Lowe, A. Richter, J.P. Burrows, and W.H. Schroeder. 2002. Antarctic springtime depletion of atmospheric mercury, *Environ. Sci. Technol.*, 36, 1238–1244.

Eisler, R. 2000. Mercury. In *Handbook of Chemical Risk Assessment: Health Hazards to Humans, Plants, and Animals,* Vol. 1, p. 313–409. Lewis Publishers, Boca Raton, FL.

Eisler, R. 2004. Mercury hazards from gold mining to humans, plants, and animals, *Rev. Environ. Contam. Toxicol.*, 181, 139–198.

Elhassani, S.B. 1983. The many faces of methylmercury poisoning, *J. Toxicol.*, 19, 875–906.

Falter, R. and R.D. Wilken. 1998. Isotope experiments for the determination of abiotic mercury methylation potential of a River Rhine sediment, *Vom Wasser*, 90, 217–232.

Fitzgerald, W.F. 1986. Cycling of mercury between the oceans and the atmosphere. In P. Buat-Menard (Ed.), *The Role of Air-Sea Exchange in Geochemical Cycling,* p. 363–408. D. Reidel Publ., Boston.

Fitzgerald, W.F. and T.W. Clarkson. 1991. Mercury and monomethylmercury: present and future concerns, *Environ. Health Perspect.*, 96, 159–166.

Fitzgerald, W.F., D.R. Engstrom, R.P. Mason, and E.A. Nater. 1998. The case for atmospheric mercury contamination in remote areas, *Environ. Sci. Technol.*, 32, 1–7.

Gabriel, M.C. and D.G. Williamson. 2004. Principal biogeochemical factors affecting the speciation and transport of mercury through the terrestrial environment, *Environ. Geochem. Health*, 26, 421–434.

Gailer, J., G.N. George, I.J. Pickering, S. Madden, R.C. Prince, and E.Y. Yu. 2000. Structural basis of the antagonism between inorganic mercury and selenium in mammals, *Chem. Res. Toxicol.*, 13, 1135–1142.

Gill, G.A. and K.W. Bruland. 1990. Mercury speciation in surface freshwater systems in California and other areas, *Environ. Sci. Technol.*, 24, 1392–1400.

Gilmour, C.C., E.A. Henry, and R. Mitchell. 1992. Sulfate stimulation of mercury methylation in freshwater sediments, *Environ. Sci. Technol.*, 26, 2281–2287.

Gilmour, C.C., G.S. Reidel, M.C. Ederington, J.T. Bell, J.M. Benoit, G.A. Gill, and M.C. Stordal. 1998. Methylmercury concentrations and production rates across a trophic gradient in the northern Everglades, *Biogeochemistry*, 40, 327–345.

Goyer, R.A. 1986. Toxic effects of metals. In C.D. Klaassen, M.O. Amdur, and J. Doull (Eds.), *Casarett and Doull's Toxicology, third edition,* p. 582–635. Macmillan, New York.

Greener, Y. and J.A. Kochen. 1983. Methyl mercury toxicity in the chick embryo, *Teratology*, 28, 23–28.

Grigal, D.F. 2002. Inputs and outputs of mercury from terrestrial watersheds: a review, *Environ. Rev.*, 10, 1–39.

Hamasaki, T., H. Nagawe, Y. Yoshioka, and T. Sato. 1995. Formation, distribution, and ecotoxicity of methylmetals of tin, mercury, and arsenic in the environment, *Crit. Rev. Environ. Sci. Technol.*, 25, 45–91.

Hines, M.E., M. Horvat, J. Faganeli, J.C.J. Bonzongo, T. Barkay, E.B. Major, K.J. Scott, E.A. Bailey, J.J Warwick, and W.B. Lyons. 2000. Mercury biogeochemistry in the Idrija River, Slovenia, from above the mine into the Gulf of Trieste, *Environ. Res.*, 83A, 129–139.

Hirayama, K. and A. Yasutake. 2001. *In vivo* degradation of methylmercury: its mechanism and significance in methylmercury induced neurotoxicity. In Y. Takizawa and M. Osame (Eds.), *Understanding of Minamata Disease: Methylmercury Poisoning in Minamata and Niigata, Japan,* p. 103–110. Japan Public Health Assn., Tokyo.

Hook, J.B. and W.R. Hewitt. 1986. Toxic responses of the kidney. Pages 310–329 In C.D. Klaassen, M.O. Amdur, and J. Doull (Eds.), *Casarett and Doull's Toxicology, third edition,* p. 310–329. Macmillan, New York.

Huckabee, J.W., J.W. Elwood, and S.G. Hildebrand. 1979. Accumulation of mercury in freshwater biota. In J.O. Nriagu (Ed.), *The Biogeochemistry of Mercury in the Environment.* p. 277–302. Elsevier/North-Holland Biomedical Press, NY.

Hurley, J.P., J.M. Benoit, C.L. Babiarz, M.M. Shafer, A.W. Andren, J. R. Sullivan, R. Hammond, and D. A. Webb. 1995. Influences of watershed characteristics on mercury levels in Wisconsin rivers, *Environ. Sci. Technol.*, 29, 1867–1875.

Ikemoto, T., T. Kunito, H. Tanaka, N. Baba, N. Miyazaki, and S. Tanabe. 2004. Detoxification mechanism of heavy metals in marine mammals and seabirds: interaction of selenium with mercury, silver, zinc, and cadmium in liver, *Arch. Environ. Contam. Toxicol.*, 47, 402–413.

Itano, K., S. Kawai, N. Miyazaki, R.Tatsukawa, and T. Fujiyama. 1984. Mercury and selenium levels in striped dolphins caught off the Pacific coast of Japan, *Agricul. Biol. Chem.*, 48, 1109–1116.

Iwata, H., T. Masukawa, H. Kito, and M. Hiyashi. 1982. Degradation of methylmercury by selenium, *Life Sci.*, 31, 859–866.

Jackson, T.A. 1986. Methyl mercury levels in a polluted prairie river-lake system: seasonal and site-specific variations, and the dominant influence of trophic conditions, *Can. J. Fish. Aquat. Sci.*, 43, 1873–1887.

Jackson, T.A. 1998. Mercury in aquatic ecosystems. In W.J. Langston and M.J. Bebianno (Eds.), *Metal Metabolism in Aquatic Environments*, p. 76–157. Chapman & Hall, London.

Jernelov, A. 1969. Conversion of mercury compounds. In M.W. Miller and G.G. Berg (Eds.), *Chemical Fallout, Current Research on Persistent Pesticides*, p. 68–74. Chas. C Thomas, Springfield, IL.

Kim, E.Y., K. Saeki, S. Tanabe, H. Tanaka, and R. Tatsukawa. 1996. Specific accumulation of mercury and selenium in seabirds, *Environ. Pollut.*, 94, 261–265.

Kim, J.H., P.J. Hanson, M.O. Barnett, and S.E. Lindberg. 1997. Biogeochemistry of mercury in the air-soil-plant system. In A. Sigel and H. Sigel (Eds.), *Metal Ions in Biological Systems, Vol. 34, Mercury and its Effects on Environment and Biology,* p. 185–212. Marcel Dekker, New York.

King, J.K., J.E. Kostka, M.E. Frischer, F.M. Saunders, and R.A. Jahnke. 2001. A quantitative relationship that demonstrates mercury methylation rates in marine sediments are based on the community composition and activity of sulfate-reducing bacteria, *Environ. Sci. Technol.*, 35, 2491–2496.

Koeman, J.H, W.H.M. Peeters, C.H.M. Koudstaal-Hol, P.S. Tijoe, and J.J.M. de Goeij. 1973. Mercury-selenium correlations in marine mammals, *Nature*, 254, 385–386.

Krabbenhoft, D.P. and C. Babiarz. 1992. The role of groundwater transport in aquatic mercury cycling, *Water Resour. Res.*, 28, 3119–3128.

Krabbenhoft, D.P., J.M. Benoit, C.L. Babiarz, J.P. Hurley, and A.W. Andrem. 1995. Mercury cycling in the Allequash Creek watershed, *Water Air Soil Pollut.*, 80, 425–433.

Krabbenhoft, D.P., J.P. Hurley, M.L. Olson, and L.B. Cleckner. 1998. Diel variability of mercury phase and species distributions in the Florida Everglades, *Biogeochemistry*, 40, 311–325.

Krabbenhoft, D.P., M.L. Olson, J.F. Dewild, D.W. Clow, R.S. Striegl, M.M. Dornblaster, and P. Van Metre. 2002. Mercury loading and methylmercury production and cycling in high altitude lakes from the western United States, *Water Air Soil Pollut. Focus*, 2, 233–249.

Law, R.J. 1996. Metals in marine mammals. In W.N. Beyer, G.H. Heinz, and A.W. Redmon-Norwood (Eds.), *Environmental Contaminants in Wildlife: Interpreting Tissue Concentrations,* p. 357–376. CRC Press, Boca Raton, FL.

Lansens, P., M. Leermakers, and W. Baeyens. 1991. Determination of methylmercury in fish by headspace-gas chromatography with microwave-induced-plasma detection, *Water Air Soil Pollut.*, 56, 103–115.

Lindberg, S.E., S. Brooks, C.J. Lin, K.J. Scott, M.S. Landis, R.K. Stevens, M. Goodsite, and A. Richter. 2002. Dynamic oxidation of gaseous mercury in the Arctic troposphere at polar sunrise, *Environ. Sci. Technol.*, 36, 1245–1256.

Lindberg, S.E. and W.J. Stratton. 1998. Atmospheric mercury speciation: concentrations and behavior of reactive gaseous mercury in ambient air, *Environ. Sci. Technol.*, 32, 49–57.

Lorey, P. and C.T. Driscoll. 1999. Historical trends of mercury deposition to Adirondack lakes, *Environ. Sci. Technol.*, 33, 718–722.

Martoja, R. and J.P. Berry. 1980. Identification of tiemannite as a probable product of demethylation of mercury by selenium in cetaceans: a complement to the scheme of the biological cycle of mercury, *Vie Milieu*, 30, 7–10.

Marvin-DiPasquale, M., J. Agee, C. McGowan, R.S. Oremland, M. Thomas, D.P. Krabbenhoft, and C.C. Gilmour. 2000. Methyl mercury degradation pathways — a comparison among three mercury impacted ecosystems, *Environ. Sci. Technol.*, 34, 4908–4916.

Marvin-DiPasquale, M. and R.S. Oremland. 1998. Bacterial methylmercury degradation in Florida Everglades sediment and periphyton, *Environ. Sci. Technol.*, 32, 2556–2563.

Mason, R.P. and W.F. Fitzgerald. 1991. Mercury speciation in open ocean waters, *Water Air Soil Pollut.*, 56, 779–789.

Mason, R.P., W.F. Fitzgerald, and F.M.M. Morel. 1994. The biogeochemical cycling of elemental mercury: anthropogenic influences, *Geochim. Cosmochim. Acta*, 58, 3191–3198.

Merck Index. 1976. *The Merck Index, 9th edition.* Merck and Co. Inc., Rahway, NJ.

Munthe, J., I. Wangberg, N. Pirrone, A. Iverfeld, R. Ferrara, R. Ebinghaus, X. Feng, K. Gardfeldt, G. Keeler, E. Lanzillotta, S. Lindberg, J. Lu, Y. Mamane, E. Prestbo, S. Schmolke, W.H. Schroeder, J. Sommar, F. Sprovieri, R.K. Stevens, W. Stratton, G. Tuncel, and A. Urba. 2001. Intercomparison of methods for sampling and analysis of atmospheric mercury species, *Atmospher. Environ.*, 35, 3007–3017.

Nagase, H., Y. Ose, and T. Sato. 1988. Possible methylation of inorganic mercury by silicones in the environment, *Sci. Total Environ.*, 73, 29–36.

Nakamura, K. 1994. Mercury compounds-decomposing bacteria in Minamata Bay. In Proceedings of the International Symposium on "Assessment of Environmental Pollution and Health Effects from Methylmercury," p. 198–209. October 8–9, 1993, Kumamoto, Japan. National Institute for Minamata Disease, Minamata City, Kumamoto 867, Japan.

Nigro, M. 1994. Mercury and selenium localization in macrophages of the striped dolphin, *Stenella coeruleoalba*, *J. Mar. Biol. Assoc. U.K.*, 74, 975–978.

Nigro, M. and C. Leonzio. 1996. Intracellular storage of mercury and selenium in different marine vertebrates, *Mar. Ecol. Prog. Ser.*, 135, 137–143.

Olson, B.H. and R.C. Cooper. 1976. Comparison of aerobic and anaerobic methylation of mercuric chloride by San Francisco Bay sediments, *Water Res.*, 10, 113–116.

Oremland, R.S., C.W. Culbertson, and M.R. Winfrey. 1991. Methylmercury decomposition in sediments and bacterial cultures--involvement of methanogens and sulfate reducers in oxidative demethylation, *Appl. Environ. Microbiol.*, 57, 130–140.

Pacyna, E.G., J.M. Pacyna, and N. Pirrone. 2001. European emissions of atmospheric mercury from anthropogenic sources in 1995, *Atmosph. Environ.*, 35, 2987–2996.

Pak, K.R. and R. Bartha. 1998. Mercury methylation and demethylation in anoxic lake sediments and by strictly anaerobic bacteria, *Appl. Environ. Microbiol.*, 64, 1013–1017.

Palmisano, F., N. Cardellicchio, and P.G. Zambonin. 1995. Speciation of mercury in dolphin liver: a two-stage mechanism for the demethylation accumulation process and role of selenium, *Mar. Environ. Res.*, 40, 109–121.

Pelletier, E. 1985. Mercury-selenium interactions in aquatic organisms: a review, *Mar. Environ. Res.*, 18, 111–132.

Ping, L., H. Nagasawa, K. Matsumoto, A. Suzuki, and K. Fuwa. 1986. Extraction and purification of a new compound containing selenium and mercury accumulated in dolphin liver, *Biol. Trace Elem. Res.*, 11, 185–199.

Porcella, D., J. Huckabee, and B. Wheatley (Eds.). 1995. Mercury as a global pollutant. *Water Air Soil Pollut.*, 80, 1–1336.

Ramamoorthy, S. and K. Blumhagen. 1984. Uptake of Zn, Cd, and Hg by fish in the presence of competing compartments, *Can. J. Fish. Aquat. Sci.*, 41, 750–756.

Ravichandran, M., G.R. Aiken, J.N. Ryan, and M.M. Reddy. 1999. Inhibition of precipitation and aggregation of metacinnabar (mercury sulfide) by dissolved organic matter isolated from the Florida Everglades, *Environ. Sci. Technol.*, 33, 1418–1423.

Rawson, A.J., J.P. Bradley, A. Teetsov, S.B. Rice, E.M. Haller, and G.W Patton. 1995. A role for airborne particulates in high mercury levels of some cetaceans, *Ecotoxicol. Environ. Safety*, 30, 309–314.

Regnell, O. 1990. Conversion and partitioning of radio-labelled mercury chloride in aquatic model systems, *Can. J. Fish. Aquat. Sci.*, 47, 548–553.

Reimers, R.S. and P.A. Krenkel. 1974. Kinetics of mercury adsorption and desorption in sediments, *J. Water Pollut. Control Feder.*, 46, 352–365.

Rolfhus, K.R. and W.F. Fitzgerald. 1995. Linkages between atmospheric mercury deposition and the methylmercury content of marine fish, *Water Air Soil Pollut.*, 80, 291–297.

Roulet, M. and M. Lucotte. 1995. Geochemistry of mercury in pristine and flooded ferralitic soils of a tropical rain forest in French Guiana, South America, *Water Air Soil Pollut.*, 80, 1079–1088.

Rowland, I.R. 1988. Interactions of the gut microflora and the host in toxicology, *Toxicol. Pathol.*, 16, 147–153.

Rowland, I.R., R.D. Robinson, and R.A. Doherty. 1984. Effects of diet on mercury metabolism and excretion in mice given methylmercury: role of gut flora, *Arch. Environ. Health*, 39, 401–408.

Schintu, M., F. Jean-Caurant, and J.C. Amiard. 1992. Organomercury determination in biological reference materials: application to a study on mercury speciation in marine mammals off the Faroe Islands, *Ecotoxicol. Environ. Safety*, 24, 95–101.

Schroeder, W.H. and J. Munthe. 1998. Atmospheric mercury — an overview, *Atmos. Environ.*, 32, 809–822.

Schuster, P.F., D.P. Krabbenhoft, D.L. Naftz, L.D. Cecil, M.L. Olson, and J.F. Dewild. 2002. Atmospheric mercury deposition during the last 270 years: a glacial ice core record of natural and anthropogenic sources, *Environ. Sci. Technol.*, 36, 2302–2310.

Sellers, P., C.A. Kelly, and J.W.M. Rudd. 2001. Fluxes of methylmercury to the water column of a drainage lake: the relative importance of internal and external sources, *Limnol. Ocean.*, 46, 623–631.

Sellers, P., C.A. Kelly, J.W.M. Rudd, and A.R. MacHutchon. 1996. Photodegradation of methylmercury in lakes, *Nature*, 380, 694–697.

Sellinger, M., N. Ballatori, and J.L. Boyer. 1991. Mechanism of mercurial inhibition of sodium-coupled alanine uptake in liver plasma membrane vesicles from *Raja erinacea*, *Toxicol. Appl. Pharmacol.*, 107, 369–376.

Semkin, R.G., G. Mierle, and R.J. Neureuther. 2005. Hydrochemistry and mercury cycling in a high Arctic watershed, *Sci. Total Environ.*, 342, 199–221.

Sicialiano, S.D., N.J. O'Driscoll, R. Tordon, J. Hill, S. Beauchamp, and D.R.S. Lean. 2005. Abiotic production of methylmercury by solar radiation, *Environ. Sci. Technol.*, 39, 1071–1077.

Spencer, J.N. and A.F. Voigt. 1961. Thermodynamics of the solution of mercury metal. I. Tracer determination of the solubility in various liquids, *Jour. Phys. Chem.*, 72, 464–470.

Stratton, W.J., S. Lindberg, and C.J. Perry. 2001. Atmospheric mercury speciation: laboratory and field evaluation of a mist chamber method for measuring reactive gaseous mercury, *Environ. Toxicol. Chem.*, 35, 170–177.

Suda, I., S. Totoki, T. Uchida, and H. Takahashi. 1992. Degradation of methyl and ethyl mercury into inorganic mercury by various phagocytic cells, *Arch. Toxicol.*, 66, 40–44.

Takizawa, Y. 1975. Studies on the distribution of mercury in several body organs — application of activography to examination of mercury distribution of Minamata Disease patients. In *Studies on the Health Effects of Alkylmercury in Japan*, p. 5–9. Environment Agency Japan, Tokyo.

Thompson, D.R. 1990. Metal levels in marine vertebrates. In R.W. Furness and P.S. Rainbow (Eds.), *Heavy Metals in the Marine Environment*, p. 143–182. CRC Press, Boca Raton, FL.

U.S. Environmental Protection Agency (USEPA). 1980. Ambient water quality criteria for mercury, *U.S. Environ. Protection Agen. Rep. 440/5-80-058.* Available from Natl. Tech. Infor. Serv., 5285 Port Royal Road, Springfield, VA 22161.

U.S. Environmental Protection Agency (USEPA). 1985. Ambient water quality criteria for mercury — 1984. *U.S. Environ. Protection Agen. Rep. 440/5-84-026.* 136 pp. Available from Natl. Tech. Infor. Serv., 5285 Port Royal Road, Springfield, VA 22161.

U.S. Environmental Protection Agency (USEPA). 1997. Mercury study report to Congress, U.S. Environ. Protection Agen. Publ. 452R-97--004.. Washington, D.C.

U.S. National Academy of Sciences (USNAS). 1978. An Assessment of Mercury in the Environment. Natl. Acad. Sci., Washington, D.C., 185 pp.

Weast, R.C. and M. J. Astle (Eds.). 1982. *CRC Handbook of Chemistry and Physics*. CRC Press, Boca Raton, FL.

Wiener, J.G., D.P. Krabbenhoft, G.H. Heinz, and A.M. Scheuhammer. 2003. Ecotoxicology of mercury. In D.J. Hoffman, B.A. Rattner, G.A. Burton Jr., and J. Cairns Jr. (Eds.)., *Handbook of Ecotoxicology, 2nd edition*, p. 409–436. Lewis Publ., Boca Raton, FL.

Windom, H.L. and D.R. Kendall. 1979. Accumulation and biotransformation of mercury in coastal and marine biota. In J.O. Nriagu (Ed.), *The Biogeochemistry of Mercury in the Environment,* p. 301–323. Elsevier/North-Holland Biomedical Press, NY.

Winfrey, M.R. and J.W.M. Rudd. 1990. Environmental factors affecting the formation of methylmercury in low pH lakes, *Environ. Toxicol. Chem.*, 9, 853-869.

World Health Organization (WHO). 1976. *Mercury. Environmental Health Criteria 1*. WHO, Geneva, 131 pp.

World Health Organization (WHO). 1989. *Mercury — Environmental Aspects. Environmental Health Criteria 86*. WHO, Geneva. 115 pp.

World Health Organization (WHO). 1990. *Methylmercury. Environmental Health Criteria 101*. WHO, Geneva. 144 pp.

World Health Organization (WHO). 1991. *Inorganic Mercury. Environmental Health Criteria 118*. WHO, Geneva. 168 pp.

Yang, J., T. Kunito, S. Tanabe, and N. Miyazaki. 2002. Mercury in tissues of Dall's porpoise (*Phocoenoides dalli*) collected on Sanriku coast of Japan, *Fish Sci.*, 68 (Suppl. 1), 256–259.

Yasutake, A. and K. Hirayama. 2001. Evaluation of methylmercury biotransformation using rat liver slices, *Arch. Toxicol.*, 75, 400–406.

Zhang, H. and S.E. Lindberg. 2001. Sunlight and iron(III)-induced photochemical production of dissolved gaseous mercury in freshwater, *Environ. Sci. Technol.*, 35, 928–935.

Zillioux, E.J., D.B. Porcella, and J.M. Benoit. 1993. Mercury cycling and effects in freshwater wetland ecosystems, *Environ. Toxicol. Chem.*, 12, 2245–2264.

CHAPTER 4

Mercury Poisoning and Treatment

The toxicity of mercury has been recognized since antiquity (Hook and Hewitt, 1986). No other metal demonstrates the diversity of effects caused by different biochemical forms than does mercury (Goyer, 1986). Toxicologically, there are three forms of mercury: (1) elemental mercury, (2) inorganic mercury compounds, and (3) organomercurials (Goyer, 1986). Among the various forms of mercury and its compounds, elemental mercury in the form of vapor, mercuric mercury, and methylmercury have the greatest toxicological potential (Satoh, 1995). Metallic or elemental mercury volatilizes to mercury vapor at ambient air temperatures, and most human exposure is by inhalation. Mercury vapor is lipid soluble, readily diffuses across the alveolar membranes, and concentrates in erythrocytes and the central nervous system. Inorganic mercury salts can be divalent (mercuric) or monovalent (mercurous). Gastrointestinal (GI) absorption of inorganic salts of mercury from food is less than 15.0% in mice and about 7.0% in a study of human volunteers; however, methylmercury absorption in the GI tract is 90.0 to 95.0%. Methylmercury, when compared to inorganic mercury compounds, is about 5 times more soluble in erythrocytes than in plasma, and about 250 times more abundant in hair than in blood (Goyer, 1986).

4.1 POISONING

The primary mode of action of both inorganic and organic mercury compounds is associated with interference of membrane permeability and enzyme reactions through binding of mercuric ion to sulfhydryl groups, although organomercurials can penetrate membranes more readily (Norton, 1986). Early accounts of acute and chronic mercury poisoning and their treatment follow (Anon., 1948); however, it is cautioned that treatment should be under the guidance of a physician.

4.1.1 Elemental Mercury

Metallic or elemental mercury volatilizes to mercury vapor at ambient air temperatures, and most human exposure is by way of inhalation (Goyer, 1986). The saturated vapor pressure at 20.0°C is 13.2 mg/m^3. This value far exceeds the threshold limited value (TLV) of 0.05 mg/m^3; accordingly, mercury intoxication due to inhalation of the vapor readily occurs in various occupational and environmental situations (Satoh, 1995). Mercury vapor readily diffuses across the alveolar membrane and is lipid soluble so that it has an affinity for the central nervous system and red blood cells. Metallic mercury, unlike mercury vapor, is only slowly absorbed by the GI tract (0.01%) at a rate related to the vaporization of the elemental mercury and is of negligible toxicological significance (Goyer, 1986).

Inhaled mercury vapor (Hg^o) is readily oxidized via the catalase–H_2O_2 complex and converted to Hg^{2+}, mainly in liver and erythrocytes (Satoh, 1995). Although this reaction is rapid, some Hg^o crosses the blood–brain barrier and accumulates to a greater extent than does Hg^{2+} after ionic mercury exposure. Because Hg^{2+} is reduced to Hg^o, there is probably an oxidation-reduction cycle of mercury in the body (Satoh, 1995).

In typical Hg^o vapor poisonings, excessive bronchitis and bronchiolitis occur within a few hours after heavy exposure, that is, direct inhalation of mercury vapor generated by heating metallic mercury (Satoh, 1995). This is followed by pneumonitis and respiratory distress, excitability, and tremors. If the amount inhaled is sufficiently large, renal failure will develop. In one Japanese factory producing sulfuric acid using Hg^o in the process, a few of the workers died of respiratory distress associated with renal failure (Satoh, 1995). Moderate and repeated exposure results in classical mercury poisoning (Satoh, 1995). Inhalation of mercury vapor, if not fatal, is associated with an acute, corrosive bronchitis, interstitial pneumonitis, tremors, and increased excitability (Goyer, 1986). With chronic exposure to mercury vapor, the major effects are on the central nervous system. Early signs are nonspecific and have been termed "micromercurialism" or "asthenic-vegetative syndrome" (Goyer, 1986). Micromercurialism is characterized by weakness, fatigue, anorexia, weight loss, and GI disturbance (Satoh, 1995). This syndrome is characterized clinically by at least three of the following: tremors, thyroid enlargement, increased radioiodine uptake by thyroid, tachycardia, unstable pulse, dermographism, gingivitis, changes in blood chemistry, and increased excretion of mercury in urine (Goyer, 1986). With increasing exposure, the symptoms include tremors of the fingers, eyelids, and lips, and may progress to generalized trembling of the entire body and violent chronic spasms of the arms and legs. This is accompanied by changes in personality and behavior, with loss of memory, increased excitability, severe depression, delirium, and hallucination. Another characteristic feature of mercury toxicity is severe salivation (Goyer, 1986). Tremors, increased excitability, and gingivitis have been recognized historically as the major manifestation of mercury poisoning from inhalation of mercury vapor and exposure in the fur, felt, and hat industry to mercuric nitrate (Goldwater, 1972).

Effects of mercury vapor exposure prevail long after cessation of exposure, although typical symptoms such as tremors, gingivitis, and salivation usually disappear quickly (Satoh, 1995). Residual effects due to previous exposure have been documented in workers with a peak urinary mercury concentration greater than 0.6 mg/L; neurobehavioral disturbances were observed in these workers 20 to 35 years post exposure (Satoh, 1995).

4.1.2 Inorganic Mercurials

Inorganic mercury salts can be divalent (mercuric) or monovalent (mercurous). Chronic exposure to low levels of inorganic mercury compounds is associated with psychological changes including abnormal irritability (erethismas mercurialis), colored mercury compounds in the anterior lens capsule of the eye (mercurialentis), tremors, and excessive salivation (Norton, 1986). Inorganic forms of mercury are corrosive and produce symptoms that include metallic taste, burning, irritation, salivation, vomiting, diarrhea, upper GI tract edema, abdominal pain, and hemorrhage (Rumack and Lovejoy, 1986). These effects are seen acutely and may subside with subsequent lower GI tract ulceration. Large ingestion of the mercurial salts may produce kidney ingestion, such as nephrosis, oliguria, and anuria (Rumack and Lovejoy, 1986).

4.1.2.1 Mercuric Mercury

Acute poisoning by mercurials usually occurs in the case of mercuric perchlorides, with intense gastrointestinal inflammation, vomiting, diarrhea, and extreme collapse. Treatment is usually with albumin, which forms an insoluble compound with the perchloride (Anon., 1948). Chronic poisoning, or mercurialism, is marked by tenderness of the teeth while eating and offensive breath. Later, the gums become inflamed, salivary glands are swollen and tender, and saliva pours from the mouth. The teeth may become loose and fall out. The symptoms are aggravated until the tongue and mouth ulcerate, the jaw bone necroses, hemorrhages occur in various parts of the body, and death results from anemia, septic inflammation, or exhaustion. Treatment includes administration of potassium iodide in low, repeated doses (Anon., 1948).

Chronic exposure to low levels of inorganic mercuric compounds produces tremors, excess salivation, and psychological changes characterized by irritability and excitement (Norton, 1986). Collectively, this is often described as the "mad hatter syndrome" (Rumack and Lovejoy, 1986). Mercuric mercury (Hg^{2+}) is a potentially toxic chemical, although it is poorly absorbed by the GI tract and other body parts (Satoh, 1995). Accidental or suicidal ingestion of mercuric chloride or other mercuric salts produces corrosive ulceration, bleeding, necrosis of the intestinal tract and is usually accompanied by shock and circulatory collapse (Goyer, 1986). If the patient survives the gastrointestinal damage, renal failure may occur within 24 hours owing to necrosis of the proximal tubular epithelium followed by diminished secretion of urine and kidney pathology. These may be followed by ultrastructural changes consistent with irreversible cell injury. Regeneration of the tubular lining is possible if the patient can be maintained by dialysis (Goyer, 1986).

The pathogenesis of chronic mercury kidney damage has two phases: an early phase with antibasement membrane glomerulonephritis, followed by a superimposed immune-complex glomerulonephritis (Roman-Franco et al., 1978). The pathogenesis of the nephropathy in humans appears similar, although early glomerular nephritis may progress to an interstitial immune-complex nephritis (Tubbs et al., 1982). Injection of mercuric chloride produces necrosis of kidney epithelium (Gritzka and Trump, 1968). Cellular changes include fragmentation of the plasma membrane, disruption of cytoplasmic membranes, loss of ribosomes, and mitochondrial swelling; however, all of these changes are associated with renal cell necrosis from a variety of insults. High doses of mercuric chloride are directly toxic to renal tubular lining cells and chronic low-level doses to mercuric salts may induce an immunologic glomerular disease that is the most common form of mercury-induced nephropathy. Chronic mercury-induced kidney damage seldom occurs in the absence of detectable damage to the nervous system (Gritzka and Trump, 1968). Although kidneys contain the highest concentrations of mercury following exposure to inorganic mercury salts and mercury vapor, it is emphasized that organomercurials concentrate in the posterior cortex of the brain (Goyer, 1986).

4.1.2.2 Mercurous Mercury

Mercurous (Hg^{+}) mercury compounds are unstable and easily break down to Hg^{o} and Hg^{2+} (Satoh, 1995). Mercurous compounds are less corrosive and less toxic than mercuric compounds, probably because they are less soluble (Rumack and Lovejoy, 1986).

Calomel — a mercurous chloride-containing powder with a long history of medical use — is known to be responsible for acrodynia or "pink-disease" in children when used as a teething powder (Matheson et al., 1980). This is probably a hypersensitivity response by skin to mercurous salts, producing vasodilation, hyperkeratosis, and excessive sweating. Afflicted children had fever, pink-colored rashes, swollen spleen and lymph nodes, and hyperkeratosis and swelling of fingers. Effects were independent of dose and are thought to be a hypersensitivity reaction (Matheson et al., 1980).

4.1.3 Organomercurials

Among the organomercurials, alkylmercurials — especially methylmercurials (CH_3Hg^{+}) — are the most environmentally and ecologically significant (Satoh, 1995). Methylmercury is naturally produced from inorganic mercury by microbial activity; methylmercurials are lipid soluble and readily cross blood–brain and placental barriers (Satoh, 1995). The sensitivity of the developing brain to methylmercury is due to placental transfer of lipophilic methylmercury to the central nervous system

(CNS) (Campbell et al., 1992). The blood–brain barrier is incomplete during the first year of life in humans, and methylmercury can cross this barrier during that time (Rodier, 1995). Phenylmercurials ($C_6H_5Hg^+$) and methoxyethyl mercurials ($CH_3OC_2H_4Hg^+$) have been used as fungicides and pesticides and readily transform into inorganic mercurials in living organisms with toxic properties similar to those of inorganic mercurials (Satoh, 1995).

Ingestion of organomercurials, such as ethylmercury, may produce symptoms of nausea, vomiting, abdominal pain, and diarrhea, but in most cases the main toxicity is neurologic involvement presenting with paresthesias, visual disturbances, mental disturbances, hallucinations, ataxia, hearing defects, stupor, coma, and death (Rumack and Lovejoy, 1986). Symptoms may occur for several weeks after exposure. Exposure and poisoning can occur after ingestion of mercury-contaminated seafoods, grains, or inhalation of vaporized organomercurials (Rumack and Lovejoy, 1986).

The toxic signs of alkylmercury compounds, such as methylmercury, are different than those of inorganic mercurials owing to greater penetration of the organomercurials into the brain (Norton, 1986). Methylmercury causes necrosis of the granule cell layer of the cerebellum, which is associated with carbohydrate metabolism and kidney disorders. Focal atrophy of the cortex, with sensory disturbances, ataxia, and dysarthria, is found after methylmercury intoxication. The emotional changes and autonomous nervous system involvement with inorganic mercury compounds are not seen with organomercurials (Norton, 1986).

Sensory nerve fibers are selectively damaged (Norton, 1986). The primary mode of action of both inorganic and organic mercury compounds may be interference with membrane permeability and enzyme actions by binding of mercuric ion to sulfhydryl groups. Small neurons in the CNS are more likely to be damaged than large neurons in the same area by methylmercury (Norton, 1986). The major clinical features of methylmercury toxicity are neurologic, consisting of paresthesia, ataxia, dysarthria, and deafness, appearing in that order (Roizin et al., 1977). The main pathologic features include degeneration and necrosis of neurons in focal areas of the cerebral cortex and in the granular layer of the cerebellum. Studies of both inorganic- and organic-mercury-related neuropathy show degeneration of primary sensory ganglion cells. Lesion distribution in the CNS suggests that mercury damages small nerve cells in the cerebellum and visual cortex (Roizin et al., 1977). Methylmercuric chloride, as an environmental pollutant, has produced renal damage in humans and animals through inhibition of mitochondrial and other enzyme systems (Hook and Hewitt, 1986).

4.2 MERCURY TREATMENT

Treatment of mercury-poisoned victims is complex, and should be supervised by a physician. Therapy of mercury poisoning is directed toward lowering the concentration of mercury at the critical organ or site of injury (Berlin, 1979). For the most severe cases, particularly with acute renal failure, hemodialysis together with infusion of mercury-chelating agents such as cysteine and penicillamine is warranted. For less severe cases of inorganic mercury poisoning, chelation with BAL (dimercaprol) is recommended. Chelation therapy, however, is not effective for poisoning with organomercurials. In those cases, oral administration of a nonabsorbable thiol resin that binds mercury and enhances intestinal excretion, or surgical establishment of gallbladder drainage have proven satisfactory (Berlin, 1979).

Treatment usually consists of emesis or lavage, followed by administration of activated charcoal and a saline cathartic (Rumack and Lovejoy, 1986). Cow's milk may be given to help precipitate the mercury compound. Blood and urine levels of mercury may be useful in determining whether chelating agents, such as D-penicillamine or BAL (dimercaprol) should be administered. D-penicillamine is given at 250.0 mg orally, four times daily in adults. For children, D-penicillamine is given at 100.0 mg/kg body weight daily to a maximum recommended dose of 1000 mg daily for 3 to 10 days with continuous monitoring of mercury urinary excretion. In patients unable to tolerate penicillamine, BAL can be administered at a dose of 3.0 to 5.0 mg/kg body weight (BW) every 4 h by deep intramuscular (im) injection for the first 2 days, then 2.5 to 3.0 mg/kg BW im every 6 h for 2 days, followed by 2.5 to 3.0 mg/kg BW im every 12 h for 1 week. Adverse reactions associated with BAL, such as skin eruptions (urticaria), can often be controlled with antihistamines such as diphenylhydramine. The development of renal failure contraindicates use of penicillamine because the kidney is the main excretory route for penicillamine. BAL therapy can be used cautiously despite renal failure because BAL is excreted in the bile; however, BAL toxicity, which consists of fever, rash, hypertension, and CNS stimulation must be closely monitored. Dialysis is not recommended because it does not remove chelated or free mercury (Rumack and Lovejoy, 1986).

Mercury-antagonistic and mercury-protectant drugs and compounds now include 2,3-dimercaptopropanol, polythiol resins, selenium salts, thiamin, vitamin E, metallothionein-like proteins, and sulfhydryl agents (Magos and Webb, 1979; Elhassani, 1983; Siegel et al., 1991; USPHS, 1994; Caurent et al., 1996). Thiols (R-SH), which compete with mercury for protein binding sites, are the most important antagonists of inorganic mercury salts, and have been used extensively in attempts to counteract mercury poisoning in humans (Das et al., 1982). Thiamin was the most effective of the Group VIB derivatives (which includes sulfur, selenium, and tellurium) in protecting against organomercury poisoning in higher animals (Siegel et al., 1991). The protective action of selenium (Se) against adverse or lethal effects induced by inorganic or organic mercury salts is documented for algae, aquatic invertebrates, fish, birds, and mammals (Magos and Webb, 1979; Heisinger, 1979; Chang et al., 1981; Lawrence and Holoka, 1981; Das et al., 1982; Gotsis, 1982; Satoh et al., 1985; Eisler, 1987; Goede and Wolterbeek, 1994; Paulsson and Lundbergh, 1989; USPHS, 1994; Caurent et al., 1996; Kim et al., 1996a, 1996b). For example, selenium, as sodium selenite, that was introduced into a nonacidified mercury-contaminated lake in Sweden to concentrations of 3.0 to 5.0 μg Se/L (from 0.4 μg Se/L) and sustained at this level for 3 years resulted in declines of 50.0 to 85.0% in mercury concentrations in fish muscle (Paulsson and Lundbergh, 1989). The mercury-protective effect of selenium is attributed to competition by selenium for mercury-binding sites associated with toxicity, formation of a Hg-Se complex that diverts mercury from sensitive targets, and prevention of oxidative damage by increasing the amount of selenium available to the selenium-dependent enzyme glutathione peroxidase (USPHS, 1994). In seabirds, an equivalent molar ratio of 1:1 between total mercury and selenium was found in livers of individual seabirds that contained more than 100.0 mg Hg/kg DW; this relation was unclear in other individuals, which had relatively low mercury levels (Kim et al., 1996a, 1996b). The selenium-protective mechanism in birds is explained by a strong binding between mercury and selenium, possibly by the formation of a selenocystamine–methylmercury complex ($CH_3HgSeCH_2CH_2NH_3^+$), mercury binding to selenocysteine residues ($CH_3HgSeCH_2CH(NH_3)(COO)\bullet H_2O$), the formation of insoluble mercuric selenide (HgSe), or binding of mercury to SeH residues of selenoproteins, notably metallothioneins with thiols replaced by SeH (Goede and Wolterbeek, 1994). However, high selenium concentrations in tissues of marine wading birds do not have their origin in elevated levels of mercury. The Se:Hg ratio in marine wading birds from the Wadden Sea is 32:1 and greatly exceeds the 1:1 ratio found when selenium is accumulated to detoxify mercury (Goede and Wolterbeek, 1994). In marine mammals and humans, selenium and mercury concentrations are closely related, almost linearly in a 1:1 molar ratio (Eisler, 1987). The molar ratio between mercury and selenium in marine mammals suggests that the major mechanism of detoxification is through the formation of a complex Hg–Se that leads to mercury demethylation (Caurent et al., 1996). The site of this process is the liver in which mercury appeared mainly as inorganic; whereas in the muscle, the percent of organic to total mercury was much higher. Detoxification is limited in lactating female whales, and sometimes in all the individuals of one school (Caurent et al., 1996). Selenium does not, however, protect against mercury-induced birth defects, such as cleft palate in mice (USPHS, 1994). It is clear that more research is needed on mercury protectants.

4.3 SUMMARY

Both inorganic and organomercurials interfere with membrane permeability and enzyme reactions through binding of mercuric ion to sulfhydryl groups; organomercurials usually penetrate membranes more readily. Symptoms of acute and chronic mercury poisoning caused by elemental mercury, mercuric mercury, mercurous mercury, and organomercurial compounds are listed, mechanisms of action discussed, and treatment regimes prescribed. For elemental mercury, inhalation of Hg^o is the primary toxicological route. Inhaled Hg^o vapor is readily oxidized within the body, mainly in liver and erythrocytes, and converted to Hg^{2+}. Neurobehavioral disturbances were observed in some Hg^o vapor poisoning cases 20 to 35 years after exposure. For inorganic mercuric compounds, exposure routes include inhalation and ingestion, with primary damage to the renal system. Mercurous (Hg^+) mercury compounds are unstable and degrade to Hg^o and Hg^{2+}. Mercurous compounds are less corrosive and less toxic than mercuric compounds and this could be associated with their comparatively low solubility. Among the organomercurials, methylmercury compounds (CH_3Hg^+) are the most significant toxicologically because they are produced naturally from inorganic mercury by microbial activity and are lipid soluble, thus readily crossing blood–brain and placental barriers. Ingestion is the main route of administration for methylmercurials and the primary target organs are brain and other neurologic tissues.

Treatment of mercury-poisoned victims is complex and should be supervised by a physician. Therapy is directed to lowering the mercury concentration at the critical organ or site of injury through emesis, lavage, cathartics, administration of activated charcoal and various mercury chelating agents, and — in the most severe cases — dialysis. The development of mercury-antagonistic and mercury-protectant drugs is proceeding, and some already available have been used to treat cases of inorganic mercury poisoning (thiols) and organomercurial poisoning (thiamin, and selenium-, sulfur-, and tellurium-containing drugs).

REFERENCES

Anon. 1948. Mercury, *Encylop. Brittanica*, 15, 269–272.

Berlin, M. 1979. Mercury. In L. Friberg, G.F. Nordberg, and C. Nordman (Eds.), *Handbook on the Toxicology of Metals*, p. 503–530. Elsevier/North-Holland Biomedical Press, New York.

Campbell, D, M. Gonzales, and J.B. Sullivan Jr. 1992. Mercury. Pages 824–833 in J.B. Sullivan Jr. and G.R. Krieger (Eds.), *Materials Toxicology — Clinical Principles of Environmental Health,* p. 824–833. Williams and Wilkins, Baltimore, MD.

Caurant, F., M. Navarro, and J.C. Amiard. 1996. Mercury in pilot whales: possible limits to the detoxification process, *Sci. Total Environ.*, 186, 95–104.

Chang, P.S.S., D.F. Malley, N.E. Strange, and J.F. Klaverkamp. 1981. The effects of low pH, selenium and calcium on the bioaccumulation of ^{203}Hg by seven tissues of the crayfish, *Orconectes virilis*, *Canad. Tech. Rep. Fish. Aquat. Sci.*, 1151, 45–67.

Das, S.K., A. Sharma, and G. Talukder. 1982. Effects of mercury on cellular systems in mammals — a review, *Nucleus (Calcutta)*, 25, 193–230.

Eisler, R. 1987. Mercury hazards to fish, wildlife, and invertebrates: a synoptic review, *U. S. Fish Wildl. Serv. Biol. Rep.*, 85(1.10), 1–90.

Elhassani, S.B. 1983. The many faces of methylmercury poisoning, *J. Toxicol.*, 19, 875–906.

Goede, A.A. and H.T. Wolterbeek. 1994. Have high selenium concentrations in wading birds their origin in mercury?, *Sci. Total Environ.*, 144, 247–253.

Goldwater, L.J. 1972. In *Mercury: A History of Quicksilver,* p. 270–277. York Press, Baltimore, MD.

Gotsis, O. 1982. Combined effects selenium/mercury and selenium/copper on the cell population of the alga *Dunaliella minuta*, *Mar. Biol.*, 71, 217–222.

Goyer, R.A. 1986. Toxic effects of metals. In C.D. Klaassen, M.O. Amdur, and J. Doull (Eds.), *Casarett and Doull's Toxicology, third edition,* p. 582–635. Macmillan, New York.

Gritzka, T.L. and B.F. Trump. 1968. Renal tubular lesions caused by mercuric chloride, *Am. J. Pathol.*, 52, 1225–1227.

Heisinger, J.F., C.D. Hansen, and J.H. Kim. 1979. Effect of selenium dioxide on the accumulation and acute toxicity of mercuric chloride in goldfish, *Arch. Environ. Contam. Toxicol.*, 8, 279–283.

Hook, J.B. and W.R. Hewitt. 1986. Toxic responses of the kidney. In C.D. Klaassen, M.O. Amdur, and J. Doull (Eds.), *Casarett and Doull's Toxicology, third edition,* p. 310–329. Macmillan, New York.

Kim, E.Y., K. Saeki, S. Tanabe, H. Tanaka, and R. Tatsukawa. 1996a. Specific accumulation of mercury and selenium in seabirds, *Environ. Pollut.*, 94, 261–265.

Kim, E.Y., T. Murakami, K. Saeki, and R. Tatsukawa. 1996b. Mercury levels and its chemical form in tissues and organs of seabirds, *Arch. Environ. Contam. Toxicol.*, 30, 259–266.

Lawrence, S.G. and M.H. Holoka. 1981. Effect of selenium on impounded zooplankton in a mercury contaminated lake, *Can. Tech. Rep. Fish. Aquat. Sci.*, 1151, 83–92.

Magos, L. and M. Webb. 1979. Synergism and antagonism in the toxicology of mercury. In J.O. Nriagu (Ed.), *The Biogeochemistry of Mercury in the Environment*, p. 581–599. Elsevier/North-Holland Biomedical Press, New York.

Matheson, D.S., T.W. Clarkson, and E.W. Gelfand. 1980. Mercury toxicity (acrodynia) induced by long-term injection of gamma globulin, *J. Pediatr.*, 97, 153–155.

Norton, S. 1986. Toxic responses of the central nervous system. In C.D. Klaassen, M.O. Amdur, and J. Doull (Eds.), *Casarett and Doull's Toxicology, third edition,* p. 359–386. Macmillan, New York.

Paulsson, K. and K. Lundbergh. 1989. The selenium method for treatment of lakes for elevated levels of mercury in fish, *Sci. Total Environ.*, 87/88, 495–507.

Rodier, P.M. 1995. Developing brain as a target of toxicity, *Environ. Health Perspect.*, 103 (Suppl. 6), S73–S76.

Roizin, L., H. Shiraki, and N. Grceric. 1977. *NeuroToxicology, Volume 1.* Raven Press, New York. 658 pp.

Roman-Franco, A.A., M. Twirello, B. Abini, and E. Ossi. 1978. Anti-basement membrane antibodies with antigen-antibody complexes in rabbits injected with mercuric chloride, *Clin. Immunol. Immunopathol.*, 9, 404–411.

Rumack, B.H. and F.H. Lovejoy Jr. 1986. Clinical toxicology. Pages 879–901 in C.D. Klaassen, M.O. Amdur, and J. Doull (Eds.). *Casarett and Doull's Toxicology, third edition.* Macmillan, New York.

Satoh, H. 1995. Toxicological properties and metabolism of mercury; with an emphasis on a possible method for estimating residual amounts of mercury in the body. In *Proceedings of the International Workshop on "Environmental Mercury Pollution and its Health Effects in Amazon River Basin,"* p. 106–112, Rio de Janeiro, 30 November–2 December 1994. Published by National Institute for Minamata Disease, Minamata City, Kumamoto 867, Japan.

Satoh, H., N. Yasuda, and S. Shimai. 1985. Development of reflexes in neonatal mice prenatally exposed to methylmercury and selenite, *Toxicol. Lett.*, 25, 199–203.

Siegel, B.Z., S.M. Siegel, T. Correa, C. Dagan, G. Galvez, L. Leeloy, A. Padua, and E. Yaeger. 1991. The protection of invertebrates, fish, and vascular plants against inorganic mercury poisoning by sulfur and selenium derivatives, *Arch. Environ. Contam. Toxicol.*, 20, 241–246.

Tubbs, R.R., G.N. Gephardt, J.T. McMahon, M.C. Phol, D.G. Vidt, S.A. Barenberg, and R. Valenzuela. 1982. Membraneous glomerulonephritis associated with industrial mercury exposure, *Am. J. Clin. Pathol.*, 77, 409–413.

U.S. Public Health Service (USPHS). 1994. *Toxicological Profile for Mercury (Update), TP-93/10.* U.S. PHS, Agen. Toxic Substances Dis. Registry, Atlanta, GA. 366 pp.

Part 2

Mercury Concentrations in Field Collections of Abiotic Materials, Plants, and Animals

CHAPTER 5

Mercury Concentrations in Abiotic Materials

Mercury burdens in sediments and other nonbiological materials are estimated to have increased up to five times prehuman levels, primarily as a result of anthropogenic activities (USNAS, 1978). Maximum increases are reported in freshwater and estuarine sediments and in freshwater lakes and rivers, but estimated increases in oceanic waters and terrestrial soils have been negligible (USNAS, 1978). Methylmercury accounts for a comparatively small fraction of the total mercury found in sediments, surface waters, and sediment interstitial waters of Poplar Creek, Tennessee, which was initially contaminated with mercury in the 1950s and 1960s. Mercury measurements in Poplar Creek from 1993 to 1994 showed that methylmercury accounted for 0.01% of the total mercury in sediments, 0.1% in surface waters, and 0.3% in sediment interstitial waters (Campbell et al., 1998). The residence time of mercury in nonbiological materials is variable, and depends on a number of physicochemical conditions. Estimated half-time residence values for mercury are 11 days in the atmosphere, 1000 years in terrestrial soils, 2100 to 3200 years in ocean waters, and more than 250 million years in oceanic sediments (USNAS, 1978; Boudou and Ribeyre, 1983; Clarkson et al., 1984); however, this estimate was only 1 month to 5 years for water from the contaminated Saguenay River in Quebec (Smith and Loring, 1981).

This chapter documents mercury concentrations in air, coal, pelagic clays, sediments, sewage sludge, snow, soils, suspended particulate matter, seawater, freshwater, groundwater, and sediment interstitial waters from selected geographic locales.

5.1 AIR

Most (up to 59.1%) of the mercury contributed to the atmosphere each year is from anthropogenic sources such as combustion of fossil fuels from power plants, with natural sources such as oceans, land runoff, and volcanoes contributing almost all the remainder (Table 5.1).

Atmospheric concentrations of total mercury in the northern hemisphere are about three times higher than those sampled in the southern hemisphere owing to greater sources from human activities in the comparatively industrialized and populated north (Lamborg et al., 1999). Different mercury species are dominant at different Japanese locations. For example, in 1977 to 1978, Hg^{2+} was the dominant species in air over hot springs, volcanoes, and urban centers; however, Hg^{o} was dominant in air over chloralkali plants and rural areas (Takizawa et al., 1981; Table 5.2). Enrichment of toxic metals in respirable particulate matter emissions from a coal-fired power plant in Central India is documented, especially for mercury that was enriched 4.8 times over the coal (Sharma and Pervez, 2004; Table 5.2).

Table 5.1 Contributions to Atmospheric Mercury from Natural and Anthropogenic Sources

Source (% of total)[a]	Mercury Concentration or Emission	Estimated Amount of Mercury Discharged into Global Atmosphere per Year (metric tons)	Ref.[b]
Volcanoes (1.0–1.4)	28.0–1400.0 ng/m^3	60	1–3
Land (16.7–22.7)	1.0–6.0 ng/m^3	1000	2, 4, 5
Mines (0.2–2.3)	1.0–5000.0 ng/m^3	10–100	4, 6–9
Oceans (13.3–45.4)	1.0–3.0 ng/m^3	800–2000	5, 10
Anthropogenic (33.3–59.1)	10.0–900.0 kg/year[c]	2000–2600	5, 10, 11

[a] The estimated total amount of mercury contributed to the atmosphere worldwide ranges between 4400 and 6000 metric tons annually; mercury concentrations in the atmosphere usually range between 1.0 and 2.0 ng/m^3 (Porcella, 1994; Lamborg et al., 2000; Gray, 2003).

[b] Reference: 1, Fitzgerald, 1989; 2, Varekamp and Buseck, 1986; 3, Ferrara et al., 1994; 4, Gustin et al., 1994; 5, Mason et al., 1994; 6, Ferrara et al., 1991; 7, Ferrara et al., 1998; 8, Gustin et al., 1996; 9, Gustin et al., 2000; 10, Lamborg et al., 2000; 11, USEPA, 2000.

[c] Range of mercury emissions from power plants in the United States (USEPA, 2000).

Table 5.2 Mercury Concentrations in Selected Abiotic Materials

Material (units)	Concentration	Ref.[a]
Air (ng/m^3)		
Atmosphere over open ocean; total mercury:		
Southern hemisphere, 60°S	1.0	25
Northern hemisphere	3.0	25
India; near a large coal-fired power station:		
Respirable suspended particulate matter	50.0	31
Nonrespirable suspended particulate matter	30.0	31
Flyash	28.0	31
Japan:		
Remote areas	< 5.0–20.0	12
Urban areas	85.0–100.0	12
Japan; 1977–1978; maximum values:		
Over open volcanoes (southern Japan) vs. over hot springs (northern Japan):		
Particulate mercury	2.0 vs. 1.0	27, 29
Hg^{2+}	368.0 vs. 126.0	29
CH_3Hg^+	19.0 vs. 9.0	27, 29
Hg^o	97.0 vs. 21.0	29
$(CH_3)_2Hg$	9.0 vs. 5.0	29
Urban vs. rural:		
Particulate mercury	0.0 vs. 1.0	27
Hg^{2+}	44.0 vs. 33.0	27
CH_3Hg^+	38.0 vs. 8.0	27
Hg^o	10.0 vs. 30.0	27
$(CH_3)_2Hg$	5.0 vs. 6.0	27
Over chloralkali plant:		
Particulate mercury	0.0	27
Hg^{2+}	80.0	27, 29
CH_3Hg^+	34.0	27
Hg^o	61.0	27
$(CH_3)_2Hg$	17.0	27
Siberia, 1992–1993; summer vs. winter:		
Gaseous	0.7–2.3 vs. 1.2–6.1	14
Particulate	0.005–0.02 vs. 0.02–0.09	14

Table 5.2 (continued) Mercury Concentrations in Selected Abiotic Materials

Material (units)	Concentration	Ref.[a]
Coal (mg/kg dry weight = DW)		
Bituminous	0.07	10
Lignite	0.12	10
Sub-bituminous	0.03	10
Various	mean 0.2; rarely >1.0	33
Pelagic Clays (mg/kg fresh weight)		
430 km southeast of San Diego, CA	0.39	18
Rock (μg/kg DW)		
Limestone, Pennsylvania	9.0 (4.0–14.0)	36
Shale, eastern Canada	42.0	37
Sediments, Mercury-Contaminated Areas (mg/kg DW)		
Near chloralkali plant:		
Quebec, Canada	12.0	3
Norway	250.0 (90.0–350.0)	4
Thailand	8.0–58.0	5
Near gold mining operations[b]:		
South Dakota	0.1–4.1	6
Australia	120.0	7
Ecuador (cyanide extraction facility); 1988; dry season	0.1–5.8[c]	28
Near mercury-fungicide plant:		
Denmark	22.0	8
Near offshore oil fields:		
Southeastern Brazil; 1998–1999 vs. reference locations	0.036–0.047 (0.013–0.08) vs. 0.01–0.05	21
Near acetaldehyde plant:		
Minamata Bay, Japan[b]	28.0–713.0	4
Near pulp and paper mill:		
Finland	746.0	9
Tennessee; Oak Ridge; 1993–1994; contaminated in mid-1950s to early 1960s; methylmercury vs. total mercury	Max. 0.012 vs. 0.6–140.0	15
Sediments, Uncontaminated Areas (mg/kg DW)		
Adriatic Sea	0.13–1.5	17
Lakes	0.1–0.3	12
Marine	0.05–0.08	12
Rivers:	< 0.05	12
North Central United States	0.02–0.06, max. 0.11	6
South Dakota	0.02–0.1	6
Thailand	0.03	5
Finland	0.02	5
Various lakes	Usually < 10.0, frequently < 1.0	4
Wisconsin:		
Deep precolonial strata	0.04–0.07	13
Top 15 cm	0.09–0.24	13
Sewage Sludge (mg/kg DW)		
50 publicly owned treatment works (POTW), United States	2.8	23
74 POTW, Missouri	3.9 (0.6–130.0)	23

(continued)

Table 5.2 (continued) Mercury Concentrations in Selected Abiotic Materials

Material (units)	Concentration	Ref.[a]
Snow (ng/L)		
Arctic Alaska	1.5–7.5	35
Siberia, 1992—1993:		
Total mercury	8.0–60.0	14
Methylmercury	0.1–0.25	14
Soils (μg/kg DW)		
Ribeiro Preto, Brazil; vicinity solid waste landfill site:		
2002 vs. 2003	< 25.0 vs. 50.0	22
0–500 m from site vs. 200 m from site	50.0 (20.0–80.0) vs. 50.0	22
Critical limit for Brazilian agricultural soils	< 2500.0	22
China; contaminated by mercury-containing wastewater from acetaldehyde factory; about 135 tons of mercury discharged into Zhuja River between 1970 and 2000:		
Quingshen City; high contamination area vs. low contamination area:		
Total mercury	61,400.0, max. 329,900.0 vs. 7,000.0, max. 98,300.0	26
Methylmercury	45.0, max. 65.1 vs. 5.8, max. 43.6	26
Lanchong reference site:		
Total mercury	110.0, max. 1780.0	26
Methylmercury	2.3, max. 7.0	26
Jakobstad, Finland:		
Humus:		
Rural	303.0 (116.0–393.0)	20
Urban	280.0 (150.0–1028.0)	20
Urban		
Topsoil	93.0 (11.0–2,309.0)	20
Subsoil	44.0 (< 5.0–540.0)	20
Forest soils; total mercury vs. methylmercury:		
Humus	100.0–250.0 vs. 0.2–0.5	24
Mineral horizon	15.0–30.0 vs. < 0.05	24
Jamaica; agricultural soils	221.0 (max. 830.0)	32
United Kingdom; near combustion plants; top soils (0–15 cm) vs. subsurface soils (15–30 cm):		
Coal-fired power plant with flue gas desulfurization system (FGD):		
Total Hg	297.0 vs. 86.0	16
Elemental Hg	95.0 vs. 64.0	16
Exchangeable Hg	2.0 vs. 1.0	16
Organic Hg	5.0 vs. 5.0	16
Hg sulfide	185.0 vs. 11.0	16
Residual Hg	10.0 vs. 4.0	16
Coal-fired power plant without FGD system:		
Total Hg	495.0 vs. 135.0	16
Elemental Hg	193.0 vs. 107.0	16
Exchangeable Hg	2.0 vs. 3.0	16
Organic Hg	3.0 vs. 4.0	16
Hg sulfide	245.0 vs. 16.0	16
Residual Hg	51.0 vs. 4.0	16
Waste incinerator:		
Total Hg	1470.0 vs. 2310.0	16
Elemental Hg	446.0 vs. 1950.0	16
Exchangeable Hg	9.0 vs. 3.0	16
Organic Hg	5.0 vs. 4.0	16
Hg sulfide	869.0 vs. 310.0	16
Residual Hg	136.0 vs. 33.0	16

Table 5.2 (continued) Mercury Concentrations in Selected Abiotic Materials

Material (units)	Concentration	Ref.[a]
Crematorium:		
Total Hg	392.0 vs. 680.0	16
Elemental Hg	193.0 vs. 480.0	16
Exchangeable Hg	5.0 vs. 3.0	16
Organic Hg	5.0 vs. 2.0	16
Hg sulfide	182.0 vs. 180.0	16
Residual Hg	6.0 vs. 14.0	16
Suspended Particulate Matter (mg/kg DW)		
Germany, Elbe River, 1988, mercury-contaminated by chloralkali plants:		
Total mercury	30.0; max. 150.0	11
Methylmercury	2.7	11
Reference site, total mercury	0.4	11
Water (ng/L)		
Coastal seawater	< 20.0	2
Drainage water from mercury mines; California:		
Total mercury	450,000.0	34
Methylmercury	70.0	34
Estuarine seawater	< 50.0	2
Freshwater; Ecuador; near cyanide extraction gold mining facility; 1988; dry season	2.2–1100.0[d]	28
Freshwater, surface:		
Arctic Canada	0.23–0.76	38
Arctic Russia	0.30–1.00	39
Finland	1.3–7.2	40
Lake Superior	0.49	41
High-altitude lakes, United States	1.07	42
Northern Minnesota	0.2–3.2	43
Glacial waters	10.0	2
Groundwater	50.0	2
Groundwater; southern Nevada; 190 km NW of Las Vegas; Amaragosa Desert; Aug.–Sept. 2002:		
Unfiltered	11.9 (0.4–36.7)	19
Filtered	5.4 (< 0.22–15.7)	19
Japan; Akita area; June–July 1978; maximum values		
Rain		
Total Hg	15.2	30
Inorganic Hg	14.3	30
Organic Hg	0.9	30
Lake water vs. river water		
Total Hg	13.8 vs. 21.1	30
Inorganic Hg	8.8 vs. 9.0	30
Organic Hg	5.5 vs. 12.1	30
Lake water:		
Siberia, Lake Baikal, summer 1992–1993:		
Total mercury	0.14–0.77	14
Methylmercury	max. 0.038	14
Sweden:		
Total mercury	1.4–15.1	12
Methylmercury	0.04–0.8	12
United States:		
Total mercury	0.4–10.7	12
Methylmercury	0.03–0.64	12

(continued)

Table 5.2 (continued) Mercury Concentrations in Selected Abiotic Materials

Material (units)	Concentration	Ref.[a]
Open ocean	5.3 (3.1–7.5)	1
Open ocean	< 10.0	2
Rainwater:		
Open ocean	1.0	2
Coastal ocean	10.0	2
Continents	Often > 50.0	2
Siberia, 1992–1993:		
Total mercury	3.0–20.0	14
Methylmercury	0.1–0.25	14
Sweden:		
Total mercury	7.5–89.7	12
Methylmercury	0.04–0.6	12
Rivers and lakes	10.0, max. 50.0	2
River water:		
Canada, Ottawa River:		
Total mercury	4.6–9.8	12
Methylmercury	1.6–2.8	12
Japan:		
Total mercury	19.3–25.9	12
Methylmercury	5.8–7.0	12
Siberia, 1992–1993:		
Total mercury	Max. 2.0	14
Methylmercury	Max. 0.16	14
United States, Connecticut River:		
Total mercury	45.0	12
Methylmercury	21.0	12
Seawater:		
Japan:		
Total mercury	3.2–12.5	12
Methylmercury	0.2–1.0	12
United States, New York		
Total mercury	47.0–78.0	12
Methylmercury	25.0–33.0	12
Sediment interstitial water:		
Total mercury	100.0–600.0	12, 15
Methylmercury	2.0	15

[a] Reference: 1, Nishimura and Kumagai, 1983; 2, Fitzgerald, 1979; 3, Smith and Loring, 1981; 4, Skei, 1978; 5, Suckcharoen and Lodenius, 1980; 6, Martin and Hartman, 1984; 7, Bycroft et al., 1982; 8, Kiorboe et al., 1983; 9, Paasivirta et al., 1983; 10, Chu and Porcella, 1995; 11, Wilken and Hintelmann, 1991; 12, Hamasaki et al., 1995; 13, Rada et al., 1989; 14, Meuleman et al., 1995; 15, Campbell et al., 1998; 16, Panyametheekul, 2004; 17, Vucetic et al., 1974; 18, Williams and Weiss, 1973; 19, Cizdziel, 2004; 20, Peltola and Astrom, 2003; 21, Lacerda et al., 2004; 22, Segura-Munoz et al., 2004; 23, Beyer, 1990; 24, Nater and Grigal, 1992; 25, Lamborg et al., 1999; 26, Matsuyama et al., 2004; 27, Takizawa et al., 1981; 28, Tarras-Wahlberg et al., 2000; 29, Takizawa, 1995; 30, Minagawa and Takizawa, 1980; 31, Sharma and Pervez, 2004; 32, Howe et al., 2005; 33, Finkelman, 2003; 34, Rytuba, 2003; 35, Snyder-Conn et al., 1997; 36, McNeal and Rose, 1974; 37, Cameron and Jonasson, 1972; 38, Semkin et al., 2005; 39, Coquery et al., 1995; 40, Verta et al., 1994; 41, Hurley et al., 2002; 42, Krabbenhoft et al., 2002; 43, Monson and Brezonik, 1998.

[b] This subject is covered in greater detail later.

[c] Sediment criterion for aquatic life protection in Ecuador is < 0.45 mg Hg/kg DW (Tarras-Wahlberg et al., 2000).

[d] Freshwater criterion for aquatic life protection in Ecuador is < 100.0 ng Hg/L (Tarras-Wahlberg et al., 2000).

In general, mercury concentrations were low in the atmosphere over the open ocean (1 to 3.0 ng/m^3), up to 100.0 ng/m^3 in the air of large cities, and highest (495.0 ng/m^3, 74.0% Hg^{2+}) in the atmosphere over open volcanoes (Table 5.2).

5.2 COAL

Mercury concentrations were highest in lignite coal (0.12 mg/kg DW), lowest in sub-bituminous coal (0.03 mg/kg DW), and intermediate (0.07 mg/kg DW) in bituminous coal samples measured (Table 5.2). More recent information (Finkelman, 2003) indicates that coal contains, on average, 0.2 mg Hg/kg and may contain as much as 1.0 mg/kg. Most of the mercury in coal is associated with arsenic-bearing pyrite; other forms include organically bound mercurials, elemental mercury, and mercuric sulfides and selenides. In coal samples with low pyrite, mercury selenides may be the primary form (Finkelman, 2003).

It is noteworthy that installation of available pollution control technology can significantly lower mercury concentrations in surface soils near coal-fired power plants in the United Kingdom. Thus, surface soils near a coal-fired power plant with a flue gas desulfurization (FGD) system contained 0.297 mg total Hg/kg DW vs. 0.495 mg total Hg/kg DW in a coal-fired power plant without FGD, a 40.0% reduction (Panyametheekul, 2004; Table 5.2).

5.3 SEDIMENTS

Much of the mercury that enters freshwater lakes is deposited in bottom sediments (Rada et al., 1993). Sedimentary pools of mercury in these lakes greatly exceed the inventories of mercury in water, seston, and fish, and the release of mercury from the sediments would significantly increase bioavailability and uptake. The dry weight mercury concentrations of sediments seem to underrepresent the significance of the shallow water sediments as a reservoir of potentially available mercury when compared to the mass per volume of wet sediment, which more accurately portrayed the depth distribution of mercury in Wisconsin seepage lakes (Rada et al., 1993). The increase in the mercury content of recent lake sediments in Wisconsin is attributed to increased atmospheric deposition of mercury, suggesting that the high mercury burdens measured in gamefish in certain Wisconsin lakes originated from atmospheric sources (Rada et al., 1989). Levels of mercury in sediments can be reflected by an increased mercury content in epibenthic marine fauna. For example, mercury concentrations in sediments near the Hyperion sewer outfall in Los Angeles, which ranged up to 820.0 μg/kg and decreased with increasing distance from the outfall, were reflected in the mercury content in crabs, scallops, and whelks. Concentrations of mercury were highest in organisms collected nearest the discharge, and lowest in those collected tens of kilometers away (Klein and Goldberg, 1970).

In sediments that were anthropogenically contaminated with mercury, concentrations were significantly elevated (usually > 20.0 mg/kg) when compared with uncontaminated sediments (usually < 1.0 mg/kg; Table 5.2). Significant mercury enrichment in sediments of Newark Bay, New Jersey, may represent a hazard to aquatic life (Gillis et al., 1993). In Finland, sediments near a pulp and paper mill — where mercury was used as a slimicide — contained up to 746.0 mg Hg/kg dry weight (Paasivirta et al., 1983; Table 5.2). In Florida, methylmercury in sediments from uncontaminated southern estuaries in 1995 accounted for 0.77% of the total mercury and was not correlated with total mercury or organic content of sediments (Kannan et al., 1998).

5.4 SEWAGE SLUDGE

Concentrations of total mercury in sewage sludge from 74 publicly owned treatment works in Missouri ranged from 0.6 to 130.0 mg/kg DW (Beyer, 1990; Table 5.2), this strongly indicates that sewage sludge applications to agricultural soils should be carefully monitored.

5.5 SNOW AND ICE

Total mercury concentrations in Siberian snow ranged between 8.0 and 60.0 ng/L; the maximum methylmercury concentration was 0.25 ng/L (Table 5.2).

Mercury concentrations in Arctic ice 4000 to 12,000 years ago during the precultural period were about 20.0% that of present-day concentrations; however, 13,000 to 30,000 years ago during the last glacial period, they were about five times higher than precultural levels (Vandal et al., 1993). Ice cores taken in Wyoming representing the 270-year period from 1715 to 1985 demonstrate that annual concentrations between 1715 and 1900 were usually less than 5.0 ng/L, except for volcanic eruptions in 1815 (Tambora; up to 15.0 ng/L) and 1883 (Krakatoa; up to 25.0 ng/L), and the California gold rush between 1850 and 1884 (up to 18.0 ng/L) (Atkeson et al., 2003). Between 1880 and 1985, mercury concentrations ranged up to 10.0 ng/L annually (1880 to 1950) due to industrialization and World War II manufacturing (1939 to 1945), up to 30.0 ng/L owing to the eruption of Mount St. Helena in 1980, and around 15.0 ng/L for the remainder due to increased industrialization (Atkeson et al., 2003).

5.6 SOILS

In general, soil mercury concentrations were higher in the vicinity of acetaldehyde plants, solid waste landfill sites, urban areas, coal-fired power plants, waste incinerators, and crematoriums (Table 5.2). The highest mercury concentrations recorded in soils were from receiving mercury-containing wastes of a Chinese acetaldehyde plant. These soils contained up to 329.9 mg total mercury (0.045 mg methylmercury)/kg dry weight vs. up to 1.78 mg total mercury/kg dry weight from reference sites (Matsuyama et al., 2004; Table 5.2).

Jamaican agricultural soils contained up to 0.83 mg total mercury/kg DW, mean 0.221 mg/kg DW (Table 5.2). This was far in excess of Danish and Canadian guidelines for mercury in crop soils (i.e., < 0.007 mg/kg DW) (Howe et al., 2005). Jamaican soils also exceeded Danish and Canadian proposed limits for arsenic, cadmium, copper, and chromium in agricultural soils (Howe et al., 2005).

Total mercury concentrations in surface soils near combustion plants range from 0.3 to 1.47 mg/kg DW, and in subsurface soils from 0.09 to 2.3 mg/kg DW (Panyametheekul, 2004; Table 5.2). Total mercury in both topsoils and subsoils is dominated by elemental mercury and mercuric sulfide, with increasing sulfur content in soil associated with increasing HgS. Solubility and pH conditions also influence the occurrence and distribution of mercury in soils near combustion plants (Panyametheekul, 2004). Uptake from the soil is probably a significant route for the entrance of mercury into vegetation in terrestrial ecosystems. In Italy, elevated mercury concentrations in soils near extensive cinnabar deposits and mining activities were reflected in elevated mercury concentrations in plants grown on those soils (Ferrara et al., 1991). Mercury concentrations in tissues of different species of vascular plants growing on flood plain soils at Waynesboro, Virginia, were directly related to soil mercury concentrations that ranged between < 0.2 and 31.0 mg Hg/kg DW soil (Cocking et al., 1995). In a study conducted in Fulton County, Illinois, it was shown that repeated applications of sewage sludge to land will significantly increase the concentration of mercury in surface soils (Granato et al., 1995). However, 80.0 to 100.0% of the mercury remained in the top 15 cm and was not bioavailable to terrestrial vegetation. The authors concluded that models developed by the U.S. Environmental Protection Agency overpredict the uptake rates of mercury from sludge-amended soils into grains and animal forage, and need to be modified (Granato et al., 1995).

5.7 WATER

Mercury-sensitive ecosystems are those where comparatively small inputs or inventories of total mercury (i.e., 1.0 to 10.0 g Hg/ha) result in elevated concentrations of methylmercury in natural resources. These systems are characterized by efficient conversion of inorganic mercuric mercury to methylmercury sufficient to contaminate aquatic and wildlife food webs (Brumbaugh et al., 1991; Spry and Wiener, 1991; Bodaly et al., 1997; Heyes et al., 2000; Wiener et al., 2003). Known sensitive ecosystems include surface waters adjoining wetlands (St. Louis et al., 1994; Gilmour et al., 1998); low-alkalinity or low-pH lakes (Spry and Wiener, 1991; Watras et al., 1994; Meyer et al., 1995); wetlands (St. Louis et al., 1996; Plourde et al., 1997); and flooded terrestrial areas (Kelly et al., 1997).

In southern Nevada groundwaters, mercury concentrations were usually less than 20.0 ng/L in unfiltered samples and less than 10.0 ng/L in filtered samples. Mining activities in southern Nevada have not significantly increased mercury concentrations in groundwater, as was the case in parts of northern Nevada (Cizdziel, 2004). In seawater, most authorities agree that mercury exists mainly bound to suspended particles (Jernelov et al., 1972); that the surface area of the sediment granules is instrumental in determining the final mercury content (Renzoni et al., 1973); that mercury conversion and transformation occur in the surface layer of the sediment or on suspended particles in the water (Dean, 1972; Fagerstrom and Jernelov, 1972; Jernelov et al., 1972); and, finally, that mercury-containing sediments require many decades to return to background levels under natural conditions (Langley, 1973). High concentrations of methylmercury in sub-thermocline, low-oxygen seawater were significantly and positively correlated with median daytime depth (< 200 m to > 300 m) of eight species of pelagic fishes; mean total mercury concentrations in whole fishes ranged between 57.0 and 377.0 μg/kg DW. The enhanced mercury accumulations in the marine mesopelagic compartment is attributable to diet and ultimately to water chemistry that controls mercury speciation and uptake at the base of the food chain (Monteiro et al., 1996).

Total mercury concentrations in uncontaminated natural waters (presumably unfiltered) now range from about 1.0 to 50.0 ng/L (Table 5.2). Concentrations as high as 1100.0 ng/L are reported in freshwaters near active gold mining facilities in Ecuador (Tarras-Wahlberg et al., 2000; Table 5.2), and up to 450,000.0 ng/L in drainage water from abandoned mercury mines in California (Rytuba, 2003; Table 5.2). Mercury and methylmercury from mercury mine drainage is adsorbed onto iron-rich precipitates and seasonally flushed in streams during periods of high water (Rytuba, 2003). Maximum concentrations of 89.7 ng/L in Swedish rain, 78.0 ng total mercury (47.0 ng methylmercury)/L in coastal seawater of New York, and 600.0 ng/L in sediment interstitial waters are documented (Table 5.2). Suspended particulate matter in the Elbe River, Germany, contained up to 150.0 mg total mercury (2.7 mg methylmercury)/kg dry weight as a result of mercury-contaminated wastes from a chloralkali plant (Wilken and Hintelmann, 1991; Table 5.2).

5.8 SUMMARY

Maximum concentrations of mercury recorded were less than 50.0 ng/L in uncontaminated natural waters; 60.0 ng/L in snow; 78.0 ng/L in coastal seawater; 89.7 ng/L in rain; 600.0 ng/L in groundwater; 1100.0 ng/L in freshwaters near active gold mining sites; 450,000 ng/L in drainage water from mercury mines; 495.0 ng/m^3 in air over Japanese volcanoes (74.0% as Hg^{2+}); 0.12 to 1.0 mg/kg dry weight in coal; 0.39 mg/kg fresh weight in pelagic clays; 130.0 mg/kg dry weight in Missouri, U.S., sewage sludge; 150.0 mg total mercury (2.7 mg methylmercury)/kg dry weight suspended particulate matter in water contaminated by chloralkali wastes in Germany; 329.9 mg total mercury/kg dry weight (0.045 mg methylmercury/kg dry weight) in soils contaminated by mercury-containing wastewater from a Chinese acetaldehyde plant; and 746.0 mg/kg dry weight in sediments near a Finnish pulp and paper mill where mercury was used as a slimicide.

REFERENCES

Atkeson, T., D. Axelrad, C. Pollman, and G. Keeler. 2003. Integrating Atmospheric Mercury Deposition and Aquatic Cycling in the Florida Everglades: An Approach for Conducting a Total Maximum Daily Load Analysis for an Atmospherically Derived Pollutant. Integrated Summary. Final Report. 272 pp. Avail. from Mercury and Applied Science MS6540, Florida Dept. Environmental Conservation, 2600 Blair Stone Road, Tallahassee, FL 32399-2400. See also <http://www.floridadep.org/labs/mercury/index.htm>

Beyer, W.N. 1990. Evaluating soil contamination, *U.S. Fish Wildl. Serv. Biol. Rep.*, 90(2), 1–25.

Bodaly, R.A., V.L. St. Louis, M.J. Paterson, R.J.P. Fudge, B.D. Hall, D.M. Rosenberg, and J.W.M. Rudd. 1997. In A. Sigel and H. Sigel (Eds.), *Metal Ions in Biological Systems, Vol. 34, Mercury and Its Effects on Environment and Biology*, p. 259–287. Marcel Dekker, New York.

Boudou, A. and F. Ribeyre. 1983. Contamination of aquatic biocenoses by mercury compounds: an experimental toxicological approach. In J.0. Nriagu (Ed.), *Aquatic Toxicology*, p. 73–116. John Wiley, New York.

Brumbaugh, W.G., D.P. Krabbenhoft, D.R. Helsel, J.G. Wiener, and K.R. Echols. 2001. A national pilot study of mercury contamination of aquatic ecosystems along multiple gradients: bioaccumulation in fish, *U.S. Geol. Surv., Biol. Sci. Rep. USGS/BRD/BSR 2001–0009*, 25 pp.

Bycroft, B.M., B.A.W. Coller, G.B. Deacon, D.J. Coleman, and P.S. Lake. 1982. Mercury contamination of the Lerderderg River, Victoria, Australia, from an abandoned gold field, *Environ. Pollut.*, 28A, 135–147.

Cameron, E.M. and I.R. Jonasson. 1972. Mercury in Precambrian shales of the Canadian Shield, *Geochim. Cosmochim. Acta*, 36, 985–1005.

Campbell, K.R., C.J. Ford, and D.A. Levine. 1998. Mercury distribution in Poplar Creek, Oak, Ridge, Tennessee, USA, *Environ. Toxicol Chem.*, 17, 1191–1198.

Chu, P. and D.B. Porcella. 1995. Mercury stack emissions from U.S. electric utility power plants, *Water Air Soil Pollut.*, 80, 135–144.

Cizdziel, J. 2004. Mercury concentrations in groundwater collected from wells on and near the Nevada test site, USA, *Bull. Environ. Contam. Toxicol*, 72, 202–210.

Clarkson, T.W., R. Hamada, and L. Amin-Zaki. 1984. Mercury. In J.O. Nriagu (Ed.), *Changing Metal Cycles and Human Health,* p. 285–309. Springer-Verlag, Berlin.

Cocking, D., M. Rohrer, R. Thomas, J. Walker, and D. Ward. 1995. Effects of root morphology and Hg concentration in the soil on uptake by terrestrial vascular plants, *Water Air Soil Pollut.*, 80, 1113–1116.

Coquery, M. D. Cossa, and J.M. Martin. 1995. The distribution of dissolved and particulate mercury in three Siberian estuaries and adjacent Arctic coastal waters, *Water Air Soil Pollut.*, 80, 653–664.

Dean, R.B. 1972. The case against mercury, Available from *Natl. Tech. Inform. Serv.*, Springfield VA, as PB-213-692, 1-11.

Fagerstrom, T. and A. Jernelov. 1972. Some aspects of the quantitative ecology of mercury, *Water Res.*, 6, 1193–1202.

Ferrara, R., B.E. Maserti, M. Andersson, H. Edner, P. Ragnarson, S. Svanberg, and A. Hwrnandez. 1998. Atmospheric mercury concentrations and fluxes in the Almaden District (Spain), *Atmospher. Environ.*, 32, 3897–3904.

Ferrara, R., B.E. Maserti, and R. Breder. 1991. Mercury in biotic and abiotic compartments of an area affected by a geochemical anomaly (Mt. Amiata, Italy), *Water Air Soil Pollut.*, 56, 219–233.

Ferrara, R., B.E. Maserti, A. DeLiso, R. Cioni, B. Raco, G. Taddeucci, H. Edner, P. Ragnarson, S. Svanberg, and E. Wallinder. 1994. Atmospheric mercury emission at Solfatara Volcano (Pozzuoli, Phlegraean fields, Italy), *Chemosphere*, 29, 1421–1428.

Finkleman, R.B. 2003. Mercury in coal and mercury emissions from coal combustion. In J.E. Gray (Ed.), *Geologic Studies of Mercury by the U.S. Geological Survey*, p. 9—12. USGS Circular 1248. Available from USGS Information Services, Box 25286, Denver, CO 80225.

Fitzgerald, W.F. 1979. Distribution of mercury in natural waters. In J.O. Nriagu (Ed.), *The Biogeochemistry of Mercury in the Environment*, p. 161–173. Elsevier/North-Holland Biomedical Press, New York.

Fitzgerald, W.F. 1989. Atmospheric and oceanic cycling of mercury. In J.P. Ripley and R. Chester (Eds.). *Chemical Oceanography*, p. 151–186. Academic Press, New York.

Gillis, C.A., N.L. Bonnevie, and R.J. Wenning. 1993. Mercury contamination in the Newark Bay estuary, *Ecotoxicol. Environ. Safety*, 25, 214–226.

Gilmour, C.C., G.S. Reidel, M.C. Ederington, J.T. Bell, J.M. Benoit, G.A. Gill, and M.C. Stordal. 1998. Methylmercury concentrations and production rates across a trophic gradient in the northern Everglades, *Biogeochemistry*, 40, 327–345.

Granato, T.C., R.I. Pietz, J. Gschwind, and C. Lue-Hing. 1995. Mercury in soils and crops from fields receiving high cumulative sewage sludge applications: validation of U.S. EPA's risk assessment for human ingestion, *Water Air Soil Pollut.*, 80, 1119–1127.

Gray, J.E.(Ed.). 2003. *Geologic Studies of Mercury by the U.S. Geological Survey*. USGS Circular 1248. Available from USGS Information Services, Box 25286, Denver, CO 80225.

Gustin, M.S., S.E. Lindberg, K. Austin, M. Coolbaugh, A. Vette, and H. Zhang. 2000. Assessing the contribution of natural sources to regional atmospheric mercury budgets, *Sci. Total Environ.*, 259, 61–71.

Gustin, M.S., G.E. Taylor Jr., and T.L. Leonard. 1994. High levels of mercury contamination in multiple media of the Carson River drainage basin of Nevada — implications for risk assessment, *Environ. Health Perspect.*, 102, 772–778.

Gustin, M.S., G.E. Tylor Jr., T.L. Leonard, and R.E. Keislar. 1996. Atmospheric mercury concentrations associated with geologically and anthropogenically enriched sites in central western Nevada, *Environ. Sci. Technol.*, 30, 2572–2579.

Hamasaki, T., H. Nagase, Y. Yoshioka, and T. Sato. 1995. Formation, distribution, and ecotoxicity of methylmetals of tin, mercury, and arsenic in the environment, *Crit. Rev. Environ. Sci. Technol.*, 25, 45–91.

Heyes, A., T.R. Moore, J.W.M. Rudd, and J.J Dugoua. 2000. Methyl mercury in pristine and impounded boreal peatlands, Experimental Lakes area, Ontario, *Can. J. Fish. Aquat. Sci.*, 57, 2211–2222.

Howe, A., L.H. Fung, G. Lalor, R. Rattray, and M. Vutchkov. 2005. Elemental composition of Jamaican foods. 1. A survey of five food crop categories, *Environ. Geochem. Health*, 27, 19–30.

Hurley, J.P., R.C. Back, K.R. Rolfhus, R.C. Harris, D.E. Armstrong, and R. Harris. 2002. Watershed influences on transport, fate, and bioavailability of mercury in Lake Superior. In *Proceedings and Summary Report: Workshop on the Fate, Transport and Transformation of Mercury in Aquatic and Terrestrial Environments*, p. B-18. U.S. Environ. Protect. Agen. Rep. EPA/625/R–02/005.

Jernelov, A., R. Hartung, P.B. Trost, and R.E. Bisque. 1972. Environmental dynamics of mercury. Pages 167–201 in R. Hartung and B.D. Dinman (Eds.). *Environmental Mercury Contamination*. Ann Arbor Science Publ., Ann Arbor, MI.

Kannan, K., R.G. Smith Jr., R.F. Lee, H.L. Windom, P.T. Heitmuller, J.M. Macauley, and J.K. Summers. 1998. Distribution of total mercury and methyl mercury in water, sediment, and fish from south Florida estuaries, *Arch. Environ. Contam. Toxicol.*, 34, 109–118.

Kelly, C.A., J.W.M. Rudd, R.A. Bodaly, N.P. Roulet, V.L. St. Louis, A. Heyes, T.R. Moore, S. Schiff, R. Aravena, K.J. Scott, B. Dyck, R. Harris, B. Warner, and G. Edwards. 1997. Increases in fluxes of greenhouse gases and methyl mercury following flooding of an experimental reservoir, *Environ. Sci. Technol.*, 31, 1334–1344.

Kiorboe, T., F. Mohlenberg, and H.U. Riisgard. 1983. Mercury levels in fish, invertebrates and sediment in a recently recorded polluted area (Nissum Broad, western Limfjord, Denmark), *Mar. Pollut. Bull.*, 14, 21–24.

Klein, D.H. and E.D. Goldberg. 1970. Mercury in the marine environment. *Environ. Sci. Technol.*, 4, 765–768.

Krabbenhoft, D.P., M.L. Olson, J.F. Dewild, D.W. Clow, R.G. Striegl, and M.M. Dornblaser. 2002. Mercury loading and methylmercury production and cycling in high-altitude lakes from the western United States, *Water Air Soil Pollut: Focus 2002*, 2, 233–249.

Lacerda, L.D., C.E. Rezende, A.R.C. Ovalle, and C.E.V. Carvalho. 2004. Mercury distribution in continental shelf sediments from two offshore oil fields in southeastern Brazil, *Bull. Environ. Contam. Toxicol.*, 72, 178–185.

Lamborg, C.H., W.F. Fitzgerald, J.O'Donnell, and T. Torgersen. 2000. A non-steady-state compartmental model of global-scale mercury biogeochemistry with interhemispheric atmospheric gradients, *Geochim. Cosmochim. Acta*, 66, 1105–1118.

Lamborg, C.H., K.R. Rolfhus, W.F. Fitzgerald, and G. Kim. 1999. The atmospheric cycling and air-sea exchange of mercury species in the South and equatorial Atlantic Ocean, *Deep-Sea Res.*, 46, 957–977.

Langley, D.G. 1973. Mercury methylation in an aquatic environment, *Jour. Water Pollut. Contr. Feder.*, 45, 44–51.

Martin, D.B. and W.A. Hartman. 1984. Arsenic, cadmium, lead, mercury, and selenium in sediments of riverine and pothole wetlands of the north central United States, *J. Assoc. Off. Anal. Chem.*, 67, 1141–1146.

Mason, R.P., W.F. Fitzgerald, and F.M.M. Morel. 1994. The biogeochemistry of elemental mercury — anthropogenic influences, *Geochim. Cosmochim. Acta*, 58, 3191–3198.

Matsuyama, A., Q. Liya, A. Yasutake, M. Yamaguchi, R. Aramaki, L. Xiaojie, J. Pin, L. Li, A. Yumin, and Y. Yasuda. 2004. Distribution of methylmercury in an area polluted by mercury containing wastewater from an organic chemical factory in China, *Bull. Environ. Contam. Toxicol.*, 73, 846–852.

McNeal, J.M and A.W. Rose. 1974. The geochemistry of mercury in sedimentary rocks and soils in Pennsylvania, *Geochim. Cosmochim. Acta*, 38, 1757–1784.

Meuleman, C., M. Leermakers, and W. Baeyens. 1995. Mercury speciation in Lake Baikal, *Water Air Soil Pollut.*, 80, 539–551.

Meyer, M.W., D.C. Evers, T. Daulton, and W.E. Braselton. 1995. Common loons (*Gavia immer*) nesting on low pH lakes in northern Wisconsin have elevated blood mercury content, *Water Air Soil Pollut.*, 80, 871–880.

Minagawa, K. and Y. Takizawa. 1980. Determination of very low levels of inorganic and organic mercury in natural waters by cold-vapor atomic absorption spectrometry after preconcentration on a chelating resin, *Anal. Chim. Acta*, 115, 103–110.

Monson, B.A. and P.L. Brezonik. 1998. Seasonal patterns of mercury species in water and plankton from softwater lakes in northeastern Minnesota, *Biogeochemistry*, 40, 147–162.

Monteiro, L.R., V. Costa, R.W. Furness, and R.S. Santos. 1996. Mercury concentrations in prey fish indicate enhanced bioaccumulation in mesopelagic environments, *Mar. Ecol. Prog. Ser.*, 141, 21–25.

Nater, E.A. and D.F. Grigal. 1992. Regional trends in mercury distribution across the Great Lakes states, north central USA, *Nature*, 358, 139–141.

Nishimura, H. and M. Kumagai. 1983. Mercury pollution of fishes in Minamata Bay and surrounding water: analysis of pathway of mercury, *Water Air Soil Pollut.*, 20, 401–411.

Paasivirta, J., J. Sarkka, K. Surma-Aho, T. Humppi, T. Kuokkanen, and M. Marttinen. 1983. Food chain enrichment of organochlorine compounds and mercury in clean and polluted lakes of Finland, *Chemosphere*, 12, 239–252.

Panyametheekul, S. 2004. An operationally defined method to determine the speciation of mercury, *Environ. Geochem. Health*, 26, 51–57.

Peltola, P. amd M. Astrom. 2003. Urban geochemistry: a multimedia and multielement survey of a small farm in northern Europe, *Environ. Geochem. Health*, 25, 397–419.

Plourde, Y., M. Lucotte, and P. Pichet. 1997. Contribution of suspended particulate matter and zooplankton to MeHg contamination of the food chain in midnorthern Quebec (Canada) reservoirs, *Can. J. Fish. Aquat. Sci.*, 54, 821–831.

Porcella, D.B. 1994. Mercury in the environment — biochemistry. In C.J. Watras and J.W. Huckabee (Eds.), *Mercury Pollution, Integration and Synthesis*, p. 3–19. CRC Press, Boca Raton, FL.

Rada, R.G., D.E. Powell, and J.G. Wiener. 1993. Whole-lake burdens and spatial distribution of mercury in surficial sediments in Wisconsin seepage lakes, *Can. J. Fish. Aquat. Sci.*, 50, 865–873.

Rada, R.G., J.G. Wiener, M.R. Winfrey, and D.E. Powell. 1989. Recent increases in atmospheric deposition of mercury to north-central Wisconsin lakes inferred from sediment analyses, *Arch. Environ. Contam. Toxicol.*, 18, 175–181.

Renzoni, A., E. Bacci, and L.Falciai. 1973. Mercury concentration in the water, sediments and fauna of an area of the Tyrrhenian Coast. Pages 17–45 in *6th International Symposium on Medical Oceanography*, Portoroz, Yugoslavia, September 26–30, 1973.

Rytuba, J.J. 2003. Mercury from mineral deposits and potential environmental impact, *Environ. Geol.*, 43, 326–338.

St. Louis, V.L., J.W.M. Rudd, C.A. Kelly, K.G. Beaty, N.S. Bloom, and R.J. Flett. 1994. Importance of wetlands as sources of mercury to boreal forest ecosystems, *Can. J. Fish. Aquat. Sci.*, 51, 1065–1076.

St. Louis, V.L., J.W.M. Rudd, C.A. Kelly, K.G. Beaty, R.J. Flett, and N.T. Roulet. 1996. Production and loss of methylmercury and loss of total mercury from boreal forest catchments containing different types of wetlands, *Environ. Sci. Technol.*, 30, 2719–2729.

Segura-Munoz, S.I., A. Bocio, T.M.B. Trevilato, A.M.M. Takayanagui, and J.L. Domingo. 2004. Metal concentrations in soil in the vicinity of a municipal solid waste landfill with a deactivated medical waste incineration plant, Ribeirao Preto, Brazil, *Bull. Environ. Contam. Toxicol.*, 73, 575–582.

Semkin, R.G., G. Mierle, and R.J. Neureuther. 2005. Hydrochemistry and mercury cycling in a high Arctic watershed, *Sci. Total Environ.*, 342, 199–221.

Sharma, R. and S. Pervez. 2004. A case study of spatial variation and enrichment of selected elements in ambient particulate matter around a large coal-fired power station in central India, *Environ. Geochem. Health*, 26, 373–381.

Skei, J.H. 1978. Serious mercury contamination of sediments in a Norwegian semi-enclosed bay, *Mar. Pollut. Bull.*, 9, 191–193.

Smith, J.N. and D.H. Loring. 1981. Geochronology for mercury pollution in the sediment of the Saguenay Fjord, Quebec, *Environ. Sci. Technol.*, 15, 944-951.

Snyder-Conn, E., J.R. Garbarino, G.L. Hoffman, and A. Oelkers. 1997. Soluble trace elements and total mercury in Arctic Alaskan snow, *Arctic*, 50, 201–215.

Spry, D.J. and J.G. Wiener. 1991. Metal bioavailablity and toxicity to fish in low-alkalinity lades: a critical review, *Environ. Pollut.*, 71, 243–304.

Suckcharoen, S. and M. Lodenius. 1980. Reduction of mercury pollution in the vicinity of a caustic soda plant in Thailand. *Water Air Soil Pollut.*, 13, 221–227.

Takizawa, Y. 1995. Exposure assessment — mercury in the atmosphere. In *Proceedings of the International Workshop on Environmental Mercury Pollution and its Health Effects in Amazon River Basin*, 100–105. Rio de Janeiro, 30 November–2 December 1994. Published by National Institute for Minamata Disease, Minamata City, Kumamoto 867, Japan.

Takizawa, Y., K. Minagawa, and M. Fujii. 1981. A practical and simple method in fractional determination of ambient forms of mercury in air, *Chemosphere*, 10, 801–809.

Tarras-Wahlberg, N.H., A. Flachier, G. Fredriksson, S. Lane, B. Lundberg, and O. Sangfors. 2000. Environmental impact of small scale and artisanal gold mining in southern Ecuador, *Ambio*, 29, 484–491.

U.S. Environmental Protection Agency (USEPA). 2000. Emissions and generation resource integrated database (E-GRID), *<http://www.epa.gov/airmarkets/egrid>*.

U.S. National Academy of Sciences (USNAS). 1978. *An Assessment of Mercury in the Environment*. Natl. Acad. Sci., Washington, D.C., 185 pp.

Vandal, G.M., W.F. Fitzgerald, C.F. Boutron, and J.P. Candelone. 1993. Variations of mercury deposition to Antarctica over the past 34,000 years, *Nature*, 362, 621–623.

Varekamp, J.C. and P.R. Buseck. 1986. Global mercury flux from volcanic and geothermal sources, *Appl. Geochem.*, 1, 65–73.

Verta, M., T. Matilainen, P. Porvari, M. Niemi, A. Uusi-Rauva, and N.S. Bloom. 1994. Methylmercury sources in boreal lake ecosystems. In C.J. Watras and J.W. Huckabee (Eds.), *Mercury Pollution — Integration and Synthesis*, p. 119–136. Lewis, Ann Arbor, MI.

Vucetic, T., W.B. Vernberg, and G. Anderson. 1974. Long-term annual fluctuations of mercury in the zooplankton of the east central Adriatic, *Rev. Int. Ocean. Medic.*, 33, 75–81.

Watras, C.J., N.S. Bloom, R.J.M. Hudson, S. Gherini, R. Munson, S.A. Claas, K.A. Morrison, J. Hurley, J.G. Wiener, W.F. Fitzgerald, R. Mason, G. Vandal, D. Powell, R. Rada, L. Rislove, M. Winfrey, J. Elder, D. Krabbenhoft, A. W. Andren, C. Babiarz, D.B. Porcella, and J.W. Huckabee. 1994. Sources and fates of mercury and methylmercury in Wisconsin lakes. In C.J. Watras and J.W. Huckabee (Eds.), *Mercury Pollution: Integration and Synthesis*, p. 153–177. Lewis Publ., Boca Raton, FL.

Wiener, J.G., D.P. Krabbenhoft, G.H. Heinz, and A.M. Scheuhammer. 2003. Ecotoxicology of mercury. In D.J.Hoffman, B.A. Rattner, G.A. Burton Jr., and J. Cairns Jr. (Eds.), *Handbook of Ecotoxicology, 2nd edition*, p. 409–463. Lewis Publ., Boca Raton, FL.

Wilken, R.D. and H. Hintelmann. 1991. Mercury and methylmercury in sediments and suspended particles from the River Elbe, North Germany, *Water Air Soil Pollut.*, 56, 427–437.

Williams, P.M. and H.V. Weiss. 1973. Mercury in the marine environment: concentration in sea water and in a pelagic food chain, *J. Fish. Res. Bd. Canada*, 30, 293–295.

Chapter 6

Mercury Concentrations in Plants and Animals

Information on mercury residues in field collections of living organisms is especially abundant. Elevated concentrations of mercury occur in aquatic biota from areas receiving high atmospheric depositions of mercury, or when mercury concentrations in the diet or water are elevated (Sorensen et al., 1990; Wiener et al., 1990a; Fjeld and Rognerud, 1993). Mercury levels are comparatively elevated in fish-eating fishes, birds, and mammals (Langlois et al., 1995). In general, mercury concentrations in biota were usually less than 1.0 mg/kg FW tissue in organisms collected from locations not directly affected by human use of the element. However, concentrations exceed 1.0 mg/kg — and are sometimes markedly higher — in animals and vegetation from the vicinity of chloralkali plants; agricultural users of mercury; smelters; mining operations; pulp and paper mills; factories producing mercury-containing paints, fertilizers, and insecticides; sewer outfalls; sludge disposal areas; and other anthropogenic point sources of mercury (Schmitt and Brumbaugh, 1990). In some Minnesota lakes, mercury concentrations in fish are sufficiently elevated to be potentially hazardous when ingested by mink, otters, loons, and raptors (Swain and Helwig, 1989).

An elevated concentration of mercury (i.e., > 1.0 mg/kg FW), usually as methylmercury, in any biological sample is often associated with proximity to human use of mercury. The elimination of mercury point-source discharges has usually been successful in improving environmental quality; however, elevated levels of mercury in biota may persist in contaminated areas long after the source of pollution has been discontinued (Rada et al., 1986). For example, mercury remains elevated in resident biota of Lahontan Reservoir, Nevada, which received about 7500 tons of mercury as a result of gold and silver mining operations during the period 1865 to 1895 (Cooper, 1983). It is noteworthy that some groups of organisms with consistently elevated mercury residues may have acquired these concentrations as a result of natural processes rather than from anthropogenic activities. These groups include older specimens of long-lived predatory fishes, marine mammals (especially pinnipeds), and organisms living near natural mercury-ore-cinnabar deposits. In general, concentrations of mercury in feral populations of marine vertebrates — including elasmobranchs, fishes, birds, and mammals — are clearly related to the age of the organism. Regardless of species or tissue, all data for mercury and marine vertebrates show increases with increasing age of the organism (Eisler, 1984). Factors that may account, in part, for this trend include differential uptake at various life stages, reproductive cycle, diet, general health, bioavailability of different chemical species, mercury interactions with other metals, metallothioneins, critical body parts, and anthropogenic influences (Eisler, 1984).

6.1 ALGAE AND MACROPHYTES

Concentrations of total mercury were almost always below 1.0 mg/kg dry weight in aquatic and terrestrial vegetation except for those areas where human activities have contaminated the environment with mercury (Eisler, 2000; Table 6.1). In general, mercury concentrations were highest in mosses, fungi, algae, and macrophytes under the following conditions: after treatment with mercury-containing pesticides, near smelter emissions, in sewage lagoons, near chloralkali plants, exposure to mercury-contaminated soils, and proximity to industrialized areas (Table 6.1). Moreover, samples of the marine flowering plant *Posidonia oceanica* collected near a sewer outfall in Marseilles, France, had elevated concentrations of mercury — in mg/kg dry weight — of 51.5 in leaves, 2.5 in rhizomes, and 0.6 in roots (Augier et al., 1978). Also, water hyacinth *Eichornia crassipes* from a sewage lagoon in Mississippi contained up to 70.0 mg Hg/kg DW (Chigbo et al., 1982). Both *Posidonia* and *Eichornia* may be useful in phytoremediation of mercury-contaminated aquatic environments.

Highest concentrations of mercury (90.0 mg Hg/kg FW) were found in roots of alfalfa (*Medicago sativa*) growing in soil containing 0.4 mg Hg/kg, in bark of a cherry tree (*Prunus avium*) from a factory area in Slovenia (59.0 mg/kg FW), in leaves of water hyacinth (*Eichornia crassipes*; 70.0 mg/kg DW) from a sewage lagoon, in mosses near a chloralkali plant (16.0 mg/kg FW), in fungi near a smelter (35.0 mg/kg DW), and leaves of *Posidonia oceanica* (51.5 mg/kg DW) near a sewer outfall (Table 6.1).

Certain species of macrophytes strongly influence mercury cycling. For example, *Spartina alterniflora* — a dominant salt marsh plant in Georgia estuaries — accounted for almost half the total mercury budget in that ecosystem (Windom, 1973; Gardner et al., 1975; Windom et al., 1976). Mercury entered the estuary primarily in solution, delivering about 1.5 mg annually to each square meter of salt marsh. Annual uptake of mercury by *Spartina* alone was about 0.7 mg/m^2 salt marsh. Mangrove vegetation plays a similarly important role in mercury cycling in the Florida Everglades (Lindberg and Harriss, 1974; Tripp and Harriss, 1976). These findings suggest that more research is needed on the role of higher plants in the mercury cycle.

Creation of reservoirs by enlargement of riverine lakes and flooding of adjacent lands has led to a marked rise in rates of methylmercury production by microorganisms in sediments. This process has resulted mainly from increased microbial activity via increased use of organic materials under conditions of reduced oxygen (Jackson, 1988). Increased net methylation in flooded humus and peat soils, especially in anoxic conditions, was determined experimentally and judged to be the main reason for increased methylmercury concentrations in reservoirs (Porvari and Verta, 1995).

6.2 INVERTEBRATES

In general, all species of invertebrates sampled had elevated concentrations of mercury (up to 10.0 mg/kg FW, 38.7 mg/kg DW) in the vicinity of industrial, municipal, and other known sources of mercury when compared to conspecifics collected from reference locations (Table 6.2). The finding of 202.0 mg Hg/kg FW in digestive gland of *Octopus vulgaris* (Renzoni et al., 1973) needs verification. Larvae of terrestrial insects (i.e., larvae of blowflies (*Calliphora* sp.)), play an important role in mercury cycling from feeding on beached fish carcasses (Table 6.2; Sarica et al., 2005).

Comparatively high mercury concentrations of 5.7 mg/kg FW in crayfish abdominal muscle from Lahontan Reservoir, Nevada (Table 6.2), an area heavily contaminated with mercury from gold mining operations some decades earlier is discussed in greater detail in Chapter 11; and concentrations of 41.0 mg/kg DW in sea anemones and up to 100.0 mg/kg DW in crustaceans (Table 6.2), both from the heavily-contaminated Minamata Bay, Japan, are discussed in detail in Chapter 10.

Marine bivalve molluscs can accumulate mercury directly from seawater; uptake was greater in turbulent waters than in clear waters (Raymont, 1972). Molluscs sampled before and immediately after their substrate was extensively dredged had significantly elevated tissue mercury concentrations after dredging, which persisted for at least 18 months (Rosenberg, 1977). Mercury concentrations in

Table 6.1 Mercury Concentrations in Field Collections of Selected Species of Plants

Species and Other Variables	Concentration (mg/kg)	Ref.[a]
Algae and macrophytes; marine; whole:		
Malaysia; 26 species	Max. 0.35 DW	5
Korea; 17 species	0.02–0.52 DW	6
Brown alga; *Ascophyllum nodosum*; whole; marine:		
60 cm length vs. 100–140 cm length	0.07 FW vs. 0.11 FW	4
Eikhamrane, Norway vs. Flak, Norway	0.12–1.09 DW vs. 0.02–0.03 DW	7
Transplanted from Eikhamrane to Flak for 4 months	0.04–0.95 DW	7
Lofoton, Norway	0.05–0.08 DW	8
Trondheimsfjord, Norway	0.05–0.18 DW	8
Hardangerfjord, Norway	0.05–20.0 DW	8
Marine alga, *Ceramium rubrum*; whole	0.48 FW; 3.0 DW	9
Mandarin orange, *Citrus tachibana*; Japan		
Sprayed with Hg herbicide:		
Fruit skin	0.03–0.24 FW	1
Fruit pulp	0.01–0.4 FW	1
Unsprayed:		
Skin and pulp	0.01–0.05 FW	1
Fungi, *Cortinarius* spp.; near smelter	9.5–35.0 DW	1
Moss, *Dicranum scoparium*, whole		
Tennessee:		
Exposed to fly ash	1.1 DW	1
Remote areas	0.1 DW	1
Great Smoky Mountains	0.07 DW	1
Hawaii	0.16 DW	1
Iceland	0.03 DW	1
Water hyacinth, *Eichornia crassipes*; from sewage lagoon in Bay St. Louis, Mississippi; leaves	70.0 DW	2
Lichen, *Hypogymnia physodes*; whole; Finland, 1982–83; distance, in km, from chloralkali plant:		
0–1	18.0 FW	3
1–5	2.0 FW	3
5–20	0.4 FW	3
20–100	0.3 FW	3
> 100	0.3 FW	3
Kelp, *Laminaria digitata*:		
Whole	0.13 FW; 0.79 DW	9
Whole	0.17 DW	10
Labrador tea, *Ledum* sp.; Alaska; over cinnabar deposit; stem	1.0–3.5 DW	1
Alfalfa, *Medicago sativa:*		
From soil containing 0.4 mg Hg/kg:		
Root	90.0 FW	1
Leaf	0.13–0.4 FW	1
From soil with < 0.4 mg Hg/kg:		
Leaf	0.16 FW	1
Mushrooms, 10 spp.; near mercury-contaminated sludge mounds; Niigata, Japan; November 1979:		
Total mercury	0.03–2.0 FW; 0.44–24.8 DW	21
Methylmercury	< 0.005–0.1 FW; < 0.005–1.3 DW	21
Tobacco, *Nicotiana tabacum*; leaf:		
Treated with Hg (Japan)	1.0–1.6 FW	1
Untreated (U.S.)	< 0.2 FW	1
Rice, *Oryza sativa*; grain:		
Sprayed with Hg	0.1–0.7 FW	1
Unsprayed	0.02–0.1 FW	1
Phytoplankton; whole:		
Chesapeake Bay, Maryland	0.11–0.13 DW	11
North Atlantic; offshore	0.05 FW	12

(continued)

Table 6.1 (continued) Mercury Concentrations in Field Collections of Selected Species of Plants

Species and Other Variables	Concentration (mg/kg)	Ref.[a]
West coast, Norway	Max. 1.2 DW	13
West coast, Norway	0.6–25.2 DW	14
Laver, *Porphyra umbilicalis*; whole (marine alga)	0.5 FW; 2.4 DW	15
Marine flowering plant, *Posidonia oceanica*; near sewer outfall; Marseilles, France:		
Rhizomes	2.5 DW	4
Leaves	51.5 DW	4
Roots	0.6 DW	4
Cherry, *Prunus avium*; Europe (Slovenia); bark:		
Uncontaminated areas	0.06 FW	1
High Hg in soil	6.0 FW	1
Factory area	59.0 FW	1
Saltmarsh grass, *Spartina alterniflora*; Brunswick, Georgia, U.S.:		
Whole	Max. 1.4 DW	16
Leaves and stalks	0.2 DW	16, 17
Rhizomes	0.5–0.7 DW	18
Roots	0.7–8.7 DW	18
Base of stalk	0.6–1.2 DW	18
Stalk	0.4–1.1 DW	18
Leaves	0.4–1.1 DW	18
Mosses, *Sphagnum* spp.; whole; Finland, 1982–1983; distance (km) from chloralkali plant:		
0–1	3.8 (1.5–16.0) FW	3
1–5	0.8 (0.2–2.6) FW	3
5–20	0.09 (0.04–0.2) FW	3
20–100	0.05 (0.0–0.8) FW	3
> 100	0.02 FW	3
Marine algae, *Ulva* spp.; whole:		
Firth of Tay, Scotland	6.3 FW; 25.5 DW	9
Minamata Bay, Japan	Max. 14.0 DW	19
Puget Sound, Washington state	0.005–0.01 DW	20

Note: Values shown are in mg total Hg/kg fresh weight (FW) or dry weight (DW).

[a] Reference: 1, Jenkins, 1980; 2, Chigbo et al., 1982; 3, Lodenius and Tulisalo, 1984; 4, Augier et al., 1978; 5, Sivalingam, 1980; 6, Kim, 1972; 7, Myklestad et al., 1978; 8, Haug et al., 1974; 9, Jones et al., 1972; 10, Leatherland and Burton, 1974; 11, Cocoros et al., 1973; 12, Greig et al., 1975; 13, Stenner and Nickless, 1975; 14, Skei et al., 1976; 15, Preston et al., 1972; 16, Windom et al., 1976; 17, Windom, 1975; 18, Windom, 1973; 19, Matida and Kumada, 1969; 20, Schell and Nevissi, 1977; 21, Minagawa et al., 1980.

scallops are influenced by reproductive status, sex, and inherent species differences (Table 6.2; Norum et al., 2005). Mercury–sediment–water interactions influence uptake dynamics by marine benthos. Organisms feeding in direct contact with sediments have higher overall mercury levels than those feeding above the sediment–water interface (Klemmer et al., 1976). Mercury levels in mussels along European coasts tend to reflect mercury levels in water and sediments to a greater degree than does size of mussel, season of collection, or position in the intertidal zone (De Wolf, 1975). Reduced mercury inputs to coastal areas as a result of legislation and effective enforcement actions is reflected in mercury levels of common mussels, *Mytilus edulis*, in Bergen Harbor, Norway. In 2002, mussels from Bergen Harbor contained a maximum of 0.04 mg Hg/kg FW soft parts; this was about 60.0% lower than mercury levels in mussels collected from the same area in 1993. The reduced mercury was attributed to reductions in mercury content to Bergen Harbor of municipal wastewater, urban runoff, and especially of mercury-containing dental wastes (Airas et al., 2004). In every case reported wherein mercury concentrations in molluscan soft parts exceed 1.0 mg/kg FW, it was associated with mercury pollution from human activities (Eisler 1981, 2000).

Table 6.2 Mercury Concentrations in Field Collections of Selected Species of Invertebrates

Ecosystem, Species, and Other Variables	Concentration (mg/kg)	Ref.[a]
Freshwater		
Annelids, 2 families:		
From Hg-contaminated areas	0.3–0.6 FW	1
From reference locations	0.03–0.05 FW	1
Arthropods:		
Sow bug, *Asellus* sp.; whole; Sweden:		
20 km below paper mill	1.9 FW	2
1–15 km above paper mill	0.06 FW	2
Crustaceans, 2 families:		
From mercury-contaminated areas	1.9–10.0 FW	1
From reference locations	0.06–0.56 FW	1
Insects, 8 families:		
From mercury-contaminated areas	0.5–5.0 FW	1
From reference locations	0.05–0.21 FW	1
Mayfly, *Hexagenia* sp.; whole nymphs vs. sediments; upper Mississippi River; 1989	Max. 0.013 DW vs. max. 0.16 DW	3
Stonefly, *Isoperla* sp.; whole; Sweden:		
17 km below paper mill	2.4 FW	2
15 km above paper mill	0.07 FW	2
Crayfish, *Orconectes virilis*; Ontario; whole:		
Near chloralkali plant	1.4–7.4 FW	2
Reference location	0.09–0.49 FW	2
Crayfish, *Pacifiastacus* sp.; Lahontan Reservoir, Nevada; 1981; abdomen	5.7 FW	4
Crayfish; 5 species; Ontario, Canada; muscle	0.02–0.61 FW	5
Molluscs:		
From mercury-contaminated areas	0.02–2.2 FW	1
From reference locations	0.05 FW	1
Marine		
Coelenterata; whole:		
Pelagia sp.	0.07 DW	17
Sea anemone; Minamata Bay, Japan	41.0 DW	16
Anemone, *Tealia felina*	0.86 DW	13
Annelids:		
Georgia, U.S.; 3 spp.; whole; estuaries:		
Mercury-contaminated estuary; total Hg vs. methyl Hg	0.7–4.5 DW vs. max. 0.8 DW	6
Control estuary; total Hg vs. methyl Hg	0.1–0.6 DW vs. max. 0.013 DW	6
Arthropods:		
Barnacles, *Balanus* spp.; soft parts	1.0–1.35 DW; 0.1–0.22 FW	14, 28
Blue crab, *Callinectes sapidus:*		
Muscle	0.45 DW	31
Whole	0.26 (0.02–1.5) FW; 1.3 (0.1–7.7) DW	29
Rock crab, *Cancer irroratus:*		
Muscle	0.15–0.19 FW	30
Digestive gland	0.07–1.09 FW	30
Gills	0.03 FW	30
Copepods; whole	0.11–0.27 DW	32, 33
Crustaceans:		
Marine products of commerce; edible portions:		
10 species	< 0.1 FW	34, 35
9 species	0.1–0.2 FW	35
1 species	0.2–0.3 FW	35
2 species; Minamata Bay, Japan	41.0–100.0 DW	16

(continued)

Table 6.2 (continued) Mercury Concentrations in Field Collections of Selected Species of Invertebrates

Ecosystem, Species, and Other Variables	Concentration (mg/kg)	Ref.[a]
Whole:		
5 species	0.0–0.1 FW	36
12 species	0.06–1.6 DW	37
Sand shrimp, *Crangon crangon*		
Muscle	0.19 FW	38
Whole	0.2–1.7 DW	29
Whole	0.03–0.12 FW	39
Chinese mitten crab, *Eriocheir sinensis*; San Francisco Bay, California; July–August 2002:		
Hepatopancreas:		
Total mercury	0.25 (0.04–1.03) DW	54
Methylmercury	0.036 (0.006–0.069) DW	54
Other tissues:		
Total mercury	0.15 (0.04–0.69) DW	54
Methylmercury	0.04 (0.007–0.095) DW	54
Euphausids; various species; whole	Max. 0.52 DW; max. 0.06 FW	29, 32, 40
Georgia, U.S.; 2 spp; whole; estuaries:		
Mercury-contaminated estuary; total Hg vs. methyl Hg	0.4–1.8 DW vs. max. 1.0 DW	6
Control estuary: total Hg vs. methyl Hg	0.1–0.4 DW vs. max. 0.05 DW	6
American lobster, *Homarus americanus*:		
Muscle:		
Chesapeake Bay	0.03–0.06 FW	2
NW Atlantic	0.25–1.6 DW; 0.31 FW	2, 41
Nova Scotia	0.15–1.5 FW	2
Liver	0.60 FW	41
Spiny lobster, *Nephrops norvegicus*:	2.9 FW	7
Tyrrhenian Sea; 1981; muscle	2.9 FW	7
Edible portions	0.10–0.22 FW	27
Shrimp; edible portions; total Hg vs. methyl Hg	0.77 FW vs. 0.4 FW	8
Shrimp: muscle:		
Brown shrimp, *Penaeus aztecus*; Mexico	0.06 (0.01–0.67) FW	42
Penaeus spp.	0.02–0.46 FW	42–46
Shrimp: various commercial species:		
Whole:		
Gulf of Mexico	0.03–0.09 FW	47
Persian Gulf	0.24 (0.08–0.88) FW	48
North Sea	0.04–0.18 FW	49
Texas	< 0.02 DW	42
Persian Gulf	0.005–0.012 FW	50
Muscle:		
Korea	0.08–0.17 FW	51
SE United States	0.22 DW	31
Belgium	1.3 DW	24
Shell	0.02–0.05 FW	51
Molts	1.3 DW	24
Stomatopod, *Squilla mantis*; muscle	0.12 FW	43
Echinoderms:		
Sea stars, 3 spp.; 1981; Venezuela; polluted area; gonads	3.8–8.7 DW; 0.9–1.6 FW	9
Starfish, *Asterias rubens*; whole	0.12 FW; 0.22 DW	13, 18
Starfish, *Marthasterias glacialis*; whole	0.92 DW	13
Sea urchin, *Strongylocentrotus fragilis*; gonads; total mercury vs. organic mercury	0.02–0.03 FW vs. 0.003 FW	20
Various species; whole	0.28–0.40 FW	19
Molluscs:		
British Columbia, Canada; July 1999, December 1999, and February 2000; males vs. females:		
Spiny scallop, *Chlamys hastata*:		
Gonad	0.4 DW vs. 0.2 DW	55

Table 6.2 (continued) Mercury Concentrations in Field Collections of Selected Species of Invertebrates

Ecosystem, Species, and Other Variables	Concentration (mg/kg)	Ref.[a]
Gill	0.0 DW vs. 0.9 DW	55
Mantle	0.1 DW vs. 0.08 DW	55
Muscle	0.03 DW vs. 0.03 DW	55
Pacific scallop, hybrid of Japanese scallop, *Patinopecten yessoensis* X weathervane scallop, *Patinopectin caurinus:*		
Muscle	0.03 DW vs. 0.08 DW	55
Gonad	0.2 DW vs. 0.2 DW	55
Kidney	0.2 DW vs. 0.2 DW	55
Gill	0.1 DW vs. 0.1 DW	55
Mantle	0.1 DW vs. 0.04 DW	55
Red abalone, *Haliotis rufescens:*		
Gills	0.08–0.27 DW	21
Mantle	0.02–0.33 DW	21
Digestive gland	0.12–4.64 DW	21
Foot	0.03–0.09 DW	21
From vicinity chloralkali plant; Israel; 1980–1982; soft parts:		
Gastropod, *Arcularia gibbosula*	18.2–38.7 DW	10
Bivalve, *Donax venustus*	Max. 6.4 DW	10
Bivalves, various:		
From Hg-polluted area; Denmark; deposit feeders vs. suspension feeders; soft parts	1.4–4.4 FW vs. 0.9–1.9 FW	11
Edible portions; total Hg vs. methyl Hg	0.04–0.22 FW vs. Max. 0.09 FW	8
Soft parts; 2 spp; Georgia, U.S.; estuaries; Hg-contaminated estuary vs. reference site	0.5–1.2 DW vs. 0.1–0.2 DW	6
China: coastal sites along Bohai and Huanghai Sea; commercial species; soft parts:		
Gastropods	0.03 (0.002–0.09) FW	53
Bivalves	0.01–0.08 FW	53
Quahaug, *Mercenaria mercenaria*; soft parts:		
Age 3 years	0.16 DW	13
Age 4 years	0.20 DW	13
Age 10 years	0.22 DW	13
Age 15 years	0.22 DW	13
California mussel, *Mytilus californianus*; soft parts; nationwide vs. California	< 0.4 DW vs. 0.6–2.5 DW	12
Common mussel, *Mytilus edulis*:		
Soft parts:		
Belgium	1.0 DW	2
Spain	1.5 DW	2
New Brunswick	0.1 FW	2
Netherlands	0.1–0.3 FW	2
Great Britain	0.02–0.7 FW	2
New Zealand	0.02–0.48 FW	2
Norway	Max. 0.04 FW	52
Visceral mass	0.3 FW; 1.3 DW	26
Foot muscle	0.4 FW; 0.8 DW	26
Mantle	0.9 FW; 4.3 DW	26
Gills	3.4 FW; 19.9 DW	26
Shell	0.5 DW	24
Soft parts	0.03–2.1 DW	13, 24–26
Softshell clam, *Mya arenaria*; soft parts:		
Chesapeake Bay, MD	0.01–0.05 FW	2
Nova Scotia	0.03–0.13	2
New Brunswick:		
3 km below pulp mill	0.9 FW	2
3 km below chloralkali plant	3.6 FW	2

(continued)

Table 6.2 (continued) Mercury Concentrations in Field Collections of Selected Species of Invertebrates

Ecosystem, Species, and Other Variables	Concentration (mg/kg)	Ref.[a]
Common limpet, *Patella vulgata*; soft parts	0.2 FW; 0.5 DW	26
Pen shell, *Pinna nobilis:*		
Soft parts	6.0 DW	23
Mantle and gills	1.1 DW	23
Muscle	3.1 DW	23
Nervous system	1.4 DW	23
Stomach and intestines	14.0 DW	23
Gonads	3.0 DW	23
Hepatopancreas	7.6 DW	23
Byssus gland	3 DW	23
Cuttlefish, *Sepia officinalis:*		
Gills	0.9 DW	13
Mantle	0.7 DW	13
Edible portions	0.03–0.19 FW	14, 27
Octopus, *Octopus vulgaris*; Tyrrhenian coast:		
Tentacle	0.75–2.3 FW	22
Digestive gland	15.5–202.0 FW	22
Kidney	4.0–7.5 FW	22
Gill	0.48–1.9 FW	22
Brain	1.0–1.2 FW	22
Gonad	0.4–1.0 FW	22
Tunicates:	1.0 DW	
4 species; whole	0.13–0.57 DW; 0.03–0.12 FW	13, 14, 15
Styella plicata; whole; Minamata Bay, Japan	35.0 DW	16
Terrestrial		
Lacewing, *Chrysopa carnea*; whole; Illinois; fed on Hg-treated tomato plants vs. control	0.6–31.4 FW vs. 0.0–1.1 FW	2
Blowfly, *Calliphora* sp.; feeding on brook trout (*Salvelinus fontinalis*) carcasses containing 0.145 mg total mercury/kg DW (0.07 mg methylmercury/kg DW); total mercury vs. methylmercury:		
Adults (laying)	0.045 DW vs. 0.042 DW	56
Eggs	0.038 DW vs. not detectable	56
Larvae	0.082 DW vs. 0.078 DW	56
Pupae	0.150 DW vs. 0.140 DW	56
Adults (emerging)	0.004 DW vs. 0.003 DW	56

Note: Values shown are in mg total mercury/kg fresh weight (FW) or dry weight (DW).

[a] Reference: 1, Huckabee et al., 1979; 2, Jenkins, 1980; 3, Beauvais et al., 1995; 4, Cooper, 1983; 5, Allard and Stokes, 1989; 6, Windom and Kendall, 1979; 7, Schreiber, 1983; 8, Cappon and Smith, 1982; 9, Iglesias and Panchaszadeh, 1983; 10, Hornung et al., 1984; 11, Kiorboe et al., 1983; 12, Flegal et al., 1981; 13, Leatherland and Burton, 1974; 14, Yannai and Sachs, 1978; 15, Papadopoulu et al., 1972; 16, Matida and Kumada, 1969; 17, Leatherland et al., 1973; 18, DeClerck et al., 1979; 19, Williams and Weiss, 1973; 20, Eganhouse and Young, 1978; 21, Anderlini, 1974; 22, Renzoni et al., 1973; 23, Papadopoulu, 1973; 24, Bertine and Goldberg, 1972; 25, Karbe et al., 1977; 26, Jones et al., 1972; 27, Cumont et al., 1975; 28, Barbaro et al., 1978; 29, Bernhard and Zattera, 1975; 30, Greig et al., 1977; 31, Gardner et al., 1975; 32, Martin and Knauer, 1973; 33, Tijoe et al., 1977; 34, Kumagai and Saeki, 1978; 35, Hall et al., 1978; 36, Ramos et al., 1979; 37, Stickney et al., 1975; 38, DeClerck et al., 1974; 39, Zauke, 1977; 40, Greig and Wenzloff, 1977; 41, Greig et al., 1975; 42, Reimer and Reimer, 1975; 43, Establier, 1977; 44, Tuncel et al., 1980; 45, Doi and Ui, 1975; 46, Cheevaparanapivat and Menasveta, 1979; 47, Johnson and Braman, 1975; 48, Parveneh, 1977; 49, Anon., 1978; 50, Eftekhari, 1975; 51, Won, 1973; 52, Airas et al., 2004; 53, Liang et al., 2004; 54, Hui et al., 2005; 55, Norum et al., 2005; 56, Sarica et al., 2005.

In marine crustaceans, total mercury concentrations were always less than 0.5 mg/kg FW edible tissues except in organisms collected from certain areas heavily impacted by mercury-containing industrial wastes, such as Minamata, Japan (Eisler 1981; Table 6.2). Methylmercury concentrations in hepatopancreas of Chinese mitten crabs declined with increasing crab size — possibly through molting — suggesting a mechanism for mercury excretion (Hui et al., 2005), with important implications for crab predators that select larger crabs. In echinoderms, mercury concentrations in whole organisms from nonpolluted areas are low, never exceeding 0.4 mg Hg/kg FW or 0.92 mg Hg/kg DW (Eisler, 1981).

6.3 ELASMOBRANCHS AND BONY FISHES

Data on mercury concentrations in field collections of teleosts are especially abundant, and only a few of the more representative observations are listed in Table 6.3. Examination of these and other data leads to several conclusions. First, mercury tends to concentrate in the edible flesh of finfish, with older fish containing more mercury per unit weight than younger fish (Johnels et al., 1967; Hannerz, 1968; Johnels and Westermark, 1969; Nuorteva and Hasanen, 1971; Barber et al., 1972; Cumont et al., 1972; Evans et al., 1972; Forrester et al., 1972; Alexander et al., 1973; Cross et al., 1973; Giblin and Massaro, 1973; Greichus et al., 1973; Peterson et al., 1973; Taylor and Bright, 1973; DeClerck et al., 1974; Nuorteva et al., 1975; Svansson, 1975; Hall et al., 1976a, 1976b; Matsunaga, 1978; Cutshall et al., 1978; Cheevaparanapivat and Menasveta, 1979; Chvojka and Williams, 1980). This is particularly well documented in spiny dogfish, *Squalus acanthias* (Forrester et al., 1972; Greig et al., 1977); squirefish, (*Chrysophrys auratus* (Robertson et al., 1975); European eel, *Anguilla anguilla* (Establier, 1977); European hake, *Merluccius merluccius* (Yannai and Sachs, 1978); striped bass, *Morone saxatilis* (Alexander et al., 1973); and bluefish, *Pomatomus saltatrix* (Alexander et al., 1973).

Second, most of the mercury in the fish flesh was in organic form, mainly methylmercury (Westoo, 1966, 1969, 1973; Zitko et al., 1971; Ui and Kitamuri, 1971; Kamps et al., 1972; Rivers et al., 1972; Rissanen et al., 1972; Fukai et al., 1972; Suzuki et al., 1973; Peterson et al., 1973; Gardner et al., 1975; Tamura et al., 1975; Bebbington et al., 1977; Hamada et al., 1977; Eganhouse and Young, 1978; Cheevaparanapivat and Menasveta, 1979; Chvjoka and Williams, 1980; Bloom, 1992; Hammerschmidt et al., 1999). This is because fish assimilate inorganic mercury less efficiently than methylmercury from the ambient medium and from their diet, and eliminate inorganic mercury more rapidly than methylmercury (Huckabee et al., 1979; Trudel and Rasmussen, 1997; Ribeiro et al., 1999). Maximum concentrations of total mercury in shark and fish muscle usually did not exceed 2.0 mg Hg/kg FW; however, forms of mercury with very low toxicity can be transformed into forms of very high toxicity — namely, methylmercury — through biological and other processes.

Third, levels of mercury in muscle from adult tunas, billfishes, and other marine carnivorous teleosts were higher than those in younger fishes having a shorter food chain. This indicates associations among predatory behavior, longevity, and mercury accumulation (Forrester et al., 1972; Jernelov, 1972; Peakall and Lovett, 1972; Ui, 1972; Rivers et al., 1972; Peterson et al., 1973; Ratkowsky et al., 1975; Klemmer et al., 1976; Hall et al., 1976a, 1976b; Ociepa and Protasowicki, 1976; Matsunaga, 1978; Yannai and Sachs, 1978; Eisler, 1981). Oceanic tunas and swordfish caught in the 1970s had mercury levels similar to those of museum conspecifics caught nearly 100 years earlier (Miller et al., 1972). It is speculated that mercury levels in fish were much higher 13,000 to 20,000 years ago during the last period of glaciation when ocean mercury concentrations were four to five times higher than today (Vandal et al., 1993).

Fourth, total mercury was uniformly distributed in edible muscle of finfish, demonstrating that a small sample of muscle tissue taken from any region is representative of the whole muscle tissue when used for mercury analysis (Freeman and Horne, 1973a, 1973b; Hall et al., 1976a, 1976b).

Finally, elevated levels of mercury in wide-ranging oceanic fish were not solely the consequence of human activities, but also resulted from natural concentrations (Miller et al., 1972; Greig et al., 1976; Schultz et al., 1976; Scott, 1977; Yannai and Sachs, 1978). This last point is apparently not consistent with the rationale underlying U.S. seafood guidelines regulating mercury levels in

Table 6.3 Mercury Concentrations in Field Collections of Selected Species of Sharks, Rays, and Bony Fishes

Species, Tissue, and Other Variables	Concentration (mg/kg)	Ref.[a]
Rock bass, *Ambloplites rupestris*:		
Muscle:		
Ontario	0.6–4.6 FW	1
Michigan	0.4 FW	1
Western Ontario	1.1–10.9 FW	1
Lake St. Clair	0.5–2.0 FW	1
Virginia; mercury-contaminated site vs. reference site; 1986–1987:		
Liver	2.9 FW vs. 0.1 FW	2
Muscle	1.4 FW vs. 0.17 FW	2
European eel, *Anguilla anguilla*; muscle:		
San Lucar, Spain:		
Body length < 20 cm	0.12 FW	69
Body length 30–40 cm	0.25 FW	69
Body length 60–70 cm	0.36 FW	69
Cadiz, Spain:		
Body length 20–30 cm	0.11 FW	69
Body length 30–40 cm	0.16 FW	69
Body length 40–50 cm	0.23 FW	69
Body length 60–70 cm	0.36 FW	69
Sablefish, *Anoploma fimbria*; decapitated and eviscerated:		
Bering Sea	0.04 FW	70
Southeastern Alaska	0.28 FW	70
Washington State	0.40 FW	70
Oregon	0.40 FW	70
California:		
Northern	0.26 FW	70
Central	0.47 FW	70
Southern	0.60 FW	70
Blue hake, *Antimora rostrata*; NW Atlantic; 2500 m depth; muscle:		
1880	0.51 FW	3
1970	0.34 FW	3
Freshwater drum, *Aplodinotus grunniens*; whole:		
Age 0	0.05 FW	4
Age I	0.13 FW	4
Age II	0.18 FW	4
Baltic coast; 1993–2002:		
Baltic herring, *Clupea harengus:*		
Muscle vs. liver	0.10 FW vs. 0.12 FW	119
Ovary vs. testes	0.03 FW vs. 0.03 FW	119
Lumpfish, *Cyclopterus* sp.:		
Muscle vs. liver	0.02 FW vs. 0.03 FW	119
Ovary vs. testes	0.01 Fw vs. 0.01 FW	119
European smelt, *Osmerus eperlanus:*		
Muscle vs. liver	0.12 Fw vs. 0.05 FW	119
Ovary vs. testes	0.07 FW vs. 0.03 FW	119
Four-horn sculpin, *Myoxocephalus quadricornis:*		
Muscle vs. liver	0.23 FW vs. 0.11 FW	119
Ovary vs. testes	0.06 vs. 0.05 FW	119
European flounder, *Platichthys flesus:*		
Muscle vs. liver	0.08 FW vs. 0.04 FW	119
Ovary vs. testes	0.03 FW vs. 0.03 FW	119
Eelpout, *Zoarces viviparus:*		
Muscle vs. liver	0.11 FW vs. 0.09 FW	119
Ovary vs. testes	0.03 FW vs. 0.06 FW	119

Table 6.3 (continued) Mercury Concentrations in Field Collections of Selected Species of Sharks, Rays, and Bony Fishes

Species, Tissue, and Other Variables	Concentration (mg/kg)	Ref.[a]
Porgy, *Boops* sp.; mercury-contaminated area; Tyrrhenian coast:		
Muscle	0.11 FW	71
Liver	3.8 FW	71
Kidney	4.0 FW	71
Atlantic menhaden, *Brevoortia tyrannus*; muscle and organs:	0.27–0.50 DW	72
Cusk, *Brosme brosme*:		
Muscle	0.09–0.54 FW	67
Liver	0.02–0.62 FW	67
Gills	0.02–0.14 FW	67
Kidney	0.02–0.33 FW	67
California; San Joaquin River; whole body; 1986:		
Common carp, *Cyprinus carpio*	0.1–0.5 DW; max.0.8 DW	5
Mosquitofish, *Gambusia affinis*	max. 0.5 DW	5
Bluegill, *Lepomis macrochirus*	0.1–0.3 DW; max. 0.41 DW	5
Largemouth bass, *Micropterus salmoides*	0.35–0.85 DW; max. 1.9 DW	5
Canada; drainage lakes with clear-cut, burnt, or undisturbed catchments; 1996–1997; fish muscle:		
Northern pike, *Esox lucius:*		
Clear-cut	2.9 DW	116
Burnt	3.2 DW	116
Undisturbed	1.8 DW	116
Walleye, *Stizostedium vitreum:*		
Clear-cut	2.2 DW	116
Burnt	2.3 DW	116
Undisturbed	0.9 DW	116
Yellow perch, *Perca flavescens:*		
Clear-cut	1.3 DW	116
Burnt	0.8 DW	116
Undisturbed	0.8 DW	116
Canada, Waibigoon River system; Ontario; mercury-contaminated between 1962 and 1969; samples collected 1979–1981:		
Northern pike, *Esox lucius*, whole:		
Age 0+	0.04–1.0 FW	6
Age 1+	0.09–1.2 FW	6
Yellow perch, *Perca flavescens*; yearlings; whole	0.01–0.6 FW	6
Silky shark, *Carcharhinus falciformes*; North Atlantic:		
Gonads	9.0 DW	57
Muscle	5.3 DW	57
Dusky shark, *Carcharhinus obscurus*; muscle	4.2 DW	57
Squirefish, *Chrysophrys auratus*; muscle:		
Sydney, Australia vs. Nowra, Australia; 1976:		
Total mercury	0.32 (0.08–1.7) FW vs. 0.11 (0.01–0.78) FW	7
Methylmercury	0.3 (0.25–0.32) FW vs. 0.1 (0.06–0.11) FW	7
Body length < 30 cm	0.13 FW	73
Body length 35–39 cm	0.24 FW	73
Body length 45–49 cm	0.40 FW	73
Body length > 50 cm	1.0 FW	73
Blacktail, *Diplodus sargus*; muscle:		
Mercury-polluted area	0.3–1.7 FW	8
Unpolluted area	0.04–0.64 FW	8
Haifa Bay, Israel vs. reference site; 1990	0.6 FW vs. 0.15 FW	9

(continued)

Table 6.3 (continued) Mercury Concentrations in Field Collections of Selected Species of Sharks, Rays, and Bony Fishes

Species, Tissue, and Other Variables	Concentration (mg/kg)	Ref.[a]
Twoband bream, *Diplodus vulgaris*; mercury-contaminated area; Tyrrhenian coast:		
Muscle	1.9 FW	71
Liver	17.0 FW	71
Kidney	29.8 FW	71
Elasmobranchs; muscle:		
4 species	0.06–0.48 FW	58
8 species	1.5–4.0 DW	59
England; marine species; near River Tyne, 1992:		
Muscle; 5 species	0.03–0.14 (0.006–0.43) FW	10
Stomach contents; 3 species	0.01–0.04 FW; max. 0.11 FW	10
Northern pike, *Esox lucius*; muscle:		
Sweden	0.2–9.8 FW	1
Quebec	0.3–0.8 FW	1
Norway	0.1 FW	1
Saskatchewan	0.7–10.6 FW	1
Canada (unpolluted)	0.1 FW	1
Canada (polluted)	0.5–0.7 FW	1
Lake St. Clair	2.0–3.0 FW	1
NW Ontario mining area	5.6 FW; max. 16.0 FW	1, 11
Wisconsin	0.9–1.4 FW	1
Manitoba, man-made reservoir		
Preimpoundment (1971–1973)	0.25–0.35 FW	12
Postimpoundment (1979–1982)	0.67–0.95 FW	12
Finland, southern; 1960s vs. 1982–1984	5.0–6.0 FW vs. 0.15–1.4 FW	13
Finland, northern:		
Preindustrial levels	0.18–0.33 FW	14
Lake contaminated with phenylmercury until 1967:		
1971–1974 (sediments 2.1 mg Hg/kg DW)	1.5 FW	14
1990 (sediments 0.4 mg Hg/kg DW)	0.8 FW	14
Fish:		
Kidneys, 13 species:		
Total mercury	0.72 FW	83
Inorganic mercury	0.25 FW	83
Liver:		
54 species	< 0.1–0.3 FW	82
13 species	0.3–0.9 FW	82
10 species	0.9–4.0 FW	82
3 species	5.0–20.0 FW	82
Marine products of commerce; muscle:		
Malaysia, 6 species	0.08–0.1 FW	74
New Guinea, 6 species	0.02–0.19 FW	75
Belgian coast, 5 species	0.07–0.18 FW	76
Ghana; Gulf of Guinea; 20 species; November 2003–February 2004	Means range from 0.009 FW to 0.160 FW; maximum values range from 0.010 FW to 0.191 FW	114
Persian Gulf, 5 species	0.04–0.56 FW	77
Indian Ocean, 13 species	0.07–0.16 FW	62
Turkish coast 17 species	0.002–0.13 FW	78
French coast, 6 species	0.08–0.37 FW	64
Minamata Bay, Japan, 3 species	Max. 309.1 DW[b]	79
Korea, 40 species	0.17 (0.02–0.58) FW	80
Australia, 6 species	0.03–0.75 FW	81
Various locations worldwide:		
94 species	< 0.2 FW	82
33 species	0.2–0.4 FW	82

Table 6.3 (continued) Mercury Concentrations in Field Collections of Selected Species of Sharks, Rays, and Bony Fishes

Species, Tissue, and Other Variables	Concentration (mg/kg)	Ref.[a]
9 species	0.4–0.6 FW	82
10 species	0.6–0.9 FW	82
12 species	1.0–3.0 FW	82
1 species	4.0–5.0 FW	82
Muscle; 15 species (7 freshwater, 8 marine):		
Total mercury	0.01–2.8 FW (methylmercury content of 86.0–100.0%)	15
Dimethylmercury	< 0.001 FW	15
Muscle:		
Freshwater species:		
Total mercury	0.27–1.7 FW	16
Methylmercury	Max. 1.4 FW	16
Marine species:		
Total mercury	0.11–5.7 FW	16
Methylmercury	Max. 4.5 FW	16
Protein concentrate, 13 species	0.3–0.9 DW	84
Viscera:		
Minamata, Japan; 3 species	18.0–23.0 DW	85
Korea, representative species	0.16 (0.07–0.31) FW	80
Whole:		
26 species	< 0.1 FW	62, 82
10 species	0.1–0.3 FW	82, 86, 87
Florida; edible sport fish tissues; south Florida; current vs. goal	1.28 (0.45–4.03) FW vs. < 0.5 FW	115
Mummichog, *Fundulus heteroclitus*; muscle	0.001–0.009 FW	88
Blackfish, *Gadopsis marmoratus;* muscle:		
From Hg-contaminated sediments	Max. 0.64 FW	17
From uncontaminated sediments	Max. 0.06 FW	17
Atlantic cod, *Gadus morhua:*		
Roe	0.02–0.04 DW	89
Liver oil	0.12–0.15 FW	90
Cleithrum bones; 1870s vs. 1970s	0.064 DW vs. 0.056 DW	91
Three-spined stickleback, *Gasterosteus aculeatus*; Gulf of Gdansk; whole; 1988-89	0.01–0.11 FW; max. 0.49 FW	18
Lahontan Reservoir, Nevada; 1981:		
Muscle, 5 species	Max. 2.3–3.9 FW	19
Liver, 5 species	Max. 2.4–8.3 FW	19
Heart, 4 species	Max. 1.1–2.1 FW	19
Georgia, lower coastal plain, 1976–1977, liver vs. muscle:		
Carnivores	3.0 FW vs. 1.7 FW	20
Insectivores	1.1 FW vs. 1.1 FW	20
Omnivores	1.0 FW vs. 0.9 FW	20
Great lakes, Lake Ontario; whole fish:		
Slimy sculpin, *Cottus cognatus*:		
1977	0.068 FW	20
1984	0.038 FW	20
1988	0.032 FW	20
Rainbow smelt, *Osmerus mordax*:		
1977	0.067 FW	21
1984	0.059 FW	21
1988	0.037 FW	21
Greenland; Barents Sea; summer 1991–1992; muscle; demersal fishes:		
Atlantic cod, *Gadus morhua*	(0.07–0.19) DW	22
Long rough dab, *Hippoglossoides platessoides*	0.13–0.9 (0.06–1.8) DW	22

(continued)

Table 6.3 (continued) Mercury Concentrations in Field Collections of Selected Species of Sharks, Rays, and Bony Fishes

Species, Tissue, and Other Variables	Concentration (mg/kg)	Ref.[a]
Atlantic halibut, *Hippoglossus hippoglossus*	(0.24–1.1) DW	22
Starry ray, *Raja radiata*	(0.2–0.4) DW	22
Plaice, *Pleuronectes platessa*	0.3 (0.2–0.5) DW	22
Greenland halibut, *Reinhardtius hippoglossoides*	0.25–1.2 DW; max. 2.5 DW	22
Pacific halibut, *Hippoglossus stenolepis*; muscle:		
Bering Sea	0.15 FW	92
Gulf of Alaska	0.20 FW	92
Southeast Alaska	0.26 FW	92
British Columbia	0.32 FW	92
Washington–Oregon	0.45 FW	92
Channel catfish, *Ictalurus punctatus*; muscle:		
Lake Erie	0.3–1.8 FW	1
Lake St. Clair	0.5–2.0 FW	1
Ohio	0.1–0.4 FW	1
Illinois	0.03–0.2 FW	1
Oregon	0.02–1.5 FW	1
Georgia	0.1–1.9 FW	1
Texas	0.2–2.5 FW	1
Lake Chad, Africa; December 2000; muscle; 14 species	0.007–0.074 FW	107
Pumpkinseed, *Lepomis gibbosus;* 16 lakes; Ontario, Canada; 1981; muscle	0.09–0.54 FW	23
Louisiana; Atchafalaya River; 1981; whole; 8 spp.	0.06–0.79 FW	24
Maine; 120 lakes; 10 species of gamefish; muscle; 1993–1994	0.3–0.9 (0.07–1.2) FW	25
Black marlin, *Makaira indica:*		
Muscle:		
Pacific Ocean	0.6–4.3 FW	16
NE Australia	0.5–16.5 FW	16
Papua New Guinea	5.7 FW	75
Pacific Ocean, various locations	7.3 (0.5–16.5) FW	93
Liver	10.4 (0.3–63.0) FW	93
Blue marlin, *Makaira nigricans*:		
Blood	0.3 FW	95
Central nervous system	0.59 FW	94
Gill	0.3 FW	95
Gonad	0.3–0.7 FW	94, 95
Liver:		
Total mercury vs. organic mercury	6.3 FW vs. 0.2 FW	94
Total mercury vs. methylmercury	13.4 FW vs. 0.2 FW	95
Inorganic mercury	11.0 FW	95
Muscle:		
Total mercury vs. methylmercury	Max. 14.0 FW vs. max. 0.16 FW	16
Total mercury vs. organic mercury	4.3 FW vs. o,4 FW	95
Methylmercury vs. inorganic mercury	0.4 FW vs. 2.3 FW	95
Spleen; total mercury vs. methylmercury	8.5 FW vs. 0.2 FW	95
Largemouth bass, *Micropterus salmoides*:		
Muscle:		
Texas	0.1 FW	1
Utah	0.3–7.3 FW	1
California	0.1–0.6 FW	1
Oregon	0.2–1.8 FW	1
Washington	0.1–0.3 FW	1
Georgia	0.1–5.4 FW	1
Michigan	0.2–0.9 FW	1
Illinois	0.03–1.2 FW	1
Arizona	0.3 FW	1
Florida, 1989–1992	0.04–2.0 FW	26, 27

Table 6.3 (continued) Mercury Concentrations in Field Collections of Selected Species of Sharks, Rays, and Bony Fishes

Species, Tissue, and Other Variables	Concentration (mg/kg)	Ref.[a]
Whole; Florida; 1989–1992:		
Length 20 mm	0.05 FW	27
Length 320 mm	0.32 FW	27
Dover sole, *Microstomus pacificus*:		
California; Palos Verdes Peninsula vs. Catalina Island:		
Muscle	0.05 DW vs. 0.16 DW	96
Liver	0.1 DW vs. 0.14 DW	96
Kidney	0.04 DW vs. 0.03 DW	96
Gill	0.02 Dw vs. 0.02 DW	96
Muscle	0.01–0.29 FW	63
Striped bass, *Morone saxatilis*		
Nevada; Lahontan Reservoir; 1981; single specimen, 16 years old:		
Muscle	9.5 FW	19
Heart	5.6 FW	19
Liver	23.7 FW	19
New York; muscle; variable body weight:		
< 3.2 kg	< 0.5 FW	28
3.2–5.7 kg	0.5 FW	28
> 5.7 kg	> 0.5 FW	28
Striped mullet, *Mugil cephalus:*		
Muscle:		
Mercury-contaminated site	0.39 FW	71
Reference sites	0.03–0.07 FW	97, 98
Liver:		
Mercury-contaminated site	5.4 FW	71
Reference sites	0.01–0.09 FW	98
Nationwide, U.S.; freshwater; whole fish:		
1969–1970:		
Pacific Coast and Alaska	0.26 (0.05–1.7) FW	29
Southwest	0.25 (0.05–1.7) FW	29
North central	0.08 (< 0.05–0.14) FW	29
Northeast	0.23 (< 0.05–0.08) FW	29
Southeast	0.23 (< 0.05–1.0) FW	29
1972	0.15 FW	29
1976–1977	0.11 FW; max. 0.85 FW	30, 31
1978–1979	0.11 (0.01–1.1) FW	30
1980–1981	0.11 (0.01–0.77) FW	30
1984–1985	0.10 FW; max. 0.37 FW; 85th percentile 0.17 FW	31
North Dakota and Minnesota, Red River of the North, 1994:		
Common carp, *Cyprinus carpio:*		
Liver	0.11 FW	32
Muscle	0.31 FW	32
Whole	0.18 FW	32
Channel catfish:		
Liver	0.16 FW	32
Muscle	0.18 FW	32
Whole	0.11 FW	32
Yellow perch, *Perca flavescens*; whole:		
Age 0	0.07 FW	4
Age I	0.13 FW	4
Age II	0.22 FW	4
Perch, *Perca fluviatilis*; Russia; June 1989; muscle; 350 km north of Moscow; 6 lakes; acidic vs. alkaline lakes	0.5–1.1 FW vs. 0.1–0.2 FW	33

(continued)

Table 6.3 (continued) Mercury Concentrations in Field Collections of Selected Species of Sharks, Rays, and Bony Fishes

Species, Tissue, and Other Variables	Concentration (mg/kg)	Ref.[a]
European flounder, *Platichthys flesus*; muscle; Irish Sea:		
Northern areas	0.03–0.13 (0.008–0.39) DW	34
Central areas	0.16–0.48 (0.06–1.2) DW	34
Southern areas	0.16–0.40 (0.04–2.0) DW	34
Plaice, *Pleuronectes platessa*; muscle; Liverpool Bay, U.K.; sludge disposal ground:		
Early 1970s (2.7 tons of mercury yearly)	0.5 FW	35
1991 (0.16 tons of mercury annually)	0.2 FW	35
Bluefish, *Pomatomus saltatrix*; muscle:		
Fish weight < 2.4 kg	< 0.5 FW	28
Fish weight 2.4–5.6 kg	0.5 FW	28
Fish weight > 5.6 kg	> 0.5 FW	28
Total mercury	0.21 (0.05–0.41) FW	97
Methylmercury	0.2 (0.05–0.39) FW	97
Inorganic mercury	0.012 (0.00–< 0.03) FW	97
Blue shark, *Prionace glauca*; muscle	0.55–0.74 FW	60
Round whitefish, *Prospium cylindraceum*; Saginaw Bay, Michigan; 1977–1978; fillets:		
Methylmercury	Max. 0.05 FW	36
Total mercury	Max. 0.1 FW	36
Thornback ray, *Raja clavata:*		
Blood	0.22 FW	61
Gut	0.16 FW	61
Gill	0.13 FW	61
Muscle	0.05 FW	61
Blood plasma	0.001 FW	61
Trout, *Salmo* spp.; Missouri; liver and muscle:		
1946–1950	3.0 FW	37
1973	0.1–0.3 FW	37
Brook trout, *Salvelinus fontinalis*; Adirondack lakes (15), New York; whole	< 1.0 FW	38
Lake trout, *Salvelinus namaycush*:		
Muscle:		
British Columbia	1.1–10.5 FW	1
Ontario	0.3–1.3 FW	1
Quebec	0.3–1.2 FW	1
New York; Cayuga Lake; 1991; age in years:		
1	0.19 DW	39
3	0.40 DW	39
5	0.58 DW	39
10	0.70 DW	39
12	0.75 DW	39
Whole; Lake Ontario; age 4 years:		
1977	0.24 FW	40
1984	0.15 FW	40
1988	0.12 FW	40
Red drum, *Sciaenops ocellatus*; Florida; muscle:		
All fish	0.02–3.6 FW	117
Legal-sized fish (< 689 mm total length or < 565 mm standard length)	0.17–0.30 (0.02–2.7) FW	117
Chub mackerel, *Scomber japonicus*:		
Muscle	0.06–0.15 FW	99
Liver	0.08–0.29 FW	99
Spleen	0.13–0.16 FW	99

Table 6.3 (continued) Mercury Concentrations in Field Collections of Selected Species of Sharks, Rays, and Bony Fishes

Species, Tissue, and Other Variables	Concentration (mg/kg)	Ref.[a]
Scotland; 1982–1987; sludge disposal sites vs. reference sites; muscle:		
Atlantic cod, *Gadus morhua*	0.04–0.2 FW vs. 0.05–0.07 FW	41
Flounder, *Platichthys flesus*	0.09–0.55 FW vs. 0.1–0.13 FW	41
Plaice, *Pleuronectes platessa*	0.06–0.2 FW vs. 0.03–0.1 FW	41
Whiting, *Merlangius merlangius*	0.03–0.1 FW vs. 0.06–0.08 FW	41
Scotland, Firth of Clyde; 5 species; muscle; 1991–1992	0.01–0.07 (0.01–0.25) FW	42
Lesser–spotted dogfish, *Scyliorhinus caniculus*; muscle; Irish Sea; August 1985:		
North and NW areas	0.15–2.1 FW	43
Northeast sites	0.3–2.2 FW	43
Southeast sites	0.2–5.6 FW	43
Southwest sites	0.1–2.8 FW	43
Yellowtail kingfish, *Seriola grandis*; 1977–1978; 10 km offshore from Sydney, Australia; muscle	0.15 FW; max. 1.1 FW	44
Sharks; Australia; 1980; muscle; 7 spp.:		
Carcharhinus spp.	Max. 4.3 FW	45
Sphyrna spp.	Max. 4.9 FW	45
Sharks; Florida; 1988–1992; muscle; methylmercury; 9 species:		
All species	0.88 (0.06–2.9) FW	46
4 of 9 species	1.0–2.3 FW	46
Sharks > 200 cm total length	33.0% exceeded 1.0 FW (U.S. Food and Drug Administration action level)	46
Florida state survey; retail markets; 1991	1.5 FW	46
Sharks and rays; 7 species; North Atlantic; all tissues	< 2.0 DW	57
Slovak Republic; Nitra River; September 2003; muscle:		
Barbel, *Barbus barbus*	3.4 (1.9–4.6) FW	118
Eurasian perch, *Perca fluviatilis*	4.5 (2.7–6.5) FW	118
European chub, *Leuciscus cephalus*	2.3 (1.4–3.9) FW	118
Roach, *Rutilus rutilus*	1.9 (1.5–2.2) FW	118
Reference site, all species	< 0.5 FW	118
Winghead shark, *Sphyrna blochi:*		
Ovary	Not detectable	62
Liver	0.12 FW	62
Muscle	0.21 FW	62
Spiny dogfish, *Squalus acanthias*:		
Muscle	0.2–1.1 FW	63, 64
Muscle: females vs. males	0.9 (0.09–2.6) FW vs. 0.06 (0.2–1.6) FW	65
Ovarian embryos; whole; age group 0 vs. age group I	0.015 FW vs. 0.029 FW	66
Muscle; males; 72 cm total length (TL) vs. 95 cm TL	0.5 FW vs. 1.7 FW	66
Muscle; females; 77 cm TL vs. 120 cm TL	0.5 FW vs. 2.0 FW	66
Adult females:		
Muscle	0.4–1.1 FW	67
Gills	0.1–0.6 FW	67
Kidney	0.1–1.3 FW	67
Spleen	0.03–0.6 FW	67
California dogfish, *Squalus suckleyi:*		
Maternal females:		
Muscle	0.66 FW	68
Uteral membrane	0.08 FW	68
Graafian follicles; 1 cm diameter vs. 3 cm diameter	0.047 FW vs. 0.034 FW	68
Fetus; 7.5 cm vs. 10 cm:		
Yolk–sac	0.024 FW vs. 0.009 FW	68
Whole	0.046 FW vs. 0.021 FW	68
Pup; whole	0.037 FW	68

(continued)

Table 6.3 (continued) Mercury Concentrations in Field Collections of Selected Species of Sharks, Rays, and Bony Fishes

Species, Tissue, and Other Variables	Concentration (mg/kg)	Ref.[a]
Walleye, *Stizostedion vitreum vitreum*:		
Clay Lake; Ontario, Canada; contaminated by chloralkali plant 1962–1970; muscle:		
1970	15.0 FW	112
1972	7.5 FW	112
1983	3.5 FW	112
1997–1998	2.7 FW	113
Manitoba; manmade reservoir; muscle:		
Preimpoundment (1971–1977)	0.2–0.3 FW	47
Postimpoundment (1978–1992)	0.6–0.8 FW	47
Washington State; Columbia River; 1994; muscle vs. sediments	0.11–0.44 FW vs. 0.05–2.8 DW	48
Wisconsin; 1980–1982; muscle; low pH lakes vs. high pH lakes:		
Age 4 years	0.6 FW vs. 0.2 FW	49
Age 5 years	0.8 FW vs. 0.25 FW	49
Age 7 years	1.25 (0.8–1.7) FW vs. 0.4 (0.3–0.5) FW	49
Wisconsin; 1980–1989; skin-on fillets; low buffering capacity lakes vs. high buffering capacity lakes		
Total fish length:		
20–35 cm	0.3 FW vs. 0.2 FW	50
40–45 cm	0.6 FW vs. 0.3 FW	50
50–55 cm	1.1 FW vs. 0.5 FW	50
> 65 cm	1.5 FW vs. 0.5 FW	50
Wisconsin; 1990–1991; 34 northern lakes; muscle	0.3–1.0 FW; about 50.0% exceeded the fish consumption advisory of 0.5 mg Hg/kg FW muscle set by the Wisconsin Department of Natural Resources	51
Texas; coastal bays; reference locations; muscle vs. liver:		
Spotted seatrout, *Cynoscion nebulosus*	0.07–0.44 FW vs. < 0.12–0.84 FW	111
Southern flounder, *Paralichthys lethostigma*	< 0.05–0.2 FW vs. 0.1–0.2 FW	111
Red drum, *Sciaenops ocellatus*	< 0.004–0.59 FW vs. 0.07–0.63 FW	111
Thailand; various species; muscle:		
Near chloralkali plant		
No wastewater system, 1976–1977	0.3–3.6 FW	52
With wastewater system, 1978	0.1–1.4 FW	52
Control location	0.01–0.3 FW	52
Yellowfin tuna, *Thunnus albacares*:		
Kidney	0.05–0.53 FW	104
Liver	0.07–0.18 FW	104
Methylmercury:		
Muscle	0.19–0.21 FW	103
Liver	0.11 FW	103
Spleen	0.03 FW	103
Kidney	0.07 FW	103
Muscle	0.07–1.3 FW	101
Muscle	0.16–0.24 FW	104
Muscle:		
Total mercury	Max. 0.66 FW	97
Methylmercury	Max. 0.63 FW	97
Inorganic mercury	Max. 0.017 FW	97
Tuna, *Thunnus* spp.:		
Muscle	0.74–2.34 FW	105
Canned; Turkey; total mercury vs.methylmercury	0.73 FW vs. 0.56 FW	106
Bluefin tuna, *Thunnus thynnus:*		
Canned; Italy	1.33 FW	86
Muscle	0.46–1.91 FW	101

Table 6.3 (continued) Mercury Concentrations in Field Collections of Selected Species of Sharks, Rays, and Bony Fishes

Species, Tissue, and Other Variables	Concentration (mg/kg)	Ref.[a]
Tunas; 1981; 5 spp.; muscle	1.0–6.3 FW	53
Tunas; Indian Ocean; 1985–1986; blood; total mercury vs. methylmercury:		
Yellowfin, *Thunnus albacares*	0.08 (0.003–0.27) FW vs. 0.01 (0.00–0.03) FW	54
Big-eye tuna, *Thunnus obesus*	0.8 (0.5–1.3) FW vs. 0.5 (0.2–0.7) FW	54
Tunas; various species:		
Canned	0.32–2.9 FW	109
Liver, 2 species	0.08–0.27 FW	79
Muscle:		
3 species	0.15–0.71 FW	108
4 species	0.02–0.22 FW	75
Mediterranean Sea	2.5–3.5 FW	110
"Recent"	0.44–1.53 DW	100
100-year-old museum samples	0.53–1.51 DW	100
Various teleost species; muscle:		
From unpolluted areas	0.04–0.15 FW	55
From moderately Hg-polluted areas	> 1.0 FW	55
From highly polluted areas	10.0–24.0 FW	55
Swordfish, *Xiphius gladius*; muscle:		
NW Atlantic	2.0 FW; 8.1 DW	1
Peru	1.1–1.8 FW	1
Pacific	0.5–1.7 FW	1
W. Atlantic	0.05–4.9 FW	69, 101, 102
Gibraltar Straits	1.0–2.0 FW	1
Museum specimens; circa 1909 vs. 1970s	1.36 DW vs. 0.94–5.1 DW	100
Azores; November 1987; males vs. females:		
< 125 cm length	0.4 FW vs. 0.25 FW	56
> 125 cm length	1.9 FW, max. 4.9 FW vs. 1.1 FW	56

Note: Values shown are in mg total mercury/kg fresh weight (FW) or dry weight (DW).

[a] Reference: 1, Jenkins, 1980; 2, Bidwell and Heath, 1993; 3, Barber et al., 1984; 4, Busch, 1983; 5, Saiki et al., 1992; 6, Parks et al., 1991; 7, Chvojka et al., 1990; 8, Hornung et al., 1984; 9, Krom et al., 1990; 10, Dixon and Jones, 1994; 11, Wiener and Spry, 1996; 12, Bodaly et al., 1984; 13, Rask and Metsala, 1991; 14, Lodenius, 1991; 15, Bloom, 1992; 16, Cappon and Smith, 1982; 17, Bycroft et al., 1982; 18, Falandysz and Kowalewska, 1993; 19, Cooper, 1983; 20, Halbrook et al., 1994; 21, Borgmann and Whittle, 1992; 22, Joiris et al., 1997; 23, Wren and MacCrimmon, 1983; 24, Winger and Andreason, 1985; 25, Stafford and Haines, 1997; 26, Lange et al., 1993; 27, Lange et al., 1994; 28, Alexander et al., 1973; 29, Henderson and Shanks, 1973; 30, Lowe et al., 1985; 31, Schmitt and Brumbaugh, 1990; 32, Goldstein et al., 1996; 33, Haines et al., 1992; 34, Leah et al., 1992; 35, Leah et al., 1993; 36, Miller and Jude, 1984; 37, Lloyd et al., 1977; 38, Sloan and Schofeld, 1983; 39, Gutenmann et al., 1992; 40, Borgmann and Whittle, 1991; 41, Clark and Topping, 1989; 42, Mathieson and McLuskey, 1995; 43, Leah et al., 1991; 44, Chvojka, 1988; 45, Lyle, 1984; 46, as quoted in Eisler, 2000; 47, Bodaly et al., 1984; 48, Munn and Short, 1997; 49, Wiener et al., 1990b; 50, Lathrop et al., 1991; 51, Gerstenberger et al., 1993; 52, Suckcharoen and Lodenius, 1980; 53, Schreiber, 1983; 54, Kai et al., 1988; 55, USNAS, 1978; 56, Monteiro and Lopes, 1990; 57, Windom et al., 1973; 58, Menasveta and Siriyong, 1977; 59, Gardner et al., 1975; 60, Greig and Wenzloff, 1977; 61, Pentreath, 1976; 62, Kureishy et al., 1979; 63, Childs and Gaffke, 1973; 64, Cumont et al., 1975; 66, Forrester et al., 1972; 67, Greig et al., 1977; 68, Childs et al., 1973; 69, Establier, 1977; 70, Hall et al., 1976a; 71, Renzoni et al., 1973; 72, Cocoros et al., 1973; 73, Robertson et al., 1975; 74, Babji et al., 1979; 75, Sorentino, 1979; 76, DeClerck et al., 1979; 77, Parveneh, 1979; 78, Tuncel et al., 1980; 79, Fujiki, 1963; 80, Won, 1973; 81, Chvojka and Williams, 1980; 82, Hall et al., 1978; 83, Suzuki et al., 1973; 84, Beasley, 1971; 85, Matida and Kumada, 1969; 86, Kari and Kauranen, 1978; 87, Cugurra and Maura, 1976; 88, Chernoff and Dooley, 1979; 89, Julshamn and Braekkan, 1978; 90, Van de Ven, 1978; 91, Scott, 1977; 92, Hall et al., 1976b; 93, Mackay et al., 1975; 94, Schultz et al., 1976; 95, Schultz and Crear, 1976; 96, Eganhouse and Young, 1978; 97, Bebbington et al., 1977; 98, Reimer and Reimer, 1975; 99, Doi and Ui, 1975; 100, Miller et al., 1972; 101, Peterson et al., 1973; 102, Freeman et al., 1978; 103, Ueda and Takeda, 1977; 104, Hamada et al., 1977; 105, Arima and Umemoto, 1976; 106, Sanli et al. ,1977; 107, Kidd et al., 2004; 108, Greig and Krzynowek, 1979; 109, Ganther et al., 1972; 110, Anon., 1978; 111, Sager, 2004; 112, Parks and Hamilton, 1987; 113, Latif et al., 2001; 114, Voegborlo et al., 2004; 115, Atkeson et al., 2003; 116, Garcia and Carignan, 2005; 117, Adams and Onorato, 2005; 118, Andreji et al., 2005; 119, Voigt, 2004.

[b] See Chapter 10.

comestibles and formulated in the 1970s. According to Peterson et al. (1973), when the U.S. Food and Drug Administration (USFDA) introduced safety guidelines — which eventually were instrumental in the temporary removal of all swordfish and substantial quantities of canned tuna from market — it acted essentially under the assumption that the fish product was "adulterated" by an "added substance."

It is noteworthy that muscle from two species of recreationally important fish (spotted seatrout, *Cynoscion nebulosus*; red drum, *Sciaenops ocellatus*) collected from coastal bays in Texas considered "minimally impacted" by mercury exceeded the current recommended value in the United States of 0.3 mg total Hg/kg FW muscle (Sager, 2004). And walleye (*Stizostedium vitreum vitreum*) collected from Clay Lake, Ontario — a water body heavily contaminated by mercury wastes from a chloralkali plant between 1962 when discharges began and 1970 when the plant closed — contained 2.7 mg total Hg/kg FW muscle in samples collected 28 years after plant closure, a concentration in excess of the Canadian mercury criterion of < 0.5 mg total Hg/kg FW edible fish portions (Latif et al., 2001).

Of 159 species of finfish, including sharks and rays, from coastal waters of Alaska, Hawaii, and the conterminous United States, most muscle samples had mean concentrations less than 0.3 mg total Hg/kg FW (Hall et al., 1978). However, 31 species contained more than 0.5 mg total Hg/kg FW, the "action level" set by the USFDA. These 31 species represented about 0.65% of the weight of the catch from the 159 species intended for human consumption. Extrapolation of these results indicates that less than 2.0% of the U.S. catch intended for human consumption may be in excess of the USFDA action level. Of the 31 species containing more than 0.5 mg total Hg/kg FW in muscle, ten were sharks and four were billfishes (Hall et al., 1978).

Inshore marine biota often contained higher mercury concentrations than the same or similar species collected offshore (Westoo, 1969; Jones et al., 1972; Dehlinger et al., 1973). In Sweden, marine fish caught near shore often had elevated methylmercury levels, with many values in the range of 5.0 to 10.0 mg Hg/kg FW (Westoo, 1969; Ackefors et al., 1970); concentrations above 1.0 mg Hg/kg FW in Swedish fish were usually associated with industrial discharges of mercury compounds (Westoo, 1969). In Mediterranean fishes, the mercury body burden was about twice that of conspecifics of the same size from the Atlantic Ocean (Baldi et al., 1978; Renzoni et al., 1978). It is speculated that the higher body burdens of mercury in Mediterranean species is due to the elevated natural geochemical levels of mercury in the Mediterranean. Mercury concentrations in edible muscle of teleosts from German fishing grounds seldom exceeded 0.1 mg total Hg/kg FW (Jacobs, 1977). Of the total mercury in German fish muscle, methylmercury comprised 70.0 to 98.0%, which is in general agreement with Japanese, Swedish, and other reports on this subject (Jacobs, 1977).

The efficiency of mercury transfer through natural marine food chains among lower levels was comparatively low; however, higher trophic levels (including teleosts and fish-eating birds and mammals) show marked mercury amplification (Jernelov and Lann, 1971; Huckabee and Blaylock, 1972; Cocoros et al., 1973; Stickney et al., 1975; Skei et al., 1976). The variability in concentrations is explainable, in part, by collection locale wherein samples were taken from areas receiving anthropogenic mercury wastes resulting in elevated mercury loadings in the aquatic environment and a significant increase in mercury content of endemic fauna (Johnels et al., 1967; Hearnden, 1970; Wobeser et al., 1970; Kazantzis, 1971; Zitko et al., 1971; Kleinert and Degurse, 1972; Renzoni et al., 1973). Not all investigators agreed that diet was the most important mercury-concentrating mechanism for marine teleosts. For example, Fujiki (1963) reports that mercury in suspended solids and bottom sediments were not transferred in significant amounts to squirefish (*Chrysophrys major*), that accumulation via the food chain was very low, and that dissolved methylmercury in seawater was the critical pathway for methylmercury accumulation in that species. Gardner (1978) stated that there was a positive correlation between mercury content of edible bottomfish in various U.K. fishing grounds and mean mercury concentrations of water samples from the same localities. A similar case was made for Japanese waters by Matsunaga (1978), although this link needs verification. Gardner (1978) showed that ionic mercury predominated in water; however, fish tissues contained > 80.0% methylmercury. He contended that dissolved mercury was removed rapidly from seawater by particulate matter and subsequently to sediments where methylation more readily occurred. Gardner (1978) concluded that mercury variations in fish tissues were attributed to the availability of food and its mercury content; the chemical form and concentration of dissolved mercury; the fish species and trophic level; and the growth rate, sex, and age of the animal. Changes in latitude also seem important for species distributed over a wide latitudinal range and that trends in mercury levels from these species were often opposite to that reported for other fish species from the same geographical area (Hall et al., 1976a, 1976b; Cutshall et al., 1978).

Mercury was detectable in the tissues of almost all freshwater fishes examined, with the majority of the mercury (> 80.0 to 99.0%) present as methylmercury (Huckabee et al., 1979; Chvojka, 1988; Grieb et al., 1990; Southworth et al., 1995). Methylmercury is absorbed more efficiently than inorganic mercury from water, and probably from food, and is retained longer regardless of the uptake pathway (Huckabee et al., 1979; Hill et al., 1996). Three important factors modifying mercury uptake in aquatic organisms are the age of the organism, water pH, and the dissolved organic carbon content. In fish, for example, mercury tends to accumulate in muscle tissues of numerous species of freshwater and marine fishes and to increase with increasing age, weight, or length of the fish (Eisler, 1984; Braune, 1987b; Phillips et al., 1987; Chvojka, 1988; Nicoletto and Hendricks, 1988; Cope et al., 1990; Grieb et al., 1990; Sorensen et al., 1990; Wiener et al., 1990a; Leah et al., 1992, 1993; Rask and Metsala, 1991; Lange et al., 1993, 1994; Staveland et al., 1993; Mathieson and McLusky, 1995; Joiris et al., 1997; Munn and Short, 1997; Stafford and Haines, 1997). Mercury concentrations in muscle of freshwater teleosts were significantly higher in acidic lakes than in neutral or alkaline lakes (Allard and Stokes, 1989; McMurtry et al., 1989; Cope et al., 1990; Grieb et al., 1990; Wiener et al., 1990a, 1990b; Rask and Metsala, 1991; Haines et al., 1992; Lange et al., 1993). Highest levels of mercury in fish muscle were from lakes with a pH near 5.0; liming acidic lakes resulted in as much as an 80.0% decrease in muscle mercury content after 10 years (Anderson et al., 1995). And mercury concentrations in fish muscle were positively correlated with dissolved organic carbon concentration (McMurtry et al., 1989; Sorensen et al., 1990; Wren et al., 1991; Fjeld and Rognerud, 1993). Mercury content in edible portions of all species of freshwater fish sampled from the Nitra River in the Slovak Republic in 2003 exceeded that country's allowable limit of 0.5 mg total Hg/kg FW muscle by factors of 4 to 13 times (Table 6.4); authors have recommended the posting of fish consumption advisories (Andreji et al., 2005).

In addition to age, water pH, and dissolved organic carbon, other variables known to modify mercury accumulation rates in aquatic organisms include water temperature, sediment mercury concentrations, lake size, season, diet, chemical speciation of mercury, and sex. Elevated water temperatures were associated with elevated accumulations of mercury. Rates of mercury methylation were positively dependent on water temperature, and mercury demethylation rates were inversely related to water temperature (Bodaly et al., 1993). Elevated mercury concentrations in fish muscle were positively correlated with sediment mercury concentrations (Munn and Short, 1997): a similar case is made for benthic marine invertebrates (Becker and Bigham, 1995). Mercury concentrations were inversely related to lake size in planktivorous, omnivorous, and piscivorous fishes from remote lakes in northwestern Ontario; lakes ranged in size from 89 to 35,000 surface ha and were far from anthropogenic influences (Bodaly et al., 1993). Mercury levels in muscle of marine flatfishes were higher in the spring than in the autumn (Staveland et al., 1993). In the yellow perch, *Perca flavescens*, seasonal variations in uptake rate of methylmercury and in the proportion of uptake from aqueous and food sources is attributed to seasonal variations in water temperature, body size, diet, and prey availability; methylmercury uptake was primarily from aqueous sources during the spring and fall and was dominated by food sources in the summer (Post et al., 1996). Food chain transfer of mercury from benthic invertebrates to fishes depended primarily on the consumption rate of benthivorous fishes, and secondarily to the total invertebrate mercury pools (Wong et al., 1997). In the absence of pelagic forage fish, mercury concentrations in muscle of lake trout, *Salvelinus namaycush*, are

Table 6.4 Mercury Concentrations in Selected Species of Amphibians and Reptiles

Species, Tissue, and Other Variables	Concentration (mg/kg)	Ref.[a]
Wart snake, *Achrochordus javanicus*; muscle; Papua New Guinea:		
Total mercury	1.3 FW	14
Organic mercury	1.13 FW	14
Inorganic mercury	0.17 FW	14
Cottonmouth, *Agkistrodon piscivorus*; northeastern Texas (groundwater contained 3.3 µg Hg/L, sediments 0.1–0.7 mg/kg DW); total mercury; males vs. females:		
Liver	0.85 (0.3–2.18) FW vs. 0.60 (0.20–1.8) FW	25
Kidney	0.24 (0.11–0.53) FW vs. 0.18 (0.06–0.48) FW	25
Tail muscle	0.19 (0.12–0.29) FW vs. 0.14 (0.10–0.19) FW	25
Alligator, *Alligator mississippiensis*:		
From mercury-contaminated areas; 1994–1995; Florida Everglades vs. Savannah River, South Carolina:		
Blood	No data vs. 2.2 FW	1
Dermal scutes	5.8 DW vs. 4.6 DW	1
Kidney	36.4 DW vs. no data	1
Liver	41.1 DW vs. 17.7 DW	1
Muscle	5.6 DW vs. 4.1 DW	1
Muscle; Florida:		
Statewide	0.31 (0.004–0.61) FW	6
South	2.3 (0.46–3.88) FW	7
Southeast	0.7 (0.2–2.5) FW	7
North, central	0.4 (0.1–0.9) FW	7
Everglades	3.6 (1.1–6.1) FW	8, 9
Muscle; Georgia:	0.48 FW	10
Liver; Florida:		
Everglades	40.0 (8.9–99.5) FW	9
Reference site	2.5 (0.1–16.0) FW	9
Farm-raised	0.1 (0.06–0.16) FW	9
Kidney; Florida:		
Everglades	25.9 (5.4–65.3) FW	9
Reference site	1.6 (0.16–9.6) FW	9
Farm-raised	0.09 (0.03–0.2) FW	9
Spleen; Florida:		
Everglades	3.7 (1.0–13.1) FW	9
Reference site	0.5 (0.1–1.3) FW	9
Farm-raised	0.09 (0.04–0.15) FW	9
Various tissues; Florida Everglades; max. Values:		
Heart	4.6 FW	9
Brain	2.5 FW	9
Spinal cord	2.6 FW	9
Ovary	1.3 FW	9
European toad, *Bufo bufo*; Yugoslavia; control area vs. polluted mercury mining area:		
Liver	1.5 FW vs. 21.8–25.5 FW	2, 3
Kidney	1.2 FW vs. 22.8–24.0 FW	2, 3
Lung	0.2 FW vs. 1.7 FW	2, 3
Muscle	0.2 FW vs. 2.3–2.9 FW	2, 3
Egg	0.06 FW vs. 2.3 FW	2, 3
Loggerhead sea turtle, *Caretta caretta*:		
Egg, whole	0.01 FW	4
Egg yolk:		
Georgia and South Carolina	0.02–0.09 FW	15
Florida, Georgia, North Carolina, South Carolina	0.41–1.39 FW	16
Japan	0.012 (0.008–0.016) FW	17

Table 6.4 (continued) Mercury Concentrations in Selected Species of Amphibians and Reptiles

Species, Tissue, and Other Variables	Concentration (mg/kg)	Ref.[a]
Kochi Prefecture, Japan:		
Liver	1.51 (0.25–8.15) FW	17
Kidney	0.25 (0.04–0.44) FW	17
Muscle	0.10 (0.05–0.19) FW	17
Shell	0.004 (0.002–0.005) FW	17
Whole egg	0.005 (0.004–0.007) FW	17
Atlantic green turtle, *Chelonia mydas*; muscle; Papua New Guinea		
Total mercury	0.038 FW	14
Organic mercury	0.023 FW	14
Inorganic mercury	0.015 FW	14
Snapping turtle, *Chelydra serpentina*:		
New Jersey (mercury-contaminated) vs. Maryland (reference site):		
Liver	1.3 FW vs. 0.9 FW	18
Kidney	0.6 FW vs. 0.4 FW	18
Tennessee; contaminated lake (5.9 mg Hg/kg DW sediment) vs. reference wetland:		
Kidney	1.3 FW vs. 0.4 FW	19
Muscle	0.2 FW vs. 0.1 FW	19
American crocodile, *Crocodylus acutus*; Florida Everglades; egg:		
Contents	0.09 (0.07–0.14) FW	11
Whole	0.7 FW	4
Shell	0.21 DW	12
Albumin-yolk	0.13 FW; 0.66 DW	12
Crocodile, *Crocodylus niloticus*; Zimbabwe; egg contents	0.22 (0.02–0.54) DW	13
Crocodile, *Crocodilus porosus*; Papua New Guinea; muscle:		
Total mercury	0.13 FW	14
Organic mercury	0.11 FW	14
Inorganic mercury	0.02 FW	14
Water snake, *Natrix natrix*; Yugoslavia; mud contains 0.56 mg Hg/kg DW:		
Muscle	0.42 (0.25–0.53) FW	21
Liver	0.54 (0.31–0.84) FW	21
Water snake, *Nerodia sipedon*; whole; Lake Michigan, Wisconsin	0.45 FW	22
Water snake, *Nerodia* sp.; whole; Apalachicola River, Florida; upper reaches vs. lower reaches	0.18 (0.13–0.21) FW vs. 0.29 (0.17–0.38) FW	24
Bullfrog, *Rana catesbeiana:*		
Lake St. Clair:		
Carcass	0.1 FW	2
Liver	0.3 FW	2
South Carolina: 1997; tadpoles:		
With digestive tract:		
Body	0.24 DW; 0.05 FW	5
Tail	0.05 DW; 0.01 FW	5
Whole	0.18 DW; 0.04 FW	5
Without digestive tract:		
Body without gut	0.1 DW	5
Tail	0.09 DW	5
Digestive tract	0.89 DW	5
Whole	0.14 DW	5
Pig frog, *Rana grylio*; south Florida; leg muscle; 8 locations; 2001–2002; all areas vs. Everglades National Park:		
Males	0.44 FW vs. 1.23 FW	26
Females	0.31 FW vs. 1.0 FW	26
Juveniles	0.21 FW vs. 0.37 FW	26
Leopard frog, *Rana pipiens:*		
Lake St. Clair:		
Carcass	0.1–0.2 FW	2
Liver	0.5–1.1 FW	2

(continued)

Table 6.4 (continued) Mercury Concentrations in Selected Species of Amphibians and Reptiles

Species, Tissue, and Other Variables	Concentration (mg/kg)	Ref.[a]
Florida; liver	0.1 FW	2
Frog, *Rana temporaria*; Yugoslavia; 1975:		
From Hg-mining area:		
Liver	21.0 FW	3
Kidney	16.2 FW	3
Muscle	3.4 FW	3
Egg	1.3 FW	3
From control area; all tissues	< 0.08 FW	3
Garter snake, *Thamnophis sirtalis*:		
Lake St. Clair, Canada:		
Carcass	< 0.1–0.2 FW	23
Liver	0.45–0.6 FW	23
Lake Michigan, Wisconsin; whole body	0.14–0.41 FW	22
Red-eared turtle, *Trachemys scripta*		
Texas; whole body less carapace	0.08 FW	11
Tennessee; contaminated lake vs. reference wetland:		
Kidney	1.1 FW vs. 0.15 FW	20
Muscle	0.16 FW vs. 0.03 FW	20
Monitor, *Varanus* sp.; muscle; Papua New Guinea:		
Total mercury	0.175 FW	14
Organic mercury	0.157 FW	14
Inorganic mercury	0.018 FW	14

Note: All values are in mg total mercury/kg fresh weight (FW) or dry weight (DW).

[a] Reference: 1, Yanochko et al., 1997; 2, Jenkins, 1980; 3, Terhivuo et al., 1984; 4, Hall, 1980; 5, Burger and Snodgras, 1998; 6, Delany et al., 1988; 7, Hord et al., 1990; 8, Facemire et al., 1995; 9, Heaton-Jones et al., 1997; 10, Ruckel, 1993; 11, Linder and Grillitsch, 2000; 12, Stoneburner and Kushlan, 1984; 13, Phelps et al., 1986; 14, Yoshinaga et al., 1992; 15, Hillestad et al., 1974; 16, Stoneburner et al., 1980; 17, Sakai et al., 1995; 18, Albers et al., 1986; 19, Meyers-Schone et al., 1993; 20, Meyers-Schone and Walton, 1994; 21, Srebocan et al., 1981; 22, Heinz et al., 1980; 23, Dustman et al., 1972; 24, Winger et al., 1984; 25, Rainwater et al., 2005; 26, Ugarte et al., 2005.

likely to be depressed (Futter, 1994). Trophic transfer of methylmercury is much more efficient than that of Hg^{2+} (Hill et al., 1996). Sometimes, fish pellets fed to laboratory fish can contain elevated (0.09 mg Hg/kg DW) concentrations of mercury, resulting in elevated blood mercury levels (0.06 mg Hg/L) after 10 weeks, as was the case for the Sacramento blackfish, *Orthodon microlepidotus* (Choi and Cech, 1998). Sexually mature female centrarchids had significantly higher concentrations of mercury in muscle tissue than did sexually mature males (Nicoletto and Hendricks, 1988), although this has not been reported for other aquatic species. Mercury concentrations in muscle of 14 species of freshwater fishes from Lake Chad, Africa, in December 2000, were highest in fish-eating species, three to four times lower in fish that fed upon insects and other invertebrates, and lowest in herbivores (Kidd et al., 2004). Mercury concentrations in fish muscle were higher in fish from humic lakes (Rask and Metsala, 1991), from lakes of low mineralization (Allard and Stokes, 1989), and from lakes with low concentrations of dissolved iron (Wren et al., 1991), calcium, alkalinity, chlorophyll *a*, magnesium, phosphorus, and nitrogen (Lange et al., 1993). It is noteworthy that low atmospheric depositions of selenium did not affect mercury concentrations in muscle of brown trout, *Salmo trutta* (Fjeld and Rognerud, 1993); that mercury and selenium in muscle of marine fishes were not correlated (Chvojka, 1988; Chvojka et al., 1990); and that mercury and selenium concentrations in blood of tunas were independent of each other (Kai et al., 1988).

Diet, age, logging, and forest fires were all significant factors affecting mercury concentration in fish collected from Canadian drainage lakes in 1996 to 1997, wherein muscle contained between 0.2 and 20.2 mg total Hg/kg DW (Garcia and Carignan, 2005). Mercury concentrations tended to increase with increasing fish length and were higher in fish-eating fish such as northern pike, *Esox lucius*, and walleye *Stizostedium vitreum*; concentrations were highest when surrounding forest were clear-cut or fire-impacted and may reflect increased exposure to mercury when compared to conspecifics from lakes with undisturbed watersheds (Table 6.4; Garcia and Carignan, 2005).

In adult fish, females often contain higher mercury concentrations than males, possibly because they consume more food than males in order to support the energy requirements of egg production (Nicoletto and Hendricks, 1988; Trudel et al., 2000). The increased feeding rate in females causes greater dietary uptake of methylmercury; however, the transfer to egg mass is a small fraction of the maternal body burden (Hammerschmidt et al., 1999; Johnston et al., 2001). Mercury concentrations in spiny dogfish, *Squalus acanthias*, were influenced by dogfish sex, length, and area of collection. Concentrations were higher in males, higher in specimens with body length greater than 65 cm when compared to smaller dogfish, and higher in dogfish from estuarine areas than from offshore locations (Forrester et al., 1972). Mercury levels in fetuses of the California dogfish, *Squalus suckleyi*, were 21 to 42 times lower than maternal tissues, suggesting that mercury is uniquely absent from the fetal environment and may even be selectively excluded (Childs et al., 1973).

Nationwide (U.S.) monitoring of whole freshwater fish during the period 1969 to 1981 demonstrated that the highest mercury concentrations (0.33 to 1.7 mg/kg FW) were in northern squawfish (*Ptychocheilus oregonensis*) from the Columbia River basin in the Pacific Northwest (Henderson and Shanks, 1973; Lowe et al., 1985). Elevated mercury concentrations in this piscivorous species were attributed primarily to the presence of major cinnabar deposits and with Hg use associated with mineral mining in the Columbia River basin. Northern squawfish may have a natural tendency to accumulate high concentrations of mercury in their flesh — as is well known for older specimens of long-lived predatory fishes such as tunas, billfishes, bluefish (*Pomatomus saltatrix*), striped bass (*Morone saxatilis*), northern pike (*Esox lucius*), and many species of sharks; however, mercury uptake kinetics in squawfish requires further research (Lowe et al., 1985).

In the Florida Everglades ecosystem, mercury concentrations in muscle of largemouth bass, *Micropterus salmoides*, are directly correlated with atmospheric deposition of mercury (Atkeson et al., 2003). A model was formulated that showed a positive correlation between total mercury in largemouth bass muscle (in the range 0.3 to 1.8 mg/kg FW) with atmospheric Hg^{2+} wet plus dry deposition (in the range 5.0 to 35.0 μg/m^2 deposition per annum; Atkeson et al., 2003). In the absence of changes to this ecosystem other than mercury cycling (e.g., changes in sulfur cycling, nutrient cycling, and hydrology), a reduction of about 80.0% of current total annual mercury deposition rates would be needed for the mercury concentration in muscle from a 3-year-old largemouth bass at a heavily contaminated site to be reduced to less than 0.5 mg/kg FW muscle, which is Florida's present fish consumption advisory action level. Mercury concentrations in muscle from 3-year-old bass — currently averaging 2.5 mg Hg/kg FW — are predicted to achieve 50.0% of their long-term steady-state response following sustained mercury load reduction within 10 years, and 90.0% within 30 years (Atkeson et al., 2003).

In the Florida recreational fishery for red drum, *Sciaenops ocellatus*, the current maximum size limit of less than 565 mm standard length or less than 689 mm total length is an effective filter that limits consumption of large fish containing elevated mercury concentrations (Adams and Onorato, 2005). About 94.0% of all adult red drum from waters adjacent to Tampa Bay, Florida, contain mercury levels in muscle greater than 0.5 mg/kg FW muscle — the Florida Department of Health threshold level — and 64.0% contained greater than 1.5 mg Hg/kg FW muscle, which is the Florida "no consumption" level. All fish from this area containing greater than 1.5 mg Hg/kg FW muscle were longer than the 689-mm standard length (Adams and Onorato, 2005).

Reservoir construction is thought to be a cause of elevated mercury concentrations in fish. Reservoir conditions facilitating the bioavailability of mercury include upstream flooding and leaching of terrestrial sediments, relatively high pH and conductivity of the water, high bacterial counts in the water, complete thermal mixing, low clay content, and low concentrations of sulfur and iron and magnesium oxides in bottom sediments (Lodenius, 1983; Lodenius et al., 1983; Phillips et al., 1987; Allen-Gil et al., 1995). It is hypothesized that increases in mercury levels observed in fish were due to bacterial methylation of naturally occurring mercury in the flooded soils (Bodaly et al., 1984). Methylation and transfer of methylmercury from flooded soils to suspended particulate matter and zooplankton is rapid and involves the bioaccumulation of methylmercury by phytoplankton and the ingestion of suspended soil-derived organic particles by zooplankton (Plourde et al., 1997). Suspended particulate matter and zooplankton are disproportionate contributors to methylmercury contamination of aquatic food chains in Quebec reservoirs (Plourde et al., 1997). In general, mercury levels are higher in fish from younger oligotrophic reservoirs, and lower in fish from older eutrophic reservoirs; in both situations, tissue mercury levels usually decline as the reservoirs age (Abernathy and Cumbie, 1977). Mercury concentrations greater than 0.5 mg/kg FW (but less than 1.0 mg/kg) have been reported in trout from several wilderness lakes in northern Maine (Akielaszak and Haines, 1981) and from the Adirondacks region of New York (Sloan and Schofield, 1983); these values are considerably higher than might be expected for fish inhabiting remote lakes. Elevated mercury concentrations in fish tissues were usually associated with lakes of low pH, low calcium, low dissolved organic carbon concentrations, and low water hardness and alkalinity. Enlargement of northern Manitoba lakes to form hydroelectric reservoirs caused a rise in the mercury content of native fishes owing to stimulation of mercury methylating bacteria by submerged terrestrial organic matter (Jackson, 1991). Increased organic substrates beyond a critical amount mitigated this effect via promotion of mercury demethylation and production of mercury-binding agents such as sulfides. Variability in mercury concentrations between fish species was high and was due to differences in habitat preference, metabolic rate, age, growth rate, size, biomass, diet, and excretory pathways (Jackson, 1991). Elevated mercury levels in fish flesh found after impoundment of a reservoir are predicted to decline as the reservoir ages (Anderson et al., 1995). In Labrador, Canada, mercury concentrations in muscle of omnivorous species of fishes reached background levels in 16 to 20 years; however, mercury in piscivorous species remained elevated 21 years after impoundment (Anderson et al., 1995).

6.4 AMPHIBIANS AND REPTILES

Several freshwater marshes in the Florida Everglades are contaminated with mercury, and more than 900,000 ha are currently under fish consumption advisories because of high mercury concentrations — namely, greater than 1.5 mg total Hg/kg FW muscle (Ware et al., 1990; Sundolf et al., 1994; Ugarte et al., 2005). Consumption of leg muscle of the pig frog, *Rana grylio*, from certain areas in South Florida under fish consumption advisories may present a risk to human health. Total mercury in frog leg muscle was highest (max. 2.05 mg/kg FW) from areas protected from harvest in the Everglades National Park (Ugarte et al., 2005). Total mercury burdens in frog leg muscle from most harvested areas were less than 0.3 mg/kg FW, an acceptable level; however, mean concentrations in other areas regularly harvested for human consumption were greater than 0.3 mg total Hg/kg FW muscle (Ugarte et al., 2005; Table 6.4).

Elevated concentrations of mercury in amphibian tissues were also found in frogs and toads collected near a mining area in Yugoslavia (Table 6.4). Maximum concentrations (in mg/kg fresh weight) were 2.3 in egg, 2.9 in lung, 24.0 in kidney, and 25.5 in liver; conspecifics from a reference site contained less than 0.08 mg Hg/kg fresh weight in all tissues (Table 6.4).

Highest concentrations of mercury in reptiles collected were found in tissues of the American alligator (*Alligator mississippiensis*) from the Florida Everglades (Table 6.4). Maximum concentrations (in mg Hg/kg fresh weight) were 6.1 in alligator muscle, 13.1 in spleen, 65.3 in kidney, and 99.5 in liver; other tissues contained 1.3 to 4.6 mg/kg FW (Heaton-Jones et al., 1997). Mercury concentrations in spleen, kidney, and liver tissues of farm-raised alligators were always less than 0.2 mg/kg FW (Table 6.4). Based on available data, mercury concentrations in all reptiles were highest in liver, followed by kidney, muscle, and egg, in that order; in all tissues sampled, organomercurials comprised between 60.0 and 90.0% of the total mercury (Table 6.4).

In cottonmouths, *Agkistrodon piscivorus*, from Texas, Rainwater et al. (2005) report that mercury tissue concentrations were higher in males than females for kidney and liver, and that one male 96.3 cm in length had 8.6 mg Hg/kg liver FW — the highest mercury concentration ever reported for a serpent (Clark et al., 2000; Campbell and Campbell, 2001).

6.5 BIRDS

It is generally acknowledged that mercury concentrations in avian tissues and feathers are highest in species that eat fish and other birds. Mercury contamination of prey in the diet of nestling wood storks (*Mycteria americana*), an endangered species, may represent a potential concern to the recovery of this species in the southeastern United States (Gariboldi et al., 1998). Increased concentrations of total mercury in livers of diving ducks were associated with lower weights of whole body, liver, and heart, and decreased activities of enzymes related to glutathione metabolism and antioxidant activity (Hoffman et al., 1998). In seabirds, mercury concentrations were highest in tissues and feathers of species that ate fish and benthic invertebrates and lowest in birds that ate mainly pelagic invertebrates (Braune, 1987a; Lock et al., 1992; Kim et al., 1996a). In seabirds, the relation between tissues and total mercury concentrations is frequently 7:3:1 between feather, liver, and muscle; however, there is much variability and these ratios should be treated with caution. Factors known to affect these ratios include the chemical form of mercury present in liver, the sampling date relative to the stage of the molt sequence, and the types of feathers used for analysis (Thompson et al., 1990). Mercury concentrations in feathers of wading birds collected in Florida between 1987 and 1990 were highest in older birds that consumed large fishes (Beyer et al., 1997). And wading birds whose prey base consisted of larger fish had four times more mercury in livers than did species that consumed smaller fish or crustaceans (Sundlof et al., 1994). Wading birds with minimal to moderate amounts of body fat had two to three times more mercury in liver than did birds with relatively abundant body fat reserves (Sundlof et al., 1994). Essentially all mercury in body feathers of all seabirds studied was organic mercury; however, more than 90.0% of the mercury in liver is inorganic (Thompson and Furness, 1989). Mercury residues are usually highest in kidney and liver, but total mercury contents are significantly modified by food preference and availability, and by migratory patterns (USNAS, 1978; Delbekke et al., 1984). Also, there is an inverse relation between total mercury and percent methylmercury in tissues of various avian species (Norheim et al., 1982; Karlog and Clausen, 1983), a pattern that seems to hold for all vertebrate organisms for which data is available. Diet and migration are the most important mercury modifiers in birds. For example, the higher levels of mercury in juveniles than in adults of wood ducks (*Aix sponsa*) from Tennessee were related to dietary patterns: juveniles preferred insects, whereas adults preferred pondweed tubers; mercury residues were higher in the insects than in the pondweeds (Lindsay and Dimmick, 1983).

Factors that modify mercury concentrations in birds include age, tissue, migratory patterns, diet, and season. Adults of the double-crested cormorant, *Phalacrocorax auritus,* contained elevated levels of mercury in liver and whole body when compared to nestlings: 0.3 mg Hg/kg FW in nestling liver vs. 8.0 in adults, and 0.06 in nestling whole body vs. 0.64 mg Hg/kg FW in adults (Greichus et al., 1973). Among highly migratory birds, dramatic seasonal changes in mercury content are common, and are attributed, in part, to ingestion of mercury-contaminated food. Seasonal variations and diet affect mercury concentrations in avian tissues. Seasonal variations in mercury levels are reported in livers of aquatic birds (Table 6.5), being higher in winter when birds were exclusively estuarine and drastically lower in summer when birds migrated to inland and sub-Arctic breeding grounds (Parslow et al., 1973). It is possible that the wintering populations (e.g., knots, *Calidras* spp.) might previously have accumulated mercury while molting in western European estuaries, notably on the Dutch Waddenzee (Parslow, 1973). For example, three species of knots, *Calidras* spp., contained greater than 20.0 mg Hg/kg FW liver during winter while molting in

Table 6.5 Mercury Concentrations in Field Collections of Selected Species of Birds

Species, Tissue, and Other Variables	Concentration (mg/kg)	Ref.[a]
Goshawk, *Accipiter gentilis*; Sweden; feather:		
1860–1946	2.2 FW	1
1947–1965	29.0 FW	1
1967–1969	3.1–5.1 FW	1
Finnish sparrowhawk, *Accipiter nisus*; feather:		
Finland:		
1899–1960	4.1 (2.1–7.7) DW	2
1961–1970	11.1 (2.3–42.0) DW	2
1971–1982	7.4 (1.0–29.0) DW	2
Germany: 1972–1973	4.9 (0.4–20.3) DW	2
Norway; 1976	2.0–20.0 DW	2
Sharp-shinned hawk, *Accipiter striatus*; Eastern U.S.; liver; 1991 vs. 1993:	1.0 (0.06–2.2) FW vs. 0.12 FW	3
Western grebe, *Aechmophorus occidentalis*; California:		
Lake Berryessa:		
1982; found dead; kidney	20.2 FW	4
1983; normal appearing:		
Kidney	2.5 (1.1–9.0) FW	4
Liver	5.2 (2.7–11.8) FW	4
1986; found dead:		
Kidney	3.7 (2.1–6.5) FW	4
Liver	7.9 (2.7–27.3) FW	4
Clear Lake, site of abandoned mercury mine:		
1984; liver	6.1 (3.7–9.8) FW	4
1992:		
Brain	0.3 FW	5
Muscle	1.1 FW	5
Kidney	2.1 FW	5
Liver	2.7 FW	5
Wood duck, *Aix sponsa*:		
Tennessee; 1972–1973; juveniles vs. adults:		
Liver	0.4 (0.1–1.1) FW vs. 0.2 (0.1–0.1) FW	6
Muscle	0.1 (0.05–0.4) FW vs. 0.08 (0.06–0.11) FW	6
Fat	0.1 (0.01–0.4) FW vs.0.06 (0.01–0.11) FW	6
Blue-winged teal, *Anas discors*; muscle:		
Lake St. Clair	0.1–2.3 FW	1
Ontario	3.8–10.4 FW	1
Wisconsin	0.0–0.5 FW	1
Illinois	0.05 FW	1
Mallard, *Anas platyrhynchos*:		
Whole body	< 0.5 FW	69
Liver	0.06 FW	69
Kidney	0.07 FW	69
Antarctic region; February–March 1989:		
Adelie penguin, *Pygoscelis adeliae*; muscle vs. liver	max. 0.7 DW vs. max. 2.0 DW	7
Chinstrap penguin, *Pygoscelis antarctica*; feces	1.6 DW	7
Gentoo penguin, *Pygoscelis papua*; liver	34.7 DW	7
King penguin, *Aptenodytes patagonicus*; sub-Antarctic Islands; breast feathers:		
1966–1974	2.66 DW	80
2000–2001	1.98 DW	80
Golden eagle, *Aquila chrysaetos*; Scotland; 1981–1986; unhatched eggs:		
Eastern district	Not detectable	8
Western coastal district	0.4 (0.1–1.4) DW	8

Table 6.5 (continued) Mercury Concentrations in Field Collections of Selected Species of Birds

Species, Tissue, and Other Variables	Concentration (mg/kg)	Ref.[a]
Western inland district		
25 of 27 eggs	Not detectable	8
2 of 27 eggs	0.21 DW, 0.56 DW	8
Great blue heron, *Ardea herodius:*		
Liver:		
Lake St. Clair	97.0 (14.6–175.0) FW	1
New Brunswick	4.5 FW	1
Lake Erie	0.7–4.3 FW	1
Wisconsin	0.5 (0.2–1.1) FW	1
Clear Lake, California; 1993; distance from abandoned mercury mine: 8 km vs. 23 km:		
Blood	1.3 FW vs. 1.2 FW	9
Diet	0.9 FW vs. 0.5 FW	9
Feathers	2.2 FW vs. 3.1 FW	9
Kidney	1.1 FW vs. 1.1 FW	9
Liver	1.4 FW vs. 1.3 FW	9
Great white heron, *Ardea herodias occidentalis*; Florida; 1987–1989; radiotagged and recovered soon after death; liver:		
Dead from known acute causes	1.8 (0.6–4.0) FW	10
Dead from chronic multiple causes	9.8 (2.9–59.4) FW	10
Dead birds with signs of kidney disease and gout	> 25.0 FW	10
Birds:		
Antarctic; liver:		
1977–1979		
4 spp.	0.5–1.3 FW	11
3 spp.	2.7–2.9 FW	11
1980; 5 spp	0.5–2.1 FW	12
Belgium; 1970–1981; liver; 30 spp.; aquatic birds vs. terrestrial birds	0.11–0.35 FW vs. not detectable–14.0 FW	13
Hawaii; 1980; egg; 3 spp.	0.12–0.36 FW	14
North America; feather:		
From areas with mercury-treated seed dressing; seed-eating songbirds vs. upland game birds	1.6 DW vs. 1.9 DW	15
From untreated areas: seed-eating songbirds vs. upland game birds	0.03 DW vs. 0.35 DW	15
Northwestern Ontario, Canada; from a heavily mercury-contaminated freshwater system:		
Liver:		
Scavengers	57.0 (13.8–121.0) FW	16
Fish eaters	39.0 (1.7–91.0) FW	16
Omnivores	26.0 (9.5–53.0) FW	16
Invertebrate feeders	12.0 (3.2–28.0) FW	16
Vegetarians	6.1 (1.9–28.0) FW	16
Diving ducks	Max. 175.0 FW	15
Muscle:		
Diving ducks	Max. 23.0 FW	15
Mallard, *Anas platyrhynchos*	max. 6.1 FW	15
Eagle-owl, *Bubo bubo*; Sweden; feather:		
Inland populations; 1963–1976	3.2 DW	17
Coastal populations; 1963–1976	6.5 DW	17
Coastal populations; 1829–1933 vs. 1964–1965	0.3–3.6 FW vs.12.8–41.0 FW	1
Cattle egret, *Bubulcus ibis*; eggs; Egypt; 1986; declining colony between 1977 and 1984	0.48 (0.28–0.84) DW	18
Common goldeneye, *Bucephala clangula*; Minnesota; 1981; eggs of dead hens found on clutch	0.1 (0.02–0.4) FW	19

(continued)

Table 6.5 (continued) Mercury Concentrations in Field Collections of Selected Species of Birds

Species, Tissue, and Other Variables	Concentration (mg/kg)	Ref.[a]
Bulwer's petrel, *Bulweria bulwerii*; north Atlantic Ocean; methylmercury; feathers; from museum specimens collected 1885–1994; maximum concentrations; compared to Cory's shearwater, *Calonectris diomeda borealis:*		
1885–1900	5.0 FW vs. 2.0 FW	78
1900–1931	8.0 FW vs. 3.0 FW	78
1950–1970	17.0 FW vs. 4.0 FW	78
1992–1994	23.0 FW vs. 5.5 FW	78
Dunlin, *Calidris alpina*; liver:		
September–October	1.9 (0.8–3.3) DW	70
October	1.7 (0.7–2.6) DW	70
November	3.3 (3.1–3.5) DW	70
January	6.9 (4.8–9.2) DW	70
March	10.5 DW	70
Knot, *Calidris canutus*; liver:		
August	1.0 (0.4–1.5) DW	70
September–October	1.3 (0.8–1.7) DW	70
January	7.3 (6.6–8.0) DW	70
February	9.8 (5.1–12.9) DW	70
Mid-March	14.4 (5.6–24.9) DW	70
Late March	18.5 (16.5–20.5) DW	70
California, Clear Lake (mercury-contaminated site); feathers:		
Western grebe, *Aechmophorus occidentalis*; adults	9.8 DW	20
Great blue heron, *Ardea herodias*; adults	6.1 DW	20
Turkey vulture, *Cathartes aura*; adults	1.3 DW	20
Osprey, *Pandion haliaetus:*		
Adults	20.0 DW	20
Juveniles	5.3 DW	20
Juveniles from reference sites	2.3 DW	20
Great skua, *Catharcta skua*:		
Total mercury; adults vs. chicks; feather	7.0 (1.0–32.4) FW vs. 1.3 (0.7–2.4) FW	21
Total mercury vs. inorganic mercury; adults:		
Kidney	9.7 DW vs. 5.0 DW	21
Liver	11.6 DW vs. 6.2 DW	21
Muscle	2.3 DW vs. 2.3 DW	21
Black-footed albatross, *Diomedea nigripes*; Sanriku, Japan; 1997–1998:		
Liver, whole	94.0 (36.0–150.0) FW	81
Liver nuclei, lysosomes, and mitochondria	78.0 (20.0–127.0) FW	81
Diving ducks; California; 1989; livers; Tomales Bay vs. Suisin Bay		
Greater scaup, *Aythya marila*	19.0 (5.0–66.0) DW vs. 6.0 (3.0–11.0) DW	22
Surf scoter, *Melanitta perspicillata*	19.0 (3.0–35.0) DW vs. 10.0 (5.0–21.0) DW	22
Ruddy duck, *Oxyura jamaicensis*	6.0 (4.0–9.0) DW vs. 4.0 (2.0–7.0) DW	22
England; 1963–1990; liver:		
Grey heron, *Ardea cinerea:*		
1963–1970	44.0 (20.0–96.0) FW	23
1971–1975	19.0 (8.0–49.0) FW	23
1976–1980	18.0 (7.0–46.0) FW	23
1981–1985	10.0 (2.0–39.0) FW	23
1986–1990	12.0 (4.0–34.0) FW	23
Sparrowhawk, *Accipiter nisus:*		
1963–1970	4.6 (2.6–8.3) FW	23
1971–1975	5.6 FW	23
1976–1980	3.5 FW	23

Table 6.5 (continued) Mercury Concentrations in Field Collections of Selected Species of Birds

Species, Tissue, and Other Variables	Concentration (mg/kg)	Ref.[a]
1981–1985	2.3 FW	23
1986–1990	1.0 (0.2–6.3) FW	23
Kestrel, *Falco tinnunculus*:		
1963–1970	5.8 FW	23
1976–1980	1.3 FW	23
1986–1990	0.2 FW	23
Common loon, *Gavia immer:*		
Canada, central Ontario; 24 lakes; July–August 1992		
Breeding adults; blood vs. feathers	2.1 (0.9–4.3) FW vs. 13.3 (7.6–21.0) FW	24
Chicks; blood vs. feathers	0.14 (0.04–0.6) FW vs. 2.3 (1.4–3.4) FW	24
Eastern Canada; tissues from freezer archives:		
Kidney	15.0 DW	25
Liver	19.0 DW	25
Muscle	2.9 DW	25
New England; 1990–1994; loons found dead; liver; total mercury vs. methylmercury:		
Females	46.0 (10.0–100.0) FW vs. 3.9 (2.6–5.4) FW	26
Males	56.0 (11.0–187.0) FW vs. 4.3 (2.8–9.1) FW	26
Chicks	1.4 FW vs. 0.3 FW	26
Northeastern U.S.; found dead; breast feathers:		
Adults	20.2 DW	27
Immature males	7.7 DW	27
Immature females	12.4 DW	27
U.S. and Canada; 1991–1996; summers; whole blood vs. feathers:		
Males	1.9 (0.4–7.8) FW vs. 12.7 (4.1–36.7) FW	28
Females	1.5 (0.1–6.7) Fw vs. 9.4 (2.8–21.1) FW	28
Juveniles, age 3–6 weeks	0.16 (0.03–0.78) FW vs. 3.8 (0.6–13.6) FW	28
Wisconsin; summer 1991; lakes with pH < 6.3 vs. lakes with pH > 7.0:		
Clotted blood	5.0 DW vs. 2.0 DW	29
Feathers	12.0 DW vs. 9.0 DW	29
Wisconsin:		
Adults; 1992–1993; blood; lowest quartile vs. highest quartile	0.6–0.9 FW vs. 1.9–4.2 FW	30
Adults; 1992–1993; feathers; lowest quartile vs. highest quartile	3.0–9.6 FW vs.13.0–21.0 FW	30
Eggs; 1993–1996	0.9 FW	30
Peregrine, *Falco peregrinus*; feather:		
1834–1949	2.5 DW	15
1941–1965	> 40.0 DW	15
Swedish gyrfalcon, *Falco rusticolus*; nestlings; feather; percent aquatic birds in diet:		
0.0% biomass	0.035 FW	31
4.8% biomass	0.66 FW	31
10.6% biomass	1.22 FW	31
Florida; wading birds; 1987–1990:		
Nestlings; feather:		
Roseate spoonbill, *Ajaia ajaja*	2.0 (0.4–5.7) DW	32
Great blue heron, *Ardea herodias*	3.5 (1.8–7.7) DW	32
Great egret, *Casmerodius albus*	7.1 (1.6–15.0) DW	32
Great white heron, *Ardea herodias occidentalis*; feathers vs. liver:		
Nestlings	4.7 (1.0–9.1) DW vs. (3.9–9.1) DW	32
Juveniles	6.7 (2.7–15.0) DW vs. (6.2–8.1) DW	32
Adults	8.2 (4.1–14.0) DW vs. 6.2 DW	32

(continued)

Table 6.5 (continued) Mercury Concentrations in Field Collections of Selected Species of Birds

Species, Tissue, and Other Variables	Concentration (mg/kg)	Ref.[a]
Germany; North Sea Coast; feathers; pre-1940 vs. post-1941:		
Herring gull, *Larus argentatus*:		
Adults	4.6 (1.0–7.8) FW vs. 7.9 (2.1–21.2) FW	33
Juveniles	2.0 FW vs. 4.3 FW	33
Common tern, *Sterna hirundo:*		
Adults	1.0 (0.2–2.3) FW vs. 3.5 (0.4–13.9) FW	33
Juveniles	2.0 (0.4–4.9) FW vs. 4.8 (1.5–18.4) FW	33
Germany; 1991:		
Herring gull; adults:		
Eggs	Max. 2.8 FW	34
Down	Max. 58.0 FW	34
Liver	Max. 4.0 FW	34
Black-headed gull, *Larus ridibundus*; chick; body feathers	Max. 8.0 FW	34
Common tern:		
Eggs	Max. 4.0 FW	34
Down	Max. 20.0 FW	34
Chick; body feathers	Max. 16.0 FW	34
White-tailed sea-eagle, *Haliaeetus albicilla*; Gulf of Bothnia, Finland:		
Dead birds and addled eggs	Max. 26.0 FW	71
Muscle of prey bird	0.05–0.93 FW	71
Bald eagle, *Haliaeetus leucocephalus*; egg:		
Canada, BC; 1990–1992	0.08–0.29 (0.07–0.40) FW	35
Maine (highest concentrations nationwide):		
1974	0.35–0.58 FW	36
1975	0.22–0.63 FW	36
1976	0.22–0.66 FW	36
1977	0.28–0.90 FW	36
1978	0.30 FW	36
1979	0.84–1.2 FW	36
1974–1979 vs. 1980–1984	0.39 FW vs. 0.41 FW	37
Maryland; 1973–1979 vs. 1980–1984	0.04 FW vs. 0.06 FW	37
Virginia; 1976–1979 vs. 1981–1984	0.07 FW vs. 0.08 FW	37
Wisconsin; 1976–1983	0.13–0.14 FW	37
Florida; 1991–1993:		
Adults; feathers	9 (0.1–35) FW	38
Nestlings; blood vs. feathers	0.2 (0.02–0.6) FW vs. 3.2 (0.8–14.3) FW	38
Found dead; liver:		
Adults	3.2 FW ; max. 12.2 FW	38
Subadults	2.6 (0.4–5.4) FW	38
Nestlings	0.4 (0.1–1.0) FW	38
Great Lakes region; 1985–1989; feathers:		
Adult primaries	21.0 (3.6–48.9) DW	39
Adult secondaries	23.0 (5.3–66.0) DW	39
Adult retrices	19.0 (5.0–46.0) DW	39
Adult body	21.0 (0.2–48.0) DW	39
Nestlings; all feathers	9.0 (1.5–27.0) DW	39
Black-winged stilt, *Himantopus himantopus*; Portugal; several sites:		
Egg:		
1997	0.5–0.8 (0.3–1.2) DW	68
1998	0.4–0.5 (0.05–1.4) DW	68
1999	0.6–1.3 (0.3–2.5) DW	68

Table 6.5 (continued) Mercury Concentrations in Field Collections of Selected Species of Birds

Species, Tissue, and Other Variables	Concentration (mg/kg)	Ref.[a]
Chick feather:		
1997	0.7–3.4 (0.1–10.8) FW	68
1998	0.9–2.9 (0.05–6.2) FW	68
1999	0.8–2.5 (0.1–5.2) FW	68
1998:		
Down	1.2–3.0 (0.5–4.9) FW	68
Liver	0.5–1.4 (0.2–2.4) DW	68
Carcass	0.3–0.9 (0.1–1.6) DW	68
Kenya; Lake Nakura; 1970 vs. 1990:		
Pelican, *Pelecanus onocrotalus:*		
Kidney	0.03 FW vs. 0.03 FW	40
Liver	0.02 FW vs. 0.06 FW	40
Lesser flamingo, *Phoenicopterus minor:*		
Kidney	0.10 FW vs 0.04 FW	40
Liver	0.37 FW vs. 0.26 FW	40
Herring gull, *Larus argentatus*:		
Denmark; 1975–1976; liver	0.6 (0.08–2.3) FW	41
Germany; Wadden Sea:		
Egg vs. ovary	1.4 DW vs. 1.9 DW	42
Males vs. females:		
Feather	6.4 DW vs. 4.9 DW	42
Liver	4.7 DW vs. 4.4 DW	42
Muscle	2.6 DW vs. 2.0 DW	42
Great Lakes; egg:		
1973–1976	0.22–0.72 FW	43
1981–1983	0.17–0.73 FW	43
1985	0.14–0.36 FW	43
1992	0.14–0.20 FW	43
Ontario, Canada:		
Egg	1.5–15.8 FW	1
Albumen	16.1–22.7 FW	1
Yolk	3.4–3.5 FW	1
Yugoslavia; Tyrrhenian coast; mercury-contaminated:		
Liver	6.9 FW	72
Kidney	5.3 FW	72
Heart	2.9 FW	72
Muscle	2.7 FW	72
Blood	2.4 FW	72
Spleen	2.1 FW	72
Brain	1.3 FW	72
California gull, *Larus californicus*; Lahontan Reservoir, Nevada; 1981:		
Muscle	0.4 FW	4
Liver	1.0 FW	4
Egg	0.1–0.2 FW	4
Bonaparte's gull, *Larus philadelphia*; New Brunswick, Canada; autumn, 1978–1984		
Feather parts:		
Quill	2.3 DW	45
Rachis	3.3 DW	45
Vane	3.5 DW	45
Feather groups:		
Secondaries	1.9 DW	45
Wing coverts	2.3 DW	45
Primaries	2.5 DW	45
Retrices	2.8 DW	45

(continued)

Table 6.5 (continued) Mercury Concentrations in Field Collections of Selected Species of Birds

Species, Tissue, and Other Variables	Concentration (mg/kg)	Ref.[a]
All feathers:		
Juveniles	2.0 DW	45
Second-year	2.5 DW	45
Adults	4.1 DW	45
Adults; females vs. males	4.8 DW vs. 3.5 DW	45
Franklin's gull, *Larus pipixcan*; Minnesota; 1994:		
Feathers; males vs. females	0.84 DW vs. 0.77 DW	46
Eggs	0.14 DW	46
Diet (earthworms)	0.018 DW	46
Black-headed gull, *Larus ridibundus*; Mediterranean coast of Italy:		
Liver	1.3–2.4 FW	73
Muscle	0.9–1.8 FW	73
Kidney	0.6–1.4 Fw	73
Brain	0.65 FW	73
Hooded merganser, *Lophodytes cucullatus*; Minnesota; 1981; eggs of dead hens found on clutch	0.5 (0.1–2.4) FW	19
Common merganser, *Mergus merganser*, eastern Canada; tissues from freezer archives:		
Kidney	11.0 DW	25
Liver	15.0 DW	25
Muscle	3.0 DW	25
Red-breasted merganser, *Mergus serrator:*		
Muscle	0.7–0.9 FW; 3.7–3.8 DW	76
Liver	46.0 DW	76
Black-eared kite, *Milvus migrans lineatus*; Japan; premoult (April) vs. postmoult:		
Bone	0.08 DW vs. 0.07 DW	47
Brain	1.2 DW vs. 0.6 DW	47
Feathers	2.3 DW vs. 2.1 DW	47
Heart	0.4 DW vs. 0.3 DW	47
Kidney	4.7 DW vs. 1.9 DW	47
Liver	2.4 vs. 0.9 DW	47
Lung	2.5 DW vs. 0.8 DW	47
Pectoral muscle	1.1 DW vs. 0.4 DW	47
Femoral muscle	0.5 DW vs. 0.1 DW	47
Skin	0.2 DW vs. 0.07 DW	47
Whole body	0.92 DW vs. 0.69 DW	47
Wood stork, *Mycteria americana*; nestlings; Georgia; April–June 1995; diet:		
Atlantic coast colonies:		
All prey items	0.4–0.7 DW; 0.1–0.2 FW	48
Freshwater prey only	1.1 DW; 0.3 FW	48
Inland colonies; all prey items	0.7–1.0 DW; 0.18–0.28 FW	48
Osprey, *Pandion haliaetus*:		
Canada; northern Quebec; 1989–1991; built up areas vs. natural environments:		
Adults:		
Brain	1.0 FW vs. 0.2 FW	49
Egg	0.22 FW vs. 0.18 FW	49
Feathers	58.1 DW vs. 16.5 DW	49
Kidney	5.3 FW vs. 0.9 FW	49
Liver	3.6 FW vs. 0.7 FW	49
Muscle	1.8 FW vs. 0.4 FW	49
Stomach contents	0.8 FW vs. 0.3 FW	49

Table 6.5 (continued) Mercury Concentrations in Field Collections of Selected Species of Birds

Species, Tissue, and Other Variables	Concentration (mg/kg)	Ref.[a]
Chicks:		
Blood	1.9 FW vs. 0.4 FW	49
Feathers	37.3 DW vs. 7.0 DW	49
Sweden; feather:		
1840–1940	3.5–5.0 FW	1
1940–1966	> 17.0 FW	1
U.S.; egg:		
Idaho; 1973	0.06 FW	50
Florida Everglades; 1973	0.1 (0.04–0.2) FW	50
Maryland: 1973 vs. 1986	0.05 FW vs.0.1 (0.07–0.2) FW	51
Massachusetts; 1986–1987	0.06 FW	51
New Jersey; 1978	(0.05–0.25) FW	50
Virginia; 1987	0.1 (0.05–0.2) FW	51
Eastern U.S.; 1964–1973:		
Liver	< 1.0–67.5 FW	74
Kidney	0.2–130.0 FW	74
Brown pelican, *Pelecanus occidentalis*:		
Egg:		
South Carolina	0.3–0.5 FW	1
Florida	0.4 FW	1
California	0.4 FW	1
Liver	0.75 DW	52
Kidney	0.68 DW	52
Feather	0.97 DW	52
Double-crested cormorant, *Phalacrocorax auritus:*		
Adult:		
Whole body	0.6 FW	75
Muscle	0.8 FW	75
Liver	8.0 FW	75
Kidney	1.5 FW	75
Egg minus shell	0.3 FW	75
Nestlings; whole body vs. liver	0.06 FW vs. 0.28 FW	75
White-necked cormorant, *Phalacrocorax carbo*; England; 1992–1993; eggs that failed to hatch; rapidly expanding colony	2.6 (1.4–7.7) DW	53
Ring-necked pheasant, *Phasianus colchicus*; muscle:		
Denmark	0.01 FW	1
Idaho	0.0–15.0 FW	1
Indiana	0.06 FW	1
Oregon	< 0.5 FW	1
Wyoming	0.2–0.6 FW	1
Colorado	0.04–0.6 FW	1
Utah	0.2 (0.01–2.1) FW	1
California	1.6–4.7 FW	1
Wisconsin	0.01–0.08 FW	1
Illinois	0.02–0.03 FW	1
White-faced ibis, *Plegadis chihi*; egg; Carson Lake, Nevada:		
1985	Means: 0.22–0.77 DW; 30.0% had > 1.0 DW	54
1986	Means: 0.4–1.1 DW; 55.0% had > 1.0 DW	54
Great crested grebe, *Podiceps cristata*; Sweden; feather:		
1865–1940	< 10.0 FW	1
1940–1966	> 14.0 FW	1

(continued)

Table 6.5 (continued) Mercury Concentrations in Field Collections of Selected Species of Birds

Species, Tissue, and Other Variables	Concentration (mg/kg)	Ref.[a]
Kittiwake, *Rissa tridactyla*; Helgoland Island, North Sea; 1992–1994; nestlings; found dead:		
Brain:		
Age 1 day	2.0 DW	55
Age 21–40 days	0.44 DW	55
Feathers:		
Age 1 day	4.6 DW	55
Age 6–10 days	4.0 DW	55
Age 21–40 days	2.3 DW	55
Liver vs. Kidney:		
Age 1 day	2.8 DW vs. 2.0 DW	55
Age 6–10 days	1.8 DW vs. 1.4 DW	55
Age 21–40 days	1.1 DW vs. 0.9 DW	55
Seabirds:		
Antarctic and environs; 1978–1983; 15 species; eggs	0.02–1.8 FW; max. 2.7 FW	56
North and northeast Atlantic; feathers; from pre-1930s museum specimens vs. post-1980 collections:		
Great skua, *Catharcta skua*	3.7 FW vs. 5.6 FW	57
Atlantic puffin, *Fratercula arctica*	1.8 FW vs. 4.0 FW	57
Northern fulmar, *Fulmaris glacialis*	4.0–4.4 FW vs. 1.4–2.9 FW	57
Manx shearwater, *Puffinus puffinus*	1.3–1.5 FW vs. 3.3–4.2 FW	57
Atlantic gannet, *Sula bassana*	6.0 FW vs. 7.2 FW	57
Azores; 1990–1992; feathers; adults; 7 species:		
Petrels	12.5–22.1 DW; max. 35.9 DW	58
Shearwaters	2.1–6.0 DW; max. 12.4 DW	58
Terns	2.0–2.3 DW; max. 4.0 DW	58
Barents Sea; 1993; 10 species; eggs	0.06–0.34 FW	59
Canada, New Brunswick; 9 species; 1978–1984:		
Brain	0.04–0.36 FW	60
Kidney	0.24–5.3 FW	60
Liver	0.22–7.1 FW	60
Muscle	0.05–0.61 FW	60
Feathers; 10 species	1.5–30.7 FW	61
Germany; chicks; down; 1991:		
Common tern, *Sterna hirundo*	5.9 (2.7–10.2) FW	62
Black-headed gull	1.0 (0.1–3.6) FW	62
Herring gull	1.4 (0.4–2.9) FW	62
Liver; 9 species	4.9–306.0 DW	63
New Zealand; 64 species; liver:		
Albatrosses; 8 species	17.0–295.0 DW	64
Shearwaters; 3 species	0.8–1.3 DW	64
Petrels; 19 species	0.2–140.0 DW	64
Penguins; 4 species	0.5–2.4 DW	64
Various; 30 species	max. 6.7 DW	64
Little tern, *Sterna albifrons*; feathers; chicks; Portugal		
Small chicks vs. larger chicks; 2000–2002:		
2000	5.2–6.7 FW vs. 4.0–4.2 FW	79
2001	4.0–5.8 FW vs. 3.1–4.3 FW	79
2002	5.4–11.6 FW vs. 1.9–5.5 FW	79
Mourisca estuary; 2002:		
Brood 1	15.7 FW	79
Brood 2	10.8 FW	79
Brood 3	25.1 FW	79
Brood 4	6.1 FW	79
Brood 5	3.3 FW	79
Brood 6	2.8 FW	79
Brood 7	4.6 FW	79

Table 6.5 (continued) Mercury Concentrations in Field Collections of Selected Species of Birds

Species, Tissue, and Other Variables	Concentration (mg/kg)	Ref.[a]
Common tern, *Sterna hirundo*; normal vs. abnormal feather loss:		
Blood	0.4 FW vs. 0.6 FW	77
Liver	1.1 FW vs. 2.2 FW	77
Kidney	1.0 FW vs. 1.6 FW	77
Muscle	0.7 FW vs. 1.2 FW	77
Brain	0.4 FW vs 0.9 FW	77
Feathers	1.3 DW vs. 1.8 DW	77
Atlantic gannet, *Sula bassanus*; liver; dead or dying	5.9–97.7 DW	70
Tree swallow, *Tachycineta bicolor*; eggs; St. Lawrence River Basin; 1991	0.04–0.08 FW	65
Texas; Laguna Madre; eggs; 1993–1994:		
Great blue heron, *Ardea herodias*	0.1 (0.02–0.2) FW	66
Snowy egret, *Egretta thula*	0.1 (0.05–0.2 FW)	66
Tricolored heron, *Egretta tricolor*	0.1 (0.03–0.2) FW	66
Caspian tern, *Sterna caspia*	0.6(0.4–0.8) FW	66
Texas; eggs		
Black skimmer, *Rynchops niger*; Lavaca Bay vs. Laguna Vista	0.5 (0.2–0.8) FW vs. 0.19 (0.05–0.31) FW; nest success lower at Lavaca colony	67
Forster's tern, *Sterna forsteri*; Lavaca Bay vs. San Antonio Bay	0.40 FW vs. 0.22 FW; nesting success similar	67
Mourning dove, *Zenaida macroura*; liver; Eastern United States	0.07–0.67 FW	1

Note: All values are in mg total mercury/kg fresh weight (FW) or dry weight (DW).

[a] Reference: 1, Jenkins, 1980; 2, Solonen and Lodenius, 1984; 3, Wood et al., 1996; 4, Littrell, 1991; 5, Elbert and Anderson, 1998; 6, Lindsay and Dimmick, 1983; 7, Szefer et al., 1993; 8, Newton and Galbraith, 1991; 9, Wolfe and Norman, 1998; 10, Spalding et al., 1994; 11, Norheim et al., 1982; 12, Norheim and Kjos-Hanssen, 1984; 13, Delbekke et al., 1984; 14, Ohlendorf and Harrison, 1986; 15, USNAS, 1978; 16, Fimreite, 1979; 17, Broo and Odsjö, 1981; 18, Mullie et al., 1992; 19, Zicus et al., 1988; 20, Cahill et al., 1998; 21, Thompson et al., 1991; 22, Hoffman et al., 1998; 23, Newton et al., 1993; 24, Scheuhammer et al., 1998a; 25, Scheuhammer et al., 1998b; 26, Pokras et al., 1998; 27, Burger et al., 1994; 28, Evers et al., 1998; 29, Meyer et al., 1995; 30, Meyer et al., 1998; 31, Lindberg, 1984; 32, Beyer et al., 1997; 33, Thompson et al., 1993; 34, Becker et al., 1993; 35, Elliott et al., 1996; 36, Wiemeyer et al., 1984; 37, Wiemeyer et al. ,1993; 38, Wood et al., 1996; 39, Bowerman et al., 1994; 40, Kairu, 1996; 41, Karlog and Clausen, 1983; 42, Lewis et al., 1993; 43, Koster et al., 1996; 44, Cooper, 1983; 45, Braune and Gaskin, 1987; 46, Burger and Gochfeld, 1996; 47, Honda et al,. 1986; 48, Gariboldi et al., 1998; 49, DesGranges et al., 1998; 50, Wiemeyer et al., 1988; 51, Audet et al., 1992; 52, Ohlendorf et al., 1985; 53, Mason et al., 1997; 54, Henny and Herron ,1989; 55, Wenzel et al., 1996; 56, Luke et al., 1989; 57, Thompson et al., 1992; 58, Monteiro et al., 1995; 59, Barrett et al., 1996; 60, Braune, 1987a; 61, Thompson and Furness, 1989; 62, Becker et al., 1994; 63, Kim et al., 1996a; 64, Lock et al., 1992; 65, Bishop et al., 1995; 66, Mora, 1996; 67, King et al., 1991; 68, Tavares et al., 2004; 69, Stickel et al., 1977; 70, Parslow, 1973; 71, Koivusaari et al., 1976; 72, Renzoni et al., 1973; 73, Vannucchi et al., 1978; 74, Wiemeyer et al., 1980; 75, Greichus et al., 1973; 76, Bernhard and Zattera, 1975; 77, Gochfeld, 1980; 78, Monteiro and Furness, 1997; 79, Tavares et al., 2005; 80, Scheifler et al. ,2005; 81, Ikemoto et al., 2004.

western European estuaries and less than 1.0 mg/kg during summer when birds migrated to inland Arctic and sub-Arctic breeding grounds (Parslow, 1973). Concentrations of mercury in livers of Antarctic birds reflected mercury body burdens accumulated during migration, while the birds were overwintering near industrialized areas. Concentrations were highest in species that ate higher trophic levels of prey and were especially pronounced for skuas, *Catharacta* spp.; however, significant inherent interspecies differences were evident (Norheim et al., 1982; Norheim and Kjos-Hanssen, 1984). Birds that feed on aquatic fauna show elevated mercury concentrations in tissues when compared to terrestrial raptors (Johnels and Westermark, 1969; Karppanen and Henriksson, 1970; Parslow, 1973; Greichus et al., 1973). For example, mercury was highest in liver of cormorants *Phalacrocorax* spp. and pelicans *Pelecanus* spp., with concentration factors for mercury of 14 over prey fish in body of cormorants and 6 for pelicans (Greichus et al., 1973).

The recorded value of 97.7 mg Hg/kg dry weight in liver of dead or dying gannets, *Sula bassana*, (Table 6.5) requires explanation. It is possible that mercury accumulations of that magnitude were a contributory factor to death in this instance; however, the main cause of death was attributed to poisoning by polychlorinated biphenyls (Parslow, 1973). Furthermore, large variations in mercury content in gannet liver were linked to liver size (positive correlation) and to fat content (inverse relation) aver Parslow et al. (1973). The highest values observed — 67.5 and 130.0 mg Hg/kg fresh weight in liver and kidney, respectively, of osprey *Pandion haliaetus* (Table 6.5) — are attributed to a single bird. These levels are clearly excessive, reflect high environmental exposure, and are similar to concentrations found in mercury-poisoned birds (Wiemeyer et al., 1980). Of the 18 ospreys examined, except for the aberrant observation, the highest values were 6.2 mg Hg/kg FW liver and 6.5 in kidney.

Eggs of fish-eating birds, including eggs of herons and grebes, collected near mercury point-source discharges contained abnormally high levels of mercury: 29.0% of eggs contained more than 0.5 mg Hg/kg FW, and 9.0% contained more than 1.0 mg Hg/kg FW (Faber and Hickey, 1973). Parslow (1973) concludes that the main source of mercury in estuaries is probably from the direct discharge of effluent from manufacturing and refining industries into rivers. The high levels of mercury detected in eggs of the gannet *Morus bassanus* (Table 6.5) are within the range associated with negative influence on hatchability in pheasants and other sublethal effects in mallard ducks (Fimreite et al., 1980); however, the gannets appear to reproduce normally at these levels. Eggs of the common loon (*Gavia immer)* from Wisconsin in 1993 to 1996 had 0.9 mg Hg/kg FW, which is within the range associated with reproductive failure in sensitive avian species (Meyer et al., 1998).

It is generally acknowledged that feathers contain most of the total body load of mercury, while constituting usually less than 15.0% of the weight (Parslow, 1973). Mercury excretion is mainly via the feathers in both sexes, and also in the eggs (Parslow, 1973). Mercury concentrations in feathers of little tern chicks, *Sterna albifrons*, were higher in smaller chicks than larger chicks and higher in early broods (1 to 3) than later broods (4 to 7), suggesting depletion of maternal transfer of mercury (Table 6.5; Tavares et al., 2005). In Sweden, fish-eating birds had higher levels of mercury in feathers than did terrestrial raptorial species. Ospreys, which prey almost exclusively on larger fish of about 0.3 kg, show higher mercury levels in feathers than grebes, *Podiceps cristata*, which eat smaller fish and insect larvae (Johnels and Westermark, 1969). Since larger fish contain more mercury per unit weight than smaller fish, diet must be considered an important factor to account for differences in mercury concentrations of these two fish-eating species. Based on samples from museum collections, it was demonstrated that mercury content in feathers from fish-eating birds was comparatively low in the years 1815 through 1940. However, since 1940, or the advent of the chloralkali industry (wherein mercury is used as a catalyst in the process to produce sodium hydroxide and chlorine gas from sodium chloride and water, with significant loss of mercury to the biosphere), mercury concentrations in feathers were eight times higher on average (Johnels and Westermark, 1969). Others have reported that mercury levels were elevated in feathers and other tissues of aquatic and fish-eating birds from the vicinity of chloralkali plants (Fimreite et al., 1971; Fimreite, 1974); these increased levels of mercury were detectable up to 300 km from the chloralkali plant (Fimreite and Reynolds, 1973).

Bird feathers have been used for some time as indicators of mercury loadings in terrestrial and marine environments. Feathers represent the major pathway for elimination of mercury in birds, and body feathers are useful for assessment of whole-bird mercury burdens (Furness et al., 1986; Thompson et al., 1990) with almost all mercury present as methylmercury (Thompson and Furness, 1989). The keratin in bird feathers is not easily degradable, and mercury is probably associated firmly with the disulfide bonds of keratin. Consequently, it has been possible to compare mercury contents of feathers recently sampled with those from museum birds, thereby establishing a time series (Applequist et al., 1984; Thompson et al., 1992; Monteiro and Furness, 1995; Odsjo et al., 1997). There is considerable variability in mercury content of seabird feathers. Concentrations in adults were higher than those in chicks and independent of adult age or sex (Thompson et al., 1991), and were lower in spring breeders than in autumn breeders (Monteiro et al., 1995). After the completion of molting, new feathers contained up to 93% of the mercury body burden in gulls (Braune and Gaskin, 1987). The most probable source of elevated mercury residues in feathers of the Finnish sparrowhawk (*Accipiter nisus*) was from consumption of avian granivores that had become contaminated as a result of eating seeds treated with organomercury compounds; in 1981, 5.6 tons of methoxyethylmercury compounds were used in Finnish agriculture for protection of seeds against fungi (Solonen and Lodenius, 1984). Concentrations of mercury in feathers of herring gulls (*Larus argentatus*) from the German North Sea coast were higher in adults than in juveniles and two times higher after 1940 than in earlier years (Thompson et al., 1993). A maximum of 12.0 mg/kg FW in feathers during the 1940s was recorded, and presumed due to high discharges of mercury during the World War II (1939–1945). Concentrations dropped in the 1950s, increased in the 1970s to 10.0 mg Hg/kg FW, before falling in the late 1980s. This pattern correlates well with known discharges of mercury into the Elbe and Rhine (Thompson et al., 1993). Captive Swedish eagle-owls (*Bubo bubo*), with low mercury content in feathers (< 1.0 mg/kg DW), that were introduced into coastal areas quickly reflected the high (6.5 mg/kg) mercury levels in feathers of wild eagle-owls from that region. Captive birds released into inland territories, where mercury levels were near background, did not accumulate mercury in feathers (Broo and Odsjö, 1981). Mercury levels in feathers of nestling Swedish gyrfalcons (*Falco rusticolus*) showed a better correlation with mercury levels in actual food items than with levels based on adult feathers. Mercury concentrations in feathers were higher in nestlings fed partly with aquatic bird species containing more than 0.07 mg Hg/kg in pectoral muscle than in nestlings fed willow grouse (*Lagopus lagopus*) and ptarmigan (*Lagopus mutus*), both of which contained less than 0.01 mg Hg/kg in pectoral muscle (Lindberg, 1984). In some instances there was a substantial time lag, up to 10 years, between the introduction of a pesticide, such as alkylmercury, its subsequent banning, and measurable declines of mercury in feathers of several species of Swedish raptors; this was the case for various species of *Falco, Haliaeetus, Bubo, Buteo*, and *Accipiter* (Wallin, 1984). Accordingly, a reduction in mercury content in feathers of free-living birds may be sufficient to establish an improved situation.

Methylmercury concentrations in feathers from two species of north Atlantic seabirds increased about 4.8% yearly between 1885 and 1994; increases were attributed to global increases in mercury loadings rather than to local or regional sources (Monteiro and Furness, 1997). Mercury concentrations in breast feathers of the king penguin (*Aptenodytes patagonicus*) were significantly lower in 2000 to 2001 (1.98 mg Hg/kg DW) than were feathers collected from the same colony between 1966 and 1974 (2.66 mg/kg DW), and suggests that mercury concentrations in southern-hemisphere seabirds do not increase — which conflicts with trends observed in the northern hemisphere (Scheifler et al., 2005).

Molting is a major excretory pathway for mercury (Honda et al., 1986). Down and feathers were effective excretion routes of mercury in contaminated gull and tern chicks (Becker et al., 1994). Some seabirds demethylate methylmercury in the liver and other tissues, and store mercury as an immobilizable inorganic form in the liver; species with a high degree of demethylation capacity and slow molting pattern had low mercury burdens in feathers (Kim et al., 1996b). Egg laying is an important route in reducing the female's mercury burden, especially the first egg because egg mercury levels decline with laying sequence in gulls and terns (Becker, 1992; Lewis et al., 1993). In gulls and terns, 90% of the mercury in eggs is in the form of methylmercury (Becker et al., 1993). In kittiwakes (*Rissa tridactyla*), mercury concentrations in feathers and tissues of nestlings decreased with increasing age, suggesting that egg contamination was more important in chicks than consumption of mercury-contaminated food items (Wenzel et al., 1996).

Significant downward trends in mercury liver burdens of raptors in England between 1960 and 1990 (Newton et al., 1993) is a useful indicator of the prohibitions placed on mercury discharges in that region. The significance of mercury residues in birds, however, is not yet fully understood.

For example, all eggs of the bald eagle (*Haliaeetus leucocephalus*) collected nationwide (U.S.) contained detectable levels of mercury, but the mean was 0.15 mg Hg/kg (fresh weight basis) in eggs from unsuccessful nests vs. 0.11 in eggs from successful nests (Wiemeyer et al., 1984). Many other contaminants — especially organochlorine compounds — were in eagle eggs, and several were present at levels that potentially interfere with eagle reproduction (Wiemeyer et al., 1984). It is not now possible to implicate mercury as a major cause of unsuccessful eagle reproduction. In the Great Lakes, mercury has no apparent effect on reproduction or nesting success of bald eagles (Bowerman et al., 1994). Livers of 30.0 to 80.0% of some species of wading birds — such as the great blue heron, *Ardea herodias* — contained mercury concentrations greater than 30.0 mg Hg/kg FW; these herons appeared normal although concentrations greater than 30.0 mg Hg/kg FW liver are typically associated with overt neurological signs and reproductive impairment in ducks and pheasants (Sundlof et al., 1994). If reproductive disorders are expected when concentrations in feathers of adult birds approach 9.0 mg total Hg/kg DW (Beyer et al., 1997), then mercury in southern Florida may be sufficiently high to reduce productivity of wading bird populations (Beyer et al., 1997), although this must be verified.

Many factors are known to modify mercury concentrations in tissues of the common loon (*Gavia immer*). Total mercury concentrations were higher in tissues of emaciated loons when compared with apparently healthy birds, and sometimes exceeded 100.0 mg Hg/kg DW in tissues of loons that were in poor condition (Scheuhammer et al., 1998b). There was a strong positive correlation between total mercury and selenium concentrations in livers and kidneys. As total mercury concentrations increased in liver and kidney of loons, the fraction that was methylmercury decreased. Livers and kidneys with the highest total mercury concentrations had only 5.0 to 7.0% of the total as methylmercury. Concentrations of methylmercury were always less than 10.0 mg/kg DW, regardless of total mercury concentration in liver or kidney. In contrast, methylmercury contributed 80.0 to 100.0% of the total mercury in breast muscle, which ranged between 0.7 and 35.0 mg/kg DW (Scheuhammer et al., 1998b). In general, males had higher mercury concentrations in blood and feathers than did their female mates (Scheuhammer et al., 1998a). The possible transfer of mercury to eggs by females during egg laying may account for some of the sexual discrepancy. Adult loons had higher blood mercury concentrations (up to 13 times higher) than their chicks. Adult and chick blood mercury concentrations were correlated with mercury concentrations in their fish diet (Scheuhammer et al., 1998a). Blood mercury concentrations of loon chicks near Wisconsin lakes increased with decreasing lake pH (Meyer et al., 1998).Blood and feather mercury concentrations from the same individuals were correlated, especially in loons with the highest blood mercury levels (Evers et al., 1998). Common loons had aberrant nesting behavior and low reproductive success when mercury concentrations in prey — small fish and crayfish — exceeded 0.3 mg Hg/kg FW, levels known to occur in fish from many lakes in central Ontario; up to 30.0% of Ontario lakes exceeded the mercury threshold for loon reproductive impairment (Scheuhammer and Blancher, 1994; Scheuhammer et al., 1998a). Populations of the common loon are also declining in the northeastern United States and this may be due to mercury, in part, as evidenced by the high concentrations in their feathers (9.7 to 20.2 mg Hg/kg DW), being twice that of other species (Burger et al., 1994).

Mercury–selenium interactions are significant in marine mammals and seem to be a factor in loons; however, this is not the case in marine birds. Mercury concentrations in oceanic birds were not correlated with selenium concentrations, as evidenced by values in livers of murres, *Uria* spp. and razorbills *Alca torda*, and in breast muscle of sooty terns, *Sterna fuscata* (Ohlendorf et al., 1978).

6.6 HUMANS

Elevated total mercury concentrations in various human tissues and body fluids are associated with increasing consumption of fish, use of skin-lightening creams containing mercuric ammonium chloride, and recipients or users of ethylmercury compounds as wound disinfectants; moreover, hair concentrations greater than 3.1 mg methylmercury/kg DW were found in Iraqis who died in 1971 from consuming methylmercury-contaminated wheat (Table 6.6). Additional and more detailed information on mercury concentrations in human tissues is presented in Chapter 10 and Chapter 11.

Urine mercury concentrations in women are associated with age, race, smoking, fish consumption, blood mercury concentrations, and dental fillings (Dye et al., 2005; Table 6.6). Urinary mercury concentrations in this nationwide (U.S.) cohort of women are significantly associated with the increasing number of restored mercury dental amalgam surfaces (Berglund, 1990; Skare and Engqvist, 1994; Kingman et al., 1998; Dye et al., 2005).

Elevated, but not life-threatening, accumulations of mercury in human tissues are common among coastal populations, especially in fishermen who subsist to a large degree on marine fish and shellfish (Skerfring et al., 1970; Ui and Kitamuri, 1971; Establier, 1975; Nuorteva et al., 1975; Table 6.6). Some groups, however, routinely ingest diets containing greatly elevated concentrations of total mercury without apparent effect. For example, pregnant Inuit women living in close proximity to the sea and consuming seal meat and blubber on a regular basis contained greater mercury accumulations in maternal and fetal blood and tissues than did similar populations further from the sea; however, the elevated mercury concentrations reported in Inuit infants from mothers who ate seals or fish every day of their pregnancy were reportedly far below the acknowledged toxic level (Galster, 1976).

Early estimates of mercury exposure in the general population ranged from 23.0 to 78.0 μg/day: 1.0 μg/day from air, about 2.0 μg/day from water, and 20.0 to 73.0 μg/day from food, depending on the amount of fish in the diet (USHEW, 1977). Human intake of total mercury from the diet normally ranges between 7.0 and 16.0 μg daily (Schuhmacher, et al. 1994; Richardson et al., 1995). Fish consumption accounts for much of this exposure in the form of methylmercury: 27.0% of the intake and 40.0% of the absorbed dose. Intake of inorganic mercury arises primarily from foods other than fish and is estimated at 1.8 μg daily, with 0.18 μg absorbed daily (Richardson et al,. 1995). In certain areas of India, blood mercury concentrations of people who ate fish were three to four times higher than nonfish eaters (Srinivasen and Mahajan, 1989). In some countries, mercury in dental amalgams account for 2.8 μg daily, equivalent to as much as 36.0% of the total mercury intake and 42.0% of the absorbed dose (USPHS, 1994; Richardson et al., 1995).

Total blood mercury is considered the most valid biomarker of recent methylmercury exposure (Schober et al., 2003). Some Canadian aboriginal peoples had grossly elevated blood mercury concentrations of greater than 100.0 to 660.0 μg Hg/L, although there was no definitive diagnosis of methylmercury poisoning (Wheatley and Paradis, 1995). And in a study in the United States involving older adults, age 50 to 70 years, with median blood mercury levels of 2.1 μg/L (range 0.0 to 16.0 μg/L), increasing blood mercury was associated with decreasing performance on a test of visual memory; however, increasing blood mercury level was also associated with increasing manual dexterity. The authors concluded that blood mercury concentrations in older adults are not strongly indicative of declining neurobehavioral performance (Weil et al., 2005). Among Greenland (Denmark) residents, mercury intake from maritime foods may be associated with cardiovascular disease (Pedersen et al., 2005). Mercury concentrations in blood of Greenlanders was significantly higher in those consuming traditional Greenland foods (seal, whale) when compared to those consuming European diets (Table 6.6); moreover, blood pressure increased with increasing blood mercury content (Pedersen et al., 2005).

Hair has been proposed as a diagnostic indicator of mercury exposure because it is a recognized excretory pathway, is formed in a relatively short time, and mercury content is unaffected by ongoing metabolic events. In Agra City, India, higher mercury concentrations were found in hair of vegetarians when compared to nonvegetarians; higher in hair of nonsmokers when compared to smokers; and higher in hair of male alcoholics when compared to male nonalcoholics (Sharma et al., 2004), suggesting that plantstuffs may be a primary source of mercury in that population. In

Table 6.6 Mercury Concentrations in Humans from Selected Geographic Areas

Location, Tissue, and Other Variables	Concentration (mg/kg)	Ref.[a]
Canada; aboriginal peoples; 1970–1992; methylmercury:		
Blood; 514 communities; 38,571 individuals:		
23.0%	> 0.02 FW	1
1.6%	> 0.1 FW	1
0.2%	> 0.2 FW	1
Maternal	Max. 0.086 FW	1
Umbilical cord blood; 2405 samples:		
21.8%	> 0.02 FW	1
Maximum	0.224 FW	1
Canada; diet; adults:		
Total intake, in µg daily	7.7 (= 0.11 µg/kg BW daily in 70-kg individual)	2
Absorbed dose, in µg daily	5.3 (= 0.076 µg/kg BW daily in 70-kg individual)	2
Denmark; blood; consuming traditional European diet	0.022 FW	12
Faroe Islands:		
Hair, as function of number of fish meals per week:		
None	0.8 DW; max. 2.0 DW	3
1	1.6 DW; max. 3.7 DW	3
2	2.5 DW; max. 4.7 DW	3
4	5.2 DW; max. 8.0 DW	3
Whole blood; normal vs. high fish consumers	0.001–0.008 FW vs. 0.2 FW	3
Urine; women using skin-lightening creams containing 5.0–10.0% mercuric ammonium chloride; during use vs. discontinued use	0.109 FW vs. 0.006 FW	3
Greenland; blood		
Greenlanders living in Denmark consuming European diet	0.0048 FW	12
Living in Greenland consuming European diet	0.0011 FW	12
Living in Greenland consuming traditional Greenlandic food (seal, whale)	0.0249 FW	12
Hair; total mercury vs. methylmercury:		
Japan; general population (daily intake of 45.6 µg total mercury including 4.8 µg methylmercury)	5.1 FW vs. 2.1 FW	4
Japan; residents near Oyabe River (daily intake of 84.6 µg total mercury including 15.6 µg methylmercury)	No data vs. 6.7 FW	4
Japan; heavy fish eaters living near Oyabe River (daily intake of 194.0 µg total mercury including 55.0 µg methylmercury)	17.2 FW vs. 11.2 FW	4
Japan; crew of tuna fishing boat (daily intake of 119.0 µg total mercury including 42.0 µg methylmercury)	19.9 FW vs. 12.8 FW	4
Japan; Niigata patients afflicted with Minamata disease (daily intake of 1482.0 µg total mercury)	249.0 FW vs. no data	4
U.K.; general population (daily intake of 7.5 µg total mercury)	2.9 FW vs. no data	4
India; Agra City; head hair; ages 6–60; N = 354:		
Males	0.73 (0.0–21.0) DW	5
Females	0.77 (0.0–19.5) DW	5
Vegetarians	3.9 (0.1–28.0) DW	5
Nonvegetarians	2.4 (0.1–180.5) DW	5
Males:		
Smokers	0.6 (0.05–6.9) DW	5
Nonsmokers	5.1 (0.01–13.1) DW	5
Alcoholics	4.5 (0.02–38.7) DW	5
Nonalcoholics	3.8 (0.05–17.5) DW	5
India; Steel City-Bhilai; recent mothers; breast milk vs. blood:		
Employees of steel plant:		
Age 20–25 years	0.0051 FW vs. 0.0004 FW	10
Age 25–30 years	0.0152 FW vs. 0.0067 FW	10
Age 30–35 years	0.0168 FW vs. 0.0078 FW	10
Age 35–40 years	0.0211 FW vs. 0.0123 FW	10
Age 40–45 years	0.0315 vs. 0.0167 FW	10

Table 6.6 (continued) Mercury Concentrations in Humans from Selected Geographic Areas

Location, Tissue, and Other Variables	Concentration (mg/kg)	Ref.[a]
Nonemployees, but residents of Steel City:		
Age 20–25 years	0.0008 FW vs. 0.0004 FW	10
Age 25–30 years	0.0016 FW vs. 0.0009 FW	10
Age 30–35 years	0.0058 FW vs. 0.0024 FW	10
Age 35–40 years	0.0063 FW vs. 0.0032 FW	10
Age 40–45 years	0.0028 FW vs. 0.0067 FW	10
Reference site, 100 km distant:		
Age 20–25 years	0.0003 FW vs. nondetectable	10
Age 25–30 years	Nondetectable vs. 0.0001 FW	10
Age 30–35 years	Nondetectable vs. nondetectable	10
Age 35–40 years	0.0008 FW vs. 0.0009 FW	10
Age 40–45 years	0.0009 FW vs . 0.0005 FW	10
Iraq; 1971; major poisoning incident from consumption of methylmercury-contaminated wheat (> 6000 cases); blood methylmercury:		
42.0% of cases with paresthesia; 21.0% with affected vision; 11.0% with ataxia	0.5–1.0 FW	6
60.0% with paresthesia; 53.0% with visual changes; 47.0% with ataxia; 24.0% with dysarthria	1.1–2.0 FW	6
79.0% with paresthesia; 60.0% with ataxia; 56.0% with visual changes; 25.0% with dysarthria; 13.0% with hearing defects; no deaths	2.1–3.0 FW	6
17.0% dead	3.1–4.0 FW	6
28.0% dead	4.1–5.0 FW	6
Japan; healthy adults; total mercury vs. methylmercury:		
Liver	0.64 FW vs. 0.071 FW	7
Kidney	1.02 FW vs. 0.015 FW	7
Cerebrum	0.076 FW vs. 0.009 FW	7
Cerebellum	0.078 FW vs. 0.009 FW	7
Mexico; head hair; hospital medical staff; 1991–1992:		
Toluca, Mexico; widespread use of ethylmercury compound used as wound disinfectant	32.9 (2.5–143.0) FW	8
Mexico City, Mexico; declining use of ethylmercury disinfectant	1.2 FW; Max. 12.2 FW	8
U.S.; nationwide; head hair; 1999–2000; means:		
Children	0.12 DW	9
Women, age 16–19 years	0.20 DW	9
Frequent fish consumers	0.38 DW	9
All women	12.0% with > 1.0 DW (USEPA reference dose)	9
Tunafish servings per month:		9
0–1	16.2% with > 1.0 DW	9
2–3	22.6% with > 1.0 DW	9
4+	32.9% with > 1.9 DW	9
Fish servings per month (at home or in restaurant):		
0–2	8.5% with > 1.0 DW	9
3–7	29.0% with > 1.0 DW	9
7+	50.6% with > 1.0 DW	9
Age in years:		
0–19	4.1% with > 1.0 DW	9
20–40	22.7% with > 1.0 DW	9
50+	22.0% with > 1.0 DW	9
Southeast, Midwest, District of Columbia	9.0–19.1% with > 1.0 DW	9
Northeast, West, California, Florida	26.9–34.2% with > 1.0 DW	9
New York	45.2% with > 1.0 DW	9
U.S.; nationwide; 1999–2000; women, aged 16–49 years; mean total mercury in blood (μg/L) vs. mean total mercury in urine (μg/L):		
Total	2.02 μg/L vs. 1.34 μg/L	11

(continued)

Table 6.6 (continued) Mercury Concentrations in Humans from Selected Geographic Areas

Location, Tissue, and Other Variables	Concentration (mg/kg)	Ref.[a]
Age:		
35 to 49 years	2.29 μg/L vs. 1.36 μg/L	11
16 to 19 years	1.15 μg/L vs. 1.18 μg/L	11
Pregnant:		
Yes	1.78 μg/L vs. 1.28 μg/L	11
No	2.04 μg/L vs. 1.34 μg/L	11
Smoking status:		
Smoker	1.56 μg/L vs. 0.98 μg/L	11
Nonsmoker	2.16 μg/L vs. 1.45 μg/L	11
Alcohol consumption:		
Moderate/heavy drinker	3.10 μg/L vs. 1.14 μg/L	11
Abstainer	1.63 μg/L vs. 1.48 μg/L	11
Fish/shellfish consumed in past 30 days:		
Yes	2.37 μg/L vs. 1.43 μg/L	11
No	0.67 μg/L vs. 1.07 μg/L	11
Dental visit in past 12 months:		
Yes	2.22 μg/L vs. 1.37 μg/L	11
No	1.52 μg/L vs. 1.25 μg/L	11

Note: Values are in mg total mercury/kg fresh weight (FW) or dry weight (DW), unless indicated otherwise.

[a] Reference: 1, Wheatley and Paradis, 1995; 2, Richardson et al., 1995; 3, U.S. Public Health Service, 1994; 4, Takizawa, 1993; 5, Sharma et al., 2004; 6, Takizawa, 1979a; 7, Eto et al., 1992; 8, Ohno et al., 2004; 9, Maas et al., 2004; 10, Sharma and Pervez, 2004b; 11, Dye et al., 2005; 12, Pedersen et al., 2005.

Kenya, habitual use of skin-lightening soaps containing elevated concentrations of inorganic mercury was associated with high residues of mercury in hair, tremors, lassitude, neurasthenia, and other symptoms of inorganic mercury poisoning (Harada et al., 2001). The U.S. Environmental Protection Agency proposed a Reference Dose of 1.0 mg total Hg/kg DW hair as indicative of mercury exposure, and the concentration at which women of child-bearing age are advised to stop consumption of fish that might have elevated mercury levels (Maas et al., 2004). A survey shows that hair levels in excess of 1.0 mg Hg/kg DW are related to increasing fish consumption, to age over 20 years, and to certain geographic areas, especially New York and Florida (Table 6.6).

Recent mothers who were employees of the largest steel producing plant in India, located in Steel City-Bhilai, had elevated concentrations of total mercury in breast milk and blood when compared to recent mothers who did not work at the plant but were residents of Steel City. Both groups had higher levels than recent mothers from a reference site 100 km distant (Table 6.6; Sharma and Pervez, 2004b). In Steel City, mercury concentrations in breast milk and blood increased with increasing age of the mother (Table 6.6), and this can be related to increasing respiratory exposure to mercury-contaminated dust (Sharma and Pervez, 2004a).

6.7 OTHER MAMMALS

Among nonhuman mammals, marine pinnipeds contained the highest reported concentrations of mercury in tissues (Table 6.7). Total mercury content in all tissues examined of marine mammals — including muscle, brain, blubber, kidney, and liver — generally increased with increasing age of the animal (Gaskin et al., 1972; Freeman and Horne, 1973c; Holden, 1975; Harms et al., 1978; Smith and Armstrong, 1978; Eisler, 1984). This was especially pronounced in livers of northern fur seals, *Callorhinus ursinus* (Anas, 1974); grey seals, *Halichoerus grypus* (Sergeant and Armstrong, 1973); California sea lions, *Zalophus californianus* (Martin et al., 1976); harp seals, *Pagophilus groenlandica* (Freeman and Horne, 1973c; Figure 6.1); and in various tissues of the Baikal

Table 6.7 Mercury Concentrations in Field Collections of Selected Species of Mammals

Organism, Tissue, and Other Variables	Concentration (mg/kg)	Ref.[a]
Antarctica; February–March; 1989; marine mammals:		
Leopard seal, *Hydrurga leptonyx*:		
Kidney	Max. 6.1 DW	1
Liver	Max. 18.1 DW	1
Muscle	Max. 3.2 DW	1
Stomach contents	Max. 1.2 DW	1
Weddell seal, *Leptonchyotes weddellii*:		
Kidney	Max. 15.9 DW	1
Liver	Max. 48.8 DW	1
Muscle	Max. 3.6 DW	1
Crabeater seal, *Lobodon carcinophagus*:		
Kidney	Max. 12.5 DW	1
Liver	Max. 16.3 DW	1
Muscle	Max. 6.2 DW	1
Australian fur seal, *Arctocephalus pusillus*:		
Muscle	0.9 (0.1–1.9) FW	2
Liver	62.0 (1.0–170.0) FW	2
Kidney	0.6 (0.1–1.7) FW	2
Spleen	1.3 (0.0–3.8) FW	2
Brain	0.7 (0.0–2.5) FW	2
Fur	9.6 (1.1–19.8) FW	2
Woodmouse, *Apodemus sylvaticus*; U.K.:		
From field with Hg-treated wheat seed:		
Liver	Max. 7.1 FW	3
Kidney	Max. 11.7 FW	3
From chloralkali plant area:		
Liver	Max. 0.5 FW	3
Kidney	Max. 1.3 FW	3
From control site:		
Liver	Max. 0.07 FW	3
Kidney	Max. 0.3 FW	3
Fin whale, *Balaenoptera physalus*; Spain and Iceland; 1983–1986; total vs. organic mercury; maximum concentrations		
Kidney	3.3 FW vs. 0.4 FW	4
Liver	5.4 FW vs. 1.4 FW	4
Muscle	1.2 FW vs. 0.9 FW	4
Northern fur seal, *Callorhinus ursunis:*		
Adult males vs. adult females:		
Liver	3.0–19.0 FW vs. 7.0–172.0 FW	40
Muscle	0.1–0.4 FW vs. 0.2–0.4 FW	40
Kidney	0.7 FW vs. 0.6–1.6 FW	40
Pups; liver vs. muscle	0.1–0.3 FW vs. 0.1 FW	40
Fetus; liver vs. kidney	0.4 FW vs. 0.2 FW	40
Nursing cows:		
Fur	4.9 FW	41
Blood	0.1 FW	41
Milk	0.014 FW	41
Newborn pups:		
Fur	3.7 FW	41
Blood	0.02 FW	41
Lice, *Antarctophthirus callorhini*; whole	0.22 FW	41
Pups; age 2 months:		
Fur	5.4 FW	41
Blood	0.07 FW	41
Lice	0.51–0.63 FW	41

(continued)

Table 6.7 (continued) Mercury Concentrations in Field Collections of Selected Species of Mammals

Organism, Tissue, and Other Variables	Concentration (mg/kg)	Ref.[a]
Sanriku, Japan; 1997–1998:		
Liver, whole	59.0 (7.6–121.0) FW	10
Liver nuclei, lysosomes, and mitochondria	39.0 (3.9–83.0) FW	10
Roe deer, *Capreolus capreolus*; males; Poland; 1977–1978; mercury-contaminated habitat vs. reference site:		
Muscle	0.047 FW vs. 0.013 FW	5
Liver	0.036 FW vs. 0.015 FW	5
Kidney	0.053 FW vs.0.027 FW	5
Serow (feral bovine ruminant), *Capricornis crispus*; Japan; 1981–1983:		
Fleece:		
Fawns	0.37 FW	6
Yearlings	0.38 FW	6
Adults, up to age 10 years	0.35 FW	6
Adults, age 10–17.5 years	0.44 FW	6
Other tissues	Always < 0.02 FW	6
Beaver, *Castor canadensis*; Wisconsin; 1972–1975; all tissues	< 0.09 FW	7
Red deer, *Cervus elaphus*; East Slovakia; muscle	0.1 (0.03–0.4) FW	8
Dugong, *Dugong dugon*; Australia; liver:		
Queensland, 1996–2000:		
Mature	0.3 (0.05–1.11) DW	11
Immature	0.09 (0.04–0.28) DW	11
Northern Australia	0.24 DW	62, 63
Torres Straits	< 0.1–0.22 DW	12, 61
Bearded seal, *Erignathus barbatus*; various Arctic locales:		
Barrow Strait; liver; adult vs. adolescents	79.2 FW vs. 9.4 FW	42
W. Victoria Island; mean age 8.5 years:		
Liver; total mercury vs. methylmercury	143.0 FW vs. 0.3 FW	42
Muscle	0.5 FW	42
Belcher Island; age 4.9 years		
Liver; total mercury vs. methylmercury	26.2 FW vs. 0.12 FW	42
Muscle	0.09 FW	42
Florida panther, *Felis concolor coryi:*		
1989; found dead:		
Blood	21.0 FW	9
Fur	130.0 FW	9
Liver	110.0 FW	9
1978–1991; liver:		
Southeast FL vs. Southwest FL; young panthers	25.8 FW vs. 0.3 FW	9
Southwest FL; older vs. younger panthers	14.6 FW vs. 0.3 FW	9
Whole blood; females:		
1.46 kittens per female per year	0.0–0.25 FW	9
0.167 kittens per female per year	> 0.5 FW	9
Domestic cat, *Felis domesticus*; ate fish from below chloralkali plant; NW Ontario:		
Brain	6.9–16.4 FW	3
Pancreas	4.3–4.9 FW	3
Kidney	0.8–13.4 FW	3
Liver	14.2–67.1 FW	3
Fur	121.0–392.0 FW	3
Pilot whale, *Globicephala macrorhynchus*:		
Liver	19.0–1570.0 FW	43
Liver	57.0–454.0 FW	44
Muscle	3.0–5.0 FW	43
Kidney	6.0–56.0 FW	43
Blubber	Max. 2.4 FW	44

Table 6.7 (continued) Mercury Concentrations in Field Collections of Selected Species of Mammals

Organism, Tissue, and Other Variables	Concentration (mg/kg)	Ref.[a]
Wolverine, *Gulo luscus*; British Columbia, Canada; winter 1997; liver	0.18 DW; 0.05 FW	37
Grey seal, *Halichoerus grypus*:		
Adult males vs. adult females:		
Fur	1.4–12.0 FW vs. 3.1–16.0 FW	45
Claw	5.0–9.8 FW vs. 8.4–8.6 FW	45
Liver	10.0–30.0 FW vs. 11.0–26.0 FW	45
Kidney	3.0–5.7 FW vs. 2.8–5.0 FW	45
Flipper	0.9 FW vs. 0.9 FW	45
Muscle	0.9–1.6 FW vs. 0.9–1.6 FW	45
Heart	0.4–0.8 FW vs. 0.4–0.7 FW	45
Gonad	0.2–0.4 Fw vs. 0.3–0.6 FW	45
Blubber	0.06–0.09 FW vs. 0.1 FW	45
Brain	0.3–0.4 FW vs. 0.3–0.4 FW	45
Adults vs. pups		
Blubber	0.04–0.16 FW vs. 0.02–0.06 FW	46
Muscle	0.72–2.4 FW vs. 0.17–0.50 FW	46
Liver	14.3–387.0 FW vs. 0.46–1.2 FW	46
Maternal females; inorganic mercury vs. methylmercury:		
Liver	15.0–127.0 FW vs. 3.2–28.0 FW	47
Bile	0.05–0.22 FW vs. 0.02–0.08 FW	47
Pups; liver; inorganic mercury vs. methylmercury	0.05–2.3 FW vs. 0.025–2.0 FW	47
Juveniles found dead; inorganic mercury vs. methylmercury:		
Brain	0.03–0.35 FW vs. 0.11–0.85 FW	47
Blubber	0.03–0.11 FW vs. 0.36–1.0 FW	47
Muscle	0.09–1.6 FW vs. 0.31–2.8 FW	47
Spleen	0.06–3.6 FW vs. 0.25–1.7 FW	47
Kidney	0.68–4.9 FW vs. 0.4–4.2 FW	47
Liver	0.65–60.0 FW vs. 0.65–14.0 FW	47
Adults found dead; inorganic mercury vs. methylmercury:		
Brain	0.07–17.0 FW vs. 0.2–4.2 FW	47
Blubber	0.07–0.25 FW vs. 0.15–2.3 FW	47
Muscle	1.3 FW vs. 2.5 FW	47
Spleen	18.0 FW vs. 7.8 FW	47
Kidney	2.5–14.0 FW vs. 0.7–7.8 FW	47
Liver	16.0–250.0 FW vs. 4.3–62.0 FW	47
Faroe Islands; summers 1993–1995:		
Males; mature vs. immatures:		
Liver	123.0 (46.0–199.0) FW vs. 10.0 (2.0–65.0) FW	39
Kidney	8.0 (3.0–16.0) FW vs. 2.0 (0.4–5.0) FW	39
Muscle	2.0 (0.6–4.6) FW vs. 0.5 (0.2–1.2) FW	39
Females; mature vs. immatures:		
Liver	133.0 (34.0–238.0) FW vs. 14.0 (1.0–90.0) FW	39
Kidney	4.0 (1.0–7.0) FW vs. 1.0 (0.6–4.0) FW	39
Muscle	1.0 (0.3–2.6) FW vs. 0.4 (0.2–1.0) FW	39
Liver	66.0 FW; 224.8 DW	48
Kidney	4.8 FW; 23.5 DW	48
Spleen	0.7 FW; 2.7 DW	48
Blubber	2.7 FW; 3.1 DW	48
Skin and hair	4.4 FW; 8.9 DW	48
White-beaked dolphin, *Lagenorhynchus albirostris*:		
Fascical fat	1.3 FW	50
Blubber	0.9 FW	50
Liver	19.0 FW	50

(continued)

Table 6.7 (continued) Mercury Concentrations in Field Collections of Selected Species of Mammals

Organism, Tissue, and Other Variables	Concentration (mg/kg)	Ref.[a]
Muscle	2.0 FW	50
Testicles	0.7 FW	50
Kidney	1.6 FW	50
Spleen	1.2 FW	50
Brain	3.0 FW	50
River otter, *Lutra canadensis*:		
Canada; Manitoba; Winnipeg River; 1979–1981; males vs. females; maximum values:		
Liver	8.9 FW vs. 3.9 FW	13
Kidney	6.5 FW vs. 1.8 FW	13
Brain	3.1 FW vs. 0.6 FW	13
Canada; Ontario; 1983–1985; trapped; English River vs. Sudbury:		
Brain	3.2 FW vs. 0.2 FW	14
Kidney	3.5 FW vs. 0.6 FW	14
Liver	3.5 FW vs. 0.9 FW	14
Muscle	1.1 FW vs. 0.3 FW	14
Canada; Ontario::		
Fur	9.6 (4.0–20.0) DW	65
Found dead; liver vs. kidney	96.0 FW vs. 58.0 FW	72
Canada; liver	1.6 FW	71
Georgia; 1976–1977; lower coastal plain vs. Piedmont:		
Fur	24.2 DW vs. 15.2 DW	15
Liver	7.5 FW vs. no data	15
Muscle	4.4 FW vs. 1.5 FW	15
Maine; fur	20.3 (1.1–33.7) DW	66
Michigan; 1987–1989; kidney vs. liver	1.5 (< 0.3–6.2) DW vs. 2.2 (< 0.3–7.0) DW	16
New York; 1982–1984; liver	1.3–2.2 FW	17
Wisconsin; 1972–1975:		
Brain	0.7 FW	7
Muscle	1.4 FW	7
Liver	3.3 FW; max. 24.0 FW	7
Kidney	8.5 FW; max. 21.0 FW	7
Fur; industrial area vs. nonindustrial area	9.5 FW; max. 63.0 FW vs. 3.8 FW	7
European otter, *Lutra lutra:*		
Liver		
Denmark; 1980–1990		
Adults	4.0 (0.9–12.0) FW	18
Juveniles	0.3 (0.03–2.0) FW	18
Subadults	1.6 (0.2–6.0) FW	18
Hungary	0.6 FW	69
Austria	1.0 FW	69
Scotland	4.7 FW	70
Fur:		
Finland	18.5 (0.7–61.3) DW	67
U.K.	18.7 (1.3–85.1) DW	68
Muscle; U.K.	1.6 FW	68
Bobcat, *Lynx rufus*; fur; Georgia, U.S.; upper coastal plain vs. lower coastal plain	13.1 DW vs. 0.9 DW	3
Marten, *Martes americana*; British Columbia, Canada; winters 1994–1996; kidney	1.0 DW; 0.31 FW	37
Mink, *Mustela vison*:		
Canada; Ontario; 1983–1985; English River vs. Sudbury		
Brain	0.5 FW vs. 0.4 FW	14
Kidney	2.1 FW vs. 0.6 FW	14
Liver	2.0 FW vs. 0.4 FW	14
Muscle	0.8 FW vs. 0.4 FW	14

Table 6.7 (continued) Mercury Concentrations in Field Collections of Selected Species of Mammals

Organism, Tissue, and Other Variables	Concentration (mg/kg)	Ref.[a]
Canada; Manitoba; Winnipeg River; 1979–1981; males vs. females; maximum values:		
Liver	9.9 FW vs. 10.7 FW	13
Kidney	6.4 FW vs. 8.1 FW	13
Brain	2.4 FW vs. 2.1 FW	13
Canada; eastern Quebec:		
Kidney	36.0 FW	38
Liver	64.0 FW	38
New York; 1982–1984; eight locations; liver	0.0–3.0 (0.6–6.0) FW	17
Tennessee; Oak Ridge vs. reference site; fur; adults:	104.0 DW vs. max. 14.7 DW	19
Wisconsin; 1972–1975:		
Brain	0.5 FW	7
Muscle	1.3 FW	7
Liver	2.1 FW; max. 17.0 FW	7
Kidney	2.3 FW; max. 12.0 FW	7
Fur; industrialized area vs. nonindustrialized area	10.5 FW; max. 41.0 FW vs. 3.0 FW	7
Norway; winter 1989–1990; Arctic coast; marine mammals; four species; maximum values:		
Grey seal, *Halichoerus grypus:*		
Brain	2.0 FW	20
Kidney	16.0 FW	20
Liver	48.3 FW	20
Harp seal, *Pagophilus groenlandica*:		
Brain	0.1 FW	20
Kidney	0.4 FW	20
Liver	1.1 FW	20
Ringed seal, *Phoca hispida:*		
Brain	0.4 FW	20
Kidney	0.5 FW	20
Liver	0.7 FW	20
Harbor seal, *Phoca vitulina:*		
Brain	2.0 FW	20
Kidney	8.7 FW	20
Liver	16.0 FW	20
Dall's porpoise, *Phocoenoides dalli*; Sanriku, Japan; 1997–1998:		
Liver, whole	3.8 (0.6–7.1) FW	10
Liver nuclei, lysosomes, and mitochondria	1.7 (0.2–4.2) FW	10
White-tailed deer, *Odocoileus virginianus*; Alabama; 1992–1993; kidney vs. liver	0.5 (0.3–0.7) FW vs. 0.1 (0.05–0.1) FW	21
Muskrat, *Ondatra zibethicus*:		
Tennessee; Oak Ridge vs. reference sites; fur:		
Adults	4.0 (0.8–23.0) DW vs. 0.1–0.2 (0.03–0.6) DW	19
Juveniles	1.6 DW vs. max. 0.2 DW	19
Wisconsin; 1972–1975; all tissues	< 0.06 FW	7
Sheep, *Ovis aries*; grazing for 23 months on Hg-contaminated field:		
Diet (grass); winter vs. summer	6.5 DW vs. 1.9 DW	22
Lung	max. 4.0 FW	22
Kidney	max. 3.1 FW	22
Liver	max. 2.4 FW	22
Brain	max. 1.1 FW	22
Muscle	< 1.0 FW	22
Harp seal, *Pagophilus groenlandica*:		
Muscle	0.34 FW	49
Liver	5.1 FW	49

(continued)

Table 6.7 (continued) Mercury Concentrations in Field Collections of Selected Species of Mammals

Organism, Tissue, and Other Variables	Concentration (mg/kg)	Ref.[a]
Adult females vs. pups:		
Fur	3.2 FW vs. 1.7 FW	45
Claws	3.7 FW vs. 1.8 FW	45
Liver	4.6 FW vs. 0.5 FW	45
Flipper	0.5 FW vs. 0.2 FW	45
Muscle	0.5 FW vs. 0.2 FW	45
Heart	0.3 FW. vs. 0.2 FW	45
Gulf of St. Lawrence; 1984; mother vs. pup less than 3 weeks old:		
Kidney	0.8 FW; 3.5 DW vs. 0.29 FW; 1.2 DW	23
Liver	10.4 FW; 34.7 DW vs. 0.32 FW; 1.1 DW	23
Muscle	0.38 FW; 1.3 DW vs. 0.14 FW; 0.51 DW	23
Mother's milk	0.0065 (0.0026–0.010) FW	23
Ringed seal, *Phoca hispida*:		
Various Arctic locales		
Aston Bay; age unknown; liver vs. muscle	19.3 FW vs. 0.44 FW	42
Barrow Strait; age 10 years:		
Liver; total mercury vs. methylmercury	16.1 FW vs. 0.9 FW	42
Muscle	0.9 FW	42
Pond Inlet; age 5 years:		42
Liver; total mercury vs. methylmercury	3.8 FW vs. 0.5 FW	42
Muscle	0.17 FW	42
Cape Parry; age 1.3 years; liver vs. muscle	1.0 FW vs. 0.11 FW	42
W. Victoria Island; age 13 years:		
Liver; total mercury vs. methylmercury	27.5 FW vs. 1.0 FW	42
Muscle	0.7 FW	42
Claw	1.1–3.7 FW	45
Liver	27.5 FW	52
Liver	14.0–300.0 FW	53
Liver	0.64 FW	54
Liver	210.0 FW	49
Muscle	0.72 FW	52
Kidney	2.8–5.2 FW	53
Saimaa ringed seal, *Phoca hispida saimensis:*		
Muscle	1.3–6.1 FW	53
Liver	72.0–210.0 FW	53
Kidney	1.9–13.0 FW	53
Blubber	0.14–0.46 FW	53
Baikal seal, *Phoca sibirica*; Lake Baikal; Russia; 1992:		
All seals:		
Kidney	1.8 (0.6–3.6) FW	24
Liver	2.3 (0.2–9.1) FW	24
Muscle	0.2 (0.1–0.7) FW	24
Male; age 19.5 years:		
Brain, intestine, fat, bone, skin	0.01–0.1 FW	24
Liver, pancreas, spleen, lung, stomach, heart, muscle, whole	0.11–1.0 FW	24
Fur	4.3 FW	24
Female; age 13.5 years:		
Liver	1.2 FW	24
Kidney	1.7 FW	24
Fur	3.0 FW	24
Other tissues	< 1.0 FW	23
Juveniles vs. adults:		
Fur:	2.1 FW vs. 2.5 FW	25
Kidney	1.0 FW vs. 2.1 FW	25
Liver	0.7 FW vs. 2.8 FW	25
Muscle	0.1 FW vs. 0.3 FW	25

Table 6.7 (continued) Mercury Concentrations in Field Collections of Selected Species of Mammals

Organism, Tissue, and Other Variables	Concentration (mg/kg)	Ref.[a]
Harbor seal, *Phoca vitulina*:		
Liver:		
California	269.0 (81.0–700.0) FW	3, 40
Oregon	0.3–68.0 FW	3, 40
Washington	1.3–60.0 FW	3, 40
Pribilof Islands	0.6–9.0 FW	3, 40
Maine	0.5–7.9 FW	57
New Brunswick, Canada	1.7–50.9	57
Found dead	225.0–765.0 FW	55
Liver	1.5–160.0 FW	54
Liver	257.0–326.0 FW	55
Liver	110.0 FW	56
Liver	8.9 FW	46
Blubber	0.04 FW	46
Muscle	1.0–10.0 FW	54
Muscle	0.71 FW	46
Fur	1.6 FW	46
Harbor porpoise, *Phocoena phocoena*:		
North Sea; 1987–1990; maximum values; juveniles vs. adults:		
Kidney	6.0 DW vs. 23.0 DW	26
Liver	6.0 DW vs. 504.0 DW	26
Muscle	3.0 DW vs. 24 DW	26
Males vs. females:		
Muscle	0.8 (0.2–1.9) FW vs. 1.0 (0.3–2.6) FW	58
Liver	0.9–18.3 FW vs. 0.6–91.3 FW	58
Blubber	0.7 (0.5–0.9) FW	50
Liver	22.0 (1.5–69.0) FW	50
Muscle	1.9 (0.8–3.2) FW	50
Muscle	0.16 FW	54
Liver	28.0 FW	54
Sperm whale; *Physeter macrocephalus*; southern Australia; 1976; muscle:		
Breeding vs. nonbreeding females	10 (8–12) DW vs. 6 (0.8–10) DW	27
All females	7 (0.8–12) DW	27
Males	6 (0.9–12) DW	27
Raccoon, *Procyon lotor*:		
Alabama; 1992–1993; kidney vs. liver	0.24 FW vs. 0.41 FW	28
California; Clear Lake; 1993; distance from abandoned mercury mine; 8 km vs. 23 km:		
Blood	0.4 FW vs. 0.2 FW	29
Fur	22.0 FW vs. 4.0 FW	29
Liver	3.3 FW vs. 7.0 FW	29
Wisconsin; 1975:		
Brain	< 0.02 FW	7
Muscle	0.08 FW	7
Kidney	1.4 FW	7
Liver	2.0 FW	7
Fur	3.8 FW	7
Giant otter, *Pteronura brasiliensis*; Rio Negro, Brazil; found dead; 2002–2003:		
Fur	2.9–3.7 DW	64
Kidney	1.1–4.6 FW	64
Liver	1.5–4.3 FW	64
Muscle	0.17 FW	64
Gray squirrel, *Sciurus carolinensis*; fur; Florida; 1974:		
Rural areas	0.43 FW	3
Urban areas; age 0–1 years vs. age > 2 years	1.0 (0.1–7.0) FW vs. 3 (0.3–9.0) FW	3

(continued)

Table 6.7 (continued) Mercury Concentrations in Field Collections of Selected Species of Mammals

Organism, Tissue, and Other Variables	Concentration (mg/kg)	Ref.[a]
Seals, two species; body length 84–114 cm vs. body length 155–254 cm:		
Liver	1.6 FW vs. 178.0 FW	51
Brain	0.14 FW vs. 0.34 FW	51
Blubber	0.03 FW vs. 0.05 FW	51
Dolphin, *Stenella attenuata*; eastern tropical Pacific Ocean; 1977–1985:		
Blood	0.4 FW	30
Brain	2.0 FW	30
Blubber	7.6 FW	30
Kidney	5.6 FW	30
Liver	62.3 FW; max. 217.5 FW	30
Muscle	2.2 FW	30
Pancreas	6.6 FW	30
Striped dolphin, *Stenella coeruleoalba*:		
Adults; Japan; 1977–1980; total mercury vs. methylmercury:		
Muscle	15.2 FW vs. 5.3 FW	31
Liver	205.0 FW vs. 7.0 FW	31
Kidney	14.7 FW vs. 3.2 FW	31
Whole body:		
Age 1 year	0.8 FW vs. 0.4 FW	32
Age 3 years	1.8 FW vs. 1.0 FW	32
Age 4 years	3.0 Fw vs. 1.5 FW	32
Age 14 years	4.5 FW vs. 2.6 FW	32
Age 20 years	10.7 FW vs. 3.5 FW	32
Found stranded on French Atlantic coast vs. Mediterranean coasts; 1972–1980:		
Intestine	Max. 23.6 FW	33
Kidney	7.0 FW vs. 30.0 FW; max. 179.0 FW	30, 33
Liver	52.0 FW vs. 346.0 FW; max. 1,544.0 FW	30, 33
Melon fat	0.5 FW vs. 2.0 FW	30
Muscle	4.0 FW vs. 28.0 FW; max. 81.0 FW	30, 33
Stomach	Max. 32.0 FW	33
Wild boar, *Sus scrofa scrofa*; East Slovakia; muscle	0.02 (0.0–0.1) FW	8
Red fox, *Vulpes vulpes*:		
Georgia; U.S.; fur; upper coastal plain vs. lower coastal plain	2.3 DW vs. 0.5 DW	3
Wisconsin; 1972–1975; fur vs. other tissues	0.6 FW vs. < 0.14 FW	7
Brown bear, *Ursus arctos*; Slovak Republic; 1988–1990:		
Fat	Max. 0.06 FW	35
Kidney	0.2 FW; max. 0.9 FW	35
Liver	0.04 FW; max. 0.8 FW	35
Muscle	0.004 FW; max. 0.04 FW	35
Polar bear, *Ursus maritimus*:		
Alaska; 1972; total mercury; young vs. adults		
Northern area:		
Liver	22.4 FW vs. 38.1 FW	36
Muscle	0.15 FW vs. 0.19 FW	36
Western area:		
Liver	3.9 FW vs. 4.8 FW	36
Muscle	0.04 FW vs. 0.04 FW	36
Greenland; adults; fur		
NW Greenland; 1978–1989	8.0 (5.0–14.0) DW	34
Eastern Greenland; 1984–1989	4.6 (2.5–8.8) DW	34
Svalbard; 1980; recently molted	2.0 (1.0–4.6) DW	34
Whales; 3 species; muscle	0.06–0.97 FW	49

Table 6.7 (continued) Mercury Concentrations in Field Collections of Selected Species of Mammals

Organism, Tissue, and Other Variables	Concentration (mg/kg)	Ref.[a]
California sea lion, *Zalophus californianus:*		
Mother vs. pup:		
Liver	73.0–1,026 DW vs. 0.9–16.0 DW	3
Kidney	4.0–43.0 DW vs. 0.6–6.7 DW	3
Healthy animals vs. sick animals (leptospirosis):		
Liver	74.1 FW vs. 161.3 FW	59
Muscle	1.2 FW vs. 1.6 FW	59
Mothers with normal pups vs. mothers with premature pups:		
Liver	747.0 DW vs. 204.0 DW	60
Kidney	28.4 DW vs. 7.1 DW	60
Pups; normal vs. born prematurely:		
Liver	9.6 DW vs. 1.8 DW	60
Kidney	4.6 DW vs. 0.9 DW	60

Note: Values are in mg total mercury/kg fresh weight (FW) or dry weight (DW), unless indicated otherwise.

[a] Reference: 1, Szefer et al., 1993; 2, Bacher, 1985; 3, Jenkins, 1980; 4, Sanpera et al., 1993; 5, Krynski et al., 1982; 6, Honda et al., 1987; 7, Sheffy and St. Amant, 1982; 8, Kacmar and Legath, 1991; 9, Roelke et al., 1991; 10, Ikemoto et al., 2004; 11, Haynes et al., 2005; 12, Haynes and Kwan, 2001; 13, Kucera, 1983; 14, Wren et al., 1986; 15, Halbrook et al., 1994; 16, Ropek and Neely, 1993; 17, Foley et al., 1988; 18, Mason and Madsen, 1992; 19, Stevens et al., 1997; 20, Skaare et al., 1994; 21, Khan and Forester, 1995; 22, Edwards and Pumphery, 1982; 23, Wagemann et al., 1988; 24, Watanabe et al., 1996; 25, Watanabe et al., 1998; 26, Joiris et al., 1991; 27, Cannella and Kitchener, 1992; 28, Khan et al., 1995; 29, Wolfe and Norman, 1998; 30, Andre et al., 1991a; 31, Itano et al., 1984a; 32, Itano et al., 1984b; 33, Andre et al., 1991b; 34, Born et al., 1991; 35, Zilincar et al., 1992; 36, Lentfer and Galster, 1987; 37, Harding, 2004; 38, Fortin et al., 2001; 39, Bustamente et al., 2004; 40, Anas, 1974; 41, Kim et al., 1974; 42, Smith and Armstrong, 1978; 43, Gaskin et al., 1974; 44, Stoneburner, 1978; 45, Freeman and Horne, 1973c; 46, Sergeant and Armstrong, 1973; 47, Van de Ven et al., 1979; 48, Jones et al., 1972; 49, Holden, 1973; 50, Andersen and Rebsdorff, 1976; 51, Holden, 1975; 52, Smith and Armstrong, 1975; 53, Kari and Kauranen, 1978; 54, Harms et al., 1978; 55, Koeman et al., 1973; 56, Roberts et al., 1976; 57, Gaskin et al., 1973; 58, Gaskin et al., 1972; 59, Buhler et al., 1975; 60, Martin et al., 1976; 61, Gladstone, 1996; 62, Denton and Breck, 1981; 63, Denton et al., 1980; 64, Fonseca et al., 2005; 65, Evans et al., 1998; 66, Evers et al., 2002; 67, Hyvarinen et al., 2003; 68, Mason et al., 1986; 69, Gutleb et al., 1998; 70. Mason and Reynolds, 1988; 71, Evans et al., 2000; 72, Wren, 1985.

seal, *Phoca sibirica* (Watanabe et al., 1998). Unlike teleosts, a high percentage of the mercury in seal liver occurs in the inorganic form (Harms et al., 1978). Moreover, most of the mercury in seal liver is biologically unavailable through complexation with selenium in a 1:1 atomic ratio (Koeman et al., 1975; Martin et al., 1976; Smith and Armstrong, 1978). Total mercury concentrations were usually highest in livers of marine mammals, intermediate in muscle, and lowest in blubber (Eisler, 1981). With some exceptions, mercury content in meat, blubber, and especially liver of adult and newborn marine mammals exceeded mercury safety guidelines established by regulatory agencies for foodstuffs. The relatively high concentrations appeared to be a result of natural processes rather than of anthropogenic activities, and probably did not represent a significant risk to pinniped health (Eisler, 1981). In one case, liver from an older grey seal contained 387.0 mg Hg/kg FW, a level substantially in excess of levels found toxic to humans (Sergeant and Armstrong, 1973). However, in contrast to fish, a high percentage of the total mercury occurs in the inorganic form in seal and whale liver (Harms et al., 1978). Inuit sled dogs, subsisting largely on seal meat, contained elevated levels of mercury in liver, up to 11.5 mg/kg FW, without apparent harm (Smith and Armstrong, 1975).

Mercury in pinniped muscle, unlike liver, was mostly methylmercury in both mothers and pups; pups acquired most of their mercury during gestation (Wagemann et al., 1988). The percentage of methylmercury in any tissue from any marine mammal appears to be inversely correlated with total mercury content (Gaskin et al., 1972, 1974; Sergeant and Armstrong, 1973; Buhler et al., 1975; Koeman et al., 1975; Smith and Armstrong, 1975). For example, Gaskin et al. (1972) found that

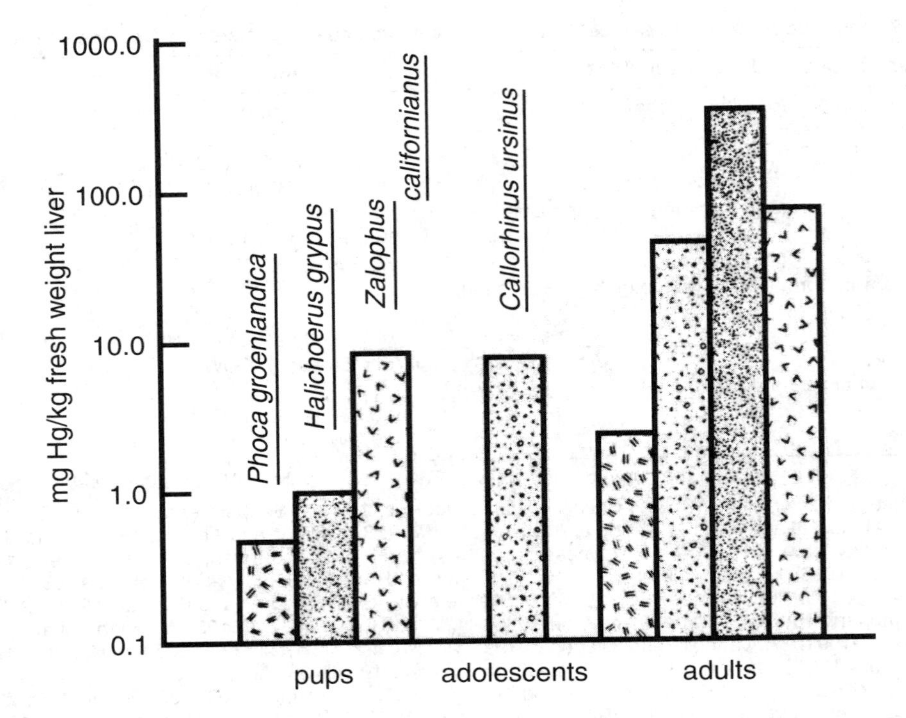

Figure 6.1 Mercury concentrations in livers of four species of pinniped mammals: harbor seal, *Phoca groenlandica*; grey seal, *Halichoerus grypus*; California sea lion, *Zalophus californianus*; and northern fur seal, *Callorhinus ursinus*.

liver of harbor seals from Maine contained a maximum of 7.8 mg total Hg/kg FW vs. 50.9 from those collected from New Brunswick, Canada; methylmercury accounted for 13.0 to 37.0% of total mercury in Maine (U.S.) seal livers but only 2.0 to 11.0% in Canadian seals. Among healthy California sea lions, *Zalophus californianus*, Buhler et al. (1975) reported concentrations of total mercury in tissues, in mg/kg FW and percent methylmercury, respectively, as follows: liver 74.0 and 3.7%; kidney 7.0 and 17.2%; muscle 1.2 and 88.6%; and heart 0.59 and 88.1%.

Many factors are known to modify mercury accumulation and retention in marine pinniped mammals, including diet, age of mammal, sex, general health, proximity to urban areas, selenium residues, and migrations through areas of high tectonic activity. Diet, for example, is an important concentrating mechanism in seals. Grey seals (*Halichoerus grypus*), hooded seals (*Cystophora cristata*), and harbor seals (*Phoca vitulina*), which feed on large fish and cephalopods, contained up to 10 times more mercury in their tissues than harp seals (*Pagophilus groenlandicus*), which feed on small pelagic fish and crustaceans (Sergeant and Armstrong, 1973). Although concentrations of copper, zinc, cadmium, and lead in muscle and liver tissues of prey fish did not differ significantly from corresponding organs of marine mammals, mercury concentrations were considerably higher in liver of whales and seals than in fish (Harms et al., 1978).

Mercury and selenium concentrations in livers of marine pinnipeds seem to be positively correlated (Koeman et al., 1973, 1975; Martin et al., 1976). Koeman et al. (1973) state that selenium protects these species by completely binding to sub-cellular S sites, the presumed location of mercury's toxic action. Both ringed (*Phoca hispida*) and bearded (*Erignathus barbatus*) seals accumulate high levels of naturally occurring mercury in their livers; however, in the large sample of seals that were examined, there was no indication of mercury intoxication (Smith and Armstrong, 1978). The presence of selenium in a 1:1 atomic ratio with mercury in seal liver indicates a biochemical binding process. The mechanism of detoxification by selenium is not understood with certainty, but it appears that when selenium is ingested along with mercury, some mechanism operates in seals that causes both metals to combine and become immobilized in the liver (Smith and Armstrong, 1978).

The mechanisms to account for mercury accumulation in pinnipeds are similar to those reported by Itano et al. (1984a, 1984b, 1984c) for the striped dolphin (*Stenella coeruleoalba*). Itano and co-workers showed that tissue concentrations of mercury in striped dolphins increased with increasing age of the animal, reaching a plateau in 20 to 25 years; were highest in liver, although muscle accounted for about 90.0% of the total body mercury burden; were present in the methylated form in fetal and suckling stages, but the proportion of methylmercury decreased over time with no absolute increase after age 10 years; were excreted slowly by all developmental stages, and slowest in older dolphins (resulting in higher accumulations); and were correlated strongly with selenium concentrations in all age groups. It is probable that inorganic mercury and selenium were complexed in a 1:1 molar ratio, in a form biologically unavailable to marine mammals (and probably other mammals), thereby significantly decreasing the risk of mercury toxicosis to individuals with grossly elevated mercury body burdens (Eisler, 1984, 1985). The Hg:Se ratio was close to 1.0 in adults of four species of Norwegian seals, provided that tissue mercury concentrations were greater than 15 mg/kg FW (Skaare et al., 1994). Total mercury in livers of pinniped mothers — but not pups — was correlated positively with selenium (Wagemann et al., 1988). In grey seals, *Halichoerus grypus*, mercury concentrations were higher in liver, kidney, and muscle of mature males and females when compared to immature individuals; this was especially pronounced in liver, where maximum mercury concentrations of 199.0 mg/kg FW and 238.0 mg/kg FW were recorded in mature males and females, respectively (Table 6.7). A strong correlation between cadmium, mercury, and zinc in kidney suggests the presence of a detoxification process involving metallothionein proteins; another strong and positive correlation between mercury and selenium and a molecular Hg:Se ratio close to 1.0 in liver of grey seals suggests a demethylation process leading to the formation of mercuric selenide granules (Bustamente et al., 2004). Large colonies of pinnipeds and, to a lesser extent, marine birds along the western coast of the United States may make mercury available to California mussels (*Mytilus californianus*) through fecal elimination of large amounts of mercury, resulting in abnormally high mercury levels in mussels from several west coast sites (Flegal et al., 1981).

Increasing mercury concentrations in tissues of marine mammals were also associated with poor health due to leptospirosis (Buhler et al., 1975), with proximity to urbanized areas (Anas, 1974; Roberts et al., 1976) and with starvation (Jones et al., 1976). Accumulations were usually highest in adult females, then adult males (Gaskin et al., 1972); placental transfer of mercury to developing pups is low (Freeman and Horne, 1973c; Kim et al., 1974; Jones et al., 1976). Methylmercury concentrations in seal pups were lower than that of their mothers. Moreover, the seal fetus does not show a preference for mercury over that of the mother's tissues, suggesting that seals may possess enzyme systems capable of demethylating methylmercury (Freeman and Horne, 1973c).

A source of mercury in livers of pilot whales (*Globicephala macrorhyncha*) may be due to volcanic activity in areas of mercury ore deposits and through which whales migrate; or exposure over time rather than food web (Gaskin et al., 1974). If this is correct, the elevated mercury levels in pilot whales of St. Lucia probably reflect accumulation resulting form existence in a tectonically active region with a higher than average environmental level of mercury, with exposure to a probably small fraction of air-transported mercury from outside the region, possibly industrial in origin.

Mercury concentrations in mammals other than pinnipeds are modified by age, sex, sexual condition, diet, season of collection, and other variables. Increasing concentrations of total and organic mercury in muscle and liver were observed with increasing age of fin whales (*Balaenoptera physalus*; Sanpera et al., 1993) and striped dolphins (Andre et al., 1991a), and in livers of white-tailed deer (*Odocoileus virginianus*; Khan and Forester, 1995), otters (Mason and Madsen, 1992), and the endangered Florida panther (*Felis concolor coryi*; Roelke et al., 1991). However, in harbor porpoises (*Phocoena phocoena*), total mercury — but not methylmercury — increased in tissues with increasing age (Joiris et al., 1991). Pregnant or lactating sperm whales (*Physeter macrocephalus*) had significantly higher mercury concentrations in muscle than nonbreeding females (Cannella and Kitchener, 1992). In river otters (*Lutra canadensis*), mercury concentrations in liver and kidney were higher in males than in females (Ropek and Neely, 1993). No sexual differences in liver mercury concentrations were evident in white-tailed deer (Khan and Forester, 1995) or European otters (*Lutra lutra*; Mason and Madsen, 1992). Polar bears (*Ursus maritimus*) probably obtain mercury from ringed seals (*Phoca hispida*), their main food (Lentfer and Galster, 1987). Polar bear cubs had lower concentrations of mercury in fur than did yearlings or adults, and the low concentrations in adult fur in summer is attributed to molting (Born et al., 1991). Florida panthers found dead contained as much as 110.0 mg total Hg/kg FW liver, a level found lethal to feral cats (*Felis cattus*) in Minamata, Japan. Florida panthers feed primarily on raccoons that contain as much as 3.0 mg Hg/kg FW (Roelke et al., 1991), but it is not known if this is the source of the elevated liver mercury in panthers.

Among furbearers in the Wisconsin River drainage system, mercury burdens were higher in fish-eating than in herbivorous species: that is, river otter > mink (*Mustela vison*) > raccoon (*Procyon lotor*) > red fox (*Vulpes fulva*) > muskrat (*Ondatra zibethicus*) > beaver (*Castor canadensis*) (Sheffy and St. Amant, 1982). In general, fur contained the highest mercury levels, followed by liver, kidney, muscle, and brain, in that order (Table 6.7; Sheffy and St. Amant, 1982). Mercury levels in fish-eating furbearers collected from the Wisconsin River basin paralleled mercury levels in fish, crayfish, and bottom sediments from that system; levels in all compartments were highest about 30 km downstream from an area that supported 16 pulp and paper mills and a chloralkali plant (Sheffy and St. Amant, 1982). Mercury concentrations in tissues of minks and otters trapped from various locations in Ontario between 1983 and 1985 varied by as much as sixfold (Table 6.7); mercury levels in fish and crayfish from the study areas followed a similar pattern (Wren et al., 1986). Mink and river otter accumulated about ten times more mercury than did predatory fishes from the same drainage areas, suggesting that these furbearers can serve as sensitive indicators of mercury, even at very low levels of mercury contamination (Kucera, 1983). The shorttail shrew, *Blarina brevicauda*, from certain mercury-contaminated sites in Tennessee have extremely high concentrations in kidney (38.8 mg Hg/kg FW) and may be ingesting nephrotoxic levels of mercury through the diet (8.8 mg Hg/kg FW ration; Talmage and Walton, 1993).

In the serow (*Capriocornis crispus*), a free-ranging bovine ruminant, about 40.0% of the total mercury body burden was in the fleece, at 0.37 mg Hg/kg FW fleece (Honda et al., 1987). Domestic sheep (*Ovis* sp.), allowed to graze for 23 months on grass contaminated with mercury (up to 6.5 mg/kg dry weight) caused by atmospheric emissions of a nearby chloralkali site, retained about 0.1% of the total mercury taken in by ingestion and inhalation, although residues in flesh were negligible (Edwards and Pumphery, 1982). It was concluded that contamination of grass as a result of atmospheric discharges of inorganic mercury from chloralkali sites causes no hazard, either directly to grazing animals or indirectly to humans who might ultimately consume their flesh.

6.8 SUMMARY

Mercury concentrations in worldwide field collections of algae and higher plants, invertebrates, fishes, amphibians, reptiles, birds, and mammals are listed; factors known to modify mercury accumulations and their significance are discussed. Aquatic and terrestrial plants usually contain less than 1.0 mg total mercury/kg dry weight (DW), except in the vicinity of naturally occurring cinnabar deposits and various anthropogenic activities such as smelters, sewage lagoons, chloralkali plants, newly formed reservoirs, and applications of mercury-containing agricultural chemicals. Some vegetation from impacted environments had up to 70.0 mg Hg/kg DW and 59.0 mg Hg/kg fresh weight (FW); these species may be useful in mercury phytoremediaton removal from mercury-contaminated sites. Invertebrates collected near industrial, municipal, and other known sources of mercury contained up to 10.0 mg Hg/kg FW and 38.7 mg Hg/kg DW; conspecifics from reference sites usually had less than 0.5 mg Hg/kg FW. Passage of environmental legislation and effective enforcement were considered instrumental in reducing mercury concentrations in mussels from Bergen Harbor, Norway, by 60.0% in 10 years.

In bony fishes and elasmobranchs, mercury tends to accumulate in muscle, mainly as methylmercury. Accumulations in muscle and other tissues increases with increasing age of the fish, and is highest in carnivorous species. Concentrations exceeding 2.0 mg Hg/kg FW in muscle of some wide-ranging oceanic fishes — such as tuna, marlin, and swordfish — were common, owing to a combination of human activities and natural processes. For most fish products sold, muscle mercury concentrations were less than 0.3 mg Hg/kg FW; however, about 2.0% of the total catch landed may contain more than 0.5 mg Hg/kg FW.

Amphibians usually contained less than 0.5 mg Hg/kg FW in various organs; however, frogs collected near a mercury mine in the former Yugoslavia had 24.0 mg Hg/kg FW in kidney and 25.5 mg Hg/kg FW in liver. In reptiles, most of the mercury in tissues was organic mercury; concentrations were highest in liver, kidney, muscle, and egg, in that order. Alligators from the mercury-contaminated Florida Everglades had the highest mercury concentrations recorded in reptiles: 6.1 mg/kg FW in muscle (which is eaten locally), 65.3 mg/kg FW in kidney, and 99.5 mg Hg/kg FW in liver. Farm-raised alligators in Florida always had less than 0.2 mg Hg/kg FW in tissues. In birds, the highest mercury concentrations measured were 187.0 mg/kg FW in liver of male loons found dead in New England; 175.0 mg/kg FW in liver of great blue herons from mercury-contaminated Lake. St. Clair; 175.0 mg/kg FW in liver of diving ducks from a mercury-contaminated ecosystem; 130.0 mg/kg FW in kidney of ospreys from the eastern United States; 96.0 mg/kg FW in liver of grey herons from England; 306.0 mg/kg DW in liver of seabirds; 295.0 mg/kg DW in albatross liver and 140.0 mg/kg DW in petrel liver, both species collected in New Zealand. Elevated concentrations of mercury in avian tissues were associated with low organ and body weight, inhibited reproduction, and decreased activities of enzymes related to glutathione metabolism and antioxidant activities. Feathers usually had the highest mercury concentrations, followed by liver, then muscle, often in the ratio of 7:3:1; however, almost all mercury in feathers is organic and in liver it is inorganic mercury. Many factors modify mercury concentrations in avian tissues, including food preference and availability, migration patterns, age and sex, mercury loadings in the immediate biosphere, season of collection, proximity to industrialized areas, inherent species differences, molting stage, general health, and interactions with selenium.

In humans, increasing mercury concentrations in hair and tissues are primarily a result of increasing consumption of fish and shellfish, and to a lesser degree to dental amalgams. Pregnant aboriginal women who routinely consume seal meat and blubber with high mercury concentrations throughout pregnancy had elevated mercury concentrations in maternal and fetal blood and other tissues without apparent effect on the fetus or the resultant infant; however, this requires verification. Total mercury concentrations in hair greater than 1.0 mg/kg DW are now considered indicative of mercury exposure and is the recommended concentration at which women of child-bearing age should restrict consumption of fish, according to the U.S. Environmental Protection Agency. However, more than 45.0% of New Yorkers and 34.0% of Floridians now exceed 1.0 mg Hg/kg DW hair, suggesting reexamination of this value. Mercury poisoning in humans is associated with total mercury concentrations in hair greater than 249.0 mg/kg FW and methylmercury concentrations in blood greater than 3.1 mg/L. Among nonhuman mammals, mercury concentrations were highest (143.0 to 765.0 mg total Hg/kg FW) in tissues of marine pinnipeds, especially in livers of older seals and sea lions; accumulations were not attributed to anthropogenic activities and did not seem to pose a significant threat to pinniped health or to human consumers. Among land mammals, livers of the endangered Florida panther had up to 110.0 mg Hg/kg FW, possibly from consumption of mercury-contaminated raccoons.

REFERENCES

Abernathy, A.R. and P.M. Cumbie. 1977. Mercury accumulation by largemouth bass (*Micropterus salmoides*) in recently impounded reservoirs, *Bull. Environ. Contam. Toxicol.*, 17, 595–602.

Ackefors, H., G. Lofroth, and C.G. Rosen. 1970. A survey of the mercury pollution problem in Sweden with special reference to fish, *Oceanogr. Mar. Bio. Annu. Rev.*, 8, 203–224.

Adams, D.H. and G.V. Onorato. 2005. Mercury concentrations in red drum, *Sciaenops ocellatus*, from estuarine and offshore waters of Florida, *Mar. Pollut. Bull.*, 50, 291–300.

Airas, S., A. Duinker, and K. Julshamn. 2004. Copper, zinc, arsenic, cadmium, mercury, and lead in blue mussels (*Mytilus edulis*) in the Bergen Harbor area, western Norway, *Bull. Environ. Contam. Toxicol.*, 73, 276–284.

Akielaszak, J.J. and T.A. Haines. 1981. Mercury in the muscle tissue of fish from three northern Maine lakes, *Bull. Environ. Contam. Toxicol.*, 27, 201–208.

Albers, P.H., L. Sileo, and B.M. Mulhern. 1986. Effects of environmental contaminants on snapping turtles of a tidal wetland, *Arch. Environ. Contam. Toxicol.*, 15, 39–49.

Aldington, R. and D. Ames (translators). 1968. *New Larousse Encyclopedia of Mythology.* Hamlyn Publ., New York. 500 pp.

Alexander, J.E., J. Foehrenbach, S. Fisher, and D. Sullivan. 1973. Mercury in striped bass and bluefish, *N.Y. Fish Game J.*, 20, 147–151.

Allard, M. and P.M. Stokes. 1989. Mercury in crayfish species from thirteen Ontario lakes in relation to water chemistry and smallmouth bass (*Micropterus dolomieui*) mercury, *Can. J. Fish. Aquat. Sci.*, 46, 1041–1046.

Allen-Gil, S.M., D.J. Gilroy, and L.R. Curtis. 1995. An ecoregion approach to mercury bioaccumulation by fish in reservoirs, *Arch. Environ. Contam. Toxicol.*, 28, 61–68.

Anas, R.E. 1974. Heavy metals in northern fur seals, *Callorhinus ursinus*, and harbor seals, *Phoca vitulina richardi*, *U.S. Dept. Comm. Fish. Bull.*, 72, 133–137.

Anderlini, V. 1974. The distribution of heavy metals in the red abalone, *Haliotis rufescens*, on the California coast, *Arch. Environ. Contam. Toxicol.*, 2, 253–265.

Andersen, S.H. and A. Rebsdorff. 1976. Polychlorinated hydrocarbons and heavy metals in the harbour porpoise (*Phocoena phocoena*) and whitebeaked dolphin (*Lagenorhynchus albirostris*) from Danish waters, *Aquatic Mammals*, 4, 14–20.

Anderson, M.R., D.A. Scruton, U.P. Williams, and J.F. Payne. 1995. Mercury in fish in the Smallwood Reservoir, Labrador, twenty one years after impoundment, *Water Air Soil Pollut.*, 80, 927–930.

Andre, J.M., A. Boudou, and F. Ribeyre. 1991a. Mercury accumulation in Delphididae, *Water Air Soil Pollut.*, 56, 187–201.

Andre, J., A. Boudou, F. Ribeyre, and M. Bernhard. 1991b. Comparative study of mercury accumulation in dolphins (*Stenella coeruleoalba*) from French Atlantic and Mediterranean coasts, *Sci. Total Environ.*, 104, 191–209.

Andreji, J., I. Stranai, P. Massanyi, and M. Valent. 2005. Concentrations of selected metals in muscle of various fish species, *J. Environ. Sci. Health*, 40, 899–912.

Anon. 1978. Selected pollution profiles: North Atlantic, North Sea, Baltic Sea, and Mediterranean Sea, *Ambio*, 7, 75–78.

Applequist, H., S. Asbirk, and I. Drabaek. 1984. Mercury monitoring: mercury stability in bird feathers, *Mar. Pollut. Bull.*, 15, 22–24.

Arima, S. and S. Umemoto. 1976. Mercury in aquatic organisms. II. Mercury distribution in muscles of tunas and swordfish, *Bull. Japan. Soc. Sci. Fish.*, 42, 931–937.

Atkeson, T., D. Axelrad, C. Pollman, and G. Keeler. 2003. Integrating Atmospheric Mercury Deposition and Aquatic Cycling in the Florida Everglades: An Approach for Conducting a Total Maximum Daily Load Analysis for an Atmospherically Derived Pollutant. Integrated Summary. Final Report. 272 pp. Available from Mercury and Applied Science MS6540, Florida Dept. Environmental Conservation, 2600 Blair Stone Road, Tallahassee, FL 32399-2400. See also <http://www.floridadep.org/labs/mercury/index.htm>

Audet, D.J., D.S. Scott, and S.N. Wiemeyer. 1992. Organochlorines and mercury in osprey eggs from the eastern United States, *J. Raptor Res.*, 26, 219–224.

Augier, H., G. Gilles, and G. Ramonda. 1978. Recherche sur la pollution mercurielle du milieu maritime dans la region de Marseille (Mediterranee, France). 1. Degre de contamination par le mercure de la phanerograme marine *Posidonia oceanica* delile a proximite du complexe portuaire et dans la zone de reject due grand collecteur disgouts de la ville de Marseille, *Environ. Pollut.*, 17, 269–285.

Babji, A.S., M.S. Embong, and W.W. Woon. 1979. Heavy metal contents in coastal water fishes of West Malaysia, *Bull. Environ. Contam. Toxicol.*, 23, 830–836.

Bacher, G.J. 1985. Mercury concentrations in the Australian fur seal *Arctocephalus pusillus* from SE Australian waters, *Bull. Environ. Contam. Toxicol.*, 35, 490–495.

Baldi, F., A. Renzoni, and M. Bernhard. 1978. Mercury concentrations in pelagic fishes (anchovy, mackerel and sardine) from the Italian coast and Strait of Gibraltar, *IVes J. Etud. Pollut.*, 251–254.

Barbaro, A., A. Francescon, B. Polo, and M. Bilio. 1978. *Balanus amphitrite* (Cirripedia: Thoracica) — a potential indicator of fluoride, copper, lead, chromium, and mercury in North Adriatic lagoons, *Mar. Biol.*, 46, 247–257.

Barber, R.T., P.J. Whaling, and D.M. Cohen. 1984. Mercury in recent and century-old deep-sea fish, *Environ. Sci. Technol.*, 18, 552–555.

Barber, R.T., A. Vijayakumar, and F.A. Cross. 1972. Mercury concentrations in recent and ninety-year old benthopelagic fish, *Science*, 178, 636–639.

Barrett, R.T., J.U. Skaare, and G.W. Gabrielsen. 1996. Recent changes in levels of persistent organochlorines and mercury in eggs of seabirds from the Barents Sea, *Environ. Pollut.*, 92, 13–18.

Beasley, T.M. 1971. Mercury in selected fish protein concentrates, *Environ. Sci. Technol.*, 5, 634–635.

Beauvais, S.L., J.G. Wiener, and G.J. Atchison. 1995. Cadmium and mercury in sediment and burrowing mayfly nymphs (*Hexagenia*) in the upper Mississippi River, USA, *Arch. Environ. Contam. Toxicol.*, 28, 178–183.

Bebbington, G. N., N.J. Mackay, R. Chvojka, R.J. Williams, A. Dunn, and E.H. Auty. 1977. Heavy metals, selenium and arsenic in nine species of Australian commercial fish, *Austral. J. Mar. Freshwater Res.*, 28, 277–286.

Becker, D.S. and G.N. Bigham. 1995. Distribution of mercury in the aquatic food web of Onondaga Lake, New York, *Water Air Soil Pollut.*, 80, 563–571.

Becker, P.H. 1992. Egg mercury levels decline with the laying sequence in charadriiformes, *Bull. Environ. Contam. Toxicol.*, 48, 762–767.

Becker, P.H., R.W. Furness, and D. Henning. 1993. The value of chick feathers to assess spatial and interspecific variation in the mercury contamination of seabirds, *Environ. Monitor. Assess.*, 28, 255–262.

Becker, P.H., D. Henning, and R.W. Furness. 1994. Differences in mercury contamination and elimination during feather development in gull and tern broods, *Arch. Environ. Contam. Toxicol.*, 27, 162–167.

Berglund, A. 1990. Estimation by a 24-hour study of the daily dose of intra-oral mercury vapor inhaled after release from dental amalgam, *J. Dent. Res.*, 70, 233–237.

Bernhard, M. and A. Zattera. 1975. Major pollutants in the marine environment. In E.A. Pearson and R. Frangipane (Eds.), *Marine Pollution and Marine Waste Disposal,* p. 195–300. Pergamon, Elmsford, NY.

Bertine, K.K. and E.D. Goldberg. 1972. Trace elements in clams, mussels, and shrimp, *Limnol. Ocean.*, 17, 877–884.

Beyer, W.N., M. Spalding, and D. Morrison. 1997. Mercury concentrations in feathers of wading birds from Florida, *Ambio*, 26, 97–100.

Bidwell, J.R. and A.G. Heath. 1993. An *in situ* study of rock bass (*Ambloplites rupestris*) physiology: effect of season and mercury contamination, *Hydrobiologia*, 264, 137–152.

Bishop, C.A., M.D. Koster, A.A. Chek, D.J.T. Hussell, and K. Jock. 1995. Chlorinated hydrocarbons and mercury in sediments, red-winged blackbirds (*Agelaius phoeniceus*) and tree swallows (*Tachycineta bicolor*) from wetlands in the Great Lakes-St. Lawrence River basin, *Environ. Toxicol. Chem.*, 14, 491–501.

Bloom, N.S. 1992. On the chemical form of mercury in edible fish and marine invertebrate tissue, *Can. J. Fish. Aquat. Sci.*, 49, 1010–1017.

Bodaly, R.A., R.E. Hecky, and R.J.P. Fudge. 1984. Increases in fish mercury levels in lakes flooded by the Churchill River diversion, northern Manitoba, *Can. J. Fish. Aquat. Sci.*, 41, 682–691.

Bodaly, R.A., J.W.M. Rudd, and R.J.P. Fudge. 1993. Mercury concentrations in fish related to size of remote Canadian Shield lakes, *Can. J. Fish. Aquat. Sci.*, 50, 980–987.

Borgmann, U. and D.M. Whittle. 1991. Contaminant concentration trends in Lake Ontario lake trout (*Salvelinus namaycush*): 1977 to 1988, *J. Great Lakes Res.*, 17, 368–381.

Borgmann, U. and D.M. Whittle. 1992. DDE, PCB, and mercury concentration trends in Lake Ontario rainbow smelt (*Osmerus mordax*) and slimy sculpin (*Cottus cognatus*): 1977–1988, *J. Great Lakes Res.*, 18, 298–308.

Born, E.W., A. Renzoni, and R. Dietz. 1991. Total mercury in hair of polar bears (*Ursus maritimus*) from Greenland and Svalbard, *Polar Res.*, 9, 113–120.

Bowerman, W.W.IV., E.D. Evans, J.P. Giesy, and S. Postupalsky. 1994. Using feathers to assess risk of mercury and selenium to bald eagle reproduction in the Great Lakes region, *Arch. Environ. Contam. Toxicol.*, 27, 294–298.

Braune, B.M. 1987a. Comparison of total mercury levels in relation to diet and molt for nine species of marine birds, *Arch. Environ. Contam. Toxicol.*, 16, 217–224.

Braune, B.M. 1987b. Mercury accumulation in relation to size and age of Atlantic herring (*Clupea harengus harengus*) from the southwestern Bay of Fundy, Canada, *Arch. Environ. Contam. Toxicol.*, 16, 311–320.

Braune, B.M. and D.E. Gaskin. 1987. Mercury levels in Bonaparte's gulls (*Larus philadelphia*) during autumn molt in the Quoddy region, New Brunswick, Canada, *Arch. Environ. Contam. Toxicol.*, 16, 539–549.

Broo, B. and T. Odsjo. 1981. Mercury levels in feathers of eagle-owls *Bubo bubo* in a captive, a reintroduced and a native wild population in SW Sweden, *Holarctic Ecol.*, 4, 270–277.

Buhler, D.R., R.R. Claeys, and B.R. Mate. 1975. Heavy metal and chlorinated hydrocarbon residues in California sea lions (*Zalophus californianus californianus*), *J. Fish. Res. Bd. Canada*, 32, 2391–2397.

Burger, J. and M. Gochfeld. 1996. Heavy metal and selenium levels in Franklin's gull (*Larus pipixcan*) parents and their eggs. *Arch. Environ. Contam. Toxicol.*, 30, 487–491.

Burger, J., M. Pokras, R. Chafel, and M. Gochfeld. 1994. Heavy metal concentrations in feathers of common loons (*Gavia immer*) in the northeastern United States and age differences in mercury levels, *Environ. Monitor. Assess.*, 30, 1–7.

Burger, J. and J. Snodgrass. 1998. Heavy metals in bullfrog (*Rana catesbeiana*) tadpoles: effects of depuration before analysis, *Environ. Toxicol. Chem.*, 17, 2203–2209.

Busch, W.D.N. 1983. Decline of mercury in young fishes from western Lake Erie between 1970–71 and 1974, *Prog. Fish–Cult.*, 45, 202–206.

Bustamente, P., C.F. Morales, B. Mikkelsen, M.Dam, and F. Caurant. 2004. Trace element bioaccumulation in grey seals *Halichoerus grypus* from the Faroe Islands, *Mar. Ecol. Prog. Ser.*, 267, 291–304.

Bycroft, B.M., B.A.W. Coller, G.B. Deacon, D.J. Coleman, and P.S. Lake. 1982. Mercury contamination of the Lerderderg River, Victoria, Australia, from an abandoned gold field, *Environ. Pollut.*, 28A, 135–147.

Cahill, T.M., D.W. Anderson, R.A. Elbert, B.P. Perley, and D.R. Johnson. 1998. Elemental profiles in feather samples from a mercury-contaminated lake in central California, *Arch. Environ. Contam. Toxicol.*, 35, 75–81.

Campbell, K.R. and T.S. Campbell. 2001. The accumulation and effects of environmental contaminants on snakes: a review, *Environ. Monitor. Assess.*, 70, 253–301.

Cannella, E.G. and D.J. Kitchener. 1992. Differences in mercury levels in female sperm whale, *Physeter macrocephalus* (Cetacea: Odontoceti), *Austral. Mammal.*, 15, 121–123.

Cappon, C.J. and J.C. Smith. 1982. Chemical form and distribution of mercury and selenium in edible seafood, *J. Anal. Toxicol.*, 6, 10–21.

Cheevaparanapivat, V. and P. Menasveta. 1979. Total and organic mercury in marine fish of the upper Gulf of Thailand, *Bull. Environ. Contam. Toxicol.*, 23, 291–299.

Chernoff, B. and J.K. Dooley. 1979. Heavy metals in relation to the biology of the mummichog, *Fundulus heteroclitus*, *J. Fish Biol.*, 14, 309328.

Chigbo, F.E., R.W. Smith, and F.L. Shore. 1982. Uptake of arsenic, cadmium, lead and mercury from polluted waters by the water hyacinth *Eichornia crassipes*, *Environ. Pollut.*, 27A, 31–36.

Childs, E.A. and J.N. Gaffke. 1973. Mercury content in Oregon ground fish, *U.S. Dept. Comm. Fish. Bull.*, 71, 713–717.

Childs, E.A., J.N. Gaffke, and D.L. Crawford. 1973. Exposure of dogfish shark feti to mercury, *Bull. Environ. Contam. Toxicol.*, 9, 276–280.

Choi, M.H., and J.J. Cech, Jr. 1998. Unexpectedly high mercury level in pelleted commercial fish feed, *Environ. Toxicol. Chem.*, 17, 1979–1981.

Chvojka, R. 1988. Mercury and selenium in axial white muscle of yellowtail kingfish from Sydney, Australia, *Mar. Pollut. Bull.*, 19, 210–213.

Chvojka, R. and R.J. Williams. 1980. Mercury levels in six species of Australian commercial fish, *Austral. J. Mar. Freshwater Res.*, 31, 469–474.

Chvojka, R., R.J. Williams, and S. Fredrickson. 1990. Methyl mercury, total mercury, and selenium in snapper from two areas of the New South Wales coast, Australia, *Mar. Pollut. Bull.*, 21, 570–573.

Clark, D.R. Jr., J.W. Bickham, D.L. Baker, and D.F. Cowman. 2000. Environmental contaminants in Texas, USA, wetland reptiles: evaluation using blood samples, *Environ. Toxicol. Chem.*, 19, 2259–2265.

Clark, G. and G. Topping. 1989. Mercury concentrations in fish from contaminated areas in Scottish waters, *J. Mar. Biol. Assoc. U.K.*, 69, 437–445.

Cocoros, G.P., H. Cahn, and W, Siler. 1973. Mercury concentrations in fish, plankton and water from three western Atlantic estuaries, *J. Fish Biol.*, 5, 641–647.

Cooper, J.J. 1983. Total mercury in fishes and selected biota in Lahontan Reservoir, Nevada: 1981, *Bull. Environ. Contam. Toxicol.*, 31, 9–17.

Cope, W.G., J.G. Wiener, and R.G. Rada. 1990. Mercury accumulation in yellow perch in Wisconsin seepage lakes: relation to lake characteristics, *Environ. Toxicol. Chem.*, 9, 931–940.

Cross, F.A., L. H. Hardy, N.Y. Jones, and R.T. Barber. 1973. Relation between total body weight and concentrations of manganese, iron, copper, zinc and mercury in white muscle of bluefish (*Pomatomus saltatrix*) and bathyl-demersal fish (*Antimora rostrata*), *J. Fish. Res. Bd. Canada*, 30, 1287–1291.

Cugurra, F. and G. Maura. 1976. Mercury content in several species of marine fish, *Bull. Environ. Contam. Toxicol.*, 15, 568–573.

Cumont, G., G. Gilles, F. Bernard, M.B. Briand, G. Stephan, G. Ramonda, and G. Guillou. 1975. Bilan de la contamination des poissons de mer par le mercure a l'occasion d'un controle portant sur 3 annees, *Ann. Hyg. L. Fr. Med. et Nut.*, 11 (1), 17–25.

Cumont, G. G. Viallex, H. Lelievre, and P. Bobenreith. 1972. Mercury contamination of fish of the sea, *Rev. Int. Ocean. Med.*, 28, 95–127.

Cutshall, N.H., J.R. Naidu, and W.G. Pearcy. 1978. Mercury concentrations in Pacific hake (*Merluccius productus*) (Ayres), as a function of length and latitude, *Science*, 200, 1489–1491.

DeClerck, R., R. Vanderstappen, and W. Vyncke. 1974. Mercury content of fish and shrimps caught off the Belgian coast, *Ocean Manage.*, 2, 117–126.

DeClerck, R., R. Vanderstappen, W. Vyncke, and P. Van Hoeyweghen. 1979. La teneur en metaux lords dans les organismes marins provenant de la capture accesssoire de la peche coteire belge, *Rev. de l'Agric.*, 3(32), 739–801.

Dehlinger, P., W.F. Fitzgerald, S.Y. Feng, D.F. Paskausky, M.W. Garvine, and W.F. Bohlen. 1973. Determination of budgets of heavy metal wastes in Long Island Sound, Annual Rept., Parts I and II, University of Connecticut, Marine Sciences Institute, Groton, CT.

Delany, M.F., J.U. Bell, and S.F. Sundlof. 1988. Concentrations of contaminants in muscle of the American alligator in Florida, *J. Wildl. Dis.*, 24, 62–66.

Delbekke, K., C. Joiris, and G. Decadt. 1984. Mercury contamination of the Belgian avifauna 1970–1981, *Environ. Pollut.*, 7B, 205–221.

Denton, G.R.W. and W.G. Breck. 1981. Mercury in tropical marine organisms from north Queensland, *Mar. Pollut. Bull.*, 12, 116–121.

Denton, G.R.W., H. Marsh, G.E. Heinsohn, and C. Burden-Jones. 1980. The unusual metal status of the dugong *Dugong dugon*, *Mar. Biol.*, 57, 201–219.

DesGranges, J.L., J. Rodrigue, B. Tardiff, and M. Laperle. 1998. Mercury accumulation and biomagnification in ospreys (*Pandion haliaetus*) in the James Bay and Hudson Bay regions of Quebec, *Arch. Environ. Contam. Toxicol.*, 35, 330–341.

De Wolf, P. 1975. Mercury content of mussels from West European coasts, *Mar. Pollut. Bull.*, 6, 61–63.

Dixon, R. and B. Jones. 1994. Mercury concentrations in stomach contents and muscle of five fish species from the north east coast of England, *Mar. Pollut. Bull.*, 28, 741–745.

Doi, R. and J. Ui. 1975. The distribution of mercury in fish and its form of occurrence. In P.A. Krenkel (Ed.), *Heavy Metals in the Aquatic Environment,* p. 197–221. Pergamon, Elmsford, NY.

Dustman, E.H., L.F. Stickel,and J.B. Elder. 1972. Mercury in wild animals at Lake St. Clair 1970. In R. Hartung and R.D. Dinman (Eds.), *Environmental Mercury Contamination,* p. 46–52. Ann Arbor Press, Ann Arbor, MI.

Dye, B.A., S.E. Schober, C.F. Dillon, R.L. Jones, C. Fryar, M. McDowell, and T.H. Sinks. 2005. Urinary mercury concentrations associated with dental restorations in adult women aged 16–49 years: United States, 1999–2000, *Occup. Environ. Med.*, 62, 368–375.

Edwards, P.R. and N.W.J. Pumphery. 1982. Ingestion and retention of mercury by sheep grazing near a chlor-alkali plant, *J. Sci. Food Agric.*, 33, 237–243.

Eftekhari, M. 1975. Teneur en mercure de quelques crevettes du Golfe Persique, *Science Peche. Bull. Inst. Peches Marit.*, 250, 9–10.

Eganhouse, R.P. and D.R. Young. 1978. Total and organic mercury in benthic organisms near a major submarine wastewater outfall system, *Bull. Environ. Contam. Toxicol.*, 19, 758–766.

Eisler, R. 1981. *Trace Metal Concentrations in Marine Organisms.* Pergamon, Elmsford, NY. 687 pp.

Eisler, R. 1984. Trace metal changes associated with age of marine vertebrates, *Biol. Trace Elem. Res.*, 6, 165–180.

Eisler, R. 1985. Selenium hazards to fish, wildlife, and invertebrates: a synoptic review, *U.S. Fish Wildl. Serv. Biol. Rep.*, 85(1.5), 1–57.

Eisler, R. 2000. Mercury. In *Handbook of Chemical Risk Assessment: Health Hazards to Humans, Plants, and Animals. Vol. 1, Metals,* p. 313–409. Lewis Publishers, Boca Raton, FL.

Elbert, R.A. and D.W. Anderson. 1998. Mercury levels, reproduction, and hematology in western grebes from three California lakes, USA, *Environ. Toxicol. Chem.*, 17, 210–213.

Elliott, J.E., R.J. Norstrom, and G.E.J. Smith. 1996. Patterns, trends, and toxicological significance of chlorinated hydrocarbon and mercury contaminants in bald eagle eggs from the Pacific coast of Canada, 1990–1994, *Arch. Environ. Contam. Toxicol.*, 31, 354–367.

Establier, R. 1975. Contenido en mercurio de las anguillas (*Anguilla anguilla*) de la desembocadura del rio Guadalquiver y esteros de les salines de la zona de Cadiz, *Invest. Pesquera*, 39(1), 249–255.

Establier, R. 1977. Estudio de la contaminacion marine por metals pesados y sus effectos biologicos, *Inform. Tecn. Inst. Invest. Pesqueras*, 47, 1–36.

Eto, K., S. Oyanagi, Y. Itai, H. Tokunaga, Y. Takizawa, and I. Suda. 1992. A fetal type of Minamata Disease. An autopsy case report with special reference to the nervous system, *Molecul. Chem. Neuropathol.*, 16, 171–186.

Evans III, E.C., T.J. Peeling, A.E. Murchison, and Q.D. Stephen-Hassard. 1972. A Proximate Biological Survey of Pearl Harbor, Oahu, Available from Natl. Tech. Infor. Serv., Springfield, VA as AD-744 233, Rep. NUC TP 290, 1–65.

Evans, R.D., E.M. Addison, J.Y Villeneuve, K.S. MacDonald, and D.G. Joachim. 1998. An examination of spatial variation in mercury concentrations in otter (*Lutra canadensis*) in south-central Ontario, *Sci. Total Environ.*, 213, 239–245.

Evans, R.D., E.M. Addison, J.Y. Villeneuve, K.S. MacDonald, and D.G. Joachim. 2000. Distribution of inorganic and methylmercury among tissues in mink (*Mustela vison*) and otter (*Lutra canadensis*), *Environ. Res.*, 84. 133–139.

Evers, D.C., J.D. Kaplan, M.W. Meyer, P.S. Reaman, W.E. Braselton, A. Major, N. Burgess, and A.M. Scheuhammer. 1998. Geographic trend in mercury measured in common loon feathers and blood, *Environ. Toxicol. Chem.*, 17, 173–183.

Evers, D.C., D. Yates, and L. Savoy. 2002. *Developing a Mercury Exposure Profile for Mink and River Otter in Maine*. Maine Report 2002–10, Biodiversity Research Institute, Falmouth, ME.

Faber, R.A. and J.J. Hickey. 1973 Eggshell thinning, chlorinated hydrocarbons and mercury in inland aquatic bird eggs, 1969 and 1970, *Pestic. Monitor. J.*, 7, 27–36.

Facemire, C., T. Augspurger, D. Bateman, M. Brim, P. Conzelmann, S. Delchamps, E. Douglas, L. Inmon, K. Looney, F. Lopez, G. Masson, D. Morrison, N. Morse, and A. Robison. 1995. Impacts of mercury contamination in the southeastern United States, *Water Air Soil Pollut.*, 80, 923–926.

Falandysz, J. and M. Kowalewska. 1993. Mercury concentration of stickleback *Gasterosteus aculeatus* from the Gulf of Gdansk, *Bull. Environ. Contam. Toxicol.*, 51, 710–715.

Fimreite, N. 1974. Mercury contamination of aquatic birds in northwestern Ontario, *Jour. Wildl. Manage.*, 38, 120–131.

Fimreite, N. 1979. Accumulation and effects of mercury on birds. In J.O. Nriagu (Ed.), *The Biogeochemistry of Mercury in the Environment,* p. 601–607. Elsevier/North-Holland Biomedical Press, New York.

Fimreite, N., W.N. Holsworth, J.A. Keith, P.A. Pearce, and I.M. Gruchy. 1971. Mercury in fish and fish-eating birds near sites of industrial contamination in Canada, *Canad. Field Natur.*, 85, 211–220.

Fimreite, N., N. Kveseth, and E.M. Brevik. 1980. Mercury, DDE, and PCBs in eggs from a Norwegian gannet colony, *Bull. Environ. Contam. Toxicol.*, 24, 142–144.

Fimreite, N. and L.M. Reynolds. 1973. Mercury contamination of fish in northwestern Ontario, *J. Wildl. Manage.*, 37, 62–68.

Fjeld, E. and S. Rognerud. 1993. Use of path analysis to investigate mercury accumulation in brown trout (*Salmo trutta*) in Norway and the influence of environmental factors, *Can. J. Fish. Aquat. Sci.*, 50, 1158–1167.

Flegal, A.R., M. Stephenson, M. Martin, and J.H. Martin. 1981. Elevated concentrations of mercury in mussels (*Mytilus californianus*) associated with pinniped colonies, *Mar. Biol.*, 65, 45–48.

Foley, R.E., S.J. Jackling, R.J. Sloan, and M.K. Brown. 1988. Organochlorine and mercury residues in wild mink and otter: comparison with fish, *Environ. Toxicol. Chem.*, 7, 363–374.

Fonseca, F.R.D., O. Malm, and H.F. Waldemarin. 2005. Mercury levels in tissues of giant otters (*Pteronura brasiliensis*) from the Rio Negro, Pantanal, Brazil, *Environ. Res.*, 98, 368–371.

Forrester, C.R., K.S. Ketchen, and C.C. Wong. 1972. Mercury content of spiny dogfish (*Squalus acanthias*) in the Strait of Georgia, British Columbia, *J. Fish. Res. Bd. Canada*, 29, 1487–1490.

Fortin, C., G. Beauchamp, M. Dansereau, N. Lariviere, and D. Belanger. 2001. Spatial variation in mercury concentrations in wild mink and river carcasses from the James Bay Territory, Quebec, Canada, *Arch. Environ. Contam. Toxicol.*, 40, 121–127.

Freeman, H.C. and D.A. Horne. 1973a. The total mercury and methylmercury content of the American eel (*Anguilla rostrata*), *J. Fish. Res. Bd. Canada*, 30, 454–456.

Freeman, H.C. and D.A. Horne. 1973b. Sampling the edible muscle of the swordfish (*Xiphius gladius*) for total mercury analysis, *J. Fish. Res. Bd. Canada*, 30, 1251–1252.

Freeman, H.C. and D.A. Horne. 1973c. Mercury in Canadian seals, *Bull. Environ. Contam. Toxicol.*, 10, 172–180.

Freeman, H.C., G. Shum, and J.F. Uthe. 1978. The selenium content in swordfish (*Xiphias gladius*) in relation to total mercury content, *J. Environ. Sci. Health*, A13(3), 235–240.

Fujiki, M. 1963. Studies on the course that the causative agent of Minamata disease was formed, especially on the accumulation of the mercury compound in the fish and shellfish of Minamata Bay, *J. Kumamoto Med. Soc.*, 37, 494–521.

Fukai, S., K. Tanaka, S. Kanno, and T. Ukita. 1972. Improvements in the determination method of methyl mercury in fish tissues and the ratio of methyl mercury to total mercury in fish tissues, *Adv. Water Pollut. Res., Sixth Int. Conf.*, Jerusalem, Israel, June 8–23, 1972, 819–827.

Furness, R.W., S.J. Muirhead, and M. Woodburn. 1986. Using bird feathers to measure mercury in the environment: relationships between mercury content and moult, *Mar. Pollut. Bull.*, 17, 27–30.

Futter, M.N. 1994. Pelagic food-web structure influences probability of mercury contamination in lake trout (*Salvelinus namaycush*), *Sci. Total Environ.*, 145, 7–12.

Galster, W. A. 1976. Mercury in Alaskan Eskimo mothers and infants, *Environ. Health Perspec.*, 15, 135–140.

Ganther, H.E., C. Goudie, M.L. Sunde, M.J. Kopecky, P. Wagner, S.H. Oh, and W.G. Hoekstra. 1972. Selenium: relation to decreased toxicity of methylmercury added to diets containing tuna, *Science*, 175, 1122–1124.

Garcia, E. and R. Carignan. 2005. Mercury concentrations in fish from forest harvesting and fire-impacted Canadian boreal lakes compared using stable isotopes of nitrogen, *Environ. Toxicol. Chem.*, 24, 685–693.

Gardner, D. 1978. Mercury in fish and waters of the Irish Sea and other United Kingdom fishing grounds, *Nature*, 272, 49–51.

Gardner, W.S., H.L. Windom, J.A. Stephens, F.E. Taylor, and R.R. Stickney. 1975. Concentrations of total mercury in fish and other coastal organisms: implications to mercury cycling, In F.G. Howell, J.B. Gentry, and M.H. Smith (Eds.), *Mineral Cycling in Southeastern Ecosystems, p. 268–278.* U.S. Energy Res. Dev. Admin. Available as CONF-740513 from NTIS, U.S. Dept. of Commerce, Springfield, VA.

Gariboldi, J.C., C.H. Jagoe, and A.L. Bryan Jr. 1998. Dietary exposure to mercury in nestling wood storks (*Mycteria americana*) in Georgia, *Arch. Environ. Contam. Toxicol.*, 34, 398–405.

Gaskin, D.E., R. Frank, M. Holdrinet, K. Ishida, C.J. Walton, and M. Smith. 1973. Mercury, DDT, and PCB in harbour seals (*Phoca vitulina*) from the Bay of Fundy and Gulf of Maine, *J. Fish. Res. Bd. Canada*, 30, 471–475.

Gaskin, D.E., K. Ishida, and R. Frank. 1972. Mercury in harbour porpoises (*Phocoena phocoena*) from the Bay of Fundy region, *J. Fish. Res. Bd. Canada,* 29, 1644–1646.

Gaskin, D.E., G.J.D. Smith, P.W. Arnold, M.V. Louisy, R. Frank, M. Holdrinet, and J.W. McWade. 1974. Mercury, DDT, dieldrin, and PCB in two species of Odontoceti (Cetacea) from St. Lucia, Lesser Antilles, *J. Fish. Res. Bd. Canada,* 31, 1235–1239.

Gerstenberger, S.L., J. Pratt-Shelley, M.S. Beattie, and J.A. Dellinger. 1993. Mercury concentrations of walleye (*Stizostedion vitreum vitreum*) in 34 northern Wisconsin lakes, *Bull. Environ. Contam. Toxicol.*, 50, 612–617.

Giblin, F.J. and E.J. Massaro. 1973. Pharmacodynamics of methylmercury in rainbow trout (*Salmo gairdneri*): tissue uptake, distribution, and excretion, *Toxicol. Appl. Pharmacol.*, 24, 81–91.

Gladstone, W. 1996. Trace metals in sediments, indicator organisms and the traditional seafoods of the Torres Strait, *Great Barrier Reef Marine Park Authority*, Townsville, Australia.

Gochfeld. M. 1980. Tissue distribution of mercury in normal and abnormal young common terns, *Mar. Pollut. Bull.*, 11, 362–366.

Goldstein, R.M., M.E. Brigham, and J.C. Stauffer. 1996. Comparison of mercury concentrations in liver, muscle, whole bodies, and composites of fish from the Red River of the North, *Can. J. Fish. Aquat. Sci.*, 53, 244–252.

Greichus, Y.A., A. Greichus, and R.J. Emerick. 1973. Insecticides, polychlorinated biphenyls, and mercury in wild cormorants, pelicans, their eggs, food, and environment, *Bull. Environ. Contam. Toxicol.*, 9, 321–328.

Greig, R.A. and J. Krzynowek. 1979. Mercury concentrations in three species of tunas collected from various oceanic waters, *Bull. Environ. Contam. Toxicol.*, 22, 120–127.

Greig, R. and D. Wenzloff. 1977. Final report on heavy metals in small pelagic finfish, euphasid crustaceans and apex predators, including sharks, as well as on heavy metals and hydrocarbons (C_{15+}) in sediments collected at stations in and near deepwater dumpsite 106. In *Baseline Report of the Environmental Conditions on Deepwater Dumpsite 106, Vol. III, Contaminant Inputs and Chemical Characteristics*, p. 547–564. U.S. Dept. Commerce, NOAA, Rockville, MD.

Greig, R.A., D.R. Wenzloff, and J.B. Pearce. 1976. Distribution and abundance of heavy metals in finfish, invertebrates, and sediments collected at a deepwater disposal site, *Mar. Pollut. Bull,*, 7, 185–187.

Greig, R.A., D. Wenzloff, and C. Shelpuk. 1975. Mercury concentrations in fish, North Atlantic offshore waters — 1971, *Pestic. Monitor. J.*, 9, 15–20.

Greig, R.A., D. Wenzloff, C. Shelpuk, and A. Adams. 1977. Mercury concentrations in three species of fish from North Atlantic offshore waters, *Arch. Environ. Contam. Toxicol.*, 5, 315–323.

Grieb, T. M., C. T. Driscoll, S. P. Gloss, C. L. Schofield, G. L. Bowie, and D. B. Porcella. 1990. Factors affecting mercury accumulation in fish in upper Michigan peninsula, *Environ. Toxicol. Chem.*, 9, 919–930.

Gutenmann, W.H., J.G. Ebel, Jr., H.T. Kuntz, K.S. Yourstone, and D.J. Lisk. 1992. Residues of *p,p′*-DDE and mercury in lake trout as a function of age. *Arch. Environ. Contam. Toxicol.*, 22, 452–455.

Gutleb, A.C., A. Kranz, G. Nechay, and A. Toman. 1998. Heavy metal concentrations in livers and kidneys of the otter (*Lutra lutra*) from central Europe, *Bull. Environ. Contamin. Toxicol.*, 60, 273–279.

Haines, T. A., V. Komov, and C. H. Jagoe. 1992. Lake acidity and mercury content in Darwin National Reserve, Russia, *Environ. Pollut.*, 78, 107–112.

Halbrook, R.S., J.H. Jenkins, P.B. Bush, and N.D. Seabolt. 1994. Sublethal concentrations of mercury in river otters: monitoring environmental contamination, *Arch. Environ. Contam. Toxicol.*, 27, 306–310.

Hall, A.S., F.M. Teeny, and E. Gauglitz, Jr. 1976a. Mercury in fish and shellfish of the northeast Pacific. II. Sablefish, *Anoploma fimbria, U.S. Dept. Comm. Fish. Bull.*, 74, 791–799.

Hall, A.S., F.M. Teeny, L.G. Lewis, W.H. Hardman, and E.J. Gauglitz Jr. 1976b. Mercury in fish and shellfish of the northeast Pacific. I. Pacific halibut, *Hippoglossus stenolepis, U.S. Dept. Comm. Fish. Bull,* 74, 783–789.

Hall, R.A., E.G. Zook, and G.M. Meaburn. 1978. National Marine Fisheries Service survey of trace elements in the fishery resource, *U.S. Dept. Commerce NOAA Tech. Rep.*, NMFS SSRF-721, 313 pp.

Hall, R.J. 1980. Effects of environmental contaminants on reptiles: a review, *U.S. Fish Wildl. Serv. Spec. Sci. Rep.-Wildl.*, 228, 1–12.

Hamada, M., Y. Inamasu, and T. Ueda. 1977. On mercury and selenium in tuna fish tissues. III. Mercury distribution in yellowfin tuna, *Shimonoski Univ. Fish.*, 25, 213–220.

Hammerschmidt, C.R., J.G. Wiener, B.E. Frazier, and R.G. Rada. 1999. Methylmercury content of eggs in yellow perch related to maternal exposure in four Wisconsin lakes, *Environ. Sci. Technol.*, 33, 999–1003.

Hannerz, L. 1968. Experimental investigations on the accumulation of mercury in water organisms, *Rep. Inst. Freshwater Res. Drotting.*, 48, 120–176.

Harada, M., S. Nakachi, K. Tasaka, S. Sakashita, K. Muta, K. Yamagida, R. Doi, T. Kizaki, and T. Ohno. 2001. Wide use of skin-lightening soap may cause mercury poisoning in Kenya, *Sci. Total Environ.*, 269, 183–187.

Harding, L.E. 2004. Environmental contaminants in wild martens (*Martes americana*) and wolverines (*Gulo luscus*), *Bull. Environ. Contam. Toxicol.*, 73, 98–105.

Harms, U., H.E. Drescher, and E. Huschenbeth. 1978. Further data on heavy metals and organochlorines in marine mammals for German coastal waters, *Meeresforsch*, 26, 153–161.

Haug, A., S. Melsom, and S. Omang. 1974. Estimation of heavy metal pollution in two Norwegian fjord areas by analysis of the brown alga *Ascophyllum nodosum*, *Environ. Pollut.*, 7, 179–192.

Haynes, D., S. Carter, C. Gaus, J. Muller, and W. Dennison. 2005. Organochlorine and heavy metal concentrations in blubber and liver tissue collected from Queensland (Australia) dugong (*Dugong dugon*), *Mar. Pollut. Bull.*, 51, 361–369.

Haynes, D. and D. Kwan. 2001. *Trace Metal Concentrations in the Torres Strait Environment and Traditional Seafood Species, 1997–2000.* Torres Strait Regional Authority, Thursday Island, Queensland, Australia.

Hearnden, E.H. 1970. Mercury pollution, *Fish. Canada*, 22 (10), 3–6.

Heaton-Jones, T.G., B.L. Homer, D.L. Heaton-Jones, and S.F. Sundlof. 1997. Mercury distribution in American alligators (*Alligator mississippiensis*) in Florida, *J. Zoo Wildl. Med.*, 28, 62–70.

Heinz, G.H., S.D. Haseltine, R.J. Hall, and A.J. Krynitsky. 1980. Organochlorine and mercury residues in snakes from Pilot Island and Spider Islands, Lake Michigan — 1978, *Bull. Environ. Contam. Toxicol.*, 25, 738–743.

Henderson, C. and W.E. Shanks. 1973. Mercury concentrations in fish. In D.R. Buhler (Ed.), *Mercury in the Western Environment,* p. 45–58. Oregon State University, Corvallis.

Henny, C.J. and G.B. Herron. 1989. DDE, selenium, mercury, and white-faced ibis reproduction at Carson Lake, Nevada, *J. Wildl. Manage.*, 53, 1032–1045.

Hill, W.R., A.J. Stewart, and G.E. Napolitano. 1996. Mercury speciation and bioaccumulation in lotic primary producers and primary consumers, *Can. J. Fish. Aquat. Sci.*, 53, 812–819.

Hillestad, H.O., R.J. Reimold, R.R. Stickney, H.L. Windom, and J.H. Jenkins. 1974. Pesticides, heavy metals, and radionuclide uptake in loggerhead sea turtles from South Carolina and Georgia, *Herpetol. Rev.*, 5, 75.

Hoffman, D.J., H.M. Ohlendorf, C.M. Marn, and G.W. Pendleton. 1998. Association of mercury and selenium with altered glutathione metabolism and oxidative stress in diving ducks from the San Francisco Bay region, USA, *Environ. Toxicol. Chem.*, 17, 167–172.

Holden, A.V. 1973. Mercury in fish and shellfish, a review, *J. Food Technol.*, 8, 1–25.

Holden, A.V. 1975. The accumulation of oceanic contaminants in marine mammals, *Rapp. P. -v. Reun. Cons. Int. Explor. Mer*, 169, 353–361.

Honda, K., H. Ichihashi, and R. Tatsukawa. 1987. Tissue distribution of heavy metals and their variations with age, sex, and habitat in Japanese serows (*Capricornus crispus*), *Arch. Environ. Contam. Toxicol.*, 16, 551–561.

Honda, K., T. Nasu, and R. Tatsukawa. 1986. Seasonal changes in mercury accumulation in the black-eared kite, *Milvus migrans lineatus*, *Environ. Pollut.*, 42A, 325–334.

Hord, L.J., M. Jennings, and A. Brunell. 1990. Mercury contamination of Florida alligators. In *IUCN/SSC Crocodile Specialist Group 10th Working Meeting,* April 27–30, Gainesville, Florida. *Proceedings, Volume 1, World Conservation Union*, p. 229–240. Gland, Switzerland.

Hornung, H., B.S. Krumgalz, and Y. Cohen. 1984. Mercury pollution in sediments, benthic organisms and inshore fishes of Haifa Bay, Israel, *Mar. Environ. Res.*, 12, 191–208.

Huckabee, J.W. and B.G. Blaylock. 1972. Transfer of mercury and cadmium from terrestrial to aquatic ecosystems. In S.K. Dhar (Ed.), *Metal Ions in Biological Systems*, p. 125–160. Plenum, New York.

Huckabee, J.W., J.W. Elwood, and S.G. Hildebrand. 1979. Accumulation of mercury in freshwater biota. In J.O. Nriagu (Ed.), *The Biogeochemistry of Mercury in the Environment*, p. 277–302. Elsevier/North-Holland Biomedical Press, New York.

Hui, C.A., D. Rudnick, and E. Williams. 2005. Mercury burdens in Chinese mitten crabs (*Eriocheir sinensi*) in three tributaries of southern San Francisco Bay, California, USA, *Environ. Pollut.*, 133, 481–487.

Hyvarinen, H., P. Tyni, and P. Nieminen. 2003. Effects of moult, age, and sex on the accumulation of heavy metals in the otter (*Lutra lutra*) in Finland, *Bull. Environ. Contamin. Toxicol.*, 70. 278–284.

Iglesias, N. and P.E. Penchaszadeh. 1983. Mercury in sea stars from Golfo Triste, Venezuela, *Mar. Pollut. Bull.*, 14, 396–398.

Ikemoto, T., T. Kunito, H. Tanaka, N. Baba, N. Miyazaki, and S. Tanabe. 2004. Detoxification mechanism of heavy metals in marine mammals and seabirds: interaction of selenium with mercury, silver, zinc, and cadmium in liver, *Arch. Environ. Contam. Toxicol.*, 47, 402–413.

Itano, K., S. Kawai, N. Miyazaki, R. Tatsukawa, and T. Fujiyama. 1984a. Mercury and selenium levels in striped dolphins caught off the Pacific coast of Japan, *Agric. Biol. Chem.*, 48, 1109–1116.

Itano, K., S. Kawai, N. Miyazaki, R. Tatsukawa, and T. Fujiyama. 1984b. Body burdens and distribution of mercury and selenium in striped dolphins, *Agric. Biol. Chem.*, 48, 1117–1121.

Itano, K., S. Kawai, N. Miyazaka, R. Tatsukawa, and T. Fujiyama. 1984c. Mercury and selenium levels at the fetal and suckling stages of striped dolphin, *Stenella coeruleoalba*, *Agric. Biol. Chem.*, 48, 1691–1698.

Jackson, T.A. 1988. The mercury problem in recently formed reservoirs of northern Manitoba (Canada): effects of impoundment and other factors on the production of methyl mercury by microorganisms in sediments, *Can. J. Fish. Aquat. Sci.*, 45, 97–121.

Jackson, T.A. 1991. Biological and environmental control of mercury accumulation by fish in lakes and reservoirs of northern Manitoba, Canada, *Can. J. Fish. Aquat. Sci.*, 48, 2449–2470.

Jacobs, G. 1977. Gesampt — und organisch gebundener Quecksilber-gehalt in Fischen Deutschen Fangerunden (Total and organically bound mercury content in fishes from German Fishing grounds.), *Z. Lebensm. Unters — Forsch*, 164, 71–76.

Jenkins, D.W. 1980. In Biological Monitoring of Toxic Trace Metals, Volume 2, Toxic Trace Metals in Plants and Animals of the World, Part II, Mercury, p. 779–982. U.S. Environmental Protection Agency, Rep. 600/3-80-091.

Jernelov, A. 1972. Mercury — a case study of marine pollution. *The Changing Chemistry of the Oceans. Proceedings 20th Nobel Symposium*, p. 161–169. August 16–20, 1971, Goteberg, Sweden. John Wiley, New York.

Jernelov, A. and H. Lann. 1971. Mercury accumulations in food chains, *Oikos*, 22, 403–406.

Johnels, A.G. and T. Westermark. 1969. Mercury contamination of the environment in Sweden. In M.W. Miller and G.G. Berg (Eds.), *Chemical Fallout, Current Research on Persistent Pesticides,* p. 221–239. Chas C Thomas, Springfield, IL.

Johnels, A.G., T. Westermark, W. Berg, P.I. Persson, and B. Sjostrand. 1967. Pike (*Esox lucius* L.) and some other aquatic organisms in Sweden as indicators of mercury contamination of the environment, *Oikos*, 18, 323–333.

Johnson, D.L. and R.S. Braman. 1975. The speciation of arsenic and the content of germanium and mercury in members of the pelagic *Sargassum* community, *Deep Sea Res.*, 22, 503–507.

Johnston, T.A., R.A Bodaly, M.A. Latif, R.J.P. Fudge, and N.E. Strange. 2001. Intra- and inter-population variability in maternal transfer of mercury to eggs of walleye (*Stizostedion vitreum*), *Aquat. Toxicol.*, 52, 73–85.

Joiris, C.R., I.B. Ali, L. Holsbeek, M. Kanuya-Kinoti, and Y. Tekele-Michael. 1997. Total and organic mercury in Greenland and Barents Seas demersal fish, *Bull. Environ. Contam. Toxicol.*, 58, 101–107.

Joiris, C.R., L. Holsbeek, J. Bouquegneau, and M. Bossicart. 1991. Mercury contamination of the harbor porpoise *Phocoena phocoena* and other cetaceans from the North Sea and the Kattegat, *Water Air Soil Pollut.*, 56, 283–293.

Jones, A.M., Y. Jones, and W.D.P. Stewart. 1972. Mercury in marine organisms of the Tay region, *Nature*, 238, 164–165.

Jones, D., K. Ronald, D.M. Lavigne, R. Frank, M. Holdrinet, and J.F. Uthe. 1976. Organochlorine and mercury residues in the harp seal (*Pagophilus groenlandicus*), *Sci. Total Environ.*, 5, 181–195.

Julshamn, K. and O.R. Braekkan. 1978. The relation between the concentration of some main elements and the stages of maturation of ovaries in cod (*Gadus morhua*), *Fisk. Dir. Skr. Ser. Ernoering*, 1, 1–15.

Kacmar, P. and J. Legath. 1991. Mercury in some species of game. In B. Bobek, K. Perzanowski, and W. Regelin (Eds.), Global trends in wildlife management. *Trans. 18th IUGB Congress,* p. 381–384. Jagiellon University, Krakow, 1987. Swiat Press, Krakow-Warsaw.

Kai, N., T. Ueda, Y. Takeda, and A. Kataoka. 1988. The levels of mercury and selenium in blood of tunas, *Nippon Suisan Gakkaishi*, 54, 1981–1985.

Kairu, J.K. 1996. Heavy metal residues in birds of Lake Nakuru, Kenya, *African J. Ecol.*, 34, 397–400.

Kamps, L.R., R. Carr, and H. Miller. 1972. Total mercury-monomethylmercury content of several species of fish, *Bull Environ. Contam. Toxicol.*, 8, 273–279.

Karbe, L., C. Schnier, and H.O. Siewers, 1977. Trace elements in mussels (*Mytilus edulis*) from coastal areas of the North Sea and the Baltic. Multielement analyses using instrumental neutron activation analysis, *J. Radioanal. Chem.*, 37, 927–943.

Kari, T. and P. Kauranen. 1978. Mercury and selenium contents of seals from fresh and brackish water in Finland, *Bull. Environ. Contam. Toxicol.*, 19, 273–280.

Karlog, O. and B. Clausen. 1983. Mercury and methylmercury in liver tissue from ringed herring gulls collected in three Danish localities, *Nord. Vet.-Med.*, 35, 245–250.

Karppanen, E. and K. Henriksson. 1970. Mercury content of game birds in Finland, *Nordisk Medicin*, 84(35), 1097–1098.

Kazantzis, G. 1971. The poison chain for mercury in the environment, *Int. J. Environ. Stud.*, 1, 301–306.

Khan, A.T. and D.M. Forester. 1995. Mercury in white-tailed deer forage in Russel Plantation, Macon County, Alabama, *Veterin. Human Toxicol.*, 37, 45–46.

Khan, A.T., S.J. Thompson, and H.W. Mielke. 1995. Lead and mercury levels in raccoons from Macon County, Alabama, *Bull. Environ. Contam. Toxicol.*, 54, 812–816.

Kidd, K.A., G. Stern, and J. Lemoalle. 2004. Mercury and other contaminants in fish from Lake Chad, Africa, *Bull. Environ. Contam. Toxicol.*, 73, 249–256.

Kim, C.Y. 1972. Studies on the contents of mercury, cadmium, lead, and copper in edible seaweeds in Korea, *Bull. Korean Fish. Soc.*, 5(3), 88–96.

Kim, E.Y., K. Saeki, S. Tanabe, H. Tanaka, and R. Tatsukawa. 1996a. Specific accumulation of mercury and selenium in seabirds, *Environ. Pollut.*, 94, 261–265.

Kim, E.Y., T. Murakami, K. Saeki, and R. Tatsukawa. 1996b. Mercury levels and its chemical form in tissues and organs of seabirds, *Arch. Environ. Contam. Toxicol.*, 30, 259–266.

Kim, K.C., R.C. Chu, and G.P. Barron. 1974. Mercury in tissue and lice of northern fur seals, *Bull. Environ. Contam. Toxicol.*, 11, 281–284.

King, K.A., T.W. Custer, and J.S. Quinn. 1991. Effects of mercury, selenium, and organochlorine contaminants on reproduction of Forster's terns and black skimmers nesting in a contaminated Texas bay, *Arch. Environ. Contam. Toxicol.*, 20, 32–40.

Kingman, A., T. Albertini, and U. Brown. 1998. Mercury concentrations in urine and whole blood associated with amalgam exposure in a U.S. military population, *J. Dent. Res.*, 77, 461–467.

Kiorboe, T., F. Mohlenberg, and H.U. Riisgard. 1983. Mercury levels in fish, invertebrates and sediment in a recently recorded polluted area (Nissum Broad, western Limfjord, Denmark), *Mar. Pollut. Bull.*, 14, 21–24.

Kleinert, S.J. and P.E. Degurse. 1972. Mercury levels in Wisconsin fish and wildlife, *Dept. Nat. Res. Madison, WI, Tech. Bull.*, 52, 1–22.

Klemmer, H.W., C.S. Unninayer, and W.I. Ukobo. 1976. Mercury content of biota in coastal waters in Hawaii, *Bull. Environ.Contam.Toxicol.*, 15, 454–457.

Koeman, J.H., W.H.M. Peeters, C.H.M. Koudstaal-Hol, P.S. Tijoe, and J.J.M. de Goeij. 1973. Mercury-selenium correlations in marine mammals, *Nature*, 245, 385–386.

Koeman, J.H., W.S.M. van de Ven, J.J.M. de Goeij, P.S. Tijoe, and J.L. van Haaften. 1975. Mercury and selenium in marine mammals and birds, *Sci. Total Environ.*, 3, 279–287.

Koivusaari, J., I. Nuuja, R. Palokangas, and M.L. Hattula. 1976. Chlorinated hydrocarbons and total mercury in the prey of white tailed eagle (*Haliaeetus albicilla* L.) in the Quarken Straits of the Gulf of Bothnia, Finland, *Bull. Environ. Contam. Toxicol.*, 15, 235–241.

Koster, M.D., D.P. Ryckman, D.V.C. Weseloh, and J. Struger. 1996. Mercury levels in Great Lakes herring gull (*Larus argentatus*) eggs, 1972–1992, *Environ. Pollut.*, 93, 261–270.

Krom, M.D., H. Hornung, and Y. Cohen. 1990. Determination of the environmental capacity of Haifa Bay with respect to the input of mercury, *Mar. Pollut. Bull.*, 21, 349–354.

Krynski, A., J. Kaluzinski, M. Wlazelko, and A. Adamowski. 1982. Contamination of roe deer by mercury compounds, *Acta Theriol.*, 27, 499–507.

Kucera, E. 1983. Mink and otter as indicators of mercury in Manitoba waters, *Can. J. Zool.*, 61, 2250–2256.

Kumagai, H. and K. Saeki. 1978. Contents of total mercury, alkyl mercury and methylmercury in some coastal fish and shells, *Bull. Japan. Soc. Sci. Fish.*, 44, 807–811.

Kureishy, T.W., M.D. George, and R. Sengapta. 1979. Total mercury content in some marine fish from the Indian Ocean, *Mar. Pollut. Bull.*, 10, 357–360.

Lange, T.R., H.E. Royals, and L.L. Connor. 1993. Influence of water chemistry on mercury concentration in largemouth bass from Florida lakes, *Trans. Am. Fish. Soc.*, 122, 74–84.

Lange, T.R., H.E. Royals, and L.L. Connor. 1994. Mercury accumulation in largemouth bass (*Micropterus salmoides*) in a Florida lake, *Arch. Environ. Contam. Toxicol.*, 27, 466–471.

Langlois, C., R. Langis, and M. Perusse. 1995. Mercury contamination in northern Quebec environment and wildlife, *Water Air Soil Pollut.*, 80, 1021–1024.

Lathrop, R.C., P.W. Rasmussen, and D.R. Knauer. 1991. Mercury concentrations in walleyes from Wisconsin (USA) lakes, *Water Air Soil Pollut.*, 56, 295–307.

Latif, M.A., R.A. Bodaly, T.A. Johnstom, and R.J.P. Fudge. 2001. Effects of environmental and maternally derived methylmercury on the embryonic and larval stages of walleye (*Stizostedium vitreum*), *Environ. Pollut.*, 111, 139–148.

Leah, R.T., S.E. Collings, M.S. Johnson, and S.J. Evans. 1993. Mercury in plaice (*Pleuronectes platessa*) from the sludge disposal ground of Liverpool Bay, *Mar. Pollut. Bull.*, 26, 436–439.

Leah, R., S. Evans, and M. Johnson. 1991. Mercury in muscle tissue of lesser-spotted dogfish (*Scyliorhinus caniculus* L.) from the north-east Irish Sea, *Sci. Total Environ.*, 108, 215–224.

Leah, R.T., S.J. Evans, and M.S. Johnson. 1992. Mercury in flounder (*Platichthys flesus* L.) from estuaries and coastal waters of the north-east Irish Sea, *Environ. Pollut.*, 75, 317–322.

Leatherland, T.M. and J.D, Burton. 1974. The occurrence of some trace metals in coastal organisms with particular reference to the Solent region, *J. Mar. Biol. Assoc. U.K.*, 54, 457–468.

Leatherland, T.M., J.D. Burton, F. Culkin, M.J. McCartney, and R.J. Morris. 1973. Concentrations of some trace metals in pelagic organisms and of mercury in northeast Atlantic Ocean water, *Deep Sea Res.*, 20, 679–685.

Lentfer, J.W. and W.A. Galster. 1987. Mercury in polar bears from Alaska, *J. Wildl. Dis.*, 23, 338–341.

Lewis, S.A., P.H. Becker, and R.W. Furness. 1993. Mercury levels in eggs, tissues, and feathers of herring gulls *Larus argentatus* from the German Wadden Sea coast, *Environ. Pollut.*, 80, 293–299.

Liang, L.N., J.T. Hu, D.Y. Chen, Q.F. Zhou, B. He, and G.B. Jiang. 2004. Primary investigation of heavy metal contamination status in molluscs collected from Chinese coastal sites, *Bull. Environ. Contam. Toxicol.*, 72, 937–944.

Lindberg, E. and C. Harriss. 1974. Mercury enrichment in plant detritus, *Mar. Pollut. Bull.*, 5, 93–95.

Lindberg, P. 1984. Mercury in feathers of Swedish gyrfalcons, *Falco rusticolus*, in relation to diet, *Bull. Environ. Contam. Toxicol.*, 32, 453–459.

Linder, G. and B. Grillitsch. 2000. Ecotoxicology of metals. In D.W. Sparling, G. Linder, and C.A. Bishop (Eds.), *Ecotoxicology of Amphibians and Reptiles,* p. 325–459. SETAC Press, Pensacola, FL.

Lindsay, R.C. and R.W. Dimmick. 1983. Mercury residues in wood ducks and wood duck foods in eastern Tennessee, *J. Wildl. Dis.*, 19, 114–117.

Littrell, E.E. 1991. Mercury in western grebes at Lake Berryessa and Clear Lake, California, *Calif. Fish Game*, 77, 142–144.

Lloyd, E.T., W.T. Schrenk, and J.O. Stoffer. 1977. Mercury accumulation in trout of southern Missouri, *Environ. Res.*, 13, 62–73.

Lock, J.W., D.R. Thompson, R.W. Furness, and J.A. Bartle. 1992. Metal concentrations in seabirds of the New Zealand region, *Environ. Pollut.*, 75, 289–300.

Lodenius, M. 1983. The effects of peatland drainage on the mercury contents of fish, *Suo*, 34, 21–24.

Lodenius, M. 1991. Mercury concentrations in an aquatic ecosystem during twenty years following abatement of the pollution source, *Water Air Soil Pollut.*, 56, 323–332.

Lodenius, M., A. Seppanen, and M. Herrnanen. 1983. Accumulation of mercury in fish and man from reservoirs in northern Finland, *Water Air Soil Pollut.*, 19, 237–246.

Lodenius, M. and E. Tulisalo. 1984. Environmental mercury contamination around a chlor-alkali plant, *Bull. Environ. Contam. Toxicol.*, 32, 439–444.

Lowe, T.P., T.W. May, W.G. Brumbaugh, and D.A. Kane. 1985. National Contaminant Biomonitoring Program: concentrations of seven elements in freshwater fish, 1978–1981, *Arch. Environ. Contam. Toxicol.*, 14, 363–388.

Luke, B.G., G.W. Johnstone, and E.J. Woehler. 1989. Organochlorine pesticides, PCBs and mercury in Antarctic and Subantarctic seabirds, *Chemosphere*, 19, 2007–2021.

Lyle, J.M. 1984. Mercury concentrations in four carcharhinid and three hammerhead sharks from coastal waters of the Northern Territory, *Austral. J. Mar. Freshwater Res.*, 35, 441–451.

Maas, R.P., S.C. Patch, and K.R. Sergent. 2004. *A Statistical Analysis of Factors Associated with Elevated Hair Mercury Levels in the U.S. Population: An Interim Progress Report*. Tech. Rep. 04–136, Environ. Qual. Inst., Univ. North Carolina, Ashville. 13 pp.

Mackay, N.J., M.N. Kazacos, R.J. Williams, and M.I. Leedow. 1975. Selenium and heavy metals in black marlin, *Mar. Pollut. Bull.*, 6, 57–60.

Martin, J.H., P.D. Elliot, V.C. Anderlini, D. Girvan, S.A. Jacobs, R.W. Risebrough, R.L. Delong, and W.G. Gilmartin. 1976. Mercury-selenium-bromine imbalance in premature parturient California sea lions, *Mar. Biol.*, 35, 91–104.

Martin, J.H. and G.A. Knauer. 1973. The elemental composition of plankton, *Geochim. Cosmochim. Acta*, 37, 1639–1653.

Mason, C.F., G. Ekins, and J.R. Ratford. 1997. PCB congeners, DDE, dieldrin, and mercury in eggs from an expanding colony of cormorants (*Phalacrocorax carbo*), *Chemosphere*, 34, 1845–1849.

Mason, C.F., N.I. Last, and S.M. Macdonald. 1986. Mercury, cadmium, and lead in British otters, *Bull. Environ. Contamin. Toxicol.*, 37, 844–849.

Mason, C.F. and A.B. Madsen. 1992. Mercury in Danish otters (*Lutra lutra*), *Chemosphere*, 25, 865–867.

Mason, C.F. and P. Reynolds. 1988. Organochlorine residues and metals in otters from the Orkney Islands, *Mar. Pollut. Bull.*, 19, 80–81.

Mathieson, S. and D.S. McLusky. 1995. Inter-species variation of mercury in skeletal muscle of five fish species from inshore waters of the Firth of Clyde, Scotland, *Mar. Pollut. Bull.*, 30, 283–286.

Matida, Y. and H. Kumada. 1969. Distribution of mercury in water, bottom mud and aquatic organisms of Minamata Bay, the River Agano and other water bodies in Japan, *Bull. Freshwater Fish. Res. Lab. (Tokyo)*, 19(2), 73–93.

Matsunaga, K. 1978. Concentration of mercury in marine animals, *Bull. Fac. Fish. Hokkaido Univ.*, 29(1), 70–74.

McMurtry, M.J., D.J. Wales, W.A. Scheider, G.L. Beggs, and P.E. Dimond. 1989. Relationship of mercury concentrations in lake trout (*Salvelinus namaycush*) and smallmouth bass (*Micropterus dolomieui*) to the physical and chemical characteristics of Ontario lakes, *Can. J. Fish. Aquat. Sci.*, 46, 426–434.

Menasveta, P. and R. Siriyong. 1977. Mercury content of several predacious fish in the Andaman Sea, *Mar. Pollut. Bull.*, 8, 200–204.

Meyer, M.W., D.C. Evers, T. Daulton, and W.E. Braselton. 1995. Common loons (*Gavia immer*) nesting on low pH lakes in northern Wisconsin have elevated blood mercury content, *Water Air Soil Pollut.*, 80, 871–880.

Meyer, M.W., D.C. Evers, J.J. Hartigan, and P.S. Rasmussen. 1998. Patterns of common loon (*Gavia immer*) mercury exposure, reproduction, and survival in Wisconsin, USA, *Environ. Toxicol. Chem.*, 17, 184–190.

Meyers-Schone, L., L.R. Shugart, J.J. Beauchamp, and B.T. Walton. 1993. Comparison of two freshwater turtle species as monitors of radionuclide and chemical contamination: DNA damage and residue analysis, *Environ. Toxicol. Chem.*, 12, 1487–1496.

Meyers-Schone, L. and B.T. Walton. 1994. Turtles as monitors of chemical contaminants in the environment, *Rev. Environ. Contam. Toxicol.*, 135, 93–153.

Miller, G.E., P.M. Grant, R. Kishore, F.J. Steinkruger, F.S. Rowland, and V.P. Guinn. 1972. Mercury concentrations in museum specimens of tuna and swordfish, *Science*, 175, 1121–1122.

Miller, T.J. and D.J. Jude. 1984. Organochlorine pesticides, PBBS, and mercury in round whitefish fillets from Saginaw Bay, Lake Huron, 1977–1978, *J. Great Lakes Res.*, 10, 215–220.

Minigawa, K., T. Sasaki, Y. Takizawa, R. Tamura, and T. Oshina. 1980. Accumulation route and chemical form of mercury in mushroom species. *Bull. Environ. Contam. Toxicol.*, 25, 382–388.

Monteiro, L.R. and R.W. Furness. 1995. Seabirds as monitors of mercury in the marine environment, *Water Air Soil Pollut.*, 90, 851–870.

Montiero, L.R. and R.W. Furness. 1997. Accelerated increase in mercury contamination in North Atlantic mesopelagic food chains as indicated by time series of seabird feathers, *Environ. Toxicol. Chem.*, 16, 2489–2493.

Monteiro, L.R., R.W. Furness, and A.J. del Novo. 1995. Mercury levels in seabirds from the Azores, mid-north Atlantic Ocean, *Arch. Environ. Contam. Toxicol.*, 28, 304–309.

Monteiro, L.R. and H.D. Lopes. 1990. Mercury content of swordfish, *Xiphias gladius*, in relation to length, weight, age, and sex, *Mar. Pollut. Bull.*, 21, 293–296.

Mora, M.A. 1996. Organochlorines and trace elements in four colonial waterbird species nesting in the lower Laguna Madre, Texas, *Arch. Environ. Contam. Toxicol.*, 31, 533–537.

Mullie, W.C., A. Massi, S. Focardi, and A. Renzoni. 1992. Residue levels of organochlorines and mercury in cattle egret, *Bubulcus ibis*, eggs from the Faiyum oasis, Egypt, *Bull. Environ. Contam. Toxicol.*, 48, 739–746.

Munn, M.D. and T.M. Short. 1997. Spatial heterogeneity of mercury bioaccumulation by walleye in Franklin D. Roosevelt Lake and the upper Columbia River, Washington, *Trans. Am. Fish. Soc.*, 126, 477–487.

Myklestad, S., I. Eide, and S. Melsom. 1978. Exchange of heavy metals in *Ascophyllum nodosum* (L.) Le Jod. *in situ* by means of transplanting experiments, *Environ. Pollut.*, 16, 277–284.

Newton, I. and E.A. Galbraith. 1991. Organochlorines and mercury in the eggs of golden eagles *Aquila chrysaetos* from Scotland, *Ibis*, 133, 115–120.

Newton, I., I. Wyllie, and A. Asher. 1993. Long-term trends in organochlorine and mercury residues in some predatory birds in Britain, *Environ. Pollut.*, 79, 143–151.

Nicoletto, P.F. and A.C. Hendricks. 1988. Sexual differences in accumulation of mercury in four species of centrarchid fishes, *Can. J. Zool.*, 66, 944–949.

Norheim, G. and B. Kjos-Hanssen. 1984. Persistent chlorinated hydrocarbons and mercury in birds caught off the west coast of Spitsbergen, *Environ. Pollut.*, 33A, 143–152.

Norheim, G., L. Some, and G. Holt. 1982. Mercury and persistent chlorinated hydrocarbons in Antarctic birds from Bouvetoya and Dronning Maud Land, *Environ. Pollut.*, 28A, 233–240.

Norum, U., V.W.M. Lai., and W.R. Cullen. 2005. Trace element distribution during the reproductive cycle of female and male spiny and Pacific scallops, with implications for biomonitoring, *Mar. Pollut. Bull.*, 50, 175–184.

Nuorteva, P. and E. Hasanen. 1971. Observations on the mercury content of *Myoxocephalus quadricornis* (L.) (Teleostei, cottidae) in Finland, *Ann. Zool. Fennici*, 8, 331–335.

Nuorteva, P., E. Hasanen, and S.L. Nuorteva. 1975. The effectiveness of the Finnish anti-mercury measurements in the moderately polluted area of Hameenkyro, *Ymparisto ja Terveys*, 6(8), 611–635.

Ociepa, A. and M. Protasowicki. 1976. A relationship between total mercury content and a kind of food in some chosen Pacific fish species, *Mar. Fish. Fd. Tech.*, 60, 83–87.

Odsjo, T., A. Bignert, M. Olsson, L. Asplund, U. Eriksson, L. Haggberg, K. Litzen, C. de Wit, C. Rappe, and K. Aslund. 1997. The Swedish environmental specimen bank — application in trend monitoring of mercury and some organohalogenated compounds, *Chemosphere*, 34, 2059–2066.

Ohlendorf, H.M., D.W. Anderson, D.E. Boellstorff, and B.M. Mulhern. 1985. Tissue distribution of trace elements and DDE in brown pelicans, *Bull. Environ. Contam. Toxicol.*, 35, 183–192.

Ohlendorf, H.M. and C.S. Harrison. 1986. Mercury, selenium, cadmium and organochlorines in eggs of three Hawaiian seabird species, *Environ. Pollut.*, 11B, 169–191.

Ohlendorf, H.M., R.W. Risebrough, and K. Vermeer. 1978. Exposure of marine birds to environmental pollutants, *U.S. Fish. Wildl. Serv. Wildlife Rep.*, 9, 1–40.

Ohno, H., R. Doi, Y. Kashima, S. Murae, T. Kizaki, Y. Hitomi, N. Nakano, and M. Harada. 2004. Wide use of merthiolate may cause mercury poisoning in Mexico, *Bull. Environ. Contam. Toxicol.*, 73, 777–780.

Papadopoulu, C. 1973. The elementary composition of marine invertebrates as a contribution to the sea pollution investigation. In *Proc. MAMBO meeting*, p. 1–18. Castellabate, Italy, June 18–22, 1973.

Papadopoulu, C., I. Hadzistelios, and A.P. Grimanis. 1972. Trace element uptake by *Cynthia claudicans* (Savigny), *Greek Limnol. Ocean.*, 11, 651–663.

Parks, J.W., C. Curry, D. Romani, and D.D. Russell. 1991. Young northern pike, yellow perch and crayfish as bioindicators in a mercury contaminated watercourse, *Environ. Monitor. Assess.*, 16, 39–73.

Parks, J.W. and A.L. Hamilton. 1987. Accelerating recovery of the mercury-contaminated Wabigoon/English River system, *Hydrobiologia*, 145–188.

Parslow, J.L.F. 1973. Mercury in waders from the Wash, *Environ. Pollut.*, 5, 295–304.

Parslow, J.L.F., D.J. Jefferies, and H.M. Hanson. 1973. Gannet mortality incidents in 1972, *Mar. Pollut. Bull.*, 4, 41–43.

Parvaneh, V. 1977. A survey of the mercury content of the Persian Gulf shrimp, *Bull. Environ. Contam. Toxicol*, 18, 778–782.

Parvaneh, V. 1979. An investigation on the mercury contamination of Persian Gulf fish, *Bull. Environ. Contam. Toxicol.*, 23, 357–359.

Peakall, D.B. and R.J. Lovett. 1972. Mercury: its occurrence and effects in the ecosystem, *Bioscience*, 22(1) 20–25.

Pedersen, E.B., M.E. Jorgensen, M.B. Pedersen, C. Sigaard, T.B. Sorensen, G. Mulvad, J.C. Hansen, G. Asmund, and H. Skjoldborg. 2005. Relationship between mercury in blood and 24-h ambulatory blood pressure in Greenlanders and Danes, *Am. J. Hypertension*, 18, 612–618.

Pentreath, R.J. 1976. The accumulation of mercury by the thornback ray, *Raja clavata*, *J. Exp. Mar. Biol. Ecol.*, 25, 131–140

Peterson, C.L., W.L. Klawe, and G.D. Sharp. 1973. Mercury in tunas: a review, *U.S. Dept. Comm. Fish. Bull*, 71, 603–613.

Phelps, R.J., S. Focardi, C. Fossi, C. Leonzio, and A. Renzoni. 1986. Chlorinated hydrocarbons and heavy metals in crocodile eggs from Zimbabwe, *Trans. Zimbabwe Sci. Assoc.*, 63, 8–15.

Phillips, G.R., P.A. Medvick, D.R. Skaar, and D.E. Knight. 1987. Factors Affecting the Mobilization, Transport, and Bioavailability of Mercury in Reservoirs of the Upper Missouri River Basin. U.S. Fish Wildl. Serv. Tech. Rep. 10. 64 pp.

Plourde, Y., M. Lucotte, and P. Pichet. 1997. Contribution of suspended particulate matter and zooplankton to MeHg contamination of the food chain in midnorthern Quebec (Canada) reservoirs, *Can. J. Fish. Aquat. Sci.*, 54, 821–831.

Pokras, M.A., C. Hanley, and Z. Gordon. 1998. Liver mercury and methylmercury concentrations in New England common loons (*Gavia immer*), *Environ. Toxicol. Chem.*, 17, 202–204.

Porvari, P. and M. Verta. 1995. Methylmercury production in flooded soils: a laboratory study, *Water Air Soil Pollut.*, 80, 765–773.

Post, J.R., R. Vandenbos, and D.J. McQueen. 1996. Uptake rates of food-chain and waterborne mercury by fish: field measurements, a mechanistic model, and an assessment of uncertainties, *Can. J. Fish. Aquat. Sci.*, 53, 395–407.

Preston, A., D.F. Jeffries, J.W.R. Dutton, B.R. Harvey, and A.K. Steele. 1972. British isles coastal waters: the concentrations of selected heavy metals in sea water, suspended matter and biological indicators — a pilot study, *Environ. Pollut.*, 3, 69–82.

Rada, R.G., J.E. Findley, and J.G. Wiener. 1986. Environmental fate of mercury discharged into the upper Wisconsin River, *Water Air Soil Pollut.*, 29, 57–76.

Rainwater, T.R., K.D. Reynolds, J.E. Canas, G.P. Cobb, T.A. Anderson, S.T. McMurry, and P.M. Smith. 2005. Organochlorine pesticides and mercury in cottonmouths (*Agkistrodon piscivorus*) from northeastern Texas, USA, *Environ. Toxicol. Chem.*, 24, 665–673.

Ramos, A., M. de Campos, and A.E. Olszyna-Marzys. 1979. Mercury contamination of fish in Guatemala, *Bull. Environ. Contam. Toxicol.*, 22, 488–493.

Rask, M. and T.R. Metsala. 1991. Mercury concentrations in northern pike, *Esox lucius* L., in small lakes of Evo area, southern Finland, *Water Air Soil Pollut.*, 56, 369–378.

Ratkowsky, D.A., T.G. Dix, and K.C. Wilson. 1975. Mercury in fish in the Derwent Estuary, Tasmania, and its relation to the position of the fish in the food chain, *Austral. J. Mar. Freshwater Res.*, 26, 223–231.

Raymont, J.E.G. 1972. Some aspects of pollution in Southampton waters, *Proc. Roy. Soc. London, Ser B.*, 180(1061), 451–468.

Reimer, A.A. and R.D. Reimer. 1975. Total mercury in some fish and shellfish along the Mexican coast, *Bull. Environ. Contam. Toxicol,*, 14, 105–111.

Renzoni, A., E. Bacci, and L. Falciai. 1973. Mercury concentration in the water, sediments and fauna of an area of the Tyrrhenian Coast. In *6th International Symposium on Medical Oceanography*, Portoroz, Yugoslavia, September 26–30, 1973, p. 17–45.

Renzoni, A., M. Bernhard, R. Sara, and M. Stoeppler. 1978. Comparison between the Hg body burden of *Thynnus thunnus* from the Mediterranean and the Atlantic, *IVes J. Etud. Pollut.*, 255–260.

Ribeiro, C.A., C. Rouleau, E. Pelletier, C. Audet, and H. Tjalve. 1999. Distribution kinetics of dietary methylmercury in the Arctic charr (*Salvelinus alpinus*), *Environ. Sci. Technol.*,33, 902–907.

Richardson, M., M. Mitchell, S. Coad, and R. Raphael. 1995. Exposure to mercury in Canada: a multimedia analysis, *Water Air Soil Pollut.*, 80, 21–30.

Rissanen, K., J. Erkama,and J.K. Miettinen. 1972. Experiments on microbiological methylation of mercury (2+) ion by the mud and sludge under aerobic and anaerobic conditions. In M. Ruivo (Ed.), *Marine Pollution and Sea Life,* p. 289–292. Fishing Trading News (books), London.

Rivers, J.B, J.E. Pearson, and C.D. Schultz. 1972. Total and organic mercury in marine fish, *Bull. Environ. Contam. Toxicol.*, 8, 257–266.

Roberts, T.M., P.B. Heppleston, and R.D. Roberts. 1976. Distribution of heavy metals in tissues of the common seal, *Mar. Pollut. Bull.*, 7, 194–196.

Robertson, T., G.D. Waugh, and J.C.M. Mol. 1975. Mercury levels in New Zealand snapper, *Chrysophrys auratus*, *New Zeal. J. Mar. Freshwater Res.*, 9, 265–272.

Roelke, M.E., D.P. Schultz, C.F. Facemire, and S.F. Sundlof. 1991. Mercury contamination in the free-ranging endangered Florida panther (*Felis concolor coryi*), *Proc. Am. Assoc. Zoo Veterin.*, 1991, 277–283.

Ropek, R.M. and R.K. Neely. 1993. Mercury levels in Michigan river otters, *Lutra canadensis*, *J. Freshwat. Ecol.*, 8, 141–147.

Rosenberg, R. 1977. Effects of dredging operations on estuarine benthic macrofauna, *Mar. Pollut. Bull.*, 8, 102–104.

Ruckel, S.W. 1993. Mercury concentrations in alligator meat in Georgia, *Proc. Southeast. Assoc. Fish Wildl. Agency*, 47, 287–291.

Sager, D.R. 2004. Mercury in tissues of selected estuarine fishes from minimally impacted bays of coastal Texas, *Bull. Environ. Contam. Toxicol.*, 72, 149–156.

Saiki, M.K., M.R. Jennings, and T.W. May. 1992. Selenium and other elements in freshwater fishes from the irrigated San Joaquin valley, California, *Sci. Total Environ.*, 126, 109–137.

Sakai, H. H. Ichihaschi, H. Suganuma, and R. Tatsukawa. 1995. Heavy metal monitoring in sea turtles using eggs, *Mar. Pollut. Bull.*, 30, 347–353.

Sanli, Y. A. Fouassin, and A. Noirfalise. 1977. Mercury total et methylmercure dans ces conserves de poissons provenant de Turquie, *Arch. Belges Med. Soc., Hyg., Med. Trav. Med. Legale*, March 1977 (3), 161–167.

Sanpera, C., R. Capelli, V. Minganti, and L. Jover. 1993. Total and organic mercury in north Atlantic fin whales. Distribution pattern and biological related changes, *Mar. Pollut. Bull.*, 26, 135–139.

Sarica, J., M. Amyot, J. Bey, and L. Hare. 2005. Fate of mercury accumulated by blowflies feeding on fish carcasses, *Environ. Toxicol. Chem.*, 24, 526–529.

Scheifler, R., M. Gauthier-Clerc, C. Le Bohec, N. Crini, M. Coeurdassier, P.M. Badot, P. Giraudoux, and Y. Le Maho. 2005. Mercury concentrations in king penguin (*Aptenodytes patagonicus*) feathers at Crozet Islands (sub-Antarctic): temporal trend between 1966–1974 and 2000–2001, *Environ. Toxicol. Chem.*, 24, 125–128.

Schell, W.R. and A. Nevissi. 1977. Heavy metals from waste disposal in Central Puget Sound, *Environ. Sci. Technol.*, 11, 887–893.

Scheuhammer, A.M., C.M. Atchison, A.H.K. Wong, and D.C. Evers. 1998a. Mercury exposure in breeding common loons (*Gavia immer*) in central Ontario, Canada, *Environ. Toxicol. Chem.*, 17, 191–196.

Scheuhammer, A.M. and P.J. Blancher. 1994. Potential risk to common loons (*Gavia immer*) from methylmercury exposure in acidified lakes, *Hydrobiologia*, 279/280, 445–455.

Scheuhammer, A.M., A.H K. Wong, and D. Bond. 1998b. Mercury and selenium accumulation in common loons (*Gavia immer*) and common mergansers (*Mergus merganser*) from eastern Canada, *Environ. Toxicol. Chem.*, 17, 197–202.

Schmitt, C.J. and W.G. Brumbaugh. 1990. National Contaminant Biomonitoring Program: concentrations of arsenic, cadmium, copper, lead, mercury, selenium, and zinc in U.S. freshwater fish, 1976–1984, *Arch. Environ. Contam. Toxicol.*, 19, 731–747.

Schober, S.E., T.H. Sinks, and R.L. Jones. 2003. Blood mercury levels in U.S. children and women of childbearing age, 1999–2000, *J. Am. Med. Assoc.*,289, 1667–1674.

Schreiber, W. 1983. Mercury content of fishery products: data from the last decade, *Sci. Total Environ.*, 31, 283–300.

Schuhmacher, M., J. Batiste, M.A. Bosque, J.L. Domingo, and J. Corbella. 1994. Mercury concentrations in marine species from the coastal area of Tarrgona Province, Spain. Dietary intake of mercury through fish and seafood consumption, *Sci. Total Environ.*, 156, 269–273.

Schultz, C.D. and D. Crear. 1976. The distribution of total and organic mercury in seven tissues of the Pacific blue marlin, *Makaira nigricans*, *Pacific Sci.*, 30, 101–107.

Schultz, C.D., D. Crear, J.B. Pearson, J.B. Rivers, and J.W. Hylin. 1976. Total and organic mercury in Pacific blue marlin, *Bull. Environ. Contam. Toxicol.*, 15, 230–234.

Scott, J.S. 1977. Back-calculated fish lengths and Hg and Zn levels from recent and 100-year-old cleithrum bones from Atlantic cod (*Gadus morrhua*), *J. Fish. Res. Bd. Canada*, 34, 147–150.

Sergeant, D.E. and F.A.J. Armstrong. 1973. Mercury in seals from eastern Canada, *J. Fish. Res. Bd. Canada*, 30, 843–846.

Sharma, R. and S. Pervez. 2004a. A case study of spatial variation and enrichment of selected elements in ambient particulate matter around a large coal-fired power station in central India, *Environ. Geochem. Health*, 26, 373–381.

Sharma, R. and S. Pervez. 2004b. Toxic metals status in human blood and breast milk samples in an integrated steel plant environment in central India, *Environ. Geochem. Health*, 27, 39–45.

Sharma, R., L.C. Singh, S. Tanveer, S. Verghese, and P.A. Kumar. 2004. Trace element contents in human head hair of residents from Agra City, India, *Bull. Environ. Contam. Toxicol.*, 72, 530–534.

Sheffy, T.B. and J.R. St. Amant. 1982. Mercury burdens in furbearers in Wisconsin, *J. Wildl. Manage.*, 46, 1117–1120.

Sivalingam, P.M. 1980. Mercury contamination in tropical algal species of the Island of Penang, Malaysia, *Mar. Pollut. Bull.*, 11, 106–107.

Skaare, J.U., E. Degre, P.E. Aspholm, and K.I. Ugland. 1994. Mercury and selenium in Arctic and coastal seals off the coast of Norway, *Environ. Pollut.*, 85, 153–160.

Skare, I. and A. Engqvist. 1994. Human exposure to mercury and silver released from dental amalgam restorations, *Arch. Environ. Health*, 49, 384–394.

Skei, J.M., M. Saunders, and N.B. Price. 1976. Mercury in plankton from a polluted Norwegian fjord, *Mar. Pollut. Bull.*, 7, 34–35.

Skerfving, S., K. Hansson, and J. Lindstem. 1970. Chromosome breakage in humans exposed to methyl mercury through fish consumption, *Arch. Environ. Health*, 21, 133–139.

Sloan, R. and C.L. Schofield. 1983. Mercury levels in brook trout (*Salvelinus fontinalis*) from selected acid and limed Adirondack lakes, *Northeast. Environ. Sci.*, 2, 165–170.

Smith, T.J. and F.A.J. Armstrong. 1975. Mercury in seals, terrestrial carnivores, and principal food items of the Inuit, from Holman, N.W.T., *J. Fish. Res. Bd. Canada*, 32, 795–801.

Smith, T.J. and F.A.J. Armstrong. 1978. Mercury and selenium in ringed and bearded seal tissues from Arctic Canada, *Arctic*, 32(2), 75–84.

Solonen, T. and M. Lodenius. 1984. Mercury in Finnish sparrowhawks *Accipiter nisus*, *Ornis Fennica*, 61, 58–63.

Sorensen, J.A., G.E. Glass, K.W. Schmidt, J.K. Huber, and G. R. Rapp, Jr. 1990. Airborne mercury deposition and watershed characteristics in relation to mercury concentrations in water, sediments, plankton, and fish of eighty northern Minnesota lakes, *Environ. Sci. Technol.*, 24, 1716–1727.

Sorentino, C. 1979. Mercury in marine and freshwater fish of Papua, New Guinea, *Austral. J. Mar. Freshwater Res.*, 30, 617–623.

Southworth, G.R., R.R. Turner, M.J. Peterson, and M.A. Bogle. 1995. Form of mercury in stream fish exposed to high concentrations of dissolved inorganic mercury, *Chemosphere*, 30, 779–787.

Spalding, M.G., R.D. Bjork, G.V.N. Powell, and S.F. Sundlof. 1994. Mercury and cause of death in great white herons, *J. Wildl. Manage.*, 58, 735–739.

Srebocan, V., J. Pompe-Gotal, E. Srebocan, and V. Brmalj. 1981. Heavy metal contamination of the ecosystem Crna Mlaka. 1. Mercury, *Veterinarski Arhiv.*, 51, 319–324.

Srinivasan, M. and B.A. Mahajan. 1989. Mercury pollution in an estuarine region and its effect on a coastal population, *Inter. J. Environ. Stud.*, 35, 63–69.

Stafford, C.P. and T.A. Haines. 1997. Mercury concentrations in Maine sport fishes, *Trans. Am. Fish. Soc.*, 126, 144–152.

Staveland, G., I. Marthinsen, G. Norheim, and K. Julshamn. 1993. Levels of environmental pollutants in flounder (*Platichthys flesus* L.) and cod (*Gadus morhua* L.) caught in the waterway of Glomma, Norway. II. Mercury and arsenic, *Arch. Environ. Contam. Toxicol.*, 24, 187–193.

Stenner, R.D. and G. Nickless. 1975. Heavy metals in organisms of the Atlantic coast of S.W. Spain and Portugal, *Mar. Pollut. Bull.*, 6, 89–92.

Stevens, R.T., T.L. Ashwood, and J.M. Sleeman. 1997. Mercury in hair of muskrats (*Ondatra zibethicus*) and mink (*Mustela vison*) from the U.S. Department of Energy Oak Ridge Reservation, *Bull. Environ. Contam. Toxicol.*, 58, 720–725.

Stickel, L.F., W.H. Stickel, M.A.R. McLane, and M. Bruns. 1977. Prolonged retention of methyl mercury by mallard drakes, *Bull. Environ. Contam. Toxicol.*, 18, 393–400.

Stickney, R.R., H.L. Windom, D.B. White, and F.E. Taylor. 1975. Heavy metal concentrations in selected Georgia estuarine organisms with comparative food-habit data. In Howell, F.G., J.B. Gentry, and M.H. Smith (Eds.), *Mineral Cycling in South-eastern Ecosystems,* p. 257–267. U.S. Energy Res. Dev. Admin. Available as CONF-740513 from NTIS, U.S. Dept. Commerce, Springfield VA 22161.

Stoneburner, D.L. 1978. Heavy metals in tissues of stranded short-finned pilot whales, *Sci. Total Environ.*, 9, 293–297.

Stoneburner, D.L. and J.A. Kushlan. 1984. Heavy metal burdens in American crocodile eggs from Florida Bay Florida, USA, *J. Herpetol.*, 18, 192–193.

Stoneburner, D.L., M.N. Nicora, and E.R. Blood. 1980. Heavy metals in loggerhead sea turtle eggs (*Caretta caretta*): evidence to support the hypothesis that demes exist in the western Atlantic population, *J. Herpetol.*, 14, 171–176.

Suckcharoen, S. and M. Lodenius. 1980. Reduction of mercury pollution in the vicinity of a caustic soda plant in Thailand, *Water Air Soil Pollut.*, 13, 221–227.

Sundlof, S.F., M.G. Spalding, J.D. Wentworth, and C.K. Steible. 1994. Mercury in livers of wading birds (Ciconiiformes) in southern Florida, *Arch. Environ. Contam. Toxicol.*, 27, 299–305.

Suzuki, T., T. Miyama, and C. Toyama. 1973. The chemical form and bodily distribution of mercury in marine fish, *Bull. Environ. Contam. Toxicol.*, 10, 347–355.

Svansson, A. 1975. Physical and chemical oceanography of the Skagerrak and the Kattegat. I. Open sea conditions, *Fishery Bd, of Sweden,Inst. Mar. Res. Goteborg Rept.*, 1, 1–88.

Swain, E.B. and D.D. Helwig. 1989. Mercury in fish from northeastern Minnesota lakes: historical trends, environmental correlates, and potential sources, *J. Minnesota Acad. Sci.*, 55, 103–109.

Szefer, P., W. Czarnowski, J. Pempkowiak, and E. Holm. 1993. Mercury and major essential elements in seals, penguins, and other representative fauna of the Antarctic, *Arch. Environ. Contam. Toxicol.*, 25, 422–427.

Takizawa, Y. 1979a. Epidemiology of mercury poisoning. In J.O. Nriagu (Ed.), *The Biogeochemistry of Mercury in the Environment,* p. 325–365. Elsevier/North Holland Biomedical Press, Amsterdam.

Takizawa, Y. 1993. Overview on the outbreak of Minamata disease — epidemiological aspects. In *Proceedings of the International Symposium on Epidemiological Studies on Environmental Pollution and Health Effects of Methylmercury,* p. 3–26, October 2, 1992, Kumamoto, Japan. Published by National Institute for Minamata Disease, Kumamoto 867, Japan.

Talmage, S.S. and B.T. Walton. 1993. Food chain transfer and potential renal toxicity of mercury to small mammals at a contaminated terrestrial field site, *Ecotoxicology,* 2, 243–256.

Tamura, Y., T. Maki, H. Yamada, Y. Shimamura, S. Ochiai, S. Nishigaki, and Y. Kimura. 1975. Studies on the behavior of accumulation of trace elements in fishes. III. Accumulation of selenium and mercury in various tissues of tunas, *Tokyo Toritsu Eisei Kenyusho Nenpo*, 26, 200–204. (translation available as Dept. Sec. State Canada, Fish, Mar. Serv. No. 3994, 1977, 11 pp.).

Tavares, P.C., L.R. Monteiro, R.J. Lopes, M.E. Pereira, A.C. Duarte, and R.W. Furness. 2005. Variation of mercury contamination in chicks of little tern *Sterna albifrons* in southwest Europe: brood, age, and colony related effects, *Bull. Environ. Contam. Toxicol.*, 74, 177–183.

Tavares, P.C., L.R. Monteiro, R.J. Lopes, M.M.C. dos Santos, and R.W. Furness. 2004. Intraspecific variation of mercury contamination in chicks of black-winged stilt (*Himantopus himantopus*) in coastal wetlands from southwestern Europe, *Bull. Environ. Contam. Toxicol.*, 72, 437–444.

Taylor, D.D. and T.J. Bright. 1973. The distribution of heavy metals in reef-dwelling groupers in the Gulf of Mexico and Bahama Islands, *Texas A & M Univ., College Station, Dept. Oceanography Rept.*, TAMU-SG-73-208, 1–249.

Terhivuo, J., M. Lodenius, P. Nuortiva, and E. Tulisalo. 1984. Mercury content of common frogs (*Rana temporaria* L.) and common toads (*Bufo bufo* L.) collected in southern Finland, *Ann. Zool. Fennici*, 21, 41–44.

Thompson, D.R., P.H. Becker, and R.W. Furness. 1993. Long-term changes in mercury concentrations in herring gulls *Larus argentatus* and common terns *Sterna hirundo* from the German North Sea coast, *J. Appl. Ecol.*, 30, 316–320.

Thompson, D.R., and R.W. Furness. 1989. Comparison of the levels of total and organic mercury in seabird feathers, *Mar. Pollut. Bull.*, 20, 577–579.

Thompson, D.R., R.W. Furness, and P.M. Walsh. 1992. Historical changes in mercury concentrations in the marine ecosystem of the north and north-east Atlantic ocean as indicated by seabird feathers, *J. Appl. Ecol.*, 29, 79–84.

Thompson, D.R., K.C. Hamer, and R.W. Furness. 1991. Mercury accumulation in great skuas *Catharacta skua* of known age and sex, and its effects upon breeding and survival, *J. Appl. Ecol.*, 28, 672–684.

Thompson, D.R., F.M. Stewart, and R.W. Furness. 1990. Using seabirds to monitor mercury in marine environments. The validity of conversion ratios for tissue comparisons, *Mar. Pollut. Bull.*, 21, 339–342.

Tijoe, P.S., J.J.M. de Goeij, and M. de Bruin. 1977. Determination of trace elements in dried sea plant homogenate (SP-M-1) and in dried copepod homogenate (MA-A-1) by means of neutron activation analysis, *Interuniv. Reactor Inst., Delft, Nederlands*, Rept. 133-77-05, 1–14.

Tripp, M. and R.C. Harriss. 1976. Role of mangrove vegetation in mercury cycling in the Florida Everglades. In J.O. Nriagu (Ed.), *Environmental Biogeochemistry, Vol. 2. Metals transfer and Ecological Mass Balances*, p. 489–497. Ann Arbor Sci. Publ., Ann Arbor, MI.

Trudel, M. and J.B. Rasmussen. 1997. Modeling the elimination of mercury by fish, *Environ. Sci. Technol.*, 31, 1716–1722.

Trudel, M., A. Treblay, R. Schetagne, and J.B. Rasmussen. 2000. Estimating food consumption rates of fish using a mercury mass balance model, *Can. J. Fish. Aquat. Sci.* 57, 414–428.

Tuncel, G., G. Ramelow, and T.I. Balkas. 1980. Mercury in water, organisms and sediments from a section of the Mediterranean coast, *Mar. Pollut. Bull.*, 11, 18–22.

Ueda, T. and M. Takeda. 1977. On mercury and selenium contained in tuna fish tissues. IV. Methyl mercury level in muscles and liver of yellowfin tuna, *Bull. Japan. Soc. Sci. Fish.*, 43, 1115–1121.

Ugarte, C.A., K.G. Rice, and M.A. Donnelly. 2005. Variation of total mercury concentrations in pig frogs (*Rana grylio*) across the Florida Everglades, USA, *Sci. Total Environ.*, 345, 51–59.

Ui, J. 1972. A few coastal pollution problems in Japan. In *The Changing Chemistry of the Oceans. Proceedings 20th Nobel Symp.*, p. 171–176. Stockholm, Sweden. John Wiley, New York.

Ui, J. and S. Kitamuri. 1971. Mercury in the Adriatic, *Mar. Pollut. Bull.*, 2, 56–58.

U.S. Department of Health, Education, and Welfare (USHEW). 1977. *Occupational Diseases: a Guide to their Recognition*. USHEW Publ. 77-1811, Washington, D.C.

U.S. National Academy of Sciences (USNAS). 1978. *An Assessment of Mercury in the Environment*. Natl. Acad. Sci., Washington, D.C., 185 pp.

U.S. Public Health Service (USPHS). 1994. *Toxicological profile for mercury (update), TP-93/10*. U.S. PHS, Agen. Toxic Substances Dis. Registry, Atlanta, GA. 366 pp.

Van de Ven, W.S.M. 1978. Mercury and selenium in cod-liver oil, *Clin. Toxicol.*, 12, 579–581.

Van de Ven, W.S. M., J.H. Koeman, and A. Svenson. 1979. Mercury and selenium in wild and experimental seals, *Chemosphere*, 8, 539–555.

Vandal, G.M., W.F. Fitzgerald, C.F. Boutron, and J.P. Candelone. 1993. Variations of mercury deposition to Antarctica over the past 34,000 years, *Nature*, 362, 621–623.

Vannucchi, C., S. Sivieri, and M. Ceccanti. 1978. Residues of chlorinated naphthalenes, other hydrocarbons and toxic metals (Hg, Pb, Cd) in tissues of Mediterranean seagulls, *Chemosphere*, 7, 483–490.

Voegborlo, R.B., D.A. Baah, E.E. Kwaansa-Ansah, A.A. Adimado, and J.H. Ephraim. 2004. Mercury concentrations in fish species from the Gulf of Guinea, Ghana, *Bull. Environ. Contam. Toxicol.*, 73, 1057–1064.

Voigt, H.R. 2004. Concentrations of mercury (Hg) and cadmium (Cd), and the condition of some coastal Baltic fishes, *Environmentalica Fennica*, 21, 1–22.

Wagemann, R., R.E.A. Stewart, W.L. Lockhart, and B.E. Stewart. 1988. Trace metals and methyl mercury associations and transfer in harp seal (*Phoca groenlandica*) mothers and their pups, *Mar. Mammal Sci.*, 4, 339–355.

Wallin, K. 1984. Decrease and recovery patterns of some raptors in relation to the introduction and ban of alkyl-mercury and DDT in Sweden, *Ambio,* 13, 263–265.

Ware, J.F., H. Royals, and T. Lange. 1990. Mercury contamination in Florida largemouth bass, *Southeast. Assoc. Fish Wildl. Agen.*, 44, 5–12.

Watanabe, I., H. Ichihashi, S. Tanabe, M. Amano, N. Miyazaki, E.A Petrov, and R. Tatsukawa. 1996. Trace element accumulation in Baikal seal (*Phoca sibirica*) from the Lake Baikal, *Environ. Pollut.*, 94, 169–179.

Watanabe, I., S. Tanabe, M. Amano, N. Miyazaki, E.A. Petrov, and R. Tasukawa. 1998. Age-dependent accumulation of heavy metals in Baikal seal (*Phoca sibirica*) from the Lake Baikal, *Arch. Environ. Contam. Toxicol.*, 35, 518–526.

Weil, M., J. Bressler, P. Parsons, K. Bolla, T. Glass, and B. Schwartz. 2005. Blood mercury levels and neurobehavioral function, *J. Am. Med. Assoc.*, 293, 1875–1882.

Wenzel, C., D. Adelung, and H. Theede. 1996. Distribution and age-related changes of trace elements in kittiwake *Rissa tridactyla* nestlings from an isolated colony in the German Bight, North Sea, *Sci. Total Environ.*, 193, 13–26.

Westoo, G. 1966. Determination of methylmercury compounds in foodstuffs. I. Methylmercury compounds in fish, identification, and determination, *Acta Chem. Scand.*, 20, 2131–2137.

Westoo, G. 1969. Methylmercury compounds in animal foods. In M.W. Miller and G.G. Berg (Eds.), *Chemical Fallout — Current Research on Persistent Pesticides,* p. 75–90. Chas. C Thomas, Springfield, IL.

Westoo, G. 1973. Methylmercury as percentage of total mercury in flesh and viscera of salmon and sea trout of various ages, *Science*, 181, 567–568.

Wheatley, B. and S. Paradis. 1995. Exposure of Canadian aboriginal peoples to methylmercury, *Water Air Soil Pollut.*, 80, 3–11.

Wiemeyer, S.N., C.M. Bunck, and A.J. Krynitsky. 1988. Organochlorine pesticides, polychlorinated biphenyls, and mercury in osprey eggs — 1970–79 — and their relationships to shell thinning and productivity, *Arch. Environ. Contam. Toxicol.*, 17, 767–787.

Wiemeyer, S.N., C.M. Bunck, and C.J. Stafford. 1993. Environmental contaminants in bald eagle eggs — 1980–84 — and further interpretations of relationships to productivity and shell thickness, *Arch. Environ. Contam. Toxicol.*, 24, 213–227.

Wiemeyer, S.N., T.G. Lamont, and L.N. Locke. 1980. Residues of environmental pollutants and necropsy data for eastern United States ospreys, 1964–1973, *Estuaries*, 3, 155–167.

Wiemeyer, S.N., T.G. Lamont, C.M. Bunck, C.R. Sindelar, F.J. Gramlich, J.D. Fraser, and M.A. Byrd. 1984. Organochlorine pesticide, polychlorobiphenyl, and mercury residues in bald eagle eggs — 1969–79 — and their relationships to shell thinning and reproduction, *Arch. Environ. Contam. Toxicol.*, 13, 529–549.

Wiener, J.G., W.F. Fitzgerald, C.J. Watras, and R.G. Rada. 1990a. Partitioning and bioavailability of mercury in an experimentally acidified Wisconsin lake, *Environ. Toxicol. Chem.*, 9, 909–918.

Wiener, J.G., R.E. Martini, T.B. Sheffy, and G.E. Glass. 1990b. Factors influencing mercury concentrations in walleyes in northern Wisconsin lakes, *Trans. Am. Fish. Soc.*, 119, 862–870.

Wiener, J.G. and D.J. Spry. 1996. Toxicological significance of mercury in freshwater fish. In W.N. Beyer, G.H. Heinz, and A.W. Redmon-Norwood (Eds.), *Environmental Contaminants in Wildlife: Interpreting Tissue Concentrations,* p. 297–339. CRC Press, Boca Raton, FL.

Williams, P.M. and H.V. Weiss. 1973. Mercury in the marine environment: concentration in sea water and in a pelagic food chain, *J. Fish. Res. Bd. Canada*, 30, 293–295.

Windom, H.L. 1973. Mercury distribution in estuarine–nearshore environment, *J. Am. Soc. Civ. Engin.; Waterways, Harbors Coastal Engin. Div.*, 99(WW2), 257–264.

Windom, H. 1975. Heavy metal fluxes through salt marsh estuaries. In L.E. Cronin (ed.). *Estuarine Research, Vol. 1, Chemistry, Biology, and the Estuarine System,* p. 137–152. Academic Press, New York.

Windom, H., W. Gardner, J. Stephens, and F. Taylor. 1976. The role of methylmercury production in the transfer of mercury in a salt marsh ecosystem, *Estuar. Coastal Mar. Sci.*, 4, 579–583.

Windom, H.L. and D.R. Kendall. 1979. Accumulation and biotransformation of mercury in coastal and marine biota.n J.O. Nriagu (Ed.), *The Biogeochemistry of Mercury in the Environmen,* p. 301–323*t.* Elsevier/North-Holland Biomedical Press, New York.

Windom, H., R. Stickney, R. Smith, D. White, and F. Taylor. 1973. Arsenic, cadmium, copper, mercury, and zinc in some species of North Atlantic finfish, *J. Fish. Res. Bd. Canada*, 30, 275–279.

Winger, P.V. and J.K. Andreasen. 1985. Contaminant residues in fish and sediments from lakes in the Atachafalaya River basin (Louisiana), *Arch. Environ. Contam. Toxicol.*, 14, 579– 586.

Winger, P.V., C. Siekman, T.W. May, and W.W. Johnson. 1984. Residues of organochlorine insecticides, polychlorinated biphenyls, and heavy metals in biota from the Apalachicola River Florida 1978, *J. Assoc. Off. Anal. Chem.*, 67, 325–333.

Wobeser, G., N.O. Nielsen, R.H. Dunlop, and F.M. Atton. 1970. Mercury concentrations in tissues of fish from the Saskatchewan River, *J. Fish. Res. Bd. Canada*, 27, 830–834.

Wolfe, M. and S. Norman. 1998. Effects of waterborne mercury on terrestrial wildlife at Clear Lake: evaluation and testing of a predictive model, *Environ. Toxicol. Chem.*, 17, 214–217.

Won, J.H. 1973. The concentrations of mercury, cadmium, lead, and copper in fish and shellfish of Korea, *Bull. Korean Fish Soc.*, 6 (1,2), 1–19.

Wong, A.H.K., D.J. McQueen, D.D. Williams, and E. Demers. 1997. Transfer of mercury from benthic invertebrates to fishes in lakes with contrasting fish community structures, *Can. J. Fish. Aquat. Sci.*, 54, 1320–1330.

Wood, P.B., J.H. White, A. Steffer, J.M. Wood, C.F. Facemire, and H.F. Percival. 1996. Mercury concentrations in tissues of Florida bald eagles, *J. Wildl. Manage.*, 60, 178–185.

Wren, C.D. 1985. Probable case of mercury poisoning in a wild otter, *Lutra canadensis*, in northwestern Ontario, *Can. Field Natur.*, 99, 112–114.

Wren, C.D. and H.R. MacCrimmon. 1983. Mercury levels in the sunfish, *Lepomis gibbosus*, relative to pH and other environmental variables of precambrian shield lakes, *Can. J. Fish. Aquat. Sci.*, 40, 1737–1744.

Wren, C.D., W.A. Scheider, D.L. Wales, B.W. Muncaster, and I.M. Gray. 1991. Relation between mercury concentrations in walleye (*Stizostedion vitreum vitreum*) and northern pike (*Esox lucius*) in Ontario lakes and influence of environmental factors, *Can. J. Fish. Aquat. Sci.*, 48, 132–139.

Wren, C.D., P.M. Stokes, and K.L. Fischer. 1986. Mercury levels in Ontario mink and otter relative to food levels and environmental acidification, *Can. J. Zool.*, 64, 2854–2859.

Yannai, S. and K. Sachs. 1978. Mercury compounds in some eastern Mediterranean fishes, invertebrates, and their habitats, *Environ. Res.*, 16, 408–416.

Yanochko, G.M., C.H. Jagoe, and I.L. Brisbin Jr. 1997. Tissue mercury concentrations in alligators (*Alligator mississippiensis*) from the Florida Everglades and the Savannah River site, South Carolina, *Arch. Environ. Contam. Toxicol.*, 32, 323–328.

Yoshinaga, J., T. Suzuki, T. Hongo, M. Minagawa, R. Ohtsuka, T. Kawabe, T. Inaoka, and T. Akimichi. 1992. Mercury concentration correlates with the nitrogen stable isotope ratio in the animal food of Papuans, *Ecotoxicol. Environ. Safety*, 24, 37–45.

Zauke, G.P. 1977. Mercury in benthic invertebrates of the Elbe estuary, *Helgol. wiss. Meers.*, 19, 358–374.

Zilincar, V.J., B. Bystrica, P. Zvada, D. Kubin, and P. Hell. 1992. Die Schwermeallbelastung bei den Braunbaren in den Westkarpaten, *Z. Jagdwiss.*, 38, 235–243.

Zicus, M.C., M.A. Briggs, and R.M. Pace III. 1988. DDE, PCB, and mercury residues in Minnesota common goldeneye and hooded merganser eggs, 1981, *Can. J. Zool.*, 66, 1871–1866.

Zitko, V., B.J. Finlayson, D.J. Wildish, J.M. Anderson, and A.C. Kohler. 1971. Methylmercury in freshwater and marine fishes in New Brunswick, in the Bay of Fundy, and on the Nova Scotia Banks, *J. Fish. Res. Bd. Canada*, 28, 1285–1291.

CHAPTER 7

Mercury Concentrations in Abiotic Materials and Multitaxonomic Field Collections

Mercury data for abiotic samples as well as living organisms representative of different ecosystems taken from a single collection locale, usually at the same time by the same research group, are particularly valuable. Such data may illustrate food web biomagnification and other phenomena more readily than isolated data bits drawn over several years from disparate locales using different collection methods, various sample preparations, and noncomparable chemical methodologies for mercury analysis. One integrated data set demonstrated that mercury from point-source discharges, such as sewer outfalls and chloralkali plants, was taken up by sediments, and the sediment mercury levels were then reflected by an increased mercury content of epibenthic fauna (Klein and Goldberg, 1970; Takeuchi, 1972; Dehlinger et al., 1973; Hoggins and Brooks, 1973; Klemmer et al., 1973; Parsons et al., 1973). In another case, analysis of the effluent from the Hyperion sewer outfall in Los Angeles showed a mercury concentration slightly below 0.001 mg/L (Klein and Goldberg, 1970). Concentrations of mercury in sediment samples near this outfall were as high as 0.82 mg/kg but decreased with increasing distance from the outfall; mercury levels in epibenthic fauna, including crabs, whelks, and scallops, were also highest at stations near the discharge and lowest at stations tens of kilometers distant.

Selected data sets for mercury are presented in Table 7.1 at locations in the Adriatic Sea, Alaska, Antarctica, Brazil, Canada, China, Cuba, Florida, Greenland, India, Italy, Korea, Malaysia, New Jersey, Puerto Rico, Spain, Taiwan, Tennessee, Thailand, and Vietnam.

7.1 ASIA

In the People's Republic of China, the Sonhua River received about 6.5 metric tons of mercury annually for about 15 years in the 1960s and 1970s, resulting in water concentrations of 120.0 μg methylmercury/L vs. 4.0 to 10.0 μg/L from a reference site (Soong, 1994). Catfish — a major component in the diet of fishermen — from the Sonhua River contained 1.8 to 2.7 mg mercury/kg FW muscle in 1973 to 1974 and less than 1.0 mg/kg FW in the 1990s. In 1975, 100.0% of the fishermen had greater than 4.0 mg Hg/kg FW head hair; this was only 10.0% in 1975. In 1992, head hair of fishermen contained up to 71.2 mg Hg/kg FW, a level in excess of the current World Health Organization proposed criterion of less than 50.0 mg Hg/kg FW for human health protection (see Chapter 12). In the Zhejiang coastal area of China, baseline data were collected in the 1990s on mercury concentrations in edible species of fish, molluscs, and crustaceans, and also in seawater and surface sediments (Fang et al., 2004). All measurements (Table 7.1) appeared to be of low concern to human health.

Table 7.1 Mercury Concentrations in Multitaxonomic Collections from a Single Collection Area

Locale, Taxonomic Groups, Organism, Tissue, and Other Variables	Concentration (mg/kg)	Ref.[a]
Adriatic Sea; mercury-contaminated area vs. reference site; various seafood products of commerce; edible portions		
Total mercury	0.33 FW; max. 1.9 FW vs. 0.15 FW; max. 0.5 FW	1
Methylmercury	0.16 FW; max. 0.8 FW vs. 0.13 FW; max. 0.5 FW	1
Antarctica; Terra Nova Bay; 1989–1991		
Sediments	0.012 DW	2
Phytoplankton	0.04 DW	2
Invertebrates	0.07–0.39 (0.02–1.2) DW	2
Fish; 4 species:		
Muscle	0.3–0.8 (0.01–1.8) DW	2
Liver	0.2–0.5 (0.1–0.8) DW	2
Kidney	0.3–1.0 (0.1–2.6) DW	2
Gills	0.06–0.4 (0.01–1.0) DW	2
Gonads	0.2–0.3 (0.01–0.4) DW	2
Birds:		
Petrel, *Pagodroma nivea*; eggs vs. feathers:	0.6 DW vs. 0.5 DW	2
Skua, *Catharcta maccormicki:*		2
Eggs	1.6 DW	2
Feathers	2.9 DW	2
Guano	0.2 DW	2
Chick plumage	1.9 DW	2
Adelie penguin, *Pygoscelis adeliae*:		
Egg	0.3 DW	2
Feathers	0.8 DW	2
Guano	0.2 DW	2
Chick plumage	0.4 DW	2
Stomach contents	0.08 DW	2
Muscle	0.6 DW	2
Liver	1.6 DW	2
Kidney	1.2 DW	2
Brain	0.4 DW	2
Testis	0.4 DW	2
Weddell seal, *Leptonychotes weddelli*; adult:		
Muscle	1.8 DW	2
Liver	44.0 DW	2
Spleen	24.0 DW	2
Pancreas	1.5 DW	2
Brazil; freshwater lakes in coastal southern Brazil; one natural and two man-made; sampled September 2002–August 2003		
Natural lake:		
Water	4.6 ng/L	32
Sediment	0.058 DW	32
Rain	3.4 ng/L	32
Fish muscle		
Tambica, *Oligosarcus jenynsii* (piscivore)	0.06 (0.05–0.08) FW	32
Triara, *Hoplias malabaricus* (piscivore)	0.06 (0.001–0.09) FW	32
Lambari, *Astyanax* sp. (omnivore)	0.02 (0.01–0.06) FW	32
Acara, *Geophagus brasiliensis* (planktivore)	0.019 (0..01–0.022) FW	32
Suburban lake; Rio Grande City:		
Water	4.6 ng/L	32
Sediment	0.046 DW	32

Table 7.1 (continued) Mercury Concentrations in Multitaxonomic Collections from a Single Collection Area

Locale, Taxonomic Groups, Organism, Tissue, and Other Variables	Concentration (mg/kg)	Ref.[a]
Rain	15.8 ng/L	32
Fish muscle:		
Tambica	0.22 (0.05–0.44) FW	32
Triara	0.06 (0.01–0.10) FW	32
Lambari	0.07 (0.01–0.17) FW	32
Acara	0.02 (0.01–0.05) FW	32
Industrial Lake, Pelotas City:		
Water	3.3 ng/L	32
Sediment	0.056 DW	32
Rain	76.7 ng/L	32
Fish muscle:		
Tambica	Not found	32
Triara	0.14 FW	32
Lambari	0.13 (0.09–0.20) FW	32
Acara	0.07 (0.05–0.09) FW	32
Brazil; Madeira River; gold mining area; maximum values[b]		
Sediments	157.0 DW	3
Fish muscle	2.7 FW	3
Human hair	26.7 DW	3
Air (μg/m^3)	292.0	3
Brazil; gold mining site; 1992[b]		
Sediments	Max. 0.04 DW	4
Soil	Max. 0.04 DW	4
Fish muscle:		
Spotted catfish, *Pseudoplatystoma coruscans*	0.3 FW; Max. 1.0 FW	4
Black river piranha, *Pygocentrus nattereri*	0.3 (0.1–0.5) FW	4
Bird feather:		
Black vulture, *Coragyps atratus*	6.2 FW	4
Crested caracara, *Polyborus plancus*	6.8 FW	4
Brazil; Amazon gold mining region[b]		
Water	Max. 0.008 FW	5
Fish muscle	0.04–0.61 FW (87.0–100.0% organic mercury)	5
Humans (reference site):		
Blood	Max. 0.065 FW (max. 0.010 FW)	5
Hair	Max. 32.0 FW (max. < 2.0 FW)	5
Urine	Max. 0.156 FW (max. 0.007 FW)	5
Cattle and pigs:		
Blood	0.012–0.015 FW	5
Hair	0.1–1.3 FW	5
Canada; northern Quebec; 1989–1990		
Sediment	Max. 0.18 DW	6
Fish muscle	0.6–0.9 FW	6
Bird muscle	1.0–1.6 FW	6
Mink, *Mustela vison*; muscle	2.4 FW	6
Ringed seal, *Phoca hispida*		
Brain	0.2 FW	6
Kidney	0.2 FW	6

(continued)

Table 7.1 (continued) Mercury Concentrations in Multitaxonomic Collections from a Single Collection Area

Locale, Taxonomic Groups, Organism, Tissue, and Other Variables	Concentration (mg/kg)	Ref.[a]
Liver	5.1 FW	6
Muscle	0.3 FW	6
Beluga whale, *Delphinapterus leucus*		
Brain	2.7 FW	6
Liver	20.3 FW	6
Muscle	2.6 FW	6
Canada; Canadian Arctic; Amituk Lake		
Arctic char, *Salvelinus alpinus*; muscle vs. liver	0.57 FW vs. 1.24 FW	34
Atmospheric deposition	15.06 kg annually	35
Lake sediments	0.045 DW	35
Limestone/dolomite	0.012 DW	35
Snow	1.25–4.21 ng/L	35
Surface water	0.23–0.76 ng/L	35
China; Sonhua River region; in 1960s and 1970s, area received 6.5 tons of mercury annually, including 0.5 tons of methylmercury		
Water; 1975; vs. reference site	0.12 FW (as methylmercury) vs. 0.004–0.010 FW	20
Human hair:		
1975; 100.0% of fishermen	> 4.0 FW	20
1975; 10.0% of residents	> 4.0 FW	20
1992; fishermen	19.2 FW; max. 71.2 FW	20
Catfish muscle (fish diet):		
1973 vs. 1974	2.7 FW vs. 1.8 FW	20
1990s	> 0.4 FW to < 1.0 FW	20
China; Zhejiang coastal area		
Fish; muscle; 7 species; May 1998	0.009–0.014 (0.002–0.032) FW	19
Cephalopods; mantle; 2 species; May 1998	0.016–0.024 (0.009–0.047) FW	19
Bivalve molluscs; soft parts; 4 species; May 1998	0.013–0.023 (0.009–0.044) FW	19
Shrimp; muscle; 3 species; May 1998	0.01–0.015 (0.004–0.028) FW	19
Seawater; dissolved mercury; 1996	18.0–26.0 ng/L	19
Surface sediments; 1996	0.042–0.072 DW	19
Cuba; 1985–1987; chloralkali plant vicinity		
Mimosa tree, *Mimosa pudica* vs. soils; distance from source:		
0.0–0.5 km	2.2 DW vs. 2.6 DW	7
0.6–1.0 km	0.28 DW vs. 0.12 DW	7
3.1–5.0 km	0.04 DW vs. 0.21 DW	7
21–180 km	0.03 DW vs. 0.10 DW	7
Sea urchin, *Lytechinus variegatus*; gonads:		
Near discharge	0.38 DW	7
Transition area	0.29 DW	7
Control area	0.07 DW	7
Finland; freshwater reservoirs		
Surface sediments; 0–5 cm; various water depths; 1997–1998:		
1 m	0.02–0.06 DW	31
10 m	0.04 DW	31
24 m	0.08–0.09 DW	31
32 m	0.11 DW	31

Table 7.1 (continued) Mercury Concentrations in Multitaxonomic Collections from a Single Collection Area

Locale, Taxonomic Groups, Organism, Tissue, and Other Variables	Concentration (mg/kg)	Ref.[a]
Fish muscle; 1994–1998:		
Predators:		
Northern pike, *Esox lucius*	0.85–1.58 (0.28–3.7) FW[c]	31
Burbot, *Lota lota*	Max. 0.47 FW	31
Eurasian perch, *Perca fluviatilis*	0.52 (0.36–0.69) FW	31
Ruffe, *Gymnocephalus cerneus*	Max. 0.39 FW	31
Pikeperch, *Stizostedion lucioperca*	0.60–0.91 (0.13–1.38) FW[c]	31
Nonpredators:		
Bleak, *Alburnus alburnus*	Max. 0.05 FW	31
Bream, *Abramis brama*	Max. 0.31 FW	31
European smelt, *Osmerus eperlanus*	Max. 0.22 FW	31
Florida; 1995; southern estuaries; total mercury vs. methylmercury		
Sediments	0.02 (0.001–0.22) DW vs. Max. 0.0005 DW	8
Water, filtered	Max. 0.007 μg/L vs. max. 0.002 μg/L	8
Fish muscle; 9 species	1.4 (0.1–10.1) DW; 0.31 (0.03–2.2) FW vs. 1.05 (0.06–4.5) DW; 0.23 (0.01–1.0) FW	8
Greenland; 1983–1991		
Molluscs; 5 species; soft parts	0.01–0.02 FW	9
Crustaceans; 6 species; whole	Max. 0.33 FW	9
Fish; 10 species; liver vs. muscle	< 0.01–0.6 FW vs. 0.01–0.3 FW	9
Seabirds; 10 species:		
Kidney	0.1–2.1 FW	9
Liver	0.04–2.7 FW	9
Muscle	0.02–0.67 FW	9
Marine mammals		
Seals; 4 species:		
Kidney	0.09–3.5 FW	9
Liver	0.3–19.9 FW	9
Muscle	0.06–3.6 FW	9
Baleen whales; 1 species:		
Kidney	0.3 FW	9
Liver	0.4 FW	9
Muscle	0.16 FW	9
Toothed whales; 3 species:		
Kidney	0.18–1.4 FW	9
Liver	0.8–8.2 FW	9
Muscle	0.15–0.66 FW	9
Polar bear, *Ursus maritimus:*		
Kidney	10.8–23.2 FW	9
Liver	7.2–21.6 FW	9
Muscle	0.06–0.08 FW	9
Greenland; 1984–1987; total mercury vs. organic mercury		
Birds; liver	Max. 2.3 FW vs. 0.45 (0.1–1.5) FW	10
Seals:		
Kidney	Max. 6.4 FW vs. max. 0.98 FW	10
Liver	Max. 174.5 FW vs. 0.4 (0.1–2.1) FW	10
Muscle	Max. 1.4 FW vs. max. 1.2 FW	10

(continued)

Table 7.1 (continued) Mercury Concentrations in Multitaxonomic Collections from a Single Collection Area

Locale, Taxonomic Groups, Organism, Tissue, and Other Variables	Concentration (mg/kg)	Ref.[a]
Toothed whales:		
Kidney	Max. 2.9 FW vs. max. 0.4 FW	10
Liver	Max. 16.4 FW vs. max. 1.6 FW	10
Muscle	Max. 1.3 FW vs. max. 1.2 FW	10
Baleen whales:		
Kidney	Max. 1.1 FW vs. max. 0.11 FW	10
Liver	Max. 1.5 FW vs. max. 0.4 FW	10
Muscle	Max. 0.4 FW vs. max. 0.24 FW	10
Polar bear:		
Kidney	Max. 48.6 FW vs. max. 0.2 FW	10
Liver	Max. 23.8 FW vs. max. 0.6 FW	10
Muscle	Max. 0.1 FW vs. max. 0.07 FW	10
India; Bombay		
Sediments	Max. 5.5–7.0 DW	11
Humans; blood:		
Fish eaters	0.05–0.07 (0.02–0.13) FW	11
Nonfish eaters	0.019 (0.006–0.042) FW	11
Bombay vs. reference site:		
Fish, *Arius* sp.; muscle	1.5–2.2 DW vs. 0.5 DW	11
Prawn, *Penaeus* sp.; muscle	0.0–2.1 DW vs. 0.3 DW	11
Italy; vicinity of Monte Amiata		
Near banks of roasted Cinnabar:		
Soils	1379.0 DW	12
Pine, *Pinus nigra*		
Needles	8.1 DW	12
Branches	1.8 DW	12
Roots	0.9 DW	12
Near geothermal plant:		
Soils	18.7 DW	12
Pine, *Pinus nigra*		
Needles	0.5 DW	12
Branches and roots	0.2–0.3 DW	12
Reference site; 6–7 km from known sources:		
Soils	< 0.18 DW	12
Pine, *Pinus nigra*; all samples	< 0.07 DW	12
Italy; coast; summer; 1986–1987		
Mussel, *Mytilus galloprovincialis*; soft parts	0.01–0.07 FW	13
Snail, *Murex trunculus*; soft parts	0.03–0.15 FW	13
Fish, *Serranus* spp.; muscle	0.09–0.63 FW	13
Korea		
Air; Seoul vs. Inchon; various locations	3.5–36.8 (2.2–176.2) ng/m^3 vs. 13.1–38.9 (6.4–88.3) ng/m^3	24
Fish; marine; edible parts	0.06 (0.01–0.3) FW	24
Fish; freshwater; edible parts:		
Han River	0.16 FW	24
Nakdong River Basin	0.2 FW (from water containing 0.23 μg/L)	24
Southeast; 11 species	0.07 (0.02–0.12) FW	24

Table 7.1 (continued) Mercury Concentrations in Multitaxonomic Collections from a Single Collection Area

Locale, Taxonomic Groups, Organism, Tissue, and Other Variables	Concentration (mg/kg)	Ref.[a]
Hair, human:		
Dentists, male	8.6 (2.4–84.6) DW	24
Dental nurses, female	5.6 (2.3–33.0) DW	24
Dental supply assistants, male	5.7 (2.1–9.1) DW	24
Seoul residents; males vs. females	2.6 (0.6–8.5) DW vs. 2.1 (0.6–7.4) DW	24
Sediments, coastal	0.35 (0.08–0.75) DW	24
Shellfish, marine; edible parts	0.05 (0.01–0.15) FW	24
Soils	0.14 DW; max. 1.74 DW	24
Malaysia		
Agricultural soils; top 15 cm	0.147 (0.002–0.860) DW	22
Soils vs. crops; max. concentrations:		
Cabbage, *Brassica* sp.	0.33 DW vs. 0.002 DW	22
Cocoa, *Theobroma cacao*	0.25 DW vs. 0.039 FW	22
Corn, *Zea mays*	0.31 DW vs. 0.0004 DW	22
Mustard, *Brassica juncea*	0.27 DW vs. 0.055 FW[b]	22
Oil palm, *Elaeis guineensis*	0.23 DW vs. 0.032 DW	22
Spinach, *Spinacea olarecia*	0.17 vs. 0.000002 DW	22
Spain; Catalonia; November 1992–February 1993; edible tissues		
Means:		
Fish; 9 species	0.02–0.9 FW	14
Cephalopods; 3 species	0.003–0.27 FW	14
Crustaceans; 4 species	0.006–0.72 FW	14
Molluscs (noncephalopods); 5 species	0.001–0.019 FW	14
Maximum values; all species	0.001–1.8 FW	14
Taiwan; 1995–1996; edible tissues		
Pacific oyster, *Crassostrea gigas*	0.2 (0.03–1.3) DW	15
Fish; 5 species	1.0–2.5 (0.1–6.8) DW	15
Blue marlin, *Makaira nigricans*	10.3 (1.7–22.9) DW	15
Yellowfin tuna, *Thunnus albacares*	9.8 (8.8–10.4) DW	15
Shrimp; 2 species	2.2–2.4 (0.7–5.4) DW	15
Thailand		
Agricultural soils	0.04 (0.01–0.27) DW	23
Soils vs. crops; max. concentrations:		
Cabbage	0.14 DW vs. 0.0003 DW	23
Corn	0.063 DW vs. 0.0004 DW	23
Rice	0.22 DW vs. 0.022 DW	23
United States		
Alaska; Kuskokwim River Basin; near abandoned mercury mines:		
Fish muscle; downstream; total mercury; salmon vs. other fish	< 0.1 FW vs. max. 0.62 FW	25
Mine water; filtered vs. nonfiltered	Max. 0.05 μg/l vs. max. 2.5 μg/L	26
Stream sediments, downstream:		
Methylmercury	Max. 0.031–0.041 DW	26, 28
Total mercury	Max. 5.5 DW	27
Stream water; methylmercury	Max. 0.0012 μg/L	26

(continued)

Table 7.1 (continued) Mercury Concentrations in Multitaxonomic Collections from a Single Collection Area

Locale, Taxonomic Groups, Organism, Tissue, and Other Variables	Concentration (mg/kg)	Ref.[a]
Terrestrial vegetation:		
Methylmercury	0.011 FW	29
Total mercury	Max. 0.97 FW	29
New Jersey; Newark Bay; sediments vs. fish muscle	0.1–9.8 DW vs. 0.1–1.4 DW	16
New Jersey; 57 supermarkets/fish markets; July–October 2003		
Fish fillets:		
Croaker (Scianidae)	0.1 (0.06–0.3) FW	33
Chilean sea bass (Patagonian toothfish), *Dissotichus eleginoides*	0.4 (0.2–0.6) FW	33
Flounder, several species	0.05 (0.002–0.14) FW	33
Atlantic cod, *Gadus morhua*	0.1 (0.08–0.1) FW	33
Red snapper, *Lutjanus campechanus*	0.2 (0.2–0.3) FW	33
Bluefish, *Pomatomus saltatrix*	0.3 (0.009–0.76 FW; 32.0% > 0.3 FW; 2.0% > 0.5 FW	33
Porgy (Sparidae)	0.08 (0.02–0.2) FW	33
Tuna, mainly yellowfin tuna, *Thunnus albacares*	0.6 (0.08–2.5) FW; 62.0% > 0.3 FW; 42.0% > 0.5 FW; 26.0% > 0.75 FW	33
Whiting, *Merlangius merlangius*	0.03 (0.006–0.1) FW	33
Shellfish, edible tissues:		
Scallops, various	0.01 (0.007–0.02) FW	33
Shrimp, various	0.01 (0.002–0.02) FW	33
Oregon; Willamette Basin; 2002–2003; max. mean value		
Surface water:		
Total mercury vs. dissolved mercury	5.8 ng/L vs. 2.6 ng/L	30
Total methylmercury vs. dissolved methylmercury	0.14 ng/L vs. 0.10 ng/L	30
Fish muscle; piscivores vs. omnivores	1.63 FW vs. 0.38 FW	30
Sediments; total mercury vs. methylmercury	0.71 DW vs. 0.0009 DW	30
Puerto Rico; estuaries; 1988		
Aquatic:		
Blue crab, *Callinectes sapidus*; Shrimp, *Palaemonetes* sp.	Not detectable in any tissue or whole body	17
Fish muscle		
Tarpon, *Megalops atlantica*	0.09–0.24 FW	17
Mozambique tilapia, *Tilapia mossambica*	0.08–0.46 FW	17
Lizard, *Ameiva exsul*	Not detectable in any tissue	17
Cattle egret, *Bubulcus ibis*; pectoral muscle vs. liver	0.1 FW vs. 0.1 FW	17
Moorhen, *Gallinula chloropus:*		
Liver	0.16 FW	17
Muscle	0.12 FW	17
Whole	0.08 FW	17
Tennessee; mercury-contaminated (1950–1963) site vs. reference site; 1986–1987		
Soil	269.0 FW vs. 0.2 FW	18
Vegetation	Max. 2.0 FW vs. max. 0.2 FW	18
Earthworms	16.0 FW vs. 0.2 FW	18
Centipedes	3.4 FW vs. 0.1 FW	18
Termites	2.6 FW vs. 0.7 FW	18
White-footed mouse, *Peromyscus leucopus*; kidney	1.2 FW vs. 0.5 FW	18
Short tail shrew, *Blarina brevicauda*; kidney	39.0 FW vs. 1.0 FW	18
Vietnam		
Dalat region; humans; ages 20–35 years; head hair; weekly fish consumption of 300 g (max.) of marine fish and 400 g (max.) of freshwater fish:		
Total mercury	1.17 FW	21
Methylmercury	0.45 FW	21

Table 7.1 (continued) Mercury Concentrations in Multitaxonomic Collections from a Single Collection Area

Locale, Taxonomic Groups, Organism, Tissue, and Other Variables	Concentration (mg/kg)	Ref.[a]
Nha Trang region:		
Humans; ages 20–35 years; head hair; weekly fish consumption of 700.0 g (max.) of marine fish and 300.0 g (max.) freshwater fish:		
Total mercury	2.82 FW	21
Methylmercury	1.69 FW	21
Diet (tuna fish)		
Muscle; total mercury vs. methylmercury	0.28 FW vs. 0.14 FW	21
Liver; total mercury vs. methylmercury	0.43 FW vs. 0.39 FW	21

Note: Values are in mg total mercury/kg fresh weight (FW) or dry weight (DW), unless indicated otherwise.

[a] Reference: 1, Buzina et al., 1989; 2, Bargagli et al., 1998; 3, Malm et al., 1990; 4, Hylander et al., 1994; 5, Palheta and Taylor, 1995; 6, Langlois et al., 1995; 7, Gonzalez, 1991; 8, Kannan et al., 1998; 9, Dietz et al., 1996; 10, Dietz et al., 1990; 11, Srinivasin and Mahajan, 1989; 12, Ferrara et al., 1991; 13, Giordano et al., 1991; 14, Schuhmacher et al., 1994; 15, Han et al., 1998; 16, Gillis et al., 1993; 17, Burger et al., 1992; 18, Talmage and Walton, 1993; 19, Fang et al., 2004; 20, Soong, 1994; 21, Dung et al., 1994; 22, Zarcinas et al., 2004a; 23, Zarcinas et al., 2004b; 24, Sohn and Jung, 1993; 25, Gray et al., 2000; 26, Gray and Bailey, 2003; 27, Bailey and Gray, 1997; 28, Gray et al., 1996; 29, Bailey et al., 2002; 30, Hope and Rubin, 2005; 31, Voigt, 2000; 32, Mirlean et al., 2005; 33, Burger et al., 2005; 34, Muir and Lockhart, 1993; 35, Semkin et al., 2005.

[b] Exceeds Malaysian criterion of 0.05 mg Hg/kg as consumed.

[c] Acceptable limit in Finland is 0.5 to 1.0 mg total Hg/kg FW; however, exceedance is frequent.

In Korea, air levels were elevated (up to 176.0 ng/m^3) over industrialized areas, such as Seoul, when compared to rural areas (Sohn and Jung, 1993; Table 7.1). Mercury concentrations in edible portions of marine and freshwater fishes and marine molluscs were always less than 0.3 mg/kg FW, suggesting little risk to consumers of these products. Dentists and dental technicians had elevated hair mercury concentrations — up to 84.6 mg/kg DW — presumably from contact with elemental mercury in preparing dental amalgams.

In Malaysia, most crops did not reflect soil mercury concentrations, except mustard, *Brassica juncea*, with 0.055 mg/kg FW; this slightly exceeded the Malaysian food criterion of 0.05 mg Hg/kg as consumed (Table 7.1, Zarcinas et al., 2004a). Similar studies in Thailand indicated no accumulation of mercury in crops from agricultural soils (Table 7.1; Zarcinas et al., 2004b).

In Taiwan, it appears that a mercury contamination problem is developing among marine products of commerce, as judged by the unsatisfactorily high concentrations measured in fish, crustaceans, and oysters (Table 7.1; Han et al., 1998).

In Vietnam, it was shown that increased consumption of fish, especially tuna, was associated with elevated hair mercury concentrations in consumers, although all hair concentrations were less than 3.0 mg Hg/kg FW hair (Table 7.1; Dung et al., 1994).

7.2 BRAZIL

Atmospheric deposition is the primary route by which mercury enters freshwater systems (Meili et al., 2003); In Brazil, mercury concentrations in precipitation are closely linked with proximity to sources of mercury emissions (Mirlean et al., 2005). Atmospheric mercury depositions in Brazilian lakes are directly linked to concentrations in fish, with surface-feeding carnivores attaining the highest concentrations (Mirlean et al., 2005; Table 7.1).

Other data for Brazil are from areas where elemental mercury was used extensively to extract gold through amalgamation, with resultant widespread mercury contamination of the biosphere (Malm et al. ,1990; Hylander et al., 1994; Palheta and Taylor, 1995; Table 7.1). Concentrations of mercury in air, sediments, fish muscle, and human hair were sufficiently elevated to impact human health directly and indirectly. The subject of mercury contamination from gold mining activities in Brazil, the United States, and elsewhere is discussed in detail in Chapter 11.

7.3 CARIBBEAN REGION

Mercury contamination from a Cuban chloralkali plant was reflected in elevated soil mercury concentrations and its terrestrial vegetation, and in sediments and gonads of benthic fauna from these sediments (Table 7.1). Mercury contamination is measurable for at least 5 km from the chloralkali plant (Gonzalez, 1991). In Puerto Rico, low mercury levels were measured in crab and fish muscle samples collected from estuaries in 1988, and in terrestrial reptile and bird tissues, providing useful baseline data for that area (Table 7.1; Burger et al., 1992).

7.4 EUROPE

Mercury data from seafood products of commerce taken from the Adriatic Sea (Table 7.1) showed that methylmercury accounts for about 42.0% of the total mercury from products taken in mercury-contaminated areas, and 100.0% in seafoods from noncontaminated areas (Buzina et al., 1989).

In Italy, elevated mercury concentrations were measured in soils near a geothermal plant and near a mercury mine, and in needles of *Pinus nigra* growing on these soils (Table 7.1; Ferrara et al., 1991). Low mercury concentrations were measured in marine invertebrates from Italian coastal waters in the summers of 1986 and 1987; however, fish muscle contained up to 0.63 mg Hg/kg FW (Table 7.1, Giordano et al., 1991), which should be of concern to health authorities.

In Spain, mercury data for marine seafoods in 1992 and 1993 showed that some crustaceans and fish had greater than 0.5 mg total Hg/kg FW (Table 7.1; Schuhmacher et al., 1994), a level of concern in many nations.

7.5 INDIA

Mercury concentrations were elevated in sediments and in muscle of fish and shrimp caught near Bombay when compared to a reference site; residents of Bombay who were identified as fish consumers had elevated blood mercury concentrations (maximum 0.13 mg/L) when compared to nonfish consumers (maximum 0.042 mg/L) (Table 7.1; Srinivasin and Mahajan, 1989). In all cases, mercury concentrations were of minor concern.

7.6 NORTH AMERICA

Elevated mercury concentrations in muscle and liver tissues of Arctic char, *Salvelinus alpinus,* from Amituk Lake in the Canadian Arctic were attributed to long life span and low growth rate of char in Arctic waters coupled with mercury biomagnification (Table 7.1; Schindler et al., 1995).

Mercury data from northern Quebec in 1989 to 1990 showed concentrations up to 0.9 mg/kg FW in fish muscle, up to 1.6 mg/kg FW in avian muscle, 5.1 mg/kg FW in liver of ringed seals, and 2.6 mg/kg FW in muscle and 20.3 mg/kg FW in liver of beluga whales (Table 7.1; Langlois et al., 1995). All concentrations should be of concern to human consumers of these tissues.

In Florida, baseline data collected in 1995 from southern estuaries showed acceptable mercury concentrations in sediments and water, and unacceptable levels of total mercury (up to 2.2 mg/kg FW) and methylmercury (up to 1.0 mg/kg FW) in teleost muscle (Table 7.1; Kannan et al., 1998).

In southwestern Alaska near abandoned mercury mines, total mercury concentrations were elevated in sediments (max. 5.5 mg/kg), unfiltered mine water (2.5 μg/L), muscle from fish other than salmon (max. 0.62 mg/kg FW), and terrestrial vegetation (max. 0.97 mg/kg FW; Table 7.1).

In Tennessee, high levels of mercury (39.0 mg/kg FW) in kidneys of shrews (*Blarina brevicauda*) were probably associated with feeding on earthworms (16.0 mg Hg/kg FW) from a mercury-contaminated site (269.0 mg Hg/kg FW soil); similar data from a reference site were 1.0 mg/kg FW in kidney, 0.2 mg/kg FW in earthworms, and 0.2 mg/kg FW in soil (Table 7.1; Talmage and Walton, 1993). Elevated concentrations of mercury in water, sediments, and fish tissues in the Willamette Basin, Oregon, (Table 7.1) are attributed to mercury-containing mine wastes and atmospheric deposition (Hope and Rubin, 2005).

Burger et al. (2005), on analyzing edible marine products of commerce sold in New Jersey supermarkets during 2003 for mercury content (Table 7.1), concluded that four species that were available in more than half the supermarkets or fish markets sampled contained significantly different concentrations of mercury, that is, tuna (max. 2.5 mg total Hg/kg FW muscle), bluefish (max. 0.76), snapper (max. 0.3), and flounder (max. 0.14). Consumers selecting only for price could choose from whiting, porgy, croaker, and bluefish, all with average mercury concentrations less than 0.3 mg/kg FW (Table 7.1). Because flounders were comparatively available, modest in cost, and with very low mercury levels, authors suggest that state agencies should publicize this and similar information to their citizens, enabling them to make informed decisions about risks from fish consumption (Burger et al., 2005).

7.7 POLAR REGION

Data from Terra Nova Bay, Antarctica, (Table 7.1) show comparatively low levels of mercury in sediments, fish, and phytoplankton, and elevated levels in avian and mammalian tissues. Comparatively low mercury concentrations were measured in various species of invertebrates (up to 1.2 mg/kg DW), fish kidney (up to 2.6 mg/kg DW), and fish muscle (1.8 mg/kg DW, or about 0.2 mg/kg FW). Comparatively elevated mercury levels were measured in feathers (max. 2.9 mg/kg DW) and liver (up to 1.6 mg/kg DW in Adelie penguin); and high levels in seal tissues — up to 44.0 mg/kg DW in liver of Weddell seal and 24.0 mg/kg DW in seal pancreas (Table 7.1). These data provide useful information for Antarctic investigators.

Extensive data from Greenland (Table 7.1; Dietz et al., 1990, 1996) show low (< 0.5 mg Hg/kg FW) concentrations in molluscs, crustaceans, and teleosts; and grossly elevated (> 20.0 mg/kg FW) concentrations in kidney and liver of polar bears, and in liver of some species of seals. Intermediate values were measured in tissues of seabirds and various marine mammals. As expected, organomercurial content was inversely related to total mercury content; thus, seal liver containing 174.0 mg total Hg/kg FW contained 1.2% organomercury, seal kidney with 6.4 mg total Hg/kg FW had 15.0% organomercury, and whale muscle with 1.3 mg total Hg/kg FW had 92.0% organomercury (Table 7.1).

7.8 SUMMARY

Mercury data sets on abiotic materials and biological tissues from a single collection area, a short sampling period, and chemical analysis by the same research team are particularly useful in establishing food chain dynamics, identifying areas of probable health concerns, measuring geographic areas of impact, and in predicting future problem locations. Examples are presented for selected locations in Asia, Brazil, the Caribbean region, Europe, North America, and polar regions.

REFERENCES

Bailey, E.A. and J.E. Gray. 1997. Mercury in the terrestrial environment, Kuskokwim Mountains region, southwestern Alaska. In J.A. Dumoulin and J.E. Gray (Eds.), *Geologic Studies in Alaska by the U.S. Geological Survey, 1995,* p. 41–56. U.S. Geol. Surv. Prof. Paper 1574.

Bailey, E.A., J.E. Gray, and P.M. Theodorakos. 2002. Mercury in vegetation and soils at abandoned mercury mines in southwestern Alaska, USA, *Geochem. Explor. Environ. Anal.*, 2, 275–286.

Bargagli, R.F. Monaci, J.C. Sanchez-Hernandez, and D. Cateni. 1998. Biomagnification of mercury in an Antarctic marine coastal food web, *Mar. Ecol. Prog. Ser.*, 169, 65–76.

Burger, J., K. Cooper, J. Saliva, D. Gochfeld, D. Lipsky, and M. Gochfeld. 1992. Mercury bioaccumulation in organisms from three Puerto Rican estuaries, *Environ. Monitor. Assess.*, 22, 181–197.

Burger, J., A.H. Stern, and M. Gochfeld, 2005. Mercury in commercial fish: optimizing individual choices to reduce risk, *Environ. Health Perspect.*, 113, 266–270.

Buzina, R., K. Suboticanec, J. Vukusic, J. Sapunar, and M. Zorica. 1989. Effect of industrial pollution on seafood content and dietary intake of total and methylmercury, *Sci. Total Environ.*, 78, 45–57.

Dehlinger, P., W.F. Fitzgerald, S.Y. Feng, D.F. Paskausky, M.W. Garvine, and W.F. Bohlen. 1973. Determination of budgets of heavy meatal wastes in Long Island Sound, *Annual Rept., Parts I and II,* Univ. Connecticut, Marine Sciences Inst., Groton, CT.

Dietz, R., C.O. Nielsen, M.M. Hansen, and C.T. Hansen. 1990. Organic mercury in Greenland birds and mammals, *Sci. Total Environ.*, 95, 41–51.

Dietz, R., F. Riget, and P. Johansen. 1996. Lead, cadmium, mercury and selenium in Greenland marine animals, *Sci. Total Environ.*, 186, 67–93.

Dung, H.M., N.T. Anh, and L.T. Mua. 1994. Determination of total mercury and methyl mercury in Vietnamese head hair. In *Proceedings of the International Symposium on Assessment of Environmental Pollution and Health Effects from Methylmercury,* p. 179–195. October 8–9, 1993, Kumamoto, Japan. National Institute for Minamata Disease, Minamata City, Kumamoto 867, Japan.

Fang, J., K.X. Wang, J.L. Tang, Y.M. Wang, S.J. Ren, H.Y. Wu, and J. Wang. 2004. Copper, lead, zinc, cadmium, mercury, and arsenic in marine products of commerce from Zhejiang coastal area, China, May, 1998, *Bull. Environ. Contam. Toxicol.*, 73, 583–590.

Ferrara, R., B. E. Maserti, and R. Breder. 1991. Mercury in abiotic and biotic compartments of an area affected by a geochemical anomaly (Mt. Amiata, Italy), *Water Air Soil Pollut.*, 56, 219–233.

Gillis, C.A., N.L. Bonnevie, and R.J. Wenning. 1993. Mercury contamination in the Newark Bay estuary, *Ecotoxicol. Environ. Safety*, 25, 214–226.

Giordano, R., P. Arata, L. Ciaralli, S. Rinaldi, M. Giani, A.M. Cicero, and S. Constantini. 1991. Heavy metals in mussels and fish from Italian coastal waters, *Mar. Pollut. Bull.*, 22, 10–14.

Gonzalez, H. 1991. Mercury pollution caused by a chlor-alkali plant, *Water Air Soil Pollut.*, 56, 83–93.

Gray, J.E. and E.A. Bailey. 2003. The southwestern Alaska mercury belt. In J.E. Gray (Ed.), *Geologic Studies of Mercury by the U.S. Geological Survey,* p. 19–22. U.S. Geological Survey Circular 1248. Available from USGS, Box 25286, Denver, CO 80225.

Gray, J.E., A.L. Meier, R.M. O'Leary, C. Outwater, and P.M. Theodorakas. 1996. Environmental geochemistry of mercury deposits in southwestern Alaska — mercury contents in fish, stream-sediment, and stream-water samples. In T.E. Moore and J.A. Dumoulin (Eds.). *Geologic Studies in Alaska by the U.S. Geological Survey,1994,* p. 17–29. U.S. Geol. Surv. Bull. 2152.

Gray, J.E., P.M. Theodorakos, E.A. Bailey, and R.R. Turner. 2000. Distribution, speciation, and transport of mercury in stream sediment, stream water, and fish collected near abandoned mercury mines in southwestern Alaska, U.S.A., *Sci. Total Environ.*, 260, 21–33.

Han, B.C, W.L. Jeng, R.Y. Chen, G.T. Fang, T.C. Hung, and R.J. Tseng. 1998. Estimation of target hazard quotients and potential health risks for metals by consumption of seafood in Taiwan, *Arch. Environ. Contam. Toxicol.*, 35, 711–720.

Hoggins, F.E. and R.R. Brooks. 1973. Natural dispersion of mercury from Puhipuhi, Northland, New Zealand, *New Zeal. J. Mar. Freshwater Res.*, 7, 125–132.

Hope, B.K. and J.R. Rubin. 2005. Mercury levels and relationships in water, sediment, and fish tissue in the Willamette Basin, Oregon, *Arch. Environ. Contam. Toxicol.*, 48, 367–380.

Hylander, L.D., E.C. Silva, L.J. Oliveira, S.A. Silva, E.K. Kuntze, and D.X. Silva. 1994. Mercury levels in Alto Pantantal: a screening study, *Ambio*, 23, 478–484.

Kannan, K., R.G. Smith Jr., R.F. Lee, H.L. Windom, P.T. Heitmuller, J.M. Macauley, and J.K. Summers. 1998. Distribution of total mercury and methyl mercury in water, sediment, and fish from south Florida estuaries, *Arch. Environ. Contam. Toxicol.*, 34, 109–118.

Klein, D.H. and E.D. Goldberg. 1970. Mercury in the marine environment. *Environ. Sci. Technol.*, 4, 765–768.

Klemmer, H., S.N. Luoma, and L.S. Lau. 1973. Mercury levels in marine biota, *Government Reports Announce.*, 73(10), 76.

Langlois, C., R. Langis, and M. Perusse. 1995. Mercury contamination in northern Quebec environment and wildlife, *Water Air Soil Pollut.*, 80, 1021–1024.

Malm, O., W.C. Pfeiffer, C.M.M. Souza, and R. Reuther. 1990. Mercury pollution due to gold mining in the Madeira River Basin, Brazil, *Ambio*, 19, 11–15.

Meili, M., K. Bishop, L. Bringmark, K. Johansson, J. Munthe, H. Sverdrup, and W. de Vries. 2003. Critical levels of atmospheric pollution: criteria and concepts for operational modeling of mercury in forest and lake ecosystems, *Sci. Total Environ,*, 304, 83–106.

Mirlean, N., S.T. Larned, V. Nikora, and V.T. Kutter. 2005. Mercury in lakes and lake fishes on a conservation-industry gradient in Brazil, *Chemosphere*, 60, 226–236.

Muir, D. and W.L. Lockhart. 1993. Contaminant trends in freshwater biota. J.L. Murray and R.G. Shearer (Eds.), *Environmental Studies No. 70, Synopsis of Research Conducted under the 1992/93 Northern Contaminants Program,* p. 167–173, Catalogue No. R71-19/70-1993E. Dept. Indian Affairs and Northern Develop, Ottawa, Canada.

Palheta, D. and A. Taylor. 1995. Mercury in environmental and biological samples from a gold mining area in the Amazon region of Brazil, *Sci. Total Environ.*, 168, 63–69.

Parsons, T.R., C.A. Bawden, and W.A. Heath. 1973. Preliminary survey of mercury and other metals contained in animals from the Fraser River mudflats, *J. Fish. Res. Board Canada*, 30, 1014–1016.

Schindler, D.W., K.A. Kidd, D.C.G. Muir, and W.L. Lockhart. 1995. The effects of ecosystem characteristics on contaminant distribution in northern freshwater lakes, *Sci. Total Environ.*, 160/161, 1–17.

Schuhmacher, M., J. Batiste, M.A. Bosque, J.L. Domingo, and J. Corbella. 1994. Mercury concentrations in marine species from the coastal area of Tarrgona Province, Spain. Dietary intake of mercury through fish and seafood consumption, *Sci. Total Environ.*, 156, 269–273.

Semkin, R.G., G. Mierle, and R.J. Neureuther. 2005. Hydrochemistry and mercury cycling in a high Arctic watershed, *Sci. Total Environ.*, 342, 199–221.

Sohn, D.H. and W.T. Jung. 1993. Mercury pollution in Korea. Pages 37–43 in Proceedings of the International Symposium on Epidemiological Studies on Environmental Pollution and Health Effects of Methylmercury, October 2, 1992, Kumamoto, Japan. Published by National Institute for Minamata Disease, Kumamoto 867, Japan.

Soong, T. 1994. Epidemiological research on the health effect of residents along the Sonhua River polluted by methylmercury. In *Proceedings of the International Symposium on "Assessment of Environmental Pollution and Health Effects from Methylmercury,"* p. 165–169. October 8–9, 1993, Kumamoto, Japan. National Institute for Minamata Disease, Minamata City, Kumamoto 867, Japan.

Srinivasan, M. and B.A. Mahajan. 1989. Mercury pollution in an estuarine region and its effect on a coastal population, *Int. J. Environ. Stud.*, 35, 63–69.

Takeuchi, T. 1972. Distribution of mercury in the environment of Minamata Bay and inland Ariake Sea. Pages 79–81 in Hartung, R. and B.D. Dinman (Eds.). *Environmental Mercury Contamination*. Ann Arbor Science Publ., Ann Arbor, MI.

Talmage, S.S. and B.T. Walton. 1993. Food chain transfer and potential renal toxicity of mercury to small mammals at a contaminated terrestrial field site, *Ecotoxicology*, 2, 243–256.

Voigt, H.R. 2000. Water quality and fish in two freshwater reservoirs (Gennarby and Sysilax) on the SW coast of Finland, *Acta Univer. Carolinae Environ.*, 14, 31–59.

Zarcinas, B.A., C.F. Ishak, M.J. McLaughlin, and G. Cozens. 2004a. Heavy metals in soils and crops in southeast Asia. 1. Peninsular Malaysia, *Environ. Geochem. Health*, 26, 343–357.

Zarcinas, B.A., P. Pongsakul, M.J. McLaughlin, and G. Cozens. 2004b. Heavy metals in soils and crops in southeast Asia. 2. Thailand, *Environ. Geochem. Health*, 26, 359–371.

PART 3

Lethal and Sublethal Effects of Mercury under Controlled Conditions

CHAPTER 8

Lethal Effects of Mercury

This chapter synthesizes available literature on the lethality of inorganic and organic mercury compounds to freshwater and marine biota; the effect of route of administration of various mercurials on the survival of representative species of waterfowl, passerines, raptors, and other avian groups; and the lethality of organomercury compounds to humans, small laboratory mammals, livestock, domestic cats and dogs, and various species of wildlife.

Death is the only biological variable now measured that is considered irreversible by all investigators. Nevertheless, time of death is modified by a host of physical, chemical, biological, metabolic, and behavioral variables, and it is unfortunate that some regulatory agencies still set mercury criteria to protect natural resources and human health on the basis of death — usually concentrations producing 50.0% mortality — and some variable uncertainty factor. Mercury criteria for protection of natural resources and human health, as discussed in Chapter 12, should be based — at a minimum — on the highest dose tested or highest tissue concentration found that does not produce death, impaired reproduction, inhibited growth, or disrupted well-being.

8.1 AQUATIC ORGANISMS

Lethal concentrations of mercury salts ranged from less than 0.1 μg Hg/L to more than 200.0 μg/L for representative sensitive species of marine and freshwater organisms (Table 8.1). The lower concentrations of less than 2.0 μg/L recorded were usually associated with early developmental stages, long exposures, and flowthrough tests (Table 8.1). Among teleosts, females and larger fish were more resistant to lethal effects of mercury than were males and smaller fishes (Diamond et al., 1989). Among metals tested, mercury was the most toxic to aquatic organisms, and organomercury compounds showed the greatest biocidal potential (Eisler, 1981; Jayaprakash and Madhyastha, 1987). In general, mercury toxicity was higher at elevated temperatures (Armstrong, 1979), at reduced salinities for marine organisms (McKenney and Costlow, 1981), and in the presence of other metals such as zinc and lead (Parker, 1979). Salinity stress, for example, especially abnormally low salinities, reduced significantly the survival time of mercury-exposed isopod crustaceans (Jones, 1973), suggesting that species adapted to a fluctuating marine environment — typical of the intertidal zone — could be more vulnerable to the added stress of mercury than species inhabiting more uniformly stable environments.

8.1.1 Invertebrates

The marine ciliate protozoan *Uronema marinum*, with an LC50 (24 h) value of 6.0 μg/L, failed to develop resistance to mercury over an 18-week period (Parker, 1979). However, another marine

Table 8.1 Lethality of Inorganic and Organic Mercury Compounds to Selected Species of Aquatic Organisms

Chemical Species, Ecosystem, Taxonomic Group, Species, and Other Variables	Concentration (µg Hg/L medium)	Effect[a]	Ref.[b]
Inorganic Mercury: Freshwater			
Crustaceans			
Crayfish, *Orconectes limosus*	2.0	LC50 (30 d)	1
Daphnid *Daphnia magna*	5.0	LC50 (96 h)	1
Daphnid, *Daphnia magna*	1.3–1.8	LC50 (LT)	1
Scud, *Gammarus pseudolimnaeus*	10.0	LC50 (96 h)	1
Molluscs			
Rainbow mussel, *Villosa iris:*			
Glochidia	14.0	LC50 (72 h)	33
Glochidia	25.5	All dead in 72 h	33
Juveniles; age 2 months; not fed during exposure	99.0	LC50 (96 h)	33
Juveniles; age 2 months; not fed during exposure	234.0	All dead in 96 h	33
Juveniles; age 2 months; fed	114.0	None dead in 21 d	33
Fish			
Zebrafish, *Brachydanio rerio*; embryo-larvae	< 2.0	No deaths	16
Goldfish, *Carassius auratus*	122.0	LC50 (96 h)	27
Air-breathing catfish, *Clarias batrachus*; adults	507.0	LC50 (96 h)	18
Catfish, *Clarias lazera:*			
Adults	720.0	LC50 (96 h)	17
Adults	960.0	LC50 (24 h)	17
Mosquitofish, *Gambusia affinis*; adults	1000.0	LC77 (10 d)	19
Channel catfish, *Ictalurus punctatus*; embryo-larva:			
Static test	30.0	LC50 (10 d)	2
Flowthrough test	0.3	LC50 (10 d)	2
Largemouth bass, *Micropterus salmoides*; embryo-larva:			
Static test	188.0 (138.0–238.0)	LC50 (8 d)	2, 27
Flowthrough test	5.3	LC50 (8 d)	2
Rainbow trout, *Oncorhynchus mykiss*:			
Juveniles	155.0–200.0	LC50 (96 h)	1
Embryo-larva:			
Static test	4.7 (4.2–5.3)	LC50 (28 d)	2, 27
Flowthrough test	< 0.1	LC50 (28 d)	2
Subadults	64.0	LC50 (58 d)	20
Subadults	426.0	LC50 (24 h)	20
Brook trout, *Salvelinus fontinalis*	0.3–0.9	LC50 (LT)	1
Tench, *Tinca tinca*	1000.0	All dead in 48 h	26
Tench	100.0	None dead in 3 weeks	26
Bronze featherback, *Notopterus notopterus*	440.0	LC50 (96h)	3
Amphibians			
Blanchard's cricket frog, *Acris crepitans blanchardi*; embryo-larva	10.0 (8.5–13.0)	LC50 (72 h)	27
Kentucky small-mouthed salamander, *Ambystoma barbouri*; embryo-larva	8.0 (2.0–14.0)	LC50 (168–192 h)	27
Jefferson's salamander, *Ambystoma jeffersonianum*; embryo-larva	19.0 (16.0–21.0)	LC50 (144–192 h)	27

Table 8.1 (continued) Lethality of Inorganic and Organic Mercury Compounds to Selected Species of Aquatic Organisms

Chemical Species, Ecosystem, Taxonomic Group, Species, and Other Variables	Concentration (µg Hg/L medium)	Effect[a]	Ref.[b]
Spotted salamander, *Ambystoma maculatum*; embryo-larva	31.0 (25.0–37.0)	LC50 (144–168 h)	27
Marbled salamander, *Ambystoma opacum*; embryo-larva	103.0 (63.0–153.0)	LC50 (120–144 h)	27
Small-mouthed salamander, *Ambystoma texanum*	27.0 (21.0–33.0)	LC50 (144–168 h)	27
Anurans; 4 spp; embryo-larva	36.8–67.2	LC50 (96 h)	2
Eastern green toad, *Bufo debilis debilis*; embryo-larva	40.0 (26.0–52.0)	LC50 (72 h)	27
Fowler's toad, *Bufo woodhousei fowlerei:*			
Embryo-larva	35.0 (21.0–38.0)	LC50 (72 h)	27
Tadpole	25	LC50 (72 h)	28
Toad, *Bufo melanosticus*; tadpole	185	LC50 (96 h)	28
Red-spotted toad, *Bufo punctatus*; embryo-larva	32.0 (22.0–41.0)	LC50 (72 h)	27
Narrow-mouthed toad, *Gastrophryne carolinensis*; embryo-larva	1.0 (0.9–1.9)	LC50 (72 h)	27
Southern gray treefrog, *Hyla chrysoscelis*; embryo-larva	2.3 (1.5–3.4)	LC50 (72 h)	27
Treefrogs, *Hyla* spp.; embryo-larva; 5 species	2.4–2.8	LC50 (72–96 h)	2, 27
Frog, *Microhyla ornata*; tadpoles:			
Embryo	126.0	LC50 (96 h)	29
Tadpoles:			
Recently-metamorphosed	88.0	LC50 (96 h)	29
Age 1 week	1120.0	LC50 (96 h)	13, 30
Age 4 weeks	1430.0	LC50 (96 h)	13, 30
Spring peeper, *Pseudocris crucifer*; embryo-larva	2.3 (0.3–4.9)	LC50 (72 h)	27
Frog, *Rana breviceps*; tadpole	207.0	LC50 (96 h)	28
Bullfrog, *Rana catesbeina*; embryo-larva	6.3 (4.9–8.1)	LC50 (144–192 h)	27
Frog, *Rana cyanophlyctis*:			
Adult females	960.0	LC50 (31–65 d)	14
Adult females	4800.0	LC50 (96 h)	14
Pig frog, *Rana grylio*; embryo-larva	59.0 (32.0–109.0)	LC50 (144–192 h)	27
River frog, *Rana heckscheri:*			
Embryo-larva	55.0 (38.0–78.0)	LC50 (72 h)	27
Embryo	502.0	LC50 (96 h)	31
Adults	3252.0	LC50 (96 h)	31
Adult females	880.0	No deaths in 60 d	15
Adult females	4400.0	LC50 (96 h)	15
Pickerel frog, *Rana palustris*; embryo-larva	5.1 (4.0–6.2)	LC50 (144 h)	27
Leopard frog, *Rana pipens*; embryo-larva	7.3	LC50 (96 h)	2
Northern leopard frog, *Rana pipiens pipiens*; embryo-larva	8.4 (5.3–13.3)	LC50 (144 h)	27
Southern leopard frog, *Rana sphenocephala*; tadpoles; fed diets containing various concentrations of $HgCl_2$ for 254 d:			
Control	0.0 mg/kg FW diet	12.0% dead in 240 d	34
Low Hg diet	0.1 mg Hg/kg FW diet	None dead in 240 d	34
Medium Hg diet	0.5 mg Hg/kg FW diet	22.0% dead in 70 d; 28.0% dead in 240 d	34
High Hg diet	1.0 mg Hg/kg FW diet	28.0% dead in 240 d	
South African clawed frog, *Xenopus laevis*; tadpole	74.0	LC50 (48 h)	32

(continued)

Table 8.1 (continued) Lethality of Inorganic and Organic Mercury Compounds to Selected Species of Aquatic Organisms

Chemical Species, Ecosystem, Taxonomic Group, Species, and Other Variables	Concentration (µg Hg/L medium)	Effect[a]	Ref.[b]
Inorganic Mercury: Marine			
Protozoans			
Ciliate, *Uronema marinum*	6.0	LC50 (24 h)	4
Coelenterates			
Coral, *Porites asteroides*:			
Colonies	100.0, nominal; 37.0, measured	No deaths in 15 d	25
Colonies	500.0 (nominal); 180.0 (measured)	3 of 6 colonies dead in 72 h; remaining 3 colonies survived exposure for at least 15 d	25
Molluscs			
Softshell clam, *Mya arenaria*:			
Adults	1.0	No deaths in 168 h	5
Adults	4.0	LC50 (168 h)	5
Adults	30.0	All dead in 168 h	5
Adults	400.0	LC50 (96 h)	5
Hardshell clam, *Mercenaria mercenaria*:			
Larva	4.8	LC50 (48 h)	1
Larva	4.0	LC50 (9 d)	1
American oyster, *Crassostrea virginica*:			
Embryo	3.3	5.0% dead in 12 d	1
Larva	5.6	LC50 (48 h)	1
Adult	5.5–10.2	LC50 (48 h)	1
Pacific oyster, *Crassostrea gigas*; embryos	5.7	LC50 (48 h)	10
Common mussel, *Mytilus edulis*	5.8	LC50 (96 h)	6
Mud snail, *Nassarius obsoletus*:			
Adults	100.0	No deaths in 168 h	5
Adults	700.0	LC 50 (168 h)	5
Adults	5000.0	LC (100 (168 h)	5
Adults	32,000.0	LC 50 (96 h)	5
Slipper limpet, *Crepidula fornicata*:			
Larva	60.0	LC50 (96 h)	7
Adults	330.0	LC50 (96 h)	7
Bay scallop, *Argopecten irradians*; juveniles	89.0	LC50 (96 h)	8
Crustaceans			
Fiddler crab, *Uca pugilator*, zoea	1.8	LC50 (8 d)	1
Mysid shrimp, *Mysidopsis bahia*:			
Juveniles	3.5	LC50 (96 h)	9
Egg to egg exposure	1.8	LC50 (LT)	9
Dungeness crab, *Cancer magister*, larva	6.6	LC50 (96 h)	10
Copepod, *Acartia tonsa*; adult	10.0–15.0	LC50 (96 h)	1
Hermit crab, *Pagurus longicarpus*:			
Adults	10.0	No deaths in 168 h	5
Adults	50.0	LC50 (96 h)	5
Adults	50.0	LC50 (168 h)	5
Adults	125.0	All dead in 168 h	5
Prawn, *Penaeus indicus*:			
Postlarva	16.1	LC50 (48 h)	11
Postlarva	15.3	LC50 (96 h)	11

Table 8.1 (continued) Lethality of Inorganic and Organic Mercury Compounds to Selected Species of Aquatic Organisms

Chemical Species, Ecosystem, Taxonomic Group, Species, and Other Variables	Concentration (µg Hg/L medium)	Effect[a]	Ref.[b]
Annelids			
Polychaete, *Capitella capitata*; larva	14.0	LC50 (96 h)	1
Sandworm, *Nereis virens:*			
Adults	25.0	No deaths in 168 h	5
Adults	60.0	LC50 (168 h)	5
Adults	70.0	LC50 (96 h)	5
Adults	125.0	All dead in 168 h	5
Echinoderms			
Starfish, *Asterias rubens:*			
Adults	10.0	No deaths in 168 h	5
Adults	20.0	LC50 (168 h)	5
Adults	60.0	LC50 (96 h)	5
Adults	125.0	All dead in 168 h	5
Fish			
Haddock, *Melanogrammus aeglefinus*; larvae	98.0	LC50 (96 h)	1
Spot, *Leiostomus xanthurus*; adult	36.0 (32.0–39.0)	LC50 (96 h)	23
Tidewater silverside, *Menidia peninsulae*; larvae, age 26 days	71.0 (60.0–84.0)	LC50 (96 h)	23
Mummichog, *Fundulus heteroclitus*:			
Adults	80.0	LC50 (96 h)	5
Adults	80.0	LC50 (168 h)	5
Adults	23,000.0	LC50 (24 h)	5
Organic Mercury: Freshwater			
Planarians			
Flatworm, *Dugesia dorotocephala*:			
Adult	200.0	LC0 (10 d)	12
Adult	500.0	LC100 (5 d)	12
Crustaceans			
Daphnid, *Daphnia magna*	0.9–3.2	LC50 (LT)	1
Fish			
Rainbow trout:			
Larva	24.0	LC50 (96 h)	1
Juvenile	5.0–42.0	LC50 (96 h)	1
Subadult	34.0	LC50 (48 h)	20
Subadult	4.0	< 50.0% dead in 100 d	20
Brook trout; yearling	65.0	LC50 (96 h)	1
Air-breathing catfish, *Clarias batrachus*; adults:			
Methylmercury	430.0	LC50 (96 h)	18
Methoxyethylmercury	4300.0	LC50 (96 h)	18
Blue gourami, *Trichogaster* sp.; adults	70.0	LC50 (96 h)	22

(continued)

Table 8.1 (continued) Lethality of Inorganic and Organic Mercury Compounds to Selected Species of Aquatic Organisms

Chemical Species, Ecosystem, Taxonomic Group, Species, and Other Variables	Concentration (μg Hg/L medium)	Effect[a]	Ref.[b]
Amphibians			
Toad, *Bufo bufo japonicus*; tadpole	120.0	LC50 (48 h)	28
Toad, *Bufo melanosticus*; tadpole	56.0	LC50 (96 h)	28
Frog, *Rana breviceps*; tadpole	60.0	LC50 (96 h)	28
Organic Mercury: Marine			
Molluscs			
American oyster, *Crassostrea virginica:*			
Adults	50.0 for 19 days at 0–10°C to methylmercury or phenylmercury	Most moribund or dead	24
Adults	Survivors from above removed at day 19 and transferred to flowing mercury-free seawater	All dead within 14 days	24
Crustaceans			
Amphipod, *Gammarus duebeni*	150.0	LC50 (96 h)	1
Fish			
Mummichog, *Fundulus heteroclitus*:			
Eggs, polluted creek (sediment content of 10.3 mg Hg/kg)	1700.0	LC50 (20 min)	21
Eggs, reference site	700.0	LC50 (20 min)	21

[a] Abbreviations: LT = lifetime exposure; h = hours; d = days; min = minutes.
[b] Reference: 1, USEPA, 1980; 2, Birge et al., 1979; 3, Verma and Tonk, 1983; 4, Parker, 1979; 5, Eisler and Hennekey, 1977; 6, USEPA, 1985; 7, Thain, 1984; 8, Nelson et al., 1976; 9, Gentile et al., 1983; 10, Glickstein, 1978; 11, McClurg, 1984; 12, Best et al., 1981; 13, Jayaprakash and Madhyastha, 1987; 14, Kanamadi and Saidapur, 1991; 15, Punzo, 1993; 16, Dave and Xiu, 1991; 17, Hilmy et al., 1987; 18, Kirubagaran and Joy, 1988; 19, Diamond et al., 1989; 20, Niimi and Kissoon, 1994; 21, Khan and Weis, 1987; 22, Hamasaki et al., 1995; 23, Mayer, 1987; 24, Cunningham and Tripp, 1973; 25, Bastidas and Garcia, 2004; 26, Shah and Altindag, 2004; 27, Birge et al., 2000; 28, Paulose, 1988; 29, Ghate and Mulherkar, 1980; 30, Rao and Madyastha, 1987; 31, Punzo, 1993; 32, De Zwart and Sloof, 1987; 33, Valenti et al., 2005; 34, Unrine et al., 2004.

ciliate protozoan, *Uronema nigricans*, acquired tolerance to mercury after feeding on mercury-laden bacteria and subsequently exposed to increasing levels of mercury in solution (Berk et al., 1978). The phenomenon of acquired mercury tolerance in *U. nigricans* occurred in a single generation (Berk et al., 1978). Among coral colonies of *Porites asteroides*, the LC50 (72 h) value was 180.0 μg Hg/L, as $HgCl_2$. Death was preceded by polyp contraction during the first 8 h, color loss within 24 h, and tissue loss within 48 h (Bastidas and Garcia, 2004).

In general, salts of mercury and its organic compounds have been shown in short-term bioassays to be more toxic to marine organisms than salts of other heavy metals (Kobayashi, 1971; Conner, 1972; Schneider, 1972; Berland et al., 1976; Reish et al., 1976; Eisler and Hennekey, 1977). To oyster embryos, for example, mercury salts were more toxic than salts of silver, copper, zinc, nickel, lead, cadmium, arsenic, chromium, manganese, or aluminum (Calabrese et al., 1973); to clam embryos, mercury was the most toxic metal tested, followed, in order, by silver, zinc, nickel, and lead (Calabrese and Nelson, 1974). Glickstein (1978) reported an LC50 (48 h) value of 5.7 μg Hg/L, as inorganic mercury, to embryos of the Pacific oyster, *Crassostrea gigas*; however, embryos were relatively insensitive to mercury 24 h postfertilization, and survival was enhanced by a variety of factors, including ambient selenium concentrations.

Mercury toxicity to crustaceans was markedly influenced by developmental stage, diet, sex, salinity, tissue sensitivity, and selenium. Larvae and newly molted crustaceans were more sensitive to mercury toxicity than were adults of the same species (Wilson and Conner, 1971; Vernberg et al., 1974; Shealy and Sandifer, 1975). Starved larvae of the grass shrimp had lower survival rates than fed larvae when subjected to mercury insult (Shealy and Sandifer, 1975). Also, male adult fiddler crabs (*Uca pugilator*) were more sensitive to mercury salts than females (Vernberg et al., 1974). Lethality of mercury salts to the porcelain crab (*Petrolisthus armatus*) were most pronounced at lower salinities within the range of 7 to 35‰ (Roesijadi et al., 1974). A similar pattern was recorded for the fiddler crab, *Uca pugilator* (Vernberg et al., 1974). Adult prawns (*Leander serratus*) held in lethal solutions of mercury (50.0 mg inorganic Hg/L; 1.0 mg organic mercury/L) for 3 h contained at death 320.0 to 460.0 mg Hg/kg DW in antennary gland (Corner and Rigler, 1958). High levels of selenium (> 5.0 mg/L) increased mercury toxicity to larvae of dungeness crab, *Cancer magister*, to levels below the LC50 (96 h) value of 6.6 μg Hg/L; however, moderate selenium values of 0.01 to 1.0 mg/L tended to decrease mercury toxicity (Glickstein, 1978).

Many acute toxicity bioassays were of 96-h duration, a duration that allows the senior investigator and technicians alike the opportunity to enjoy an uninterrupted weekend. But it is clear from Table 8.1 that assays of 168-h duration produced LC50 values up to 45 times lower (more toxic) than did the 96-h assays, as was shown for mud snails. It is recommended that acute toxicity bioassays with mercury and other toxicants and estuarine fauna should consist of a minimal 10-day continuous exposure followed by a 10-day observation period (Eisler, 1970).

8.1.2 Vertebrates

Signs of acute mercury poisoning in fish, included flaring of gill covers, increased the frequency of respiratory movements, loss of equilibrium, excessive mucous secretion, darkening of coloration, and sluggishness (Armstrong, 1979; Hilmy et al., 1987). Signs of chronic mercury poisoning included emaciation (due to appetite loss), brain lesions, cataracts, diminished response to change in light intensity, inability to capture food, abnormal motor coordination, and various erratic behaviors (Armstrong, 1979; Hawryshyn et al., 1982). Total mercury concentrations in tissues of mercury-poisoned adult freshwater fish that died soon thereafter ranged (in mg/kg fresh weight) from 6.0 to 114.0 in liver, 3.0 to 42.0 in brain, 5.0 to 52.0 in muscle, and 3.0 to 35.0 in whole body (Armstrong, 1979; Wiener and Spry, 1996). Whole body concentrations up to 100.0 mg/kg FW were reportedly not lethal to rainbow trout, *Oncorhynchus mykiss*, although 20.0 to 30.0 mg/kg FW in that species were associated with reduced appetite, loss of equilibrium, and hyperplasia of gill epithelium (Niimi and Lowe-Jinde, 1984). Brook trout, *Salvelinus fontinalis*, however, showed toxic signs and death at whole body residues of only 5.0 to 7.0 mg/kg FW (Armstrong, 1979).

Some fish populations have developed a resistance to methylmercury, but only in the gametes and embryonic stage. For example, eggs of the mummichog (*Fundulus heteroclitus*), an estuarine cyprinodontiform fish, from a mercury-contaminated creek, when compared to a reference site, were more than twice as resistant to methylmercury (LC-50 values of 1.7 mg Hg/L vs. 0.7 mg Hg/L) when exposed for 20 min prior to combination with untreated sperm. Eggs from the polluted creek that were subjected to 1.0 or 2.5 mg CH_3HgCl/L produced embryos with a 5.0 to 7.0% malformation frequency vs. 32.0% malformations at 1.0 mg/L and little survival at 2.5 mg/L in the reference group (Khan and Weis, 1987). Genetic polymorphism in mosquitofish (*Gambusia* sp.) at specific enzyme loci are thought to control survival during mercury exposure (Diamond et al., 1989). In one population of mosquitofish during acute exposure to mercury, genotypes at three loci were significantly related to survival time, as was heterozygosity. However, neither genotype nor heterozygosity were related to survival in a different population of mosquitofish during acute mercury exposure (Diamond et al., 1991).

Embryo-larva tests with amphibians and inorganic mercury showed that 6 of the 21 species tested were more sensitive than rainbow trout embryo-larva tests and 15 were less sensitive; however, all 21 amphibian species were more sensitive than largemouth bass embryos (Birge et al., 2000; Table 8.1). Amphibian embryos were the most sensitive stage tested to mercury and other chemicals owing to the relatively high permeability of the egg capsule at this time (Birge et al., 2000). In general, organomercurials were 3 to 4 times more lethal than inorganic mercury compounds to amphibians when the same species and life stage were tested (Table 8.1).

Exposure pathways for adult amphibians include soils (dermal contact, liquid water uptake), water (dermal contact with surface water), air (cutaneous and lung absorption), and diet (adults are carnivores). All routes of exposure are affected by various physical, chemical, and other factors. Dietary exposure in adults, for example, is related to season of year, activity rates, food availability, consumption rate, and assimilation rates (Birge et al., 2000). Knowledge of these modifiers is necessary for adequate risk assessment of mercury as a possible factor in declining amphibian populations worldwide.

8.2 TERRESTRIAL INVERTEBRATES

Methylmercury compounds at concentrations of 25.0 mg Hg/kg in soil were fatal to all tiger worms (*Eisenia foetida*) in 12 weeks; at 5.0 mg/kg, however, only 21.0% died in a similar period (Beyer et al., 1985). Inorganic mercury compounds were also toxic to earthworms (*Octochaetus pattoni*); in 60 days, 50.0% died at soil Hg^{2+} levels of 0.79 mg/kg, and all died at 5.0 mg/kg (Abbasi and Soni, 1983).

8.3 REPTILES

Data on mercury lethality in reptiles are scarce, and those available suggest that sensitivity may be both species and age dependent (Rainwater et al., 2005). For example, juveniles of the corn snake, *Elaphe guttata*, fed diets containing 12.0 mg methylmercury/kg FW diet all died within days (Bazar et al., 2002). However, adults and offspring from treated adults of the garter snake, *Thamnophis sirtalis*, fed diets containing up to 200.0 mg methylmercury/kg FW diet all survived and showed no sign of toxicity (Wolfe et al., 1998).

8.4 BIRDS

Signs of mercury poisoning in birds include muscular incoordination, falling, slowness, fluffed feathers, calmness, withdrawal, hyporeactivity, hypoactivity, and eyelid drooping. In acute oral exposures, signs appeared as soon as 20 min post-administration in mallards, *Anas platyrhynchos*, and 2.5 h in ring-necked pheasants, *Phasianus colchicus*. Deaths occurred between 4 and 48 h in mallards and 2 and 6 days in pheasants; remission took up to 7 days (Hudson et al., 1984). In studies with coturnix, *Coturnix* sp., Hill (1981) found that methylmercury was always more toxic than inorganic mercury, and that young birds were usually more sensitive than older birds. Furthermore, some birds poisoned by inorganic mercury recovered after treatment was withdrawn, but chicks that were fed methylmercury and later developed toxic signs usually died, even if treated feed was removed. Coturnix subjected to inorganic mercury, regardless of route of administration, showed a violent neurological dysfunction that ended in death 2 to 6 h posttreatment. The withdrawal syndrome in coturnix poisoned by Hg^{2+} was usually preceded by intermittent, nearly undetectable tremors, coupled with aggressiveness toward cohorts; time from onset to remission was usually 3 to 5 days, but sometimes extended to 7 days. Coturnix poisoned by methylmercury appeared normal until 2 to 5 days posttreatment; then, ataxia and low body carriage with outstretched neck were often associated with walking. In advanced stages, coturnix lost locomotor coordination and did not recover; in mild to moderate clinical signs, recovery usually took at least 1 week (Hill, 1981).

Mercury toxicity to birds varies with the form of the element, dose, route of administration, species, sex, age, and physiological condition (Fimreite, 1979). For example, in northern bobwhite chicks fed diets containing methylmercury chloride, mortality was significantly lower when the solvent was acetone than when it was another carrier such as propylene glycol or corn oil (Spann et al., 1986). In addition, organomercury compounds interact with elevated temperatures and pesticides, such as DDE and parathion, to produce additive or more-than-additive toxicity, and with selenium to produce less-than-additive toxicity (Fimreite, 1979). Acute oral toxicities of various mercury formulations ranged between 2.2 and about 31.0 mg Hg/kg body weight for most avian species tested (Table 8.2). Similar data for other routes of administration were 4.0 to 40.0 mg/kg for diet and 8.0 to 15.0 mg/kg body weight for intramuscular injection (Table 8.2).

Residues of mercury in experimentally poisoned passerine birds usually exceeded 20.0 mg/kg FW, and were similar to concentrations reported in wild birds that died of mercury poisoning (Finley et al. 1979). In one study with the zebra finch (*Poephila guttata*), adults were fed methylmercury in the diet for 76 days at dietary levels of < 0.02 (controls), 1.0, 2.5, or 5.0 mg Hg/kg DW ration (Scheuhammer, 1988). There were no signs of mercury intoxication in any group except the high-dose group, which experienced 25.0% dead and 40.0% neurological impairment. Dead birds from the high-dose group had 73.0 mg Hg/kg FW in liver, 65.0 in kidney, and 20.0 in brain; survivors without signs had 30.0 in liver, 36.0 in kidney, and 14.0 mg Hg/kg FW in brain; impaired birds had 43.0 mg Hg/kg FW in liver, 55.0 in kidney, and 20.0 in brain (Scheuhammer, 1988).

Mercury levels in tissues of poisoned wild birds were highest (45.0 to 126.0 mg/kg FW) in red-winged blackbirds (*Agelaius phoeniceus*), intermediate in European starlings (*Sturnus vulgaris*) and cowbirds (*Molothrus ater*), and lowest (21.0 to 54.0) in common grackles (*Quiscalus quiscula*). In general, mercury residues were highest in the brain, followed by the liver, kidney, muscle, and carcass. Some avian species are more sensitive than passerines (Solonen and Lodenius, 1984; Hamasaki et al., 1995). Liver residues (in mg Hg/kg FW) in birds experimentally killed by methylmercury ranged from 17.0 in red-tailed hawks (*Buto jamaicensis*) to 70.0 in jackdaws (*Corvus monedula*); values were intermediate in ring-necked pheasants, kestrels (*Falco tinnunculus*), and black-billed magpies (*Pica pica*) (Solonen and Lodenius; Hamasaki et al., 1995). Experimentally poisoned grey herons (*Ardea cinerea*) seemed to be unusually resistant to mercury; lethal doses produced residues of 415.0 to 752.0 mg Hg/kg dry weight of liver (Van der Molen et al., 1982). However, levels of this magnitude were frequently encountered in livers from grey herons collected during a massive die-off in the Netherlands during a cold spell in 1976; the interaction effects of cold stress, mercury loading, and poor physical condition of the herons are unknown (Van der Molen et al., 1982).

8.5 MAMMALS

Mercury is easily transformed into stable and highly toxic methylmercury by microorganisms and other vectors (De Lacerda and Salomons, 1998; Eisler, 2000). Methylmercury affects the central nervous system in humans — especially the sensory, visual, and auditory areas concerned with coordination; the most severe effects lead to widespread brain damage, resulting in mental derangement, coma, and death (Clarkson and Marsh, 1982; USPHS, 1994). Methylmercury has long residence times in aquatic biota and consumption of methylmercury-contaminated fish is implicated in more than 150 deaths and more than 1000 birth defects in Minamata, Japan, between 1956 and

Table 8.2 Lethality to Birds of Mercury Administered by Oral, Dietary, or Other Routes

Route of Administration, Organism, and Mercury Formulation	Mercury Concentration and Effect	Ref.[a]
	Single Oral Dose	
Chukar, *Alectoris chukar*:		
Ethylmercury	26.9 mg/Hg kg body weight (BW); LD50 within 14 d posttreatment	1
Mallard, *Anas platyrhynchos*:		
Methylmercury	2.2–23.5 mg Hg/kg BW; LD50 within 14 d posttreatment	1
Ethylmercury	75.7 mg Hg/kg BW; LD50 within 14 d posttreatment	1
Phenylmercury	524.7 mg Hg/kg BW; LD50 within 14 d posttreatment	1
Common bobwhite, *Colinus virginianus*:		
Methylmercury	23.8 mg Hg/kg BW; LD50 within 14 d posttreatment	1
Coturnix, *Coturnix japonica*:		
Methylmercury	11.0–33.7 mg Hg/kg BW; LD50 within 14 d posttreatment	1–3
Inorganic mercury	26.0–54.0 mg Hg/kg BW; LD50 within 14 d posttreatment	2, 3
Ethylmercury	21.4 mg Hg/kg BW; LD50 within 14 d posttreatment	1
Rock dove, *Columba livia*; ethylmercury	22.8 mg Hg/kg BW; LD50 within 14 d posttreatment	1
Fulvous whistling duck, *Dendrocygna bicolor*; methylmercury	37.8 mg Hg/kg BW; LD50 within 14 d posttreatment	1
Domestic chicken, *Gallus domesticus*; phenylmercury	60.0 mg Hg/kg BW; LD50 within 14d posttreatment	4
House sparrow, *Passer domesticus*; methylmercury	12.6–37.8 mg Hg/kg BW; LD50 within 14 d posttreatment	1
Gray partridge, *Perdix perdix*; ethylmercury	17.6 mg Hg/kg BW; LD50 within 14 d posttreatment	1
Ring-necked pheasant, *Phasianus colchicus*:		
Ethylmercury	11.5 mg Hg/kg BW; LD50 within 14 d posttreatment	1
Methylmercury	11.5–26.8 mg Hg/kg BW; LD50 within 14 d posttreatment	1
Phenylmercury	65.0–101.0 mg Hg/kg BW; LD50 within 14 d posttreatment	1, 4
Prairie chicken, *Tympanuchus cupido*; ethylmercury	11.5 mg Hg/kg BW; LD50 within 14 d posttreatment	1
	Dietary Route	
Mallard; hens; methylmercury	3.0 mg Hg/kg diet; reduced duckling survival over two reproductive seasons	11
Coturnix:		
Inorganic mercury	32.0 mg Hg/kg diet between hatch and age 9 weeks; no deaths	2
Inorganic mercury	2956.0–5086.0 mg Hg/kg diet for 5 days followed by 7-day observation period; LD50	2
Inorganic mercury	14-d-old coturnix fed for 5 days on treated diet, then untreated food until remission of signs: controls, 2.0% dead; 2500.0 mg Hg/kg diet as $HgCl_2$, 13.0% dead; 3535.0 mg Hg/kg diet, 33.0% dead; 7070.0 mg/kg diet, 73.0% dead; 10,000.0 mg Hg/kg diet, 80.0% dead; 5086.0 (3743.0–6912.0) mg Hg/kg diet = LD50	15
Methylmercury	Feeding regimen with 14-d-old coturnix as above: 15.0 mg Hg/kg diet, as methylmercury chloride, no deaths; 21.0 or 42.0 mg Hg/kg diet, 6.0% dead; 60.0 mg Hg/kg diet, 73.0% dead (onset of signs on day 4, remission on day 12); 47.0 (30.0–60.0) mg Hg/kg diet = LD50	15
Inorganic mercury:		
In dry salt	500.0 mg Hg/kg diet for 28 days; LD86	6
In ethanol, methanol, or water	500.0 mg Hg/kg diet for 28 days; LD55	6
In casein premix	500.0 mg Hg/kg diet for 28 days; LD33	6

Table 8.2 (continued) Lethality to Birds of Mercury Administered by Oral, Dietary, or Other Routes

Route of Administration, Organism, and Mercury Formulation	Mercury Concentration and Effect	Ref.[a]
Methylmercury	4.0 mg Hg/kg diet between hatch and age 9 weeks; no deaths	2
Methylmercury	8.0 mg Hg/kg diet for 5 d; some deaths	3
Methylmercury	31.0–47.0 mg Hg/kg diet for 5 d followed by 7-d observation period; LD50	2
Zebra finch, *Poephila guttata*:		
Methylmercury	2.5 mg Hg/kg diet for 77 d; no deaths	12
Methylmercury	5.0 mg Hg/kg diet for 77 d; LD25	12
Ring-necked pheasant:		
Ethylmercury	4.2 mg Hg/kg diet for 70 d; no deaths[b]	5
Ethylmercury	12.5 mg Hg/kg diet for 70 d; LD50	5
Ethylmercury	37.4 mg Hg/kg diet for 28 d; LD50	5
Ethylmercury	112.0 mg Hg/kg diet for 15 d; LD50	5
Birds; 4 species:		
Methylmercury	40.0 mg Hg/kg diet for 6 to 11 d; LD33	7
Birds; 3 species:		
Methylmercury	33.0 mg Hg/kg diet for 35 d; LD8 to LD90	14
	Intramuscular Injection	
Coturnix:		
Methylmercury	Single im injection of 8.0–33.0 mg Hg/kg BW; LD50	2
Inorganic mercury	Single im injection of 15.0–50.0 mg Hg/kg BW; LD50	2
Rock dove; inorganic mercury	Daily im injections for 17 d of 10.0 mg Hg/kg BW; some deaths	8
	Yolk Sac Injection	
Chicken:		
Methylmercury	Single injection of 0.015 mg Hg/egg; some deaths	13
Methylmercury	Single injection of 0.04–0.05 mg Hg/egg; LD50	13
	Applied to Egg Surface	
Mallard:		
Methylmercury	Single dose of 0.003 mg Hg/egg; some deaths	9
Methylmercury	Single dose of 0.009 mg Hg/egg; LD50	9
	In Drinking Water	
Chicken; inorganic mercury	500.0 mg Hg/L for 3 days; some deaths	10

[a] Reference: 1, Hudson et al., 1984; 2, Hill, 1981; 3, Hill and Soares, 1984; 4, Mullins et al., 1977; 5, Spann et al., 1972; 6, El-Begearmi et al., 1980; 7, Finley et al., 1979; 8, Leander et al., 1977; 9, Hoffman and Moore, 1979; 10, Grissom and Thaxton, 1985; 11, Heinz and Locke, 1976; 12, Scheuhammer, 1988; 13, Greener and Kochen, 1983; 14, Hamasaki et al., 1995; 15, Hill and Camardese, 1986.
[b] Reduction in egg production of 55.0–80.0%; embryonic survival sharply reduced.

1960 (see Chapter 10). By 1987, more than 17,000 people had been affected by methylmercury poisoning in Japan, with 999 deaths. Worldwide, De Lacerda and Salomons (1998) estimate that mercury poisoning from ingestion of contaminated food is responsible for more than 1400 human deaths and 200,000 sublethal cases. Excess mercury in human tissues is associated with an increased risk of acute myocardial infarction (Salonen et al., 1995; Gualler et al., 2002), and increased death rate from coronary heart disease (Salonen et al., 1995) and carotid atherosclerosis (Salonen et al., 2000).

In mule deer (*Odocoileus hemionus hemionus*), after acute oral mercury poisoning was induced experimentally, signs included belching, bloody diarrhea, piloerection (hair more erect than usual), and loss of appetite (Hudson et al., 1984). The kidney is the probable critical organ in adult mammals due to the rapid degradation of phenylmercurials and methoxyethylmercurials to inorganic mercury compounds and subsequent translocation to the kidney (Suzuki, 1979), whereas in the fetus the brain is the principal target (Khera, 1979). Most human poisonings were associated with organomercury compounds used in agriculture as fungicides to protect cereal seed grain (Elhassani, 1983); judging from anecdotal evidence, many wildlife species may have been similarly afflicted. Organomercury compounds, especially methylmercury, were the most toxic mercury species tested. Among sensitive species of mammals, death occurred at daily organomercury concentrations of 0.1 to 0.5 mg/kg body weight, or 1.0 to 5.0 mg/kg in the diet (Table 8.3). Larger animals such as mule deer and harp seals (*Pagophilus groenlandica*) appear to be more resistant to mercury than were smaller mammals such as mink, cats, dogs, pigs, monkeys, and river otters (Table 8.3); the reasons for this difference are unknown but may be related to differences in metabolism and mercury detoxification rates. Tissue residues in fatally poisoned mammals (in mg Hg/kg fresh weight) were approximately 6.0 in brain, 10.0 to 55.6 in liver, 17.0 in whole body, about 30.0 in blood, and 37.7 in kidney (Eisler, 2000; Table 8.3).

The lethal effects of methylmercury in various species of mammals is influenced by ambient temperature, dietary selenium, ethanol, and especially hypertension (Tamashiro et al., 1986). Tests with a genetic strain of rat with high blood pressure showed that this strain was more sensitive to methylmercury than were control strains: they died earlier and with higher tissue mercury burdens (Table 8.3). Because hypertension and borderline hypertension is common among human inhabitants of mercury-polluted areas, with estimates as high as 56.0% among individuals 40 years old and older (Tamashiro et al., 1986), more research seems warranted on the role of hypertension as a significant health problem in methylmercury-impacted populations.

Mercury interactions with other compounds should be considered. Adverse effects on growth and survival of kits of the mink, *Mustela vison*, are reported for diets containing 1.0 mg Hg/kg ration as methylmercury and Aroclor 1254 — a polychlorinated biphenyl — at 1.0 mg/kg ration (Wren et al., 1987b).

Table 8.3 Lethality of Organomercury Compounds to Selected Mammals

Route of Administration, Organism, Dose, and Other Variables	Effect	Ref.[a]
	Oral Dose	
Domestic dog, *Canis familiaris*:		
0.1 to 0.25 mg/kg body weight (BW) during entire pregnancy	High incidence of stillbirths	1
Rat, *Rattus* sp.:		
Strain with spontaneous hypertension; age 10 weeks; oral dose of 5.0 mg methylmercury/kg BW daily for 10 consecutive days. Tail blood pressure of hypertensive rats ranged from 190–231 mm Hg vs. 130–165 mm Hg for controls	Deaths and signs of mercury intoxication in hypertensive rats evident 1 week earlier than controls, but all rats in both groups dead by day 13 following last dose; total Hg concentrations in blood and other tissues were significantly higher in hypertensive rats on days 1 and 5 following the final dose	10
Pig, *Sus* spp.:		
0.5 mg/kg BW daily during pregnancy	High incidence of stillbirths	1
Rhesus monkey, *Macaca mulatta*:		
0.5 mg/kg BW daily during days 20–30 of pregnancy	Maternally toxic and abortient	1
Mule deer, *Odocoileus heminous hemionus*:		
17.9 mg/kg BW; single oral dose	LD50	5
Harp seal, *Pagophilus groenlandica*:		
25.0 mg/kg BW daily	Dead in 20 to 26 days; blood mercury levels immediately prior to death were 26.8–30.3 mg/L	6

Table 8.3 (continued) Lethality of Organomercury Compounds to Selected Mammals

Route of Administration, Organism, Dose, and Other Variables	Effect	Ref.[a]
Dietary Route		
Domestic cat, *Felis domesticus*:		
0.25 mg/kg BW daily for 90 days; total of about 85 mg mercury	LD50 (78 d). Convulsions starting at day 68; all convulsing by day 90. Liver residues of survivors were 40.2 mg/kg FW for total mercury and 18.1 mg/kg for inorganic mercury	2
Human, *Homo sapiens*:		
Methylmercury; whole body:	No effect level = 10.0–100.0 mg whole body; toxic is 100.0–1000.0 mg whole body; 1000.0 mg and higher is lethal	11, 12
20.0 mg daily for about 50 days	Lethal	11, 12
10.0 mg daily for about 500 days	Lethal	11, 12
5.0 mg daily for about 30 days	Toxic	12
2.0 mg daily for 70 days	Toxic	12
1.0 mg daily for about 500 days	Toxic	12
0.5 mg daily for < 30 days	No observable effect	12
Mink, *Mustela vison*:		
1.0 mg/kg diet daily	Fatal to 100.0% in about 2 months	3
1.0 mg/kg diet daily for 4 months, then every other day for 4 months alternating with control diet	High mortality after 4 months when subjected to cold stress; significant mercury transfer to fetus via placenta	9
5.0 mg/kg diet daily	All dead in 30 to 37 days. Elevated mercury residues (mg/kg FW) in kidney (37.7) and liver (55.6) prior to death	3
River otter, *Lutra canadensis*:		
> 2.0 mg/kg in diet	Fatal within 7 months	4, 8
15.7 mg methylmercury/kg FW ration	Lethal. Dead animals had 33.4 mg Hg/kg FW in liver and 39.2 mg/kg FW in kidney	13
Inhalation Route		
Rat, *Rattus* sp.:		
27.0 mg/m^3 air for 1–2 h	Fatal; death by asphyxiation; lung edema; necrosis of alveolar epithelium	7
Various		
Human, *Homo sapiens*	Lethal residues in tissues, in mg/kg FW, were > 6.0 in brain, > 10.0 in liver, and > 17.0 in whole body	1

[a] Reference: 1, Khera, 1979; 2, Eaton et al., 1980; 3, Sheffy and St. Amant, 1982; 4, Kucera, 1983; 5, Hudson et al., 1984; 6, Ronald et al., 1977; 7, USPHS, 1994; 8, Ropek and Neely, 1993; 9, Wren et al., 1987a; 10, Tamashiro et al., 1986; 11, Kitamura, 1971; 12, Takizawa, 1993; 13, O'Connor and Nielsen, 1980.

8.6 SUMMARY

For all organisms tested, early developmental stages were the most sensitive, and organomercury compounds — especially methylmercurials — were more toxic than inorganic forms. Numerous biological and abiotic factors modify the lethality of mercury compounds, sometimes by an order of magnitude or more, but the mechanisms of action are not clear. Lethal concentrations of total mercury to sensitive, representative organisms varied from 0.1 to 2.0 μg/L of medium for aquatic fauna; from 2.2 to 31.0 mg/kg body weight (acute oral) and 4.0 to 40.0 mg/kg (dietary) for birds; and from 0.1 to 0.5 mg/kg body weight (daily dose) and 1.0 to 5.0 mg/kg diet for mammals.

REFERENCES

Abbasi, S.A. and R. Soni. 1983. Stress-induced enhancement of reproduction in earthworms *Octochaetus pattoni* exposed to chromium (VI) and mercury (II) — implications in environmental management, *Int. J. Environ. Stud.*, 22, 43–47.

Armstrong, F.A.J. 1979. Effects of mercury compounds on fish. In J.O. Nriagu (Ed.), *The Biogeochemistry of Mercury in the Environment,* p. 657–670. Elsevier/North-Holland Biomedical Press, New York.

Bastidas, C. and E.M. Garcia. 2004. Sublethal effects of mercury and its distribution in the coral *Porites astreoides*, *Mar. Ecol. Prog. Ser.*, 267, 133–143.

Bazar, M.A., D.A. Holtzman, B.M. Adair, and S.E. Gresens. 2002. Effects of dietary methylmercury in juvenile corn snakes (*Elaphe guttata*). Abstract PO89 in *SETAC 23rd Annual Meeting, Salt Lake City, UT, November 16–20, 2002*, 176.

Berk, S.G., A.L. Mills, D.L. Henricks, and R.R. Colwell. 1978. Effects of ingesting mercury containing bacteria on mercury tolerance and growth rate of ciliates, *Microb. Ecol.*, 4, 319–330.

Berland, B.R., D.J. Bonin, V.I. Kapkov, S. Maestrini, and D.P. Arlhac. 1976. Action toxique de quatre metaux lourds sur la croissance d'algues unicellulaires marines, *C.R. Acad. Sci. Paris*, 282D, 633–536.

Best, J.B., M. Morita, J. Ragin, and J. Best, Jr. 1981. Acute toxic responses of the freshwater planarian, *Dugesia dorotocephala*, to methylmercury, *Bull. Environ. Contam. Toxicol.*, 27, 49–54.

Beyer, W.N., E. Cromartie, and G.B. Moment. 1985. Accumulation of methylmercury in the earthworm, *Eisenia foetida*, and its effect on regeneration, *Bull. Environ. Contam. Toxicol.*, 35, 157–162.

Birge, W.J., J.A. Black, A.G. Westerman, and J.E. Hudson. 1979. The effect of mercury on reproduction of fish and amphibians. In J.O. Nriagu (Ed.), *The Biogeochemistry of Mercury in the Environment*, p. 629–655. Elsevier/North-Holland Biomedical Press, New York.

Birge, W.J., A.G. Westerman, and J.A. Spromberg. 2000. Comparative toxicology and risk assessment of amphibians. In D.W. Sparling, G. Linder, and C.A. Bishop (Eds.), *Ecotoxicology of Amphibians and Reptiles,* p. 727–791. SETAC Press, Pensacola, FL.

Calabrese, A., R.S. Collier, D.A. Nelson, and J.R. MacInnes. 1973. The toxicity of heavy metals to embryos of the American oyster *Crassostrea virginica*, *Mar. Biol.*, 18, 162—166.

Calabrese, A. and D.A. Nelson. 1974. Inhibition of embryonic development of the hard clam, *Mercenaria mercenaria*, by heavy metals, *Bull. Environ. Contam. Toxicol.*, 11, 92–97.

Clarkson, T.W. and D.O. Marsh. 1982. Mercury toxicity in man. In A.S. Prasad (Ed.), *Clinical, Biochemical, and Nutritional Aspects of Trace Elements,* Vol. 6, p. 549–568. Alan R. Liss, Inc., New York.

Conner, P.M.W. 1972. Acute toxicity of heavy metals to some marine larvae, *Mar. Pollut. Bull.*, 3, 190–192.

Corner, E.D.S. and F.H. Rigler. 1958. The modes of action of toxic agents.III. Mercuric chloride and n-amylmercuric chloride on crustaceans, *J. Mar. Biol. Assn. U.K.*, 37, 85–96.

Cunningham, P.A. and M.R. Tripp. 1973. Accumulation and depuration of mercury in the American oyster *Crassostrea virginica*, *Mar. Biol*,, 31, 321–334.

Dave, G. and R. Xiu. 1991. Toxicity of mercury, copper, nickel, lead, and cobalt to embryos and larvae of zebrafish, *Brachydanio rerio*, *Arch. Environ. Contam. Toxicol.*, 21, 126–134.

De Lacerda, L.D. and W. Salomons. 1998. *Mercury from Gold and Silver Mining: a Chemical Time Bomb?.* Springer, Berlin, 146 pp.

De Zwart, D. and W. Sloof. 1987. Toxicity of mixtures of heavy metals and petrochemicals to *Xenopus laevis*, *Bull. Environ. Contam. Toxicol.*, 38, 345–351.

Diamond, S.A., M.C. Newman, M. Mulvey, P.M. Dixon, and D. Martinson. 1989. Allozyme genotype and time to death of mosquitofish, *Gambusia affinis* (Baird and Girard), during acute exposure to inorganic mercury, *Environ. Toxicol. Chem.*, 8, 613–622.

Diamond, S.A., M.C. Newman, M. Mulvey, and S.I. Guttman. 1991. Allozyme genotype and time-to-death of mosquitofish, *Gambusia holbrooki*, during acute inorganic mercury exposure: a comparison of populations, *Aquat. Toxicol.*, 21, 119–134.

Eaton, R.D.P., D.C. Secord, and P. Hewitt. 1980. An experimental assessment of the toxic potential of mercury in ringed-seal liver for adult laboratory cats, *Toxicol. Appl. Pharmacol.*, 55, 514–521.

Eisler, R. 1970. Factors affecting pesticide-induced toxicity in an estuarine fish, *U.S. Bur. Sport Fish. Wildl. Tech. Paper*, 45, 1–20.

Eisler, R. 1981. *Trace Metal Concentrations in Marine Organisms*. Pergamon, Elmsford, NY. 687 pp.

Eisler, R. 2000. Mercury. In *Handbook of Chemical Risk Assessment: Health Hazards to Humans, Plants, and Animals. Vol. 1, Metals,* 313–409. Lewis Publishers, Boca Raton, FL.

Eisler, R., and R. J. Hennekey. 1977. Acute toxicities of Cd^{2+}, Cr^{+6}, Hg^{2+}, Ni^{2+}, and Zn^{2+} to estuarine macrofauna, *Arch. Environ. Contam. Toxicol.*, 6, 315–323.

El-Begearmi, M.M., H.E. Ganther, and M.L. Sunde. 1980. Toxicity of mercuric chloride in Japanese quail as affected by methods of incorporation into the diet, *Poult. Sci.*, 59, 2216–2220.

Elhassani, S.B. 1983. The many faces of methylmercury poisoning, *J. Toxicol.*, 19, 875–906.

Fimreite, N. 1979. Accumulation and effects of mercury on birds. In J.O. Nriagu (Ed.), *The Biogeochemistry of Mercury in the Environment,* p. 601–627. Elsevier/North-Holland Biomedical Press, New York.

Finley, M.T., W.H. Stickel, and R.E. Christensen. 1979. Mercury residues in tissues of dead and surviving birds fed methylmercury, *Bull. Environ. Contam. Toxicol.*, 21, 105–110.

Gentile, J., S.M. Gentile, G. Hoffman, J.F. Heltshe, and N. Hairston, Jr. 1983. The effects of a chronic mercury exposure on survival, reproduction and population dynamics of *Mysidopsis bahia*, *Environ. Toxicol. Chem.*, 2, 61–68.

Ghate, H.V. and L. Murherkar. 1980. Effect of mercuric chloride on embryonic development of the frog *Microhyla ornata*, *Indian J. Exp. Biol.*, 18, 1094–1096.

Glickstein, N. 1978. Acute toxicity of mercury and selenium to *Crassostrea gigas* embryos and *Cancer magister* larvae, *Mar. Biol.*, 49, 113–117.

Greener, Y. and J.A. Kochen. 1983. Methyl mercury toxicity in the chick embryo, *Teratology*, 28, 23–28.

Grissom, R.E., Jr. and J.P. Thaxton. 1985. Onset of mercury toxicity in young chickens, *Arch. Environ. Contam. Toxicol.*, 14, 193–196.

Guallar, E., M.I. Sanz–Gallardo, P.V. Veer, P. Bode, A. Aro, J. Gomez-Arcena, J.D. Kark, R.A. Riemersma, J.M. Martin-Moreno, and F.J. Kok. 2002. Mercury, fish oils, and the risk of myocardial infarction, *New Engl. J. Med.*, 347, 1747–1754.

Hamasaki, T., H. Nagawe, Y. Yoshioka, and T. Sato. 1995. Formation, distribution, and ecotoxicity of methylmetals of tin, mercury, and arsenic in the environment, *Crit. Rev. Environ. Sci. Technol.*, 25, 45–91.

Hawryshyn, C.W., W.C. Mackay, and T.H. Nilsson. 1982. Methyl mercury induced visual deficits in rainbow trout, *Can. J. Zool.*, 60, 3127–3133.

Heinz, G.H. and L.N. Locke. 1976. Brain lesions in mallard ducklings from parents fed methylmercury, *Avian Dis.*, 20, 9–17.

Hill, E.F. 1981. Inorganic and Organic Mercury Chloride Toxicity to Coturnix: Sensitivity Related to Age and Quantal Assessment of Physiological Responses, Ph.D. thesis, University of Maryland, College Park. 221 pp.

Hill, E.F. and M.B. Camardese. 1986. Lethal dietary toxicities of environmental contaminants and pesticides to coturnix, *U.S. Fish Wildl. Serv. Tech. Rep.*, 2, 1–147.

Hill, E.F. and J.H. Soares, Jr. 1984. Subchronic mercury exposure in *Coturnix* and a method of hazard evaluation, *Environ. Toxicol. Chem.*, 3, 489–502.

Hilmy, A.M., N.A. El Domiaty, A.Y. Daabees, and F.I. Moussa. 1987. Short-term effects of mercury on survival, behaviour, bioaccumulation and ionic pattern in the catfish (*Clarias lazera*), *Comp. Biochem. Physiol.*, 87C:303–308.

Hoffman, D.J. and J.M. Moore. 1979. Teratogenic effects of external egg applications of methyl mercury in the mallard, *Anas platyrhynchos*, *Teratology*, 20, 453–462.

Hudson, R.H., R.K. Tucker, and M.A. Haegele. 1984. *Handbook of Toxicity of Pesticides to Wildlife*. U.S. Fish Wildl. Serv. Resour. Publ. 153. 90 pp.

Jayaprakash, R.I. and M.N. Madhyastha. 1987. Toxicities of some heavy metals to the tadpoles of frog, *Microhyla ornata* (Dumeril and Bibron), *Toxicol. Lett.*, 36, 205–208.

Jones, M.B. 1973. Influence of salinity and temperature on the toxicity of mercury to marine and brackish water isopods (Crustacea), *Estuar. Coastal Mar. Sci.*, 1, 425–431.

Kanamadi, R.D. and S.K. Saidapur. 1991. Effect of sublethal concentration of mercuric chloride on the ovary of the frog *Rana cyanophlyctis*, *J. Herpetol.*, 25, 494–497.

Khan, A.T. and J.S. Weis. 1987. Effect of methylmercury on egg and juvenile viability in two populations of killifish *Fundulus heteroclitus*, *Environ. Res.*, 44, 272–278.

Khera, K.S. 1979. Teratogenic and genetic effects of mercury toxicity. In J.O. Nriagu (Ed.), *The Biogeochemistry of Mercury in the Environment,* p. 501–518. Elsevier/North-Holland Biomedical Press, New York.

Kirubagaran, R. and K.P. Joy. 1988. Toxic effects of three mercurial compounds on survival, and histology of the kidney of the catfish *Clarias batrachus* (L.), *Ecotoxicol. Environ. Safety*, 15, 171–179.

Kitamura, S. 1971. Determination on mercury content of inhabitants, cats, fishes, and shells in Minamata District and the mud of Minamata Bay. In M. Katsuna (Ed.). *Minamata Disease,* p. 257–266. Shuhan, Kumamoto.

Kobayashi, N. 1971. Fertilized sea urchin eggs as an indicatory material for marine pollution bioassay, preliminary experiments, *Publ. Seto Mar. Biol. Lab.*, 18, 379–406.

Kucera, E. 1983. Mink and otter as indicators of mercury in Manitoba waters, *Can. J. Zool.*, 61, 2250–2256.

Leander, J.D., D.E. McMillan, and T.S. Barlow. 1977. Chronic mercuric chloride: behavioral effects in pigeons, *Environ. Res.*, 14, 424–435.

Mayer, F.L., Jr. 1987. Acute toxicity of chemicals to estuarine organisms, U.S. Environ. Protect. Agen. Rep., EPA/600/8-87/017. 274 pp.

McClurg, T.P. 1984. Effects of fluoride, cadmium and mercury on the estuarine prawn *Penaeus indicus*, *Water SA*, 10, 40–45.

McKenney, C.L., Jr. and J.D. Costlow, Jr. 1981. The effects of salinity and mercury on developing megalopae and early crab stages of the blue crab, *Callinectes sapidus* Rathbun. In F.J. Vernberg, A. Calabrese, F.P. Thurberg, and W.B. Vernberg (Eds.), *Biological Monitoring of Marine Pollutants,* p. 241–262. Academic Press, New York.

Mullins, W.H., E.G. Bizeau, and W.W. Benson. 1977. Effects of phenyl mercury on captive game farm pheasants, *J. Wildl. Manage.*, 41, 302–308.

Nelson, D.A., A. Calabrese, B.A. Nelson, J.R. MacInnes, and D. R. Wenzloff. 1976. Biological effects of heavy metals on juvenile bay scallops, *Argopecten irradians*, in short-tern exposures, *Bull. Environ. Contam. Toxicol.*, 16, 275–282.

Niimi, A.J. and L. Lowe-Jinde. 1984. Differential blood cell ratios of rainbow trout (*Salmo gairdneri*) exposed to methylmercury and chlorobenzenes, *Arch. Environ. Contam. Toxicol.*, 13, 303–311.

Niimi, A.J. and G.P. Kissoon. 1994. Evaluation of the critical body burden concept based on inorganic and organic mercury toxicity to rainbow trout (*Oncorhynchus mykiss*), *Arch. Environ. Contam. Toxicol.*, 26, 169–178.

O'Connor, D.J. and S.W. Nielsen. 1980. Environmental survey of methylmercury levels in wild mink (*Mustela vison*) and otter (*Lutra canadensis*) from the northeastern United States and experimental pathology of methylmercurialism in the otter. In J.A. Chapman and D. Pursley (Eds.), *Proceedings of the World Furbearer Conference,* p. 1728–1745, Frostburg, MD.

Parker, J.G. 1979. Toxic effects of heavy metals upon cultures of *Uronema marinum* (Ciliophora: uronematidae), *Mar. Biol.*, 54, 17–24.

Paulose, P.V. 1988. Comparative study of inorganic and organic mercury poisoning on selected freshwater organisms, *J. Environ. Biol.*, 9, 203–206.

Punzo, F. 1993. Ovarian effects of a sublethal concentration of mercuric chloride in the river frog, *Rana heckscheri* (Anura:Ranidae), *Bull. Environ. Contam. Toxicol.*, 50, 385–391.

Rainwater, T.R., K.D. Reynolds, J.E. Canas, G.P. Cobb, T.A. Anderson, S.T. McMurry, and P.M. Smith. 2005. Organochlorine pesticides and mercury in cottonmouths (*Agkistrodon piscivorus*) from northeastern Texas, USA, *Environ. Toxicol. Chem.*, 24, 665–673.

Rao, R.I. and M.N. Madyastha. 1987. Toxicities of some heavy metals to the tadpoles of frog, *Microhyla ornata* (Dumeril and Bibron), *Toxicol. Lett.*, 36, 205–208.

Reish, D.J., J.M. Martin, F.M. Piltz, and J.Q. Word. 1976. The effects of heavy metals on laboratory populations of two polychaetes with comparisons to the water quality conditions and standards in southern California marine waters, *Water Res.*, 10, 299–302.

Roesijadi, G., S.R. Petrocelli, J.W. Anderson, B.J. Presley, and R. Sims. 1974. Survival and chloride ion regulation of the porcelain crab *Petrolisthes armatus* exposed to mercury, *Mar. Biol.*, 17, 213–217.

Ronald, K., S.V. Tessaro, J.F. Uthe, H.C. Freeman, and R. Frank. 1977. Methylmercury poisoning in the harp seal (*Pagophilus groenlandicus*), *Sci. Total Environ.*, 8, 1–11.

Ropek, R.M. and R.K. Neely. 1993. Mercury levels in Michigan river otters, *Lutra canadensis*, *J. Freshwat. Ecol.*, 8, 141–147.

Salonen, J.T., K. Seppanen, T.A. Lakka, R. Salonen, and G.A. Kaplan. 2000. Mercury accumulation and accelerated progression of carotid atherosclerosis: a population-based prospective 4-year follow-up study in men in eastern Finland, *Atherosclerosis*, 148, 265–273.

Salonen, J.T., K. Seppanen, K. Nyyssonen, H. Korpela, J. Kauhanen, M. Kantola, J. Tuomilehto, H. Esterbauer, F. Tatzber, and R. Salonen. 1995. Intake of mercury from fish, lipid peroxidation, and the risk of myocardial infarction and coronary, cardiovascular, and any death in eastern Finnish men, *Circulation*, 91, 645–655.

Scheuhammer, A.M. 1988. Chronic dietary toxicity of methylmercury in the zebra finch, *Poephila guttata*, *Bull. Environ. Contam. Toxicol.*, 40, 123–130.

Schneider, J. 1972. Lower fungi as test organisms of pollutants in sea and brackish water. The effects of heavy metal compounds and phenol on *Thraustochytrium striatum*, *Mar. Biol.*, 16, 214–225.

Shah, S.L. and A. Altindag. 2004. Hematological parameters of tench (*Tinca tinca* L.) after acute and chronic exposure to lethal and sublethal mercury treatments, *Bull. Environ. Contam. Toxicol.*, 73, 911–918.

Shealy, M.H. and P.A. Sandifer. 1975. Effects of mercury on survival and development of the larval grass shrimp, *Palaemonetes vulgaris*, *Mar. Biol.*, 33, 7–16.

Sheffy, T.B. and J.R. St. Amant. 1982. Mercury burdens in furbearers in Wisconsin, *J. Wildl. Manage.*, 46, 1117–1120.

Solonen, T. and M. Lodenius. 1984. Mercury in Finnish sparrowhawks *Accipiter nisus*, *Ornis Fennica*, 61, 58–63.

Spann, J.W., R.G. Heath, J.F. Kreitzer, and L.N. Locke. 1972. Ethyl mercury *p*-toluene sulfonanilide: lethal and reproductive effects on pheasants, *Science*, 175, 328–331.

Spann, J.W., G.H. Heinz, M.B. Camardese, E.F. Hill, J.F. Moore, and H.C. Murray. 1986. Differences in mortality among bobwhite fed methylmercury chloride dissolved in various carriers, *Environ. Toxicol. Chem.*, 5, 721–724.

Suzuki, T. 1979. Dose-effect and dose-response relationships of mercury and its derivatives. In J.O. Nriagu (Ed.). *The Biogeochemistry of Mercury in the Environment,* p. 399–431. Elsevier/North-Holland Biomedical Press, New York.

Takizawa, Y. 1993. Overview on the outbreak of Minamata disease — epidemiological aspects. In Proceedings of the International Symposium on Epidemiological Studies on Environmental Pollution and Health Effects of Methylmercury, p. 3–26, October 2, 1992, Kumamoto, Japan. Published by National Institute for Minamata Disease, Kumamoto 867, Japan.

Tamashiro, H., M. Arakaki, H. Akagi, K. Hirayama, and M.H. Smolensky. 1986. Methylmercury toxicity in spontaneously hypertensive rats (SHR), *Bull. Environ. Contam. Toxicol.*, 36, 668–673.

Thain, J.E. 1984. Effects of mercury on the prosobranch mollusc *Crepidula fornicata*: acute lethal toxicity and effects on growth and reproduction of chronic exposure, *Mar. Environ. Res.*, 12, 285–309.

U.S. Environmental Protection Agency (USEPA). 1980. Ambient water quality criteria for mercury, *U.S. Environ. Protection Agen. Rep. 440/5-80-058.* Available from Natl. Tech. Infor. Serv., 5285 Port Royal Road, Springfield, VA 22161.

U.S. Environmental Protection Agency (USEPA). 1985. Ambient water quality criteria for mercury — 1984. *U.S. Environ. Protection Agen. Rep. 440/5-84-026.* 136 pp. Available from Natl. Tech. Inform. Serv., 5285 Port Royal Road, Springfield, VA 22161.

U.S. Public Health Service (USPHS). 1994. *Toxicological profile for mercury (update), TP-93/10.* U.S. PHS, Agen. Toxic Substances Dis. Registry, Atlanta, GA. 366 pp.

Unrine, J.M., C.H. Jagoe, W.A. Hopkins, and H.A. Brant. 2004. Adverse effects of ecologically relevant dietary mercury exposure in southern leopard frog (*Rana sphenocephala*) larvae, *Environ. Toxicol. Chem.*, 23, 2964–2970.

Valenti, T.W., D.S. Cherry, R.J. Neves, and J. Schmerfeld. 2005. Acute and chronic toxicity of mercury to early life stages of the rainbow mussel, *Villosa iris* (Bivalvia: Unionidae), *Environ. Toxicol. Chem.*, 24, 1242–1246.

Van der Molen, E.J., A.A. Blok, and G.J. De Graaf. 1982. Winter starvation and mercury intoxication in grey herons (*Ardea cinerea*) in the Netherlands, *Ardea*, 70, 173–184.

Verma, S.R., and I.P. Tonk. 1983. Effect of sublethal concentrations of mercury on the composition of liver, muscles and ovary of *Notopterus notopterus*, *Water Air Soil Pollut.*, 20, 287–292.

Vernberg, W.B., P.J. DeCoursey, and J. O'Hara. 1974. Multiple environmental factor effects on physiology and behavior of the fiddler crab, *Uca pugilator*. In F.J. Vernberg and W. B. Vernberg (Eds.), *Pollution and Physiology of Marine Organisms,* p. 381–425. Academic Press, New York.

Wiener, J.G., and D.J. Spry. 1996. Toxicological significance of mercury in freshwater fish. In W.N. Beyer, G.H. Heinz, and A.W. Redmon-Norwood (Eds.). *Environmental Contaminants in Wildlife: Interpreting Tissue Concentrations*. p. 297–339. CRC Press, Boca Raton, FL.

Wilson, K.W. and P.M. Conner. 1971. The use of a continuous-flow apparatus in the study of longer-term toxicity of heavy metals, *Int. Coun. Explor. Se C.M.,* 1971/E8, 343–347.

Wolfe, M.E., S. Schwarzbach, and R.A. Sulaiman. 1998. Effects of mercury on wildlife: a comprehensive review, *Environ. Toxicol. Chem.*, 17, 146–160.

Wren, C.D., D.B. Hunter, J.F. Leatherland, and P.M. Stokes. 1987a. The effects of polychlorinated biphenyls and methylmercury, singly and in combination, on mink. I. Uptake and toxic responses, *Arch. Environ. Contam. Toxicol.*, 16, 441–447.

Wren, C.D., D.B. Hunter, J.F. Leatherland, and P.M. Stokes. 1987b. The effects of polychlorinated biphenyls and methylmercury, singly and in combination on mink. II. Reproduction and kit development. *Arch. Environ. Contam. Toxicol.*, 16, 449–454.

CHAPTER 9

Sublethal Effects of Mercury

Significant sublethal effects of mercury include an increased frequency of cancers, birth defects, and chromosomal aberrations in laboratory animals and wildlife. Adverse sublethal effects of mercurials also include growth inhibition, abnormal reproduction, histopathology, high mercury accumulations and persistence, and disrupted biochemistry, metabolism, and behavior. These — and other aspects of exposure to various mercurials by living organisms — are documented and discussed for representative species of bacteria and other microorganisms, aquatic and terrestrial plants and invertebrates, fishes, amphibians, birds, and mammals.

9.1 CARCINOGENICITY, GENOTOXICITY, AND TERATOGENICITY

The social significance of cancer, inherited effects, and birth defects cannot be underestimated. For example, during the Minamata Bay, Japan, methylmercury poisoning outbreak (discussed in detail in Chapter 10) — in which thousands of people were afflicted and many died — dozens of fetuses affected *in utero* as a result of maternal ingestion of methylmercury-contaminated food were born with mental retardation and motor disorders (Y. Harada, 1977; M. Harada, 1978; Inouye and Kajiwara, 1988c). There was much concern in the neighboring population that irreversible genetic damage could produce malformations and other adverse effects in subsequent generations. Accordingly, some parents prohibited marriages of their sons and daughters with persons from Minamata (D'Itri and D'Itri, 1977). Although studies with mice indicated that intrauterine exposure effects to methylmercury were practically nil in humans (Inouye and Kajiwara, 1988c), the stigma remains.

9.1.1 Carcinogenicity

Mercury has been assigned a weight-of-evidence classification of D, which indicates that it is not classifiable as to human carcinogenicity (USPHS, 1994). Beluga whales (*Delphinapterus leucas*) in the St. Lawrence estuary have a high incidence of cancer (Gauthier et al., 1998). Studies with isolated skin fibroblasts of beluga whales exposed to mercuric chloride or methylmercury induced a significant dose-response increase of micronucleated cells. Concentrations as low as 0.5 mg Hg^{2+}/L and 50.0 µg CH_3Hg^+/L — comparable to concentrations present in certain whales of this population — significantly induced proliferation (Gauthier et al., 1998).

9.1.2 Genotoxicity

Chromosomal aberrations and sister chromatid exchanges have been produced by mercury compounds in larvae and embryos of amphibians and fishes; in various species of bacteria, yeasts, and molds; and in cultured somatic cells of fish, insects, echinoderms, and mammals — including somatic cells of rodents, dolphins, cats, and humans (Heagler et al., 1993; De Flora et al., 1994; Kramer and Newman, 1994; USPHS, 1994; Betti and Nigro, 1996). Toxicant stress to individuals may modify genetic characteristics of mosquitofish populations. Allozyme polymorphisms existing in mosquitofish populations with no history of contaminant stress can be subject to selection and provide the basis for adaptation to anthropogenic stress (Mulvey et al., 1995). For example, metabolic differences in glucosephosphatase isomerase genotypes, or closely related loci, may be expressed in degree of environmental stress or fluctuation (Mulvey et al., 1995). Allozyme genotypes could be responsible for transient genotype effects noted in electrophoretic surveys of mercury-stressed mosquitofish populations (Lee et al., 1992).

Methylmercury induces increased mutagenic effects in killifish (*Fundulus heteroclitus*) embryos after exposure to 50.0 μg Hg/L for up to 7 days post-fertilization (Perry et al., 1988). Chromosomal aberrations were 13.0% in the treated group vs. 5.1% in controls. Mutagenic effects in killifish embryos occurred at levels below that measured in some marine sediments (Perry et al., 1988).

Genetic background may influence frequency of hydrocephaly (an abnormal increase in cerebrospinal fluid within the cranial cavity accompanied by expansion of cerebral ventricles, enlargement of the forehead, and atrophy of the brain) in mice following prenatal exposure to methylmercury (Inouye and Kajiwara, 1990a). In pregnant mice of B10.D2 strain — in which hydrocephaly is rare — a single oral dose of 10.0 mg CH_3HgCl/kg BW on one of days 14 to 17 of pregnancy resulted in a 48.0 to 75.0% frequency of gross hydrocephaly in offspring at age 30 days, vs. 5.0% in sham-treated and 4.0% in untreated controls. When the study was repeated with C57BL/10 or DBA/2 strains on day 15 of pregnancy, the incidence of hydrocephaly in dosed C57BL/10 mice was 54.0% vs. 0.8% in controls. Hydrocephaly did not develop in the DBA/2 strain, suggesting that methylmercury-induced hydrocephaly is under genetic control in mice (Inouye and Kajiwara, 1990a).

There is no evidence that mercury is genotoxic in humans. Dental amalgam has been used for more than 150 years and it is well established that patients with amalgam dental fillings are chronically exposed to mercury and that the number of amalgam fillings correlates positively with mercury levels in blood and plasma (Loftenius et al., 1998). Amalgam removal, however, did not have a measurable effect on proliferation of peripheral blood lymphocytes, suggesting that mercury contributed by amalgams was not mitogenic to lymphocytes (Loftenius et al., 1998). Mercury did not adversely affect the number or structure of chromosomes in human somatic cells in workers occupationally exposed to mercury compounds by inhalation or accidentally exposed through ingestion (USPHS, 1994). In the laboratory, however, mercury compounds often exerted genotoxic effects, especially by binding SH groups and acting as spindle inhibitors, resulting in abnormal chromosome numbers (De Flora et al., 1994).

9.1.3 Teratogenicity

Teratogenicity of methylmercury has been confirmed in mice, hamsters, and rats (Harris et al., 1972; Spyker and Smithberg, 1972; Inouye and Murakami, 1975).

Methylmercury, CH_3HgCl^+, unlike Hg^o and Hg^{2+}, is efficiently absorbed through the intestinal tract and skin, crosses the human placenta resulting in fetal blood levels in excess of the mother, with elevated risks to the fetus (Inouye, 1989). Pathological features of children's brains affected by prenatal methylmercury exposure include microcephaly, dilated lateral ventricles, abnormal neuronal migration causing derangement in the gray matter architecture, and degeneration of formed nerve cells (Inouye, 1989). Liquid metallic mercury, Hg^o, is poorly absorbed from the intestinal tract of mammals and teratogenic potential of Hg^o via this route is low; however, inhaled Hg^o vapor is efficiently absorbed (Inouye, 1989). Inorganic mercury, Hg^{2+}, is also poorly absorbed (about 2.0%) from the gastrointestinal (GI) tract, and uptake of Hg^{2+} by the fetus is low. Injected inorganic mercury salts, however, are teratogenic in laboratory animals. Inorganic mercury transfers inefficiently to the fetus but accumulates in the placenta. In mice, a significant portion of the Hg^{2+} is locked in the yolk sac; it was concluded that observed fetal defects resulted from inhibition of metabolite transport and from maternal kidney dysfunction, and not from direct action of Hg^{2+} on the fetus (Inouye, 1989).

Methylmercury enhanced the teratogenic action of mitomycin in mice, *Mus* sp., and may act additively to induce cleft palate at doses not known to produce this effect (Inouye and Kajiwara, 1988a). This conclusion was based on a laboratory study in which pregnant mice were given oral nonteratogenic doses of 0.0, 2.5, 5.0, or 10.0 mg/kg BW of methylmercuric chloride on day 9 of pregnancy and then injected intraperitoneally with a teratogenic dose (4.0 mg/kg BW) of mitomycin-C on day 10. Malformations produced by mitomycin-C alone included cervical rib and vertebral anomalies, polydactyly of the hindlimb, and tail anomalies. Combined treatment significantly increased the incidence of these malformations in a dose-dependent pattern. Cleft palate was also evident in combined treatment, but not by either chemical alone (Inouye and Kajiwara, 1988a).

Studies with mice indicated teratogenic effects in subsequent generations: that is, surviving males from methylmercury-intoxicated dams mated with untreated females produced progeny with 8.2% malformations, including exencephaly and brain hernia (Inouye et al., 1985). Subsequent experiments were conducted wherein all progeny survived parturition (Inouye and Kajiwara, 1988c). Results showed the following: F_1 females retained mercury for 2 to 3 weeks after birth following maternal exposure to methylmercury during pregnancy; greater than 85.0% of F_1 males survived to adulthood and all were fertile; and F_1 males mated with untreated females produced some offspring with growth retardation ($<$ 4.0%) but no significant difference in malformation frequency from that of controls. It was concluded that male-mediated fetal effects subsequent to intrauterine exposure to methylmercury may be practically nil in humans (Inouye and Kajiwara, 1988c).

In terms of ability to affect normal development of embryos of the horseshoe crab (*Limulus polyphemus*), mercury — as the acetate or chloride — was the most toxic metal tested, followed in decreasing order by tributyltin chloride, hexavalent chromium, cadmium, copper, lead, and zinc (Itow et al., 1998). Mercury was associated with a high frequency of segment-defective embryos and could be replicated with SH inhibitors, and by compounds inhibiting SH-SS exchange (Itow et al., 1998).

9.2 BACTERIA AND OTHER MICROORGANISMS

Mercury-tolerant strains of bacteria (Colwell et al., 1976) and protozoa (Berk et al., 1978) are reported. Mercury-resistant strains of bacteria are common. In Chesapeake Bay, for example, at least 364 strains of bacteria were isolated that were resistant to $HgCl_2$. Most were pseudomonads and almost all were from seven genera (Colwell and Nelson, 1975). Other groups of bacteria known to materially influence mercury fluxes in saline waters include strains of *Escherichia coli* that convert Hg^{2+} to Hg^o (Summers and Silver, 1972) and strains of anaerobic methanogenic bacteria that enhance the transfer of methylcobalamin to Hg^{2+} under mild reducing conditions (Wood et al., 1968). Some mercury-resistant strains were reported to degrade petroleum as well (Walker and Colwell, 1974). Mercury-resistant strains of bacteria have been recommended as bioindicators of environmental mercury contamination and as markers of methylmercury in biological samples (Baldi et al., 1991). The mercury-resistant strains of bacteria cultured or discovered may have application in mobilization or fixation of mercury from contaminated aquatic environments to the extent that polluted areas may become innocuous (Vosjan and Van der Hoek, 1972; Nelson et al., 1973; Colwell and Nelson, 1975; Colwell et al., 1976).

Microorganisms may develop resistance to mercury compounds after prolonged exposure, and this phenomenon is well-documented in procaryotes such as bacteria and blue-green algae (Ghosh et al., 2004). Bacteria resistant to mercury compounds possess two inducible enzymes — mercuric reductase and organomercurial lyase — both of which reduce Hg^{2+} to Hg^o. In eucaryotic species of fungi, mercury resistance is reported by growing the organism in media supplemented with increasing concentrations of mercury compounds, including strains of *Botrytis*, *Penicillum*, *Sclerotinia*,

Stemphylium, *Pyrenophora*, *Neurospora*, and *Cryptococcus*. In addition, mercury^{2+}-resistant yeast strains of *Rhodotorula* and *Saccharomyces* were isolated from rotting guava fruit containing 28.0 to 1067.0 μg Hg^{2+}/kg FW (Ghosh et al., 2004) (note: > 500.0 μg Hg/kg FW is in violation of the food criterion set by the World Health Organization [WHO, 1976]). Mercury was not volatilized by these yeasts, indicating that the pattern of mercury resistance in yeasts is different from that of bacteria; the high resistance of mercury-resistant yeasts toward mercury may be associated with the binding of Hg^{2+} by the yeast cell wall or cytoplasmic membrane (Ghosh et al., 2004). The marine protozoan *Uronemia nigricans*, after feeding on mercury-laden bacteria, acquired mercury tolerance within a single generation (Berk et al., 1978).

Mercury-resistant and metabolizing strains of bacteria show a wide range of metabolic rates dependent on the chemical species of mercury, the bacterial biomass (Colwell and Nelson, 1975; Nelson and Colwell, 1975), and species competition, as was the case for *Escherichia coli* and *Staphylococcus aureus* (Stutzenberger and Bennett, 1965). Mercury-resistant strains of bacteria that have been developed or discovered could form the basis of a mercury-removal technology in mercury-contaminated estuaries (Vosjan and Van der Hoek, 1972; Nelson et al., 1973; Colwell and Nelson, 1975). This topic is discussed in greater detail in Chapter 10. In one case, a mercury-resistant strain of *Pseudomonas* reduced 0.025 mg of $HgCl_2$ to elemental mercury from solutions containing 6.0 mg $HgCl_2$/L (Colwell et al., 1976). A mercury-accumulating strain of *Pseudomonas* took up 12.0 mg $HgCl_2$/kg cells every 15 min from a solution containing 6.0 mg Hg/Cl_2 (Colwell et al., 1976).

9.3 TERRESTRIAL PLANTS

Seedlings of rice (*Oryza sativa*) grown on mercury-contaminated waste soil from a chloralkali factory for 75 days showed increasing mercury concentrations over time with increasing soil mercury content. At 2.5% waste soil in gardens, rice seedlings contained 8.0 mg Hg/kg FW; at 10.0%, 15.2 mg Hg/kg; and at 17.5% waste soil in gardens, seedlings had 19.1 mg Hg/kg FW (Nanda and Mishra, 1997). Content of nucleic acids and proteins in rice shoots decreased with increasing mercury concentrations in waste soils and with time, but free amino acid content increased. An increase in the RNA:DNA ratio occurred, indicating an enhanced synthesis of RNA per molecule of DNA (Nanda and Mishra, 1997). Seedlings of spruce (*Picea abies*) exposed to solutions containing up to 0.2 mg Hg/L as inorganic mercury or methylmercury showed a dose-dependent inhibition in root growth, especially for methylmercury; dose-dependent decreases in concentrations of potassium, manganese, and magnesium were evident in roots and root tips, and increases in iron in root tips (Godbold, 1991).

Some species of mushrooms have been recommended as sentinel species of mercury contamination because of their ability to accumulate very high concentrations of mercury from the ambient air (Minagawa et al., 1980). In one study, shiitake mushrooms, *Lentinus edodus*, exposed to Hg^o vapor at 172.0 μg Hg/m^3 for up to 7 days had grossly elevated concentrations of mercury in caps and stalks. After 3 days, caps had 125.0 mg Hg/kg DW and stalks had 10.0 mg Hg/kg DW. After 7 days, these values were 310.0 to 400.0 mg/kg DW in caps and 20.0 to 30.0 mg/kg DW in stalks (Minagawa et al., 1980).

9.4 TERRESTRIAL INVERTEBRATES

Methylmercury compounds have induced abnormal sex chromosomes in the fruit fly (*Drosophila melanogaster*) (USNAS, 1978; Khera, 1979). Earthworms (*Eisenia foetida*) exposed to soil containing methylmercury concentrations of 5.0 mg Hg/kg — typical of soil Hg levels near chloralkali plants — showed a significant reduction in the number of segments regenerated after 12 weeks, and contained 85.0 mg Hg/kg on a whole-body fresh weight basis. Regeneration was normal at soil Hg levels of 1.0 mg/kg, although body burdens up to 27.0 mg/kg were recorded. It was concluded that soil contaminated with methylmercury posed a greater hazard to the predators of earthworms than to the earthworms (Beyer et al., 1985). Studies with a different species of earthworm (*Octochaetus pattoni*) and mercuric chloride demonstrated a progressive initial increase in reproduction as soil mercury levels increased from 0.0 to the 60-day lethal level of 5.0 mg/kg (Abbasi and Soni, 1983).

9.5 AQUATIC PLANTS

It is emphasized that phytotoxic effects are always more pronounced when mercury is present in the organic form than in the inorganic form (Boney et al., 1959; Boney and Corner, 1959; Ukeles, 1962; Boney, 1971; Delcourt and Mestre, 1978; Eisler, 1981, 2000). Growth inhibition was recorded in freshwater algae after exposure of 24 h to 10 days to 0.3 to 0.6 μg organomercury/L (USEPA, 1980), and in the marine alga *Scripsiella faeroense* exposed to 1.0 μg Hg^{2+}/L for 24 h (Kayser, 1976). At water concentrations of 1.0 μg Hg^{2+}/L for 24 h, there was an increased incidence of frustule abnormalities and burst thecae in two species of marine algae (Kayser, 1976; Saboski, 1977).

Aquatic plants show great variation in sensitivity to mercury insult. Using marine algae as an example, phytotoxic effects — including reduced growth, developmental abnormalities, photosynthesis inhibition, and death — ranged from 1000.0 μg Hg/L and higher for the comparatively resistant *Plumaria elegans* sporelings (Boney et al., 1959) and *Isochrysis galbana* (Davies, 1974) to 1.0 μg Hg/L for the comparatively sensitive *Scripsiella faeroense* (Kayser, 1976) and *Nitzschia* sp. (Saboski, 1977). Intermediate in sensitivity were *Macrocystis pyrifera*, with effects evident at 50.0 μg Hg/L (Clendenning and North, 1959), 18 species of unicellular algae exposed for 18 days to 5.0 to 25.0 μg Hg/L (Berland et al., 1976), and various species of *Laminaria*, *Pelvetia*, *Ascophyllum*, *Dunaliella*, *Chlamydomonas*, *Skeletonema*, and *Fucus* at 50.0 to 100.0 μg Hg/L (Hopkins and Kain, 1971; Davies, 1976; Berland et al., 1977; Delcourt and Mestre, 1978; Harrison et al., 1978; Stromgren, 1980).

Rapid accumulation of mercury, especially organomercury compounds, by various species of algae is documented (Fang, 1973; Laumond et al., 1973; Davies, 1974; Eisler, 1981, 2000). However, uptake of mercury and its compounds is modified by a number of biological, chemical, and physical factors, such as initial mercury concentration (Sick and Windom, 1975; Windom et al., 1976); water pH (Gambrell et al., 1977); exposure time (Sick and Windom, 1975); age of organism (Glooschenko, 1969); salinity of medium (Gambrell et al., 1977); biological surface area (Sick and Windom, 1975); variability in mercury detoxification mechanisms (Davies, 1976); cell density (Delcourte and Mestre, 1978); and accumulation after death (Glooschenko, 1969). Among various species of marine algae, for example, *Croomonas salina* took up 1400.0 mg Hg/kg dry weight after 48 h in solutions containing 164.0 μg Hg/L (Parrish and Carr, 1976); *Chaetoceros galvestonensis* and *Phaeodactylum tricornutum* contained 7400.0 and 2400.0 mg Hg/kg, respectively, when cultured in media containing 100.0 μg Hg/L (*Chaetoceros*) or 50.0 μg Hg/L (*Phaeodactylum*) (Hannan et al., 1973); and *Isochrysis galbana* contained up to 1000.0 mg Hg/kg after exposure to 15.0 μg Hg/L (Davies, 1974). Salt marsh plants show enhanced mercury concentrations in roots at lower salinities and higher pH (Gambrell et al., 1977).

Using artificial ecosystems, reduced initial growth of natural phytoplankton from Saanach Inlet, British Columbia, Canada, was noted after addition of 1.0 to 5.0 μg Hg/L. However, recovery to above control levels appeared to occur in 21 days (Thomas et al., 1980).

9.6 AQUATIC ANIMALS

Sublethal concentrations of mercury are known to adversely affect aquatic fauna through inhibition of reproduction (Dave and Xiu, 1991; Kanamadi and Saidapur, 1991; Kirubagaran and Joy, 1992; Khan and Weis, 1993; Punzo, 1993; Birge et al., 2000); reduction in growth rate (Kanamadi and Saidapur, 1991; Punzo, 1993; Birge et al., 2000); increased frequency of tissue histopathology (Kirubagaran and Joy, 1988, 1989; Handy and Penrice, 1993; Voccia et al., 1994); impairment of ability to capture prey (Weis and Weis, 1995a, 1995b); impairment of olfactory receptor function (Baatrup et al., 1990; Baatrup and Doving, 1990); alterations in blood chemistry (Allen, 1994; Shah and Altindag, 2004) and enzyme activities (Nicholls et al., 1989; Kramer et al., 1992); and disruption of thyroid function (Kirubagaran and Joy, 1989), chloride secretion (Silva et al., 1992), and other metabolic and biochemical functions (Nicholls et al., 1989; Angelow and Nicholls, 1991).

Aquatic animals take up CH_3Hg^+ from diet, water, and sediment, and retain mercury with continued exposure because elimination is slow relative to uptake rate (Huckabee et al., 1979; Eisler, 2000). Methylmercury concentrations in fish can be six to seven orders of magnitude higher than methylmercury concentrations in ambient surface waters (Kim and Burggraaf, 1999; Bowles et al., 2001). In aquatic invertebrates, methylmercury is more readily taken up than inorganic mercury, with accumulations highest in predators and lowest in herbivores (Tremblay et al., 1996; Lawrence and Mason, 2001; Simon and Boudou, 2001). Accumulation ratios for mercury were greater than 10,000 between seston and water (Watras et al., 1998; Bowles et al., 2001); less than 10 for predatory fish muscle and their whole prey (Cope et al., 1990; Kim and Burggraaf, 1999); and 125 to 225 for seabird feathers and their diet (Monteiro et al., 1998). In general, the accumulation of mercury by aquatic biota is rapid and depuration is slow (Newman and Doubet, 1989; Angelow and Nicholls, 1991; Wright et al., 1991; Handy and Penrice, 1993; Pelletier and Audet, 1995; Geffen et al., 1998). It is emphasized that organomercury compounds, especially methylmercury, were significantly more effective than inorganic mercury compounds in producing adverse effects and accumulations (Baatrup et al., 1990; Wright et al., 1991; Kirubagaran and Joy, 1992; Odin et al., 1995). For example, mercury accumulation studies with American oysters, *Crassostrea virginica*, show that adults exposed for 74 days to seawater solutions containing 1.0 µg inorganic Hg/L had 2.0 mg Hg/kg FW soft tissues after 20 days, and 10.0 mg/kg after 60 days (Kopfler, 1974). When the experiment was repeated using methylmercury or phenylmercury compounds, oysters contained 10.0 mg total Hg/kg FW soft tissues after 20 days and 30.0 mg/kg after 60 days (Kopfler, 1974).

9.6.1 Invertebrates

Reproduction was inhibited among the most sensitive species of aquatic animals at water concentrations of 0.03 to 1.6 µg Hg/L. For less sensitive marine invertebrates such as hydroids, protozoans, and mysid shrimp, reproduction was inhibited at concentrations between 1.1 and 2.5 µg Hg^{2+}/L; this range was 5.0 to 71.0 µg/L for more resistant species of marine invertebrates (Gentile et al., 1983).

9.6.1.1 Planarians

In the planarian (*Dugesia dorotocephala*), asexual fission was suppressed at 0.03 to 0.1 µg organomercury/L; at 80.0 to 100.0 µg methylmercury/L, behavior was modified and regeneration retarded (Best et al., 1981).

9.6.1.2 Coelenterates

Studies with coral colonies of *Porites asteroides* showed that zooxanthellae and skeleton accumulated total mercury, as $HgCl_2$, in a dose-dependent manner (Bastidas and Garcia, 2004). However, polyp tissue accumulated more mercury at 37.0 μg/L than at 180.0 μg/L, suggesting either saturation of mercury in polyps, activation of mechanisms of detoxification, or both. Most of the mercury in colonies exposed to the highest mercury concentration of 180.0 μg/L for 3 days was in zooxanthellae (89.0%), with the rest in polyps (7.0%) and skeleton (4.0%). Increasing concentrations of mercury were associated with decreasing biomass of polyps and zooxanthellae, and decreasing pigment concentration. Authors concluded that the capacity of zooxanthellae and the skeleton to concentrate mercury and the decline of zooxanthellae density support the hypothesis that polyps may divert mercury to these two compartments as a detoxifying mechanism (Bastidas and Garcia, 2004). Studies with coelenterate hydrozoans, *Campanularia flexuosa* (Moore and Stebbing, 1976; Stebbing, 1976; Moore, 1977), and sea anemones, *Bundosoma cavernata* (Kasschau et al., 1980), show that concentrations as low as 0.2 μg Hg/L had an adverse effect on lysosomal enzymes of *Campanularia*; 1.7 μg Hg/L depressed growth of *Campanularia*; and exposure of *Bundosoma* to 1.2 μg Hg/L for 7 days increased concentrations of free amino acids.

9.6.1.3 Molluscs

In the slipper limpet (*Crepidula fornicata*) — the most sensitive mollusc tested — exposure to 0.25 μg Hg^{2+}/L caused a delay in spawning, a reduction in fecundity, and after 16 weeks a reduction in the growth rate of adults (Thain, 1984). In marine molluscs exposed to water concentrations of 6.0 to 10.0 μg Hg^{2+}/L for 96 h, the feeding of adults ceased and the swimming rate of larval stages declined (Thain, 1984). Juveniles of the freshwater rainbow mussel (*Villosa iris*) showed a reduction in growth rate during exposure for 21 days to sublethal concentrations of 8.0 to 114.0 μg Hg^{2+}/L (Valenti et al., 2005).

Rapid accumulation of mercury compounds, especially organomercury compounds, by various species of filter-feeding marine bivalve molluscs is well documented (Irukayama et al., 1962a, 1962b; Fang, 1973; Laumond et al., 1973; Fowler et al., 1975; Cunningham and Tripp, 1975a, 1975b; Nelson et al., 1976; Eisler, 1987, 2000). Accumulation of organomercury complexes is especially rapid in molluscan tissues with high lipid content (Wood, 1973; Eisler, 1981). Accumulation is modified by the chemical form of mercury administered (Irukayama et al., 1962a, 1962b; Cunningham and Tripp, 1975a; Pan and Wang, 2004); water temperature (Cunningham and Tripp, 1975b; Lee et al., 1975; Nelson et al., 1977; Fowler et al., 1978); salinity of the medium (Olson and Harrel, 1973; Dillon, 1977; Dillon and Neff, 1978); presence of selenium (Fowler and Benayoun, 1976; Glickstein, 1978); sexual condition (De Wolf, 1975; Cunningham and Tripp, 1975b); tissue specificity (Cunningham and Tripp, 1975a; Fowler et al., 1975, 1978; Wrench, 1978); previous acclimatization to mercury salts (Dillon, 1977); season of the year (De Wolf, 1975); soluble protein content of organism (Wrench, 1978); and presence of mercury-resistant strains of bacteria (Colwell and Nelson, 1975; Sayler et al., 1976; Colwell et al., 1976). Marine molluscs — unlike certain species of oceanic mammals, elasmobranchs, and teleosts — accumulate mercury only when the environment is contaminated with this element as a direct result of human activities (Eisler, 1981, 2000).

Adults of the American oyster, *Crassostrea virginica*, accumulate low levels of mercury in the water column to comparatively elevated concentrations in tissues and retain the mercury for extended periods. In one study, oysters exposed to 10.0 μg Hg/L as mercuric acetate for 45 days contained 28.0 mg Hg/kg FW vs. less than 0.2 mg Hg/kg FW in controls (Cunningham and Tripp, 1973). By day 60 (15 days postexposure), soft parts of oysters contained 18.0 mg/kg FW, possibly owing to spawning. After an additional 160 days in mercury-free seawater (day 22), levels declined to 15.0 mg/kg FW during the first 18 days but remained unchanged thereafter. The authors concluded that oysters can concentrate 10.0 μg Hg/L by a factor of 2800 and that total self-purification was not achieved over a 6-month period in mercury-free seawater (Cunningham and Tripp, 1973). In another study with American oysters, Kopfler (1974) found that continuous exposure to 1.0 μg Hg/L as inorganic or organic mercury resulted in accumulations considered hazardous ($>$ 0.5 mg

Hg/kg FW) to human health. And juveniles of bay scallops, *Argopecten irradians*, held for 96 h in solutions containing 40.0 μg Hg/L contained 48.9 mg Hg/kg FW soft parts, a concentration far in excess of the 0.5 mg Hg/kg FW limit in foods considered hazardous to human health (Nelson et al., 1976).

Persistence of mercury in molluscan tissues is about intermediate between that of fish and crustaceans. Time to eliminate 50.0% of biologically assimilated mercury and its compounds (Tb 1/2) ranged from 20 to 1200 days for various species of teleosts (Huckabee et al., 1979; USNAS, 1978), 297 days for the crayfish *Astacus fluviatilis*, 435 days for mussels, and 481 days for the clam *Tapes decussatus* (USNAS, 1978). In 24-h radiotracer studies with ^{203}Hg, clams (*Tapes decussata*) contained 10 times more radiomercury than did the medium (Unlu et al., 1972). Radiomercury in clams had a half-time persistence of 10 days if accumulated via the diet, but only 5 days if taken up from the water (Unlu et al., 1972); however, retention times of radiomercury-203 in most species of marine molluscs were comparatively lengthy, being up to 1000 days for mussels (*Mytilus galloprovincialis*) (Miettinen et al., 1969, 1972). Miettinen and co-workers found that phylogenetically related species follow a similar pattern of methylmercury excretion, with half-time persistence dependent on water temperature and mode of entry into the organism. Accumulation of mercury by oysters more than doubled in the presence of mercury-resistant strains of *Pseudomonas* spp. (Colwell and Nelson, 1975; Sayler et al., 1976; Colwell et al., 1976). Accumulation of inorganic mercury and methylmercury compounds by green mussels, *Perna viridis*, was modified by chemical species, water salinity, dissolved organic carbon (DOC), and colloidal organic carbon (COC) (Pan and Wang, 2004). Methylmercury was taken up 8 times faster than inorganic mercury at 30‰. At 15‰ salinity, uptake increased 75.0% for inorganic mercury and 117.0% for methylmercury. The presence of biogenic DOC derived from diatom decomposition decreased uptake of inorganic mercury, but humic acid increased uptake. Methylmercury uptake, however, was only weakly influenced by differences in DOC quality or quantity. COC increased uptake of inorganic mercury up to sevenfold when compared to low-molecular-weight complexed mercury, but decreased the uptake of methylmercury by 42.0 to 73.0%. Authors concluded that facilitated transport and direct colloidal ingestion are involved in the uptake of mercury species under different conditions of DOC and COC, and each needs to be considered separately when measuring mercury uptake (Pan and Wang, 2004).

9.6.1.4 Crustaceans

In mysid shrimp (*Mysidopsis bahia*), a sensitive species, the abortion rate increased and population size decreased after lifetime (i.e., 28 days) exposure to 1.6 μg/L of mercury as mercuric chloride (Gentile et al., 1983).

Among marine crustaceans, variables known to modify mercury accumulation and other effects include tissue specificity, diet, feeding niche, sex, life stage, general health, temperature and salinity of the medium, duration of exposure, chemical form of mercury, and the presence of other compounds. Selected studies illustrating these points follow.

American lobsters, *Homarus americanus*, immersed in seawater containing up to 6.0 μg Hg/L for 30 days had significant accumulations of mercury in various tissues: 15.2 mg Hg/kg FW in digestive gland, up from 0.12 mg Hg/kg FW at start; 85.3 in gills, up from 0.14; and 1.0 mg Hg/kg FW in muscle, up from 0.23 (Thurberg et al., 1977). With increasing exposure of 60 days, all tissues showed mercury increases; for example, gill contained 119.5 mg Hg/kg FW (Thurberg et al., 1977). Mercury accumulations from solution were, in general, more rapid and more pronounced than were accumulations from food sources (Luoma, 1976, 1977). In an alga (1400.0 mg Hg/kg DW) to copepod food link, for example, copepods showed no impairment of egg laying or egg development, and no retention of mercury in tissues, eggs, or feces (Parrish and Carr, 1976).

Mercury effects on fiddler crabs (*Uca pugilator*) immersed in solutions containing high (180.0 μg Hg/L) sublethal concentrations were modified by life stage, sex, and thermosaline regimen. Adults were more resistant than larvae, adult females more resistant than adult males, and resistance for all stages was lowest at low temperatures and salinities (Vernberg and DeCoursey, 1977). Exposure of fiddler crabs for 3 h to 180.0 μg Hg/L, as $HgCl_2$, demonstrated that gill was the major site of mercury accumulation; lesser amounts accumulated in hepatopancreas and green gland. Survival was lower at 5°C than at 33°C. The inability to transfer mercury from gill to hepatopancreas at low temperatures could be a factor in the toxicity of mercury to fiddler crabs at low temperatures (Vernberg and Vernberg, 1972; Vernberg and O'Hara, 1972). However, negligible uptake or effects were observed in whole *Uca pugilator* after exposure for 2 weeks in seawater solutions containing 100.0 μg Hg/L (Weis, 1978). Similarly, preexposure of larvae of white shrimp *Penaeus setiferous* for 57 days to 1.0 μg Hg/L had no effect on measured parameters during subsequent mercury-stress experiments with these larvae (Green et al., 1976). Although increased toxicity of mercury salts was demonstrable to crustaceans at low salinities within the range of 7 to 35‰ (Roesijadi et al., 1974; Vernberg et al., 1974), no increase in mercury uptake was noted in shrimp *Palaemon debilis* at salinities as low as 6‰ (Luoma, 1977).

Mercury accumulation in viscera of euphausiids *Meganyctiphanes norvegica* was decreased by moderate concentrations of selenium (10.0 to 1,000.0 μg Se/L) in the medium (Fowler and Benayoun, 1976). In studies with larvae of the dungeness crab *Cancer magister*, selenium concentrations of 10.0 to 1000.0 μg/L tended to decrease mercury toxicity; however, high levels of selenium (> 5000.0 μg/L) increased mercury toxicity (Glickstein, 1978).

Larvae of the barnacle *Elminius modestus* took up 920.0 mg Hg/kg DW during exposure for 3 h in 200.0 μg Hg/L as mercuric chloride (Corner and Rigler, 1958). Accumulation of amyl mercuric chloride, $n\text{-}C_5H_{11}HgCl$, was 700.0 mg Hg/kg DW during exposure for 3 h in only 1.0 μg Hg/L solutions. Similar trends were observed for brine shrimp, *Artemia salina* (Corner and Rigler, 1958). When adults of the spiny spider crab *Maia squinado* were immersed in seawater containing added mercuric chloride, mercury concentrated in gills and other sites (Corner, 1959). Eventually, the concentration of mercury in blood rose above that in the external medium, the concentration in antennary gland above that in blood, and the crabs excreted small, but increasing, amounts of mercury in the urine. Most (95.0%) of the mercury in blood was attached to protein, with mercury concentrations in blood remaining constant for several weeks following mercury exposure. When *Maia* were poisoned with methylmercury chloride, mercury again concentrated in gills and various internal organs; however, the amount in blood was comparatively low, and none was found in urine (Corner, 1959). The marine copepod *Acartia clausi*, subjected to concentrations of 0.05 μg/L of mercury and higher, reached equilibrium with the medium in only 24 h. In that study (Hirota et al., 1983), bioconcentration factor (BCF) values for whole *Acartia* after 24-h exposures were 14,360 for inorganic mercuric ion at 0.05 μg/L and, for methylmercury, 179,200 at 0.05 μg/L, and 181,000 at 0.1 μg/L.

In the shrimp *Lysmata seticaudata*, about 45.0% of all mercury was localized in viscera, 39.0% in muscle, 15.0% in exoskeleton, and 1.8% in molts (Fowler et al., 1978). It is probable that most of the mercury was in methylated form, as was the case in tissues of blue crabs (Gardner et al., 1975), fiddler crabs (Windom et al., 1976), and other benthic crustaceans (Eganhouse and Young, 1978). It is generally agreed that organomercury compounds are accumulated more rapidly by crustaceans and to a higher degree level than inorganic compounds (D'Agostino and Finney, 1974; Fowler et al., 1978). Some evidence exists showing that iodide salts of mercury have greater biomagnification potential than chloride salts in copepods and brine shrimp (Boney et al., 1959), but this requires verification.

Mercury-tolerant strains of crustaceans are documented (Green et al., 1976; Weis, 1976; Tsui and Wang, 2005). The white shrimp (*Penaeus setiferus*), preexposed for 57 days to 1.0 μg Hg/L, did not differ from controls during either exposure or subsequent mercury stress experiments (Green et al., 1976); this observation suggested that nonsensitization or adaptation mechanisms are involved. The fiddler crab (*Uca pugilator*) seemed unusually resistant and showed negligible uptake or effects during exposure to 100.0 μg Hg/L for 2 weeks (Weis, 1976). Daphnids (*Daphnia magna*) developed tolerance to Hg^{2+} after preexposure to 0.5 and 5.1 μg Hg^{2+}/L for 4 days, followed by 4 days in mercury-free media (Tsui and Wang, 2005). These daphnids had elevated body levels of mercury and metallothioneins, and higher tolerance to mercury toxicity than controls. F_1 offspring from parents exposed to 5.1 μg/L also had mercury tolerance and about 25.0% of mercury concentrations of parents, but not elevated metallothioneins. Mercury tolerance was not evident in the F_2 generation, suggesting that tolerance would disappear quickly on transfer to mercury-free environments (Tsui and Wang, 2005), unlike *Tubifex* worms where tolerance lasted for several generations (Vidal and Horne, 2003).

9.6.1.5 Annelids

Tolerance of freshwater oligochaetes (*Tubifex tubifex*) to mercury was reportedly developed in a single generation, and inherited through several generations without sustained exposure (Vidal and Horne, 2003).

9.6.1.6 Echinoderms

Sea urchin development was inhibited at 10.0 to 23.0 μg Hg^{2+}/L, with a high proportion of abnormal plutei at 30.0 μg/L (Okubo and Okubo, 1962; Kobayashi, 1971). Inhibition was more pronounced when similar concentrations of organomercury salts were tested (Kobayashi, 1971). Arrested development of sea urchin larvae was documented at medium concentrations as low as 3.0 μg Hg^{2+}/L and exposure for 40 h (USEPA, 1980). Mercury can also inhibit motility of spermatozoa of the sea urchin, *Arbacia punctata*, although effects were reversed by EDTA (Young and Nelson, 1974).

9.6.2 Vertebrates

Growth reduction, impaired reproduction, and enzyme derangement in teleosts, and metamorphosis inhibition in amphibians were observed at medium concentrations less than 1.0 μg Hg/L.

9.6.2.1 Fishes

Reduced growth of sensitive species of teleosts is recorded at water concentrations of 0.04 to 1.0 μg Hg/L. The rainbow trout (*Oncorhynchus mykiss*) was the most sensitive species tested; growth reduction was observed after 64 days in 0.04 μg Hg/L as methylmercury, or 0.11 μg Hg/L as phenylmercury (USEPA, 1985). Growth reduction was documented in brook trout (*Salvelinus fontinalis*) alevins after exposure for 21 days to 0.79 μg organomercury/L (USEPA, 1985).

Methylmercury can impair reproduction in freshwater teleosts through inhibition of gonadal development, spawning success, and survival of embryos and larvae (McKim et al., 1976; Friedmann et al., 1996; Hammerschmidt et al., 1999, 2002; Latif et al., 2001).

In the zebrafish (*Brachydanio rerio*), hatching success was reduced at 0.1 μg Hg^{2+}/L and egg deposition was reduced at 0.8 μg/L (Armstrong, 1979). Fathead minnows (*Pimephales promelas*) exposed to 0.12 μg methylmercury/L for 3 months failed to reproduce (Birge et al., 1979). Impaired egg production and spawning of goldfish (*Carassius auratus*) was associated with a total mercury concentration of 0.76 mg/kg FW gonad; for rainbow trout, 0.49 mg total Hg/kg FW gonad was associated with impaired gamete function and a reduction in survival of early life stages (Birge et al., 2000). Impairment of testicular lipid metabolism in catfish (*Clarias batrachus*) at sublethal concentrations of inorganic and organic mercury compounds may account for the mercury-induced inhibition of steroidogenesis and spermatogenesis (Kirubagaran and Joy, 1992). Inorganic and organic mercury compounds produce different morphological effects on the micropyle of fish eggs. Reduced insemination success of killifish (*Fundulus heteroclitus*) eggs exposed to methylmercury was due to rupture of cortical vesicles and blockage of the micropyle. However, reduced insemination due to inorganic mercury was due to swelling of the micropylar lip and a decrease in micropyle diameter (Khan and Weis, 1993).

Adverse effects of mercury to fishes, in addition to those listed on reproduction and growth, have been documented at water concentrations of 0.88 to 5.0 μg/L: enzyme disruption in brook trout (*Salvelinus fontinalis*) embryos immersed for 17 days in solutions containing 0.88 μg/L, as methylmercury (USEPA, 1980); decreased rate of intestinal transport of glucose, fructose, glycine, and tryptophan in the murrel (*Channa punctatus*) at 3.0 μg Hg^{2+}/L for 30 days (Sastry et al., 1982); altered blood chemistry in striped bass (*Morone saxtilis*) at 5.0 μg Hg^{2+}/L in 60 days (Dawson, 1982); and decreased respiration in striped bass 30 days postexposure after immersion for 30 to 120 days in 5.0 μg Hg^{2+}/L (Armstrong, 1979; USEPA, 1980). In largemouth bass, elevated liver metallothioneins are indicative of elevated muscle mercury concentrations, suggesting that mercury-induced metallothioneins may be useful biomarkers of mercury exposure (Schlenk et al., 1995).

At 44.0 μg Hg^{2+}/L for 30 days, the freshwater fish *Notopterus notopterus* showed generalized metabolic derangement (Verma and Tonk, 1983). And at high sublethal concentrations of methylmercury, rainbow trout were listless and darkly pigmented, appetite was reduced, and digestion was poor (Rodgers and Beamish, 1982). Olfactory receptor function in Atlantic salmon, *Salmo salar*, is highly vulnerable to brief exposures to sublethal concentrations of mercuric chloride or methylmercury chloride. In both cases, mercury was deposited in the olfactory system along its whole length from receptor cell apices to the brain (Baatrup and Doving, 1990). Inorganic mercury inhibited the olfactory response after exposure to 10.0 μg Hg/L for 10 min, with effects reversible; however, methylmercury-induced inhibition was not reversible (Baatrup et al., 1990).

The rate of accumulation of methylmercury in fish affects toxicity (Niimi and Kissoon, 1994), with slow accumulation rates associated with higher tolerances (Wiener and Spry, 1996). Laboratory studies indicate that fish accumulate high concentrations of methylmercury directly across the gills when subjected to elevated concentrations of methylmercury in the medium (McKim et al., 1976; Niimi and Kissoon, 1994). After crossing the fish gut, CH_3Hg^+ binds to erythrocytes and is transported to cell organs and tissues via the circulatory system, readily crossing internal membranes (Harrison et al., 1990; Ribeiro et al., 1999). Eventually, most of the methylmercury is translocated to skeletal muscle, where it accumulates bound to sulfhydryl groups in protein. Skeletal muscle of teleosts appears to function as a reservoir for methylmercury (Johnels et al., 1967; Giblin and Massaro, 1973; MacLeod and Pessah, 1973; Middaugh and Rose, 1974; Pentreath, 1976b; Weisbart, 1973). Maximum concentration factors of radiomercury were reached in skeletal muscle, brain, and eye lens after 34, 56, and > 90 days, respectively; maximum values in most other tissues were reached in about 7 days (Giblin and Massaro, 1973). One pathway by which anadromous fishes, such as salmon, accumulate mercury from the medium is through the gills: up to 90.0% of the mercury taken up on the gills is subsequently bound to erythrocytes within 40 min (Olson et al., 1973; Olson and Fromm, 1973). Wiener and Spry (1996) hypothesize that methylmercury storage in muscle serves as a protective mechanism in fishes because sequestration in muscle reduces central nervous system exposure to methylmercury.

At lower trophic levels, the efficiency of mercury transfer was low through natural aquatic food chains; however, in animals of higher trophic levels, such as predatory teleosts and fish-eating birds and mammals, the transfer was markedly amplified (Eisler, 1978, 1981, 1987). High uptake and accumulation of mercury from the medium by representative species of marine and freshwater teleosts and invertebrates are documented (Kopfler, 1974; Pentreath, 1976a, 1976b; Eisler, 1978, 1981; Birge et al., 1979; Huckabee et al., 1979; USEPA, 1980, 1985; Stokes et al., 1981; Rodgers and Beamish, 1982; Hirota et al., 1983; Clarkson et al., 1984; McClurg, 1984; Niimi and Lowe-Jinde, 1984; Ramamoorthy and Blumhagen, 1984; Ribeyre and Boudou, 1984; Thain, 1984; Newman and Doubet, 1989; Angelow and Nicholls, 1991; Wright et al., 1991; Handy and Penrice, 1993). Accumulation patterns were enhanced or significantly modified by numerous biological and abiotic factors (USNAS, 1978; Eisler, 1978, 1981, 1984, 1985; USEPA, 1980, 1985; Stokes et al., 1981; Rodgers and Beamish, 1982; Clarkson et al., 1984; Ramamoorthy and Blumhagen, 1984; Ribeyre and Boudou, 1984; Ponce and Bloom, 1991; Odin et al., 1995; Choi et al., 1998). In general, the accumulation of mercury was markedly enhanced at elevated water temperatures, reduced water salinity or hardness, reduced water pH, increased age of the organism, and reduced organic matter content of the medium; in the presence of zinc, cadmium, or selenium in solution; after increased duration of exposure; and in the presence of increased nominal concentrations of protein-bound mercury. Uptake patterns were significantly modified by sex, sexual condition, prior history of exposure to mercury salts, the presence of complexing and chelating agents in solution, dietary composition, feeding niche, tissue specificity, and metabolism; however, trends were not consistent between species and it is difficult to generalize. In one example, Ribeyre and Boudou (1984) immersed rainbow trout in solutions containing 0.1 μg Hg/L, as methylmercury. After 30 days, bioconcentration factors (BCF) ranged from 28,300 for brain to 238,000 for spleen; values were intermediate for muscle (30,000), whole fish (36,000), blood (102,000), liver (110,000), kidney (137,000), and gill (163,000). These values may have been higher if exposure had extended beyond 30 days; Rodgers and Beamish (1982) showed that whole-body mercury residues in rainbow trout subjected to mercury insult continued to increase for the first 66 days before stabilizing. When mercury was presented as inorganic mercuric ion at 0.1 μg/L for 30 days, BCF values were usually lower than in trout exposed to methylmercury: 2300 for muscle; 6800 for brain; 7000 for whole trout; 14,300 for blood; 25,000 for liver; 53,000 for kidney; 68,600 for gill; and 521,000 for spleen (Ribeyre and Boudou, 1984). The high BCF values recorded for rainbow trout were probably due to efficient uptake from water, coupled with slow depuration (Rodgers and Beamish, 1982).

Mercury accumulation by two species of freshwater teleosts (eastern mosquitofish, *Gambusia holbrooki*; lake chubsucker, *Erimyzon sucetta*) in an artificial wetland ecosystem was significantly enhanced by sulfate addition (Harmon et al., 2005). Authors conclude that sulfate additions result in elevated production of methylmercury in sediment and porewater — possibly due to increased mercury methylation by sulfate-reducing bacteria — which is readily evident in fish and water with subsequent increases through the food web (Harmon et al., 2005).

Total mercury concentrations (in mg/kg FW) in tissues of adult freshwater fishes with signs of methylmercury intoxication ranged from 3.0 to 42.0 in brain, 6.0 to 114.0 in liver, 5.0 to 52.0 in muscle, and 3.0 to 35.0 in whole body (Wiener and Spry, 1996). Whole-body levels up to 100.0 mg Hg/kg were reportedly not lethal to rainbow trout, although 20.0 to 30.0 mg/kg were associated with reduced appetite, loss of equilibrium, and hyperplasia of gill epithelium (Niimi and Lowe-Jinde, 1984). However, brook trout showed toxic signs and death at whole-body residues of only 5.0 to 7.0 mg/kg (Armstrong, 1979).

Elimination of accumulated mercury, both organic and inorganic, from teleosts and other aquatic animals is a complex multicompartmental process, but appears to largely depend on its rate of biological assimilation. This rate, in turn, varies widely (20.0 to 90.0%) between species, for reasons as yet unexplained (USNAS, 1978). For example, mercury associated with dietary components that are not assimilated is eliminated rapidly with feces. The rest is absorbed across the gut and incorporated into tissues. This assimilated fraction of mercury is depurated much more slowly, at a rate positively correlated with the organism's metabolism (USNAS, 1978; Rodgers and Beamish, 1982). Route of administration is also important. Bioavailability estimates of methylmercury from orally administered doses to channel catfish (*Ictalurus punctatus*) tend to overestimate the true bioavailability (McCloskey et al., 1998), and strongly indicate that data based on this route of administration should be reexamined. Time to eliminate 50.0% of biologically assimilated mercury and its compounds (Tb 1/2) is variable. Among various species of freshwater teleosts, Tb 1/2 values (in days) were 20 for guppies *Poecilia reticulatus*, 23 for goldfish *Carassius auratus*, 100 for northern pike, and 1000 each for mosquitofish *Gambusia affinis*, brook trout, and rainbow trout (Huckabee et al., 1979). For eels (*Anguilla vulgaris*) and flounders (*Pleuronectes flesus*), these values were 1030 and 1200 days, respectively (USNAS, 1978).

There are many laboratory studies on uptake, retention, and translocation of mercury by marine teleosts. Rapid accumulation of mercury by marine teleosts is documented, especially for organomercury compounds (Hibaya and Oguri, 1961; Johnels et al., 1967; Hannerz, 1968; Hasselrot, 1968; Rucker and Amend, 1969; Bligh, 1972; Olson et al., 1973; Fang, 1973; MacLeod and Pessah, 1973; Middaugh and Rose, 1974; Kramer and Neidhart, 1975; Pentreath, 1976a, 1976b). Eggs of plaice (*Pleuronectes platessa*), for example, concentrated ambient environmental levels of mercury by a factor of 465 in 12 days; for plaice larvae, this factor was 2000 in 8 days; and adult plaice had mercury concentration factors of 600 in 64 days, with most of the mercury sequestered in muscle (Pentreath, 1976a). Adults of mullet (*Mugil auratus*) held in seawater solutions containing 100.0 μg Hg/L as $HgCl_2$ for 57 days had 2.2 mg Hg/kg wet muscle vs. 0.11 in controls (Establier et al., 1978). When mercury was in the form of methylmercury, exposure for 45 days in only 8.0 μg Hg/L produced 2.6 mg Hg/kg FW muscle in *Mugil* (Establier et al., 1978).

Accumulation of mercury and its compounds from seawater by fish can be modified by the chemical form of mercury administered (Hannerz, 1968; Renfro et al., 1974; Kramer and Neidhart, 1975; Pentreath, 1976a, 1976b; Yamaguchi et al., 2004); the mode of administration (Jarvenpaa et al., 1970; Schmidt-Nielsen et al., 1977); the presence of chelating or complexing agents in the medium (Kramer and Neidhart, 1975); the initial mercury concentration (Hannerz, 1968; Calabrese et al., 1975; Kramer and Neidhart, 1975; Koeller and Wallace, 1977; Weis and Weis, 1978); tissue specificity (Hannerz, 1968; Calabrese et al., 1975; Pentreath, 1976a, 1976b; Schmidt-Nielsen et al., 1977); salinity of the medium (Renfro et al., 1974); and length of exposure (Calabrese et al., 1975; Koeller and Wallace, 1977). In one study, accumulation of methylmercury by freshwater and seawater-adapted Japanese eels, *Anguilla japonica*, held in 10.0 μg Hg/L as methylmercury chloride for 72 h, was significantly greater in marine-adapted eels than freshwater-adapted eels for liver (2.8 mg/kg FW vs. 2.0 mg/kg FW), spleen (3.4 vs. 2.4), kidney (1.6 vs. 1.4), pancreas (1.0 vs. 0.5), bile (0.4 vs. 0.06), and blood (4.1 mg/kg FW vs. 1.9 mg/kg FW) (Yamaguchi et al., 2004). In freshwater-adapted eels, but not marine-adapted eels, methylmercury induced stimulation of GSH in both liver and kidney and this may interfere with mercury uptake in that group (Yamaguchi et al., 2004).

In the food chain algae → detritus → worm → prey fish → predator fish, mercury had a long biological half life: greater than 1000 days in predatory fish, compared with about 55 days in prey fish (Huckabee and Blaylock, 1972). The transfer efficiency of inorganic mercury from prey to predator fish was about 40.0%, but from worm to prey fish, only 12.0%; worms assimilated about 60.0% of the inorganic mercury contained in the algae-detritus compartment. Huckabee and Blaylock (1972) concluded that mercury accumulation through the food chain can account for a significant percentage of the mercury body burden in fish. Methylmercury accumulation was studied in the food chain of diatom, *Skeletonema costatum* → copepod, *Acartia clausi* → fish, *Chrysophrys major* (Fujiki et al., 1978). Diatoms held for 24 h in seawater containing 5.0 μg methylmercury/L contained 3.45 mg Hg/kg. Copepods that fed on these diatoms for 4 days had 3.14 mg Hg/kg, and fish that fed on these copepods for 10 days had 3.1 mg Hg/kg. Additional studies showed that the most effective pathway for methylmercury accumulation is not via the food chain but rather from the medium by way of the gills (Fujiki et al., 1978). In the food chain clam → eel, no mercury magnification was recorded (Tsuruga, 1963). When dogfish meal containing up to 2.3 mg total Hg/kg, of which 1.9 mg/kg was methylmercury, replaced low-mercury fish rations used in salmon culture, the flesh of the salmon contained greater than 0.5 mg Hg/kg FW within 240 days (Spinelli and Mahnken, 1976). However, when dogfish meal comprised less than half the diet, mercury levels in salmon flesh were less than 0.5 mg Hg/kg FW. Similar results were observed in sablefish fed dogfish meal (Kennedy and Smith, 1972).

The retention time of mercury in marine teleosts depends on several factors. Weisbart (1973) found that the gill, heart, and swim-bladder tissues of fish lost mercury at a faster rate than did whole animals, but that some tissues — including brain, liver, muscle, and kidney — showed no significant decrease over time. Retention times of radiomercury-203 in most marine species were comparatively lengthy, ranging up to 267 days for the fish *Serranus scriba* (Miettinen et al., 1969, 1972). Miettinen and co-workers showed that phylogenetically related species follow a similar pattern of methylmercury excretion, with the biological half-life dependent on water temperature and mode of entry into organisms; it is longer after intramuscular injection than after peroral administration. A study by Amend (1970) of juvenile sockeye salmon indicated that fish treated repeatedly with mercurials during their freshwater phase (to control parasites) accumulated and retained elevated levels of mercury for several months; but after 4 years at sea, the returning salmon contained normal levels of mercury in all tissues examined. Fish heavily contaminated by methylmercury, both naturally and through laboratory-administered feeding studies, reportedly eliminated the accumulated mercury to safe levels for human consumption in 3 to 7 weeks (shorter depuration periods were unsuccessful) when maintained in a mercury-free media and fed a mercury-free diet (Kikuchi et al., 1978). In the case of *Serranus scriba*, mercury-contaminated fish eliminated mercury from gill and digestive tract when placed in clean seawater for 10 days, but mercury content was essentially unchanged in liver, kidney, and muscle tissues (Radoux and Bouquegneau, 1979).

At sublethal concentrations, methylmercury can impair the ability of fish to avoid predators, and can interfere with their ability to locate, capture, and consume prey (Weis and Weis, 1995a, 1995b; Fjeld et al., 1998; Samson et al., 2001). Long-term dietary exposure of teleosts to methylmercury is associated with impaired coordination, diminished appetite, inhibited swimming activity, starvation, and sometimes death (Wiener and Spry, 1996). Mercury-tolerant strains of fish (Weis, 1984) are reported.

Reasons to account for mercury adaptation of the estuarine cyprinodontiform teleost *Fundulus heteroclitus* to both methylmercury and inorganic mercury remain unclear (Weis, 1984).

9.6.2.2 Amphibians

Dietary exposure of amphibians in sites receiving mercury mainly via atmospheric deposition — estimated to range from 1.5 to 3.0 mg Hg/kg FW ration — may be sufficient to adversely affect survival, growth, and development (Unrine et al., 2004; Unrine and Jagoe, 2004). Studies with tadpoles of the southern leopard frog, *Rana sphenocephala*, fed mercury-containing diets for 254 days showed that about 28.0% died at the highest concentrations fed of 0.5 and 1.0 mg Hg^{2+}/kg FW ration; malformation rates were dose related, with 5.0% in controls, 5.6% in the 0.1 mg Hg/kg FW diet, 11.1% in the 0.5 mg/kg diet, and 27.8% in the 1.0 mg/kg diet. Malformations included dose-related scoliosis. Arrested growth and development, and tail resorption were also positively dose related. Total mercury body burdens were also dose–related, with about 50.0 μg Hg/kg DW whole body in controls, 100.0 Hg/kg DW in the low mercury diet, 250.0 Hg/kg DW in the medium diet, and about 450.0 μg Hg/kg DW in the 1.0 mg Hg^{2+}/kg FW diet. In all cases, methylmercury accounted for less than 20.0 μg/kg dry body weight (Unrine et al., 2004). Dietary accumulation of mercury and methylmercury under these conditions is not governed by simple partitioning processes and bioaccumulation factors may be limited as predictors of dietary uptake of methylmercury and inorganic mercury (Unrine and Jagoe, 2004).The leopard frog (*Rana pipiens*) did not metamorphose during exposure to 1.0 μg methylmercury/L for 4 months (USEPA, 1985). In the South African clawed frog, *Xenopus laevis*, a total mercury concentration of 0.49 mg/kg FW gonad was associated with impaired gamete function and reduction in survival of early life stages (Birge et al., 2000). At 0.65 mg inorganic mercury/L, there was a significant decrease in ovary weight of the river frog *Rana heckscheri* (Punzo, 1993).

The significance of mercury concentrations in amphibian tissues is not known with certainty and requires additional research for satisfactory risk assessment. Linder and Grillitsch (2000) recommend the following areas for study: acclimatization and adaption to mercury; mercury remobilization during periods of metamorphosis, hibernation, estivation, and reproduction; critical organ concentrations; and biomarkers of adverse mercury effects. These studies should also consider the influence of exposure duration and dose, mercury speciation, and mercury interaction with other metals (Linder and Grillitsch, 2000).

9.7 BIRDS

Sublethal effects of mercury on birds, administered by a variety of routes, included adverse effects on growth, development, reproduction, blood and tissue chemistry, metabolism, and behavior; histopathology and bioaccumulation were also noted. The dietary route is the most extensively studied pathway of avian mercury intake, especially in the mallard, *Anas platyrhynchos*.

Accumulation and retention of mercury in mallards from mercury-laden diets were investigated by Stickel et al. (1977), Finley and Stendell (1978), Finley et al. (1979), and Heinz (1979, 1980a). Ducklings of mallards fed diets containing 8.0 mg Hg/kg diet for 2 weeks contained (in mg Hg/kg fresh weight (FW)) 16.5 in liver, 17.6 in kidney, 9.1 in carcass, and 4.4 in whole body (Stickel et al., 1977). After 16 weeks on a mercury-free diet, values in all tissues decreased more or less progressively to 25.0% of the 2-week values. In another study, adults of the black duck (*Anas rubripes*) were fed diets containing 3.0 mg Hg/kg (Finley and Stendell, 1978). Final maximum concentrations (in mg Hg/kg FW) were 23.1 in liver, 65.7 in feathers, 3.8 in brain, 4.5 in muscle, and 16.0 in kidney. Breeding pairs fed this diet, when compared to controls, produced fewer eggs, reduced egg hatch, and lower duckling survival. Brains of dead ducklings from mercury-insulted parents contained 3.2 to 7.0 mg Hg/kg FW; brain lesions were also evident (Finley and Stendell, 1978). Methylmercury in the diet of game-farm and wild strains of mallards at 0.5 mg Hg/kg diet for three generations had little measurable effect — aside from slightly increased tissue levels of mercury — when compared to controls (Heinz, 1979, 1980a). Mercury-fed ducks weighed about the same as control ducks, but laid fewer eggs and produced fewer ducklings. Eggshell thinning in mallards might be associated with increasing concentrations of dietary methylmercury (Heinz, 1980b).

The dietary concentration of 0.5 mg Hg/kg dry weight (equivalent to about 0.1 mg/kg fresh weight) in the form of methylmercury was fed to three generations of mallards (Heinz, 1979). Females laid a greater percentage of their eggs outside nest boxes than did controls, and also laid fewer eggs and produced fewer ducklings. Ducklings from parents fed methylmercury were less responsive than controls to tape-recorded maternal calls, but were hyperresponsive to a fright stimulus in avoidance tests (Heinz, 1979). Mallard hens fed diets containing 3.0 mg Hg/kg ration as methylmercury for two reproductive seasons produced eggs with elevated mercury concentrations (5.5 to 7.2 mg Hg/kg FW); hatched ducklings from this group had brain lesions of the cerebellum and reduced survival (Heinz and Locke, 1976).

Adult female mallards fed diets containing 1.0 or 5.0 mg Hg/kg ration, as methylmercury chloride, produced eggs that contained 1.4 mg Hg/kg FW (1.0 mg/kg ration) or 8.7 mg Hg/kg FW (5.0 mg Hg/kg diet); breast muscle had 1.0 or 5.3 mg Hg/kg FW; the addition of 5.0 mg DDE/kg diet did not affect mercury retention in breast muscle or eggs (Heinz, 1987). Lesions in the spinal cord were the primary effect in adult female mallards fed diets containing 1.5 or 2.8 mg Hg/kg DW ration as methylmercury (Pass et al., 1975).

The tissues and eggs of ducks and other species of birds collected in the wild have sometimes contained levels of mercury equal to, or far exceeding, those associated with reproductive and behavioral deficiencies in domestic mallards (e.g., 9.0 to 11.0 mg/kg in feathers and greater than 2.0 mg/kg in other tissues); therefore, it is possible that reproduction and behavior of wild birds have been modified by methylmercury contamination (Heinz, 1979). Tissue mercury residues of wild-strain mallards and game-farm mallards were not significantly different after the birds were fed diets containing 0.5 mg Hg/kg as methylmercury for extended periods, indicating that game-farm mallards are suitable substitutes for wild mallards in toxicological evaluations (Heinz, 1980a).

Mercury exposure by immersion and oral administration have caused reproductive and behavioral modifications. Brief immersion of mallard eggs in solutions of methylmercury resulted in a significant incidence of skeletal embryonic aberrations at dosages of 1.0 μg Hg/egg, and higher; no increases in embryonic malformations were noted at 0.3 μg Hg/egg (Hoffman and Moore, 1979).

Interaction effects of mercury with other metals must be considered. Pekin duck (*Anas platyrhynchos*) — a color variant of the mallard — age 6 months, fed a diet containing 8.0 mg Hg/kg ration as methylmercury chloride for 12 weeks, had kidney histopathology; damage effects were exacerbated when diets also contained 80.0 mg lead acetate/kg, 80.0 mg cadmium chloride/kg, or both (Rao et al., 1989). Studies with mallard adults show significant interaction effects of mercury and selenium. In one 10-week study, mallard adults were fed diets containing 10.0 mg Hg/kg DW ration as methylmercury chloride, 10.0 mg Se/kg as seleno-DL-methionine, or a mixture containing 10.0 mg Hg/kg plus 10.0 mg Se/kg (Heinz and Hoffman, 1998). One of the 12 adult mallards fed the 10.0 mg Hg/kg diet and 8 others suffered paralysis of the legs; however, none of the 12 males on the mixture diet became sick. Both selenium and mercury diets lowered duckling production through reduced hatching success and survival; the mixture diet was worse than either Hg or Se alone. Controls produced 7.6 ducklings/female, females fed 10.0 mg Se/kg produced 2.8, females fed 10.0 mg Hg/kg produced 1.1, and the mixture diet resulted in 0.2 ducklings/female. Deformity frequency was 6.1% in control ducklings, 16.4% for the mercury group, 36.2% for the selenium group, and 73.4% for the group fed the methylmercury and selenomethionine mixture. The presence of mercury in the diet enhanced selenium storage; however, Se did not enhance Hg storage (Heinz and Hoffman, 1998). In another study with mallard adult males fed diets as outlined in the Heinz and Hoffman (1998) study, the investigators analyzed blood, liver, and brain samples for clinical and biochemical alterations after 10 weeks (Hoffman and Heinz, 1998). The 10.0 mg Hg/kg ration group had decreased hematocrit and hemoglobin, and decreased activities of various enzymes involved in glutathione metabolism. Selenium in combination with methylmercury partially or totally alleviated effects of mercury on various glutathione activities. The ability of seleno-DL-methionine to restore the glutathione status involved in antioxidative defense mechanisms may be important in protecting against the toxic effects of methylmercury (Hoffman and Heinz, 1998).

The dietary route of administration is the most extensively studied pathway of avian mercury intake. Domestic chickens fed diets containing as little as 50.0 μg/kg of mercury, as methylmercury, contained elevated total mercury (2.0 mg/kg fresh weight) residues in liver and kidney after 28 weeks; at 150.0 μg/kg diet, residues ranged from 1.3 to 3.7 mg/kg in heart, muscle, brain, kidney, and liver, in that general order; at 450.0 μg/kg in diets, residues in edible chicken tissues (3.3 to 8.2 mg/kg) were considered hazardous to human consumers, although no overt signs of mercury toxicosis were observed in the chickens (March et al., 1983). High inorganic mercury levels (500.0 mg/L) in drinking water of chickens decreased growth rate as well as food and water consumption, and elevated hemoglobin, hematocrit, and erythrocyte content within 3 days (Grissom and Thaxton, 1985).

Dietary concentrations of 1.1 mg total Hg/kg have been associated with kidney lesions in juvenile starlings (*Sturnus vulgaris*) and with elevated residues in the liver (6.5 mg/kg dry weight) and kidney (36.3 mg/kg), after exposure for 8 weeks (Nicholson and Osborn, 1984). In American black ducks (*Anas rubripes*) fed diets containing 3.0 mg Hg/kg as methylmercury for 28 weeks, reproduction was significantly inhibited; tissue residues were elevated in kidney (16.0 mg/kg fresh weight) and liver (23.0 mg/kg); and brain lesions characteristic of mercury poisoning were present (Finley and Stendell, 1978). Japanese quail (*Coturnix japonica*) fed diets containing 8.0 mg Hg/kg of inorganic mercury for 3 weeks had depressed gonad weights; those fed 3.0 mg/kg inorganic mercury or 1.0 mg/kg methylmercury for 9 weeks showed alterations in brain and plasma enzyme activities (Hill and Soares, 1984). Grossly elevated tissue residues of 400.0 mg/kg in feathers and 17.0 to 130.0 mg/kg in other tissues were measured in gray partridge (*Perdix perdix*) after dietary exposure of 20.0 to 25.0 mg total Hg/kg for 4 weeks (McEwen et al., 1973).

Reduced reproductive ability was noted in gray partridges ingesting 640.0 µg Hg (as organomercury)/kg body weight daily for 30 days (McEwen et al., 1973); similar results were observed in ring-necked pheasants (Spann et al., 1972; Mullins et al., 1977).

Behavioral alterations were noted in rock doves (*Columba livia*) given 3.0 mg inorganic Hg/kg body weight daily for 17 days (Leander et al., 1977) or 1.0 mg/kg body weight of methylmercury for 5 weeks (Evans et al., 1982). Observed behavioral changes in posture and motor coordination of pigeons were permanent after the brain accumulated greater than 12.0 mg Hg/kg fresh weight, and were similar to the "spastic paralysis" observed in wild crows during the Minamata, Japan, outbreak of the 1950s, although both species survived for years with these signs (Evans et al., 1982).

Tissue concentrations greater than 15.0 mg Hg/kg FW brain and greater than 30.0 to 40.0 mg/kg FW liver or kidney are associated with neurological impairment (Scheuhammer, 1988). Mercury residues of 0.79 to 2.0 mg/kg in egg, and 5.0 to 40.0 mg/kg in feathers, are linked to impaired reproduction in various bird species (Spann et al., 1972; USNAS, 1978; Heinz, 1979; Fimreite, 1979; Solonen and Lodenius, 1984). Residues in eggs of 1.3 to 2.0 mg Hg/kg FW were associated with reduced hatching success in white-tailed sea-eagles (*Haliaeetus albicilla*), the common loon (*Gavia immer*), and in several seed-eating species (Fimreite, 1979); this range was 0.90 to 3.1 mg/kg for ring-necked pheasants (Spann et al., 1972) and 0.79 to 0.86 mg/kg for mallards (Heinz, 1979). Residues of 5.0 to 11.0 mg Hg/kg in feathers of various species of birds have been associated with reduced hatch of eggs and with sterility (USNAS, 1978). Sterility was observed in the Finnish sparrow hawk (*Accipiter nisus*) at mercury concentrations of 40.0 mg/kg in feathers (Solonen and Lodenius, 1984). Great skuas (*Catharacta skua*) fed mercury-contaminated prey excrete mercury via feathers (Bearhop et al., 2000). Chicks of the common tern (*Sterna hirundo*) from a colony in Long Island, New York, with abnormal feather loss, had significantly elevated mercury levels in blood and liver (Gochfeld, 1980); however, the linkage of feather loss to mercury contamination requires further examination.

Interaction effects of mercury with other contaminants, such as herbicides and pesticides, could intensify hazards to avian populations (Mullins et al., 1977). For example, a striking parallel exists between levels of mercury and of DDT and its metabolites in tissues of birds of prey, suggesting the existence of common ecotoxicological mechanisms (Delbekke et al., 1984; Wiemeyer et al., 1984); additional research is clearly needed.

9.8 MAMMALS

Mercury has no known physiological function (USEPA, 1985). In humans and other mammals, it causes teratogenic, mutagenic, and carcinogenic effects; the fetus is the most sensitive life stage (USNAS, 1978; Chang, 1979; Khera, 1979; USEPA, 1980, 1985; Elhassani, 1983; Greener and Kochen, 1983; Clarkson et al., 1984; Inouye et al., 1985; Naruse et al., 1993). Methylmercury irreversibly destroys the neurons of the CNS. Frequently, a substantial latent period intervenes between the cessation of exposure to mercury and the onset of signs and symptoms; this interval is usually measured in weeks or months, but sometimes in years (Clarkson et al., 1984). Alterations in open-field operant and swimming behaviors, learning deficits, and depression in spontaneous locomotor activity have all been demonstrated in mice and rats after *in utero* exposure to methylmercury (Inouye et al., 1985; Table 9.1). Methylmercury disrupts amino acid metabolism in neural tissues of male mice; alterations were associated with specific neural cell dysfunction and appearance of muscular incoordination (Hirayama et al., 1985). Mercury can cause measurable brain developmental deficits in children exposed to even relatively low levels prior to birth (Maas et al., 2004). At high sublethal doses in humans, mercury causes cerebral palsy, gross motor and mental impairment, speech disturbances, blindness, deafness, microcephaly, intestinal disturbances, tremors, and tissue pathology (Chang, 1979; USEPA, 1980, 1985; Elhassani, 1983; Clarkson et al.,

Table 9.1 Sublethal Effects of Mercury Compounds Administered to Humans and Selected Species of Mammals

Organism, Dose, and Other Variables	Effect	Ref.[a]
Cotton top marmoset monkey, *Callithrix jacchus*; age 1.5–3.0 years; given oral doses by gavage of 1.0 mg methylmercury chloride/kg BW (0.8 mg Hg/kg BW) twice weekly in honey for 7 to 10 weeks until signs of mercury intoxication evident (i.e., unsteady gait, difficulty in resting)	Histopathological examination of brain showed severe neuronal and proliferation of astroglia in midbrain; less severe neuronal degeneration in cerebral cortex Whole blood at 5 weeks (no evidence of intoxication) had 3.5 mg Hg/kg FW. At 7–10 weeks, when all were symptomatic, maximum total mercury concentrations in tissues, in mg/kg FW, were 9.8 in whole blood, and 0.021–0.024 in cerebrum, cerebellum, brainstem, and eyeball lens	16
Guinea pig, *Cavia* spp.; pregnant animals given single oral dose of 7.5 mg Hg, as methylmercury chloride, per animal (= 9.4–15.0 mg/kg BW) on either days 21, 28, 35, 42, or 49 of gestation Killed on day 63 and fetus removed. Mercury content of maternal and fetal tissues measured; pathological examination of fetal brain	About half of all litters were aborted 4–6 days post-treatment Mercury strongly elevated in maternal kidneys (18.6–73.4 mg Hg/kg FW kidney in dams, 1.1–4.6 in fetuses, < 0.05 in untreated fetal and maternal controls); however, whole blood level of Hg in fetuses was always higher than that of dams; at 9 weeks, fetal blood contained 3.9 mg Hg/kg FW vs. 3.0 in dams Fetal hair treated at 6 or 7 weeks of gestation had 535.0 and 515.0 mg Hg/kg FW, respectively, at week 9; no hair data available from dams Brain weight of fetuses treated at 21, 28, and 35 days was reduced and histopathology evident Striking degenerative changes of the cerebrum evident in fetuses from dams treated at 42 days; prominent cerebral degeneration in fetuses of dams treated on day 49 of gestation	21
Cat, *Felis domesticus*:		
250.0 μg organomercury/kg BW daily on days 10 to 58 of gestation; oral route	Increased incidence of anomalous fetuses	1
Human, *Homo sapiens*:		
Adult; 50.0 μg organomercury/day	Risk of paresthesia, 0.3% (burning-prickling sensation of skin)	2
Adult; 200.0 μg organomercury/day	Risk of paresthesia, 8.0%	2
Adult; 1,000.0 μg organomercury/day	Risk of paresthesia, 50.0%	2
Adults; exposed to air containing 1.0 to 30.0 μg Hg°/m³; inspiration through nose and expiration through mouth:		
1.0–3.0 μg/m³	Respiratory absorption (RA) of 82.5 (74.0–92.0)%	18
4.0–6.0 μg/m³	RA of 88.8 (76.6–100.0)%	18
10.0–11.0 μg/m³	RA of 85.2 (75.5–99.2)%	18
20.0–30.0 μg/m³	RA of 87.7 (79.9–95.9)%	18
Adult; 50.0 μg organomercury/m³ air for 8 h	Mercury content, in μg/L, of 280.0 in blood and 100.0 in urine	9
Adult; 50.0 μg Hg as elemental mercury vapor/m³ air; chronic exposure for at least 1 year	Nonspecific symptoms; mercury content, in μg/L, of 35.0 in blood and 150.0 in urine	14
Adult; 100.0 μg organomercury/m³ air for 8 h	Mercury content, in μg/L, of 500.0 in blood and 350.0 in urine	9
Adult; 100.0–200.0 μg Hg as elemental mercury vapor/m³ air; chronic exposure for at least 1 year	Tremors; mercury content, in μg/L, of 70.0–140.0 in blood and 300.0–600.0 in urine	14
Whole blood mercury concentration:		
< 10.0–100.0 μg/L	Increased tremors	9
10.0–20.0 μg/L	Increased prevalence of abnormal psychomotor scores	9
> 15.0 μg/L	Disturbances in tests on verbal intelligence and memory	9

Table 9.1 (continued) Sublethal Effects of Mercury Compounds Administered to Humans and Selected Species of Mammals

Organism, Dose, and Other Variables	Effect	Ref.[a]
Urine mercury concentration:		
0.0–510.0 µg/L	Short-term memory loss	9
5.0–1,000.0 µg/L	Increased tremor frequency and reaction time	9
20.0–450.0 µg/L	Increased motor and sensory nerve latency	9
> 56.0 µg/L	Disturbances in tests on verbal intelligence and memory	9
Adult; various	Symptoms of poisoning evident at residues of 1200.0 µg Hg/L blood, 2000.0 to 3000.0 µg/kg whole body, or 3400.0 µg/kg hair	6
Infant; various	Severe effects at blood levels of 3000.0 µg Hg/L	8
Cynomolgus monkey, *Macaca fasicularis:*		
Adults; age 7–10 years; exposed *in utero*; maternal doses of 0.0, 50.0, 70.0, or 90.0 µg methylmercury/kg BW daily resulted in blood mercury levels in treated infants of 1.0–2.5 mg/L; offspring were conditioned and tested for ability to respond to a lit button for an apple juice award and other tasks	*In utero* exposure to mercury did not produce significant adverse effects on adult performance, although gender differences may interact with methylmercury on certain behaviors	11
Female adults; daily dose of methylmercury chloride in apple juice for 150 days equivalent to daily doses of 0.0, 0.4, 4.0, or 50.0 µg methylmercury/kg BW	All groups seemed normal at day 150; no differences in blood or serum cholinesterase activity	12
Cynomolgus monkeys, *Macaca* spp.; various	Visual upset at blood mercury levels of 1200.0 to 4000.0 µg/L, or brain levels of 6000.0 to 9000.0 µg Hg/kg; tremors at blood levels of 2000.0 to 10,000.0 µg Hg/L; kidney pathology at brain levels of 1500.0 µg Hg/kg	6
Rhesus monkey, *Macaca mulatta*; 16.0 µg organomercury/kg body weight (BW) daily on days 20 to 30 of pregnancy	No effect on reproduction	1
Monkeys; fed diets equivalent to 30.0 mg methylmercury/kg BW daily for 2 years	Asymptomatic	20
Mice, *Mus* sp.:		
Various strains	Residues of 2000.0 to 5000.0 µg/kg hair or > 10,000.0 µg/kg brain associated with motor incoordination and decreaed swimming ability; no observable effect at < 2000.0 µg/kg hair	6
Various; males given 5.0 mg CH_3HgCl/kg BW for 20 days	Altered levels of amino acids in neural tissues; specifically, increased taurine, aspartate, glycine, and glutamate, and decreased γ-butyric acid	19
	Alterations associated with specific neural cell dysfunction and muscular incoordination	
BALB/cAJcl strain; injected iv with 2.0 mg Hg/kg BW of either CH_3HgCl or $HgCl_2$ on one of days 9–17 of pregnancy and killed 24 h post injection; maternal blood, plasma, fetuses and membranes analyzed for total mercury	Methylmercury readily transfers to fetus; the fetal concentration increased sharply with fetal age at injection and reached higher levels than maternal blood (2.8 mg/kg FW) on days 17 and 18 of gestation	22
	Inorganic mercury was inefficiently transferred to fetus (< 0.2 mg/kg FW), but accumulated in the fetal membrane (4.0 mg/kg FW); Hg^{2+} was blocked in the proximal wall of the yolk sac	
C3H/HeN strain; pregnant; single oral dose of 20.0 mg methylmercuric chloride/kg BW on days 13, 14, 15, 16, or 17 of pregnancy	Behavioral effects of prenatal methylmercury exposure included impaired righting movement, low tail position, abnormal flexion of hind limbs, and reduced locomotor activity	15
Newborns were foster mothered, but weaning rate was low	At 10–12 weeks, neuropathology included reduced brain weight and abnormal brain development	

(continued)

Table 9.1 (continued) Sublethal Effects of Mercury Compounds Administered to Humans and Selected Species of Mammals

Organism, Dose, and Other Variables	Effect	Ref.[a]
C57BL/6N strain; given single oral dose on day 13 of pregnancy of 2.5, 5.0, 10.0, or 20.0 mg CH_3HgCl/kg BW; killed on day 14 or day 18; brain mercury concentrations; fetus vs. dam:		
Day 14:		
2.5 mg/kg	2.1 FW vs. 0.4 FW	17
5.0 mg/kg	4.5 FW vs. 1.3 FW	17
10.0 mg/kg	7.5 FW vs. 2.2 FW	17
20.0 mg/kg	18.0 FW vs. 5.5 FW	17
Day 18:		
2.5 mg/kg	1.8 FW vs. 0.9 FW	17
5.0 mg/kg	3.2 FW vs. 1.3 FW	17
10.0 mg/kg	6.5 FW vs. 3.5 FW	17
20.0 mg/kg	12.0 FW vs. 7.0 FW	17
Mink, *Mustela vison*:		
1.0 μg/kg ration as methylmercury	Reproduction normal; growth rate of kits normal	10
1100.0 mg/kg in diet	Residues of 7100.0 to 9300.0 μg/kg in brain; signs of poisoning	5
Harp seal, *Pagophilus groenlandica*; 250.0 μg organomercury/kg BW daily for 90 days in diet	Increased incidence of anomalous fetuses	3, 4
Brown Norway rat, *Rattus norvegicus*; adult males gavaged twice weekly for 19 weeks with methylmercury chloride equivalent to 0.8, 8.0 or 80.0 μg Hg/kg BW daily	Intratesticular testosterone in the high dose group reduced 44.0% and sperm count 17.0%; negative relation between fertility and testicular Hg content; testicular Hg concentrations, in μg/kg FW, were 1.0 in the controls, 10.0 in the 0.8 μg/kg BW daily group, 107.0 in the 8.0 μg/kg group, and 1670.0 in the high dose group	13
Laboratory white rat, *Rattus* sp.; organomercury compounds:		
50.0 μg/m³ air, 4 h daily for 7 days	Impaired spatial learning; increased locomotor activity	9
500.0 μg/kg BW daily; oral route	Reduced fertility	1
2,000.0 μg/kg in diet (as Pacific blue marlin); gestation through postnatal day 16	Adverse behavioral changes in offspring	6
13,300.0 to 50,000.0 μg/kg BW daily for 5 days; subcutaneous injection	Impaired cutaneous sensitivity and hearing up to 1 year posttreatment	7

[a] Reference: 1, Khera, 1979; 2, Clarkson et al., 1984; 3, Ronald et al., 1977; 4, Ramprashad and Ronald, 1977; 5, Kucera, 1983; 6, Suzuki, 1979; 7, Wu et al., 1985; 8, Elhassani, 1983; 9, USPHS, 1994; 10, Wren et al., 1987b; 11, Gilbert et al., 1994; 12, Petruccioli and Turillazzi, 1991; 13, Friedmann et al., 1998; 14, World Health Organization, 1976; 15, Inouye et al., 1985; 16, Matsumura et al., 1993; 17, Inouye et al., 1986; 18, Oikawa et al., 1982; 19, Hirayama et al., 1985; 20, Takizawa, 1979; 21, Inouye and Kajiwara, 1988b; 22, Inouye and Kajiwara, 1990b.

1984; USPHS, 1994). Pathological and other effects of mercury may vary from organ to organ, depending on factors such as the effective toxic dose in the organ, the compound involved and its metabolism within the organ, the duration of exposure, and the other contaminants to which the animal is concurrently exposed (Chang, 1979). Many compounds — especially salts of selenium — protect humans and other animals against mercury toxicity, although their mode of action is not clear (USNAS, 1978; Chang, 1979; USEPA, 1980, 1985; Eisler, 1987). Adverse effects of organomercury compounds to selected species of mammals have been recorded at administered doses of 0.25 mg Hg/kg body weight daily, dietary levels of 1.1 mg/kg, and blood Hg levels of 1.2 mg/L (Table 9.1).

Retention of mercury by mammalian tissues is longer for organomercury compounds (especially methylmercury) than for inorganic mercury compounds (USNAS, 1978; Clarkson and Marsh, 1982;

Elhassani, 1983; Clarkson et al., 1984). Excretion of all mercury species follows a complex, variable, multicompartmental pattern; the longer-lived chemical mercury species have a biological half-life that ranges from about 1.7 days in human lung to 1.36 years in whole body of various pinnipeds. In humans, increased urinary excretion and blood levels of mercury were observed in human volunteers who used phenylmercuric borate solutions or lozenges intended for treatment of throat infections (USPHS, 1994).

Organomercurials were readily transferred across placental membranes of rats, mice, and hamsters whereas inorganic mercury compounds were blocked in the placenta (Suzuki et al., 1967; Gale and Hanlon, 1976; Holt and Webb, 1986). Because human embryos lack a yolk-sac placenta, results of these studies with inorganic mercurials should be conducted with suitable experimental animals (Inouye and Kajiwara, 1990b). Hook and Hewitt (1986) have reviewed the nephrotoxic effects of mercury in humans and mammals and aver that all forms of mercury administered by a variety of routes can accumulate in the kidney to produce renal damage and sometimes acute renal failure. The basic biochemical mechanism by which mercury produces renal damage is unclear, but may include inhibition of oxidative pathways and general disruption of the plasma membrane. Mercury vapor has a greater predilection for the central nervous system (CNS) than does inorganic mercury salts, but less than organic forms of mercury (Goyer, 1986). Kidneys contain the greatest concentrations of mercury following exposure to inorganic salts of mercury and mercury vapor, whereas organomercurials have greater affinity for the brain, especially the posterior cortex (Goyer, 1986).

Respiratory absorption of elemental mercury in air (1.0 to 30.0 μg Hg^{o}/m^{3}) by humans ranged from 74.0 to 100.0% when inhaled through the nose and exhaled through the mouth; for dogs, this value was 25.0% (Oikawa et al., 1982). However, respiratory absorption in humans was only about 20.0% when inspiration and expiration was through mouth only (Oikawa et al., 1982). Two human males exposed to radiolabeled methylmercury as $CH_3{}^{203}HgCl$ via inhalation had variable half-life retention times (Tb 1/2) of mercury. One subject had a Tb 1/2 value of 103 days for the early period of observation (days 1 to 44 postadministration); and 39 days thereafter; the second had a Tb 1/2 value of 107 to 122 days for the entire postadministration observation period (Uchiyama et al., 1976).

Organomercury uptake and retention in rats is modified by metallothioneins, carbon tetrachloride, and various amino acids (Hirayama, 1980, 1985; Sato et al., 1981; Sato and Takizawa, 1983; Fujiyama et al., 1994). Methylmercury induces metallothionein synthesis in rat liver (Sato et al., 1981). Subcutaneous doses of 19.0 mg CH_3HgCl/kg BW daily for 3 days caused a threefold increase of hepatic zinc and about a fourfold increase in cysteine bound to metallothioneins, but no increase in mercury bound to hepatic metallothioneins (Sato et al., 1981). Carbon tetrachloride, at 0.5 mL/kg BW delivered intraperitoneally every 48 h for 6 days, enhanced initial uptake of methylmercury (single subcutaneous injection of 10.0 mg/kg BW) in rat brain, liver, kidney, and muscle; results suggest that CCl_4-induced liver damage is responsible for the retention time of mercury in rat tissues (Sato and Takizawa, 1983). Methylmercury uptake in brain of Wistar-strain male rats, *Rattus* sp., was influenced by certain amino acids. It was enhanced threefold by L-cysteine; this effect was depressed by L-phenylalanine and L-isoleucine, which are neutral amino acids, but not by L-lysine (basic) or L-glutamic acid (acidic; Hirayama, 1980). Cysteine appears to be an important factor in methylmercury uptake and retention in rat brain but not other tissues, suggesting that the methylmercury–cysteine complex readily penetrates the blood–brain barrier transport system (Hirayama, 1985). Cysteine accelerates the intestinal absorption of methylmercury (Hirayama, 1975), and increases methylmercury uptake in rat brain (Hirayama 1980, 1985; Aschner and Clarkson, 1988; Aschner, 1989), in brain microvessels (Aschner and Clarkson, 1989), and in astroglia (Aschner et al., 1990). Because methylmercury easily reacts with cysteine to form a conjugate similar to methionine, the conjugate is suggested to be taken up by the cells through the L-neutral amino acid carrier system (Fujiyama et al., 1994). Studies with cultured rat astroglia suggest that conjugation with glutathione is a major pathway for methylmercury efflux in rat brain cells and that elevation in cellular glutathione levels has application in cases of methylmercury poisoning in promoting accelerated elimination of methylmercury from critical tissues (Fujiyama et al., 1994).

Increasing glutathione levels were positively correlated with increasing methylmercury resistance of rat PC12 cells and decreasing methylmercury accumulations (Miura et al., 1994). In that study, methylmercury-resistant sublines of rat cells (PC12) were obtained by repeated exposure to stepwise increased concentration of methylmercury; one subline (PC12/TM) was eight to ten times more resistant to methylmercury than were parent PC12 cells. PC12/TM cells accumulated less methylmercury than did parent PC12 cells. The intracellular glutathione level in PC12/TM cells was four times higher than that of PC12 cells. Pretreatment of PC12/TM cells with buthionine sulfoximine reduced the glutathione level to that of parent cells and increased the sensitivity of the cells to methylmercury without affecting methylmercury accumulation (Miura et al., 1994).

Mercury granules were detected in brains from rats exposed to $HgCl_2$ but not CH_3HgCl, suggesting that brain mercury granules are inorganic mercury (Suda et al., 1989). In that study, Wistar rats were dosed daily with subcutaneous injections of 5.0 mg Hg/kg BW for up to 12 consecutive days with either inorganic or organic mercury. With methylmercury after 8 days, brain contained 12.5 mg total Hg/kg FW, of which 0.23 mg/kg was inorganic mercury. After 12 days, these values were 23.6 mg total Hg/kg FW and 0.32 mg/kg inorganic mercury. Mercury granules were detected in brain only after eight daily doses. Histochemical analysis of rat and of mouse brain after rodents had been subjected to mercury-intoxicated rodents showed that the lowest concentration of inorganic mercury in brain with mercury granules was 0.12 mg/kg FW in $HgCl_2$-treated rats, 0.14 mg/kg in CH_3HgCl-treated rats, and 0.12 mg/kg in CH_3HgCl-treated mice; these concentrations were similar to the levels of inorganic mercury in brain from human autopsy cases from Minamata, Japan (Suda et al., 1989).

Mercury can alter the metabolism of essential elements in rats. In one study, rats given a high sublethal dose of methylmercury chloride or mercuric chloride showed rapid changes in certain elements within 24 h postinjection, as measured by inductively coupled plasma emission spectrography (Muto et al., 1991). Zinc in brain and kidney was highest in the methylmercury group; copper was highest in liver of the inorganic mercury group. Total mercury in brain of the $HgCl_2$ group was 0.049 mg/kg FW 3 h postinjection; for the methylmercury group, it was 0.95 mg/kg FW; and for the controls, 0.02 mg/kg FW. For liver, these values were 0.2, 3.4, and 0.013 mg/kg FW, respectively. Total mercury in kidney was 4.7 mg/kg FW in the $HgCl_2$ group, 14.9 for the methylmercury group, and 0.02 for the controls (Muto et al., 1991).

There is evidence that marine teleosts and mammals contain high levels of methylmercury and selenium, and that selenium in tuna and marlin can protect against methylmercury neurotoxicity (Kari and Kauranen, 1978; Ohi et al., 1980; Eisler, 1981). In one study (Ohi et al., 1980), weanling rats were given diets for 12 weeks containing 17.5 methylmercury/kg FW diet plus either 0.3 or 0.6 mg selenium/kg FW. The mercury and selenium originated from flesh of a marine sea bass (*Sebastes iracundus*) or sperm whale (*Physeter catodon*), or from sodium selenite. Selenium in blood, brain, and spinal cord of rats was positively correlated with neurological protection (tail rotation, hind limb paralysis), while total mercury and methylmercury in these tissues were negatively correlated. Fish muscle provided greater protection against mercury neurotoxicity than did whale meat (Ohi et al., 1980).

The developing central nervous system, particularly the cerebellum, of fetuses and neonates is a primary target of mercury toxicity (Bayo et al., 2004). Mercury neurotoxicity in cerebellar granular cells from transgenic mice has been demonstrated using heat shock protein 70 as a biomarker. Effects were observed at 0.2 mg/L and higher of inorganic mercury and 0.06 mg/L and higher of methylmercury (Bayo et al., 2004).

In the C57B/6N strain of mice, mercury concentrates in fetal tissues, especially brain, when compared to their dams given 2.5 to 20.0 mg CH_3HgCl/kg BW on day 13 of pregnancy and killed on day 14 or 18 (Inouye et al., 1986). The ratio was especially high at the highest dose tested of

20.0 mg/kg BW, as this was a toxic dose and brain weight was reduced. The concentration of methylmercury in the fetal brain was 1.6 to 4.9 times higher than in maternal brain. At the lowest dose tested of 2.5 mg/kg BW, mercury concentrations were higher in male fetuses, but this was not evident at higher doses (Inouye et al., 1986). Studies with guinea pigs, *Cavia* spp., showed that developmental disturbances of the fetal brain — including abnormal neuronal migration — resulted when dams were exposed to methylmercury in early pregnancy, and focal degeneration of the cerebral cortex when dams were exposed in later pregnancy (Inouye and Kajiwara, 1988b). In Minamata, where pregnant women were chronically exposed to methylmercury throughout pregnancy, both developmental disturbances and focal degeneration of neurons might be induced in the same fetal brain (Inouye and Kajiwara, 1988b).

Dose-dependent effects of methylmercury on mouse development *in vitro* include decreases in heart beat, axial rotation, blood circulation, yolk sac diameter, crown-rump length, and number of somites (Naruse et al., 1991). Dose-dependent abnormalities of the mouse embryo *in vitro* include increases in growth retardation, neural tube closure, hind limb hypoplasia, edema, stunted head, and eye hypoplasia (Naruse et al., 1991). For mercuric chloride, dose-dependent effects and abnormalities in mice were similar to those of methylmercury, but at lower frequencies and at higher doses (Naruse et al., 1991). Fetal Minamata disease caused by exposure to methylmercury during mid- and late-gestation periods is associated with cerebral palsy, blindness, deafness, microcephaly, and other adverse effects (Naruse et al., 1993). Because human brain development in late gestation seems comparable to that of postnatal stages in rodents, a study was conducted using artificial rearing systems for infant rats wherein nutrient factors were excluded, methylmercury administered, and brain architecture studied. Artificially reared rats given 40.0 mg methylmercury chloride/kg BW on day 6 after birth were killed 3 days later. These dosed rats, when compared to controls, had decreased brain weight and increased degeneration of nerve cells in the inner granule layer of the cerebellum (Naruse et al., 1993). This system shows promise for simulating mercury effects in human embryos.

Glutathione (GSH) is a major cytosolic, low-molecular-weight sulfhydryl compound that acts as a protective agent against methylmercury (Hirayama et al., 1987, 1990, 1991; Naganuma et al., 1988). The high affinity of methylmercury for the sulfhydryl group suggests that the fate of methylmercury is closely related to GSH metabolism. In the case of C57BL/6N female mice challenged by a sublethal oral dose of methylmercury (41.0 mg/kg BW), GSH levels in liver decreased 16.0%, and in blood 20.0%, 24 h after dosing (Yasutake and Hirayama, 1994). GSH half-time persistence in liver (74 min) was reduced by 17.0% from controls (89 min) but increased by 28.0% in kidney (46 min) from controls (36 min). Methylmercury-induced alterations of GSH metabolism in mouse tissue might reflect a defense mechanism against methylmercury insult (Yasutake and Hirayama, 1994).

Low dietary protein levels (7.5%) affected methylmercury and glutathione metabolism in C57BL/6N male mice when compared to a normal protein diet of 24.8% (Adachi et al., 1992). Mice fed the low protein diet had about four times less mercury in urine than those fed a normal protein diet for 7 days, although fecal mercury levels were the same in both groups; diets of both groups contained 5.14 mg CH_3HgCl/kg. Tissue mercury levels in the low protein group were significantly higher than those in the normal protein group, except for liver. Liver glutathione in the low protein diet group was lower than the high protein group, but other tissues were the same. Efflux rate of liver glutathione was significantly lower in the low protein diet group, but efflux rates of renal glutathione were the same in both groups (Adachi et al., 1992). When methylmercury-treated mice were injected with acivicin — a specific inhibitor of γ-glutamyltranspeptidase — the urinary mercury levels increased by 60-fold in the low protein diet group and by 36-fold in the high protein diet group. Authors aver that mice fed diets low in protein would show decreased urinary excretion of mercury via increased retention of mercury metabolites in kidney, and that dietary protein status — which could modulate thiol metabolism — is important in determining the fate of methylmercury (Adachi et al., 1992). Additional studies by Adachi et al. (1994) suggest that the mechanism by which dietary protein levels affect the fate of methylmercury in mice is related to sulfur amino acids. In those studies, a low protein diet (LPD) supplemented by methionine and cysteine was fed to methylmercury-dosed mice for 5 days. These mice had increased levels of brain mercury and liver mercury that was further enhanced by the amino acid supplemented diet (ASD). Mercury levels in kidney, blood, and plasma decreased with ASD feeding. The urinary mercury level that decreased with LPD feeding was recovered and far exceeded control levels with ASD feeding. Insufficiency of sulfur amino acids in mice fed an LPD could account for observed changes in the fate of methylmercury; changes in neutral amino acid transport caused by LPD feeding is also involved in the fate of methylmercury. Although the increased brain uptake of mercury would enhance the neurotoxic action of methylmercury, the stimulated renal elimination would reduce whole body methylmercury toxicity, including neural tissues. These opposite effects might be related to the alteration in sulfur amino acid metabolism caused by excessive mercury levels (Adachi et al., 1994).

Methylmercury in combination with ionizing radiation can affect developing mouse cerebellum (Inouye et al., 1991). In one study, pregnant mice given a low oral dose of methylmercury (9.0 mg Hg/kg BW) on day 17 of gestation delivered pups that contained 4.0 to 8.0 mg Hg/kg FW cerebellum. Male pups were subsequently exposed to x-radiation of 0.125 or 0.5 Gy on the day following birth. Cell death in the external granular layer (EGL) of pup cerebellum were similar in groups exposed to 0.125 Gy alone or in combination with mercury. In the high radiation dose of 0.5 Gy, 14% of these cells were killed by radiation, regardless of methylmercury exposure. Restoration of the EGL was slightly retarded by methylmercury. Authors concluded that methylmercury does not modify radiation-induced cell death in the EGL, but does retard tissue restoration from the damage (Inouye et al., 1991).

Mercury transfer and biomagnification through mammalian food chains is well documented (Galster, 1976; USNAS, 1978; Eaton et al., 1980; Eisler, 1981, 2000; Huckabee et al., 1981; Sheffy and St. Amant, 1982; Kucera, 1983; Clarkson et al., 1984; Wren, 1986; Wren et al., 1987a) but considerable variation exists. Among terrestrial mammals, for example, herbivores such as mule deer, moose (*Alces alces*), caribou (*Rangifer tarandus*), and various species of rabbits usually contained less than 1.0 mg Hg/kg fresh weight in liver and kidney, but carnivores such as the marten (*Martes martes*), polecat (*Mustela putorius*), and red fox (*Vulpes vulpes*) frequently contained more than 30.0 mg/kg (USNAS, 1978). The usually higher mercury concentrations in fish-eating furbearers than in herbivorous species seemed to reflect the amounts of fish and other aquatic organisms in the diet. In river otter (*Lutra canadensis*) and mink (*Mustela vison*) from the Wisconsin River drainage system, mercury levels paralleled those recorded in fish, crayfish, and bottom sediments at that location. Highest mercury levels in all samples were recorded about 30 km downstream from an area that supported 16 pulp and paper mills and a chloralkali plant; residues were highest in the fur, followed by the liver, kidney, muscle, and brain (Sheffy and St. Amant, 1982). Mercury exposure has been linked to population declines of mink in Georgia, North Carolina, and South Carolina owing to elevated mercury concentrations in kidney in mink from these areas (25.0 mg/kg FW) when compared to conspecifics from reference areas (4.0 mg/kg FW) (Osowski et al., 1995). The higher mercury levels in mink kidney were toxic to this species in laboratory studies (Wobeser et al., 1976). Dietary methylmercury exposure posed a moderate risk to female mink (Moore et al., 1999) and otters (Sample and Suter, 1999) in Tennessee.

In marine mammals, more than 90.0% of the mercury content is inorganic; however, enough methylmercury occurs in selected tissues to result in the accumulation of high tissue concentrations of methylmercury in humans and wildlife consuming such meat (Clarkson et al., 1984). The liver of the ringed seal (*Phoca hispida*) normally contains 27,000.0 to 187,000.0 μg Hg/kg fresh weight, and is a traditional and common food of the coastal Inuit people (Eaton et al., 1980). Although levels of total mercury in hair (109,000.0 μg/kg) and blood (37.0 μg/L) of Inuits were grossly elevated, no symptoms of mercury poisoning were evident in the coastal Inuits. Similar high concentrations have been reported for Alaskan Eskimo mothers who, during pregnancy, ate seal oil twice a day and seal meat or fish from the Yukon-Kuskokwim Coast every day (Galster, 1976). Despite the extremely high total mercury content of seal liver, only the small organomercury component was absorbed and appeared in the tissues. Cats fed a diet of seal liver (26,000.0 μg Hg/kg fresh weight) for 90 days showed no neurologic or histopathologic signs (Eaton et al., 1980). It seems that the toxic potential of seal liver in terms of accumulated tissue levels in cats (up to 862.0 μg total Hg/L blood, and 7600.0 μg total Hg/kg hair) is better indicated by the organomercury fraction in seal liver than by the concentration of total mercury (Eaton et al., 1980).

Under laboratory conditions, gray seals (*Halichoerus grypus*) dosed with methylmercury showed a time-related increase in both mercury and selenium in liver and kidney; however, in other tissues examined (brain, thyroid, blood), only the concentrations of mercury increased (Van de Ven et al., 1979). Atomic ratios of mercury and selenium were near 1.0 in wild seals, as expected, but this ratio exceeded 1.0 in seals fed additional methylmercury (Van de Ven et al., 1979). Mercury excretion rates in ringed seals, *Pusa hispida*, followed a biphasic or polyphasic pattern. The fastest excreted component took 20 days for complete elimination, and the slowest excreted component fraction, which comprised 45.0% of all mercury, took 500 days for 50.0% elimination (Tillander et al., 1972). In a 90-day controlled feeding study, harp seals were fed 0.25 or 25.0 mg methylmercury/kg body weight (BW) daily (Ronald et al., 1977). At 0.25 mg/kg BW daily, tissue mercury concentrations at 90 days(in mg Hg/kg FW) ranged between 42.7 and 82.5 in liver, kidney, and muscle; between 13.1 and 25.0 in adrenal glands, claws, brain spleen, lung, small intestine, heart, and blood; 1.6 in hair; and 0.2 in blubber. At 25.0 mg methylmercury/kg BW daily, one seal died on day 20 and another on day 26. Blood mercury values in these animals shortly before death were 26.8 and 30.3 mg/L. In addition, these seals were diagnosed with toxic hepatitis, uremia, and renal failure (Ronald et al., 1977). Histopathological damage was also evident (Ramprashad and Ronald, 1977). In the low dose group, 5.0% of the sensory cells in the middle cochlea of the ear were damaged (24.0% in the 25.0 mg/kg group), as well as sensory hair cells along the full length of the cochlea in both groups (Ramprashad and Ronald, 1977). It is noteworthy that dead harbor seals, *Phoca vitulina*, contained 9.9 to 31.0 mg total Hg/kg FW in brain (Koeman et al., 1973). These brain mercury concentrations were of the same order as those in brain tissue of animals of various species poisoned experimentally by methylmercury (Koeman et al., 1975). In view of the numerous reported strandings and beachings of many species of large marine mammals, the significance of the findings of Ronald and co-workers and those of Koeman and colleagues should not be discounted.

9.9 SUMMARY

Mercury is a known mutagen, teratogen, and carcinogen. At comparatively low concentrations in vertebrate animals, it adversely affects reproduction, growth, development, behavior, blood and serum chemistry, motor coordination, vision, hearing, histology, and metabolism. Mercury has a high potential for bioaccumulation and biomagnification, and is slow to depurate. Organomercury compounds were more effective in producing adverse effects than were inorganic mercury compounds; however, effects were significantly enhanced or ameliorated by numerous biotic and nonbiological modifiers. For sensitive aquatic species, adverse effects on growth and reproduction were observed at water concentrations of 0.03 to 0.1 μg Hg/L. For sensitive species of birds, adverse effects — mainly on reproduction — were associated with daily intake of 640.0 μg Hg/kg body weight, dietary levels of 50.0 to 100.0 μg Hg/kg fresh weight diet, and with total mercury concentrations greater 5.0 mg Hg/kg fresh weight feather, and greater than 0.9 mg/kg FW egg. Sensitive nonhuman mammals showed significant adverse effects of mercury when daily intakes were greater than 250.0 μg/kg body weight, when dietary levels were greater than 1.1 mg/kg fresh weight diet, or when tissue concentrations exceeded 1.1 mg/kg fresh weight.

REFERENCES

Abbasi, S.A. and R. Soni. 1983. Stress-induced enhancement of reproduction in earthworms *Octochaetus pattoni* exposed to chromium (VI) and mercury (II) — implications in environmental management, *Int. J. Environ. Stud.*, 22, 43–47.

Adachi, T., A. Yasutake, and K. Hirayama. 1992. Influence of dietary protein levels on the fate of methylmercury and glutathione metabolism in mice, *Toxicology*, 72, 17–26.

Adachi, T., A. Yasutake, and K. Hirayama. 1994. Influence of dietary levels of protein and sulfur amino acids on the fate of methylmercury in mice, *Toxicology*, 93, 225–234.

Allen, P. 1994. Changes in the haematological profile of the cichlid *Oreochromis aureus* (Steindachner) during acute inorganic mercury intoxication, *Comp. Biochem. Physiol.*, 108C, 117–121.

Amend, D.F. 1970. Retention of mercury by salmon, *Prog. Fish-Cult.*, 12, 192–194.

Angelow, R.V. and D.M. Nicholls. 1991. The effect of mercury exposure on liver mRNA translatability and metallothionein in rainbow trout, *Comp. Biochem. Physiol.*, 100C, 439–444.

Armstrong, F.A.J. 1979. Effects of mercury compounds on fish. In J.O. Nriagu (Ed.), *The Biogeochemistry of Mercury in the Environment*, p. 657–670. Elsevier/North-Holland Biomedical Press, New York.

Aschner, M. 1989. Brain, kidney and liver ^{203}Hg-methylmercury uptake in the rat: relationship to the neutral amino acid carrier, *Pharmacol. Toxicol.*, 65, 17–20.

Aschner, M. and T.W. Clarkson. 1988. Uptake of methylmercury in the rat brain: effects of amino acids, *Brain Res.*, 452, 31–39.

Aschner, M. and T.W. Clarkson. 1989. Methyl mercury uptake across bovine brain capillary endothelial cells *in vitro*: the role of amino acids, *Pharmacol. Toxicol.*, 64, 293–299.

Aschner, M., N.B. Eberle, S. Goderie, and H.K. Kimelberg. 1990. Methylmercury uptake in rat primary astrocyte cultures: the role of the neutral amino acid transport system, *Brain Res.*, 521, 221–228.

Baatrup, E. and K.B. Doving. 1990. Histochemical demonstration of mercury in the olfactory system of salmon (*Salmo salar* L.) following treatments with dietary methylmercuric chloride and dissolved mercuric chloride, *Ecotoxicol. Environ. Safety*, 20, 277–289.

Baatrup, E., K.B. Doving, and S. Winberg. 1990. Differential effects of mercurial compounds on the electro-olfactogram (EOG) of salmon (*Salmo salar* L.), *Ecotoxicol. Environ. Safety*, 20, 269–276.

Baldi, F., F. Semplici, and M. Filippelli. 1991. Environmental applications of mercury resistant bacteria, *Water Air Soil Pollut.*, 56, 465–475.

Bastidas, C. and E.M. Garcia. 2004. Sublethal effects of mercury and its distribution in the coral *Porites astreoides*, *Mar. Ecol. Prog. SER.*, 267, 133–143.

Bayo, M., M.A. Serra, and C.A. Clerici. 2004. *In vitro* detection of neuronal stress induced by mercury compounds in cerebellar granule cells from hsp70/hGH transgenic mice, *Bull. Environ. Contam. Toxicol.*, 72, 62–69.

Bearhop, S., G.D. Ruxton, and R.W. Furness. 2000. Dynamics of mercury in blood and feathers of great skuas, *Environ. Toxicol. Chem.*, 19, 1638–1643.

Berk, S.G., A.L. Mills, D.L. Henricks, and R.R. Colwell. 1978. Effects of ingesting mercury containing bacteria on mercury tolerance and growth rate of ciliates, *Microb. Ecol.*, 4, 319–330.

Berland, B.R., D.J. Bonin, O.J. Guerin-Ancey, V.I. Kapkov, and D.P. Arlhac. 1977. Action de metaux lourds a des doses subletales sur les caracteristiques de la croissnce chez la diatomee *Skeletonema costatum*, *Mar. Biol.*, 42, 17–30.

Berland, B.R., D.J. Bonin, V.I. Kapkov, S. Maestrini, and D.P. Arlhac. 1976. Action toxique de quatre metaux lourds sur la croissance d'algues unicellulaires marines, *C.R. Acad. Sci. Paris*, 282D, 633–536.

Best, J.B., M. Morita, J. Ragin, and J. Best, Jr. 1981. Acute toxic responses of the freshwater planarian, *Dugesia dorotocephala*, to methylmercury, *Bull. Environ. Contam. Toxicol.*, 27, 49–54.

Betti, C. and M. Nigro. 1996. The Comet assay for the evaluation of the genetic hazard of pollutants in cetaceans: preliminary results on the genotoxic effects of methyl-mercury on the bottle-nosed dolphin (*Tursiops truncatus*) lymphocytes *in vitro*, *Mar. Pollut. Bull.*, 32, 545–548.

Beyer, W.N., E. Cromartie, and G.B. Moment. 1985. Accumulation of methylmercury in the earthworm, *Eisenia foetida*, and its effect on regeneration, *Bull. Environ. Contam. Toxicol.*, 35, 157–162.

Birge, W.J., J.A. Black, A.G. Westerman, and J.E. Hudson. 1979. The effect of mercury on reproduction of fish and amphibians. In J.O. Nriagu (Ed.), *The Biogeochemistry of Mercury in the Environment*, p. 629–655. Elsevier/North-Holland Biomedical Press, New York.

Birge, W.J., A.G. Westerman, and J.A. Spromberg. 2000. Comparative toxicology and risk assessment of amphibians. In D.W. Sparling, G. Linder, and C.A. Bishop (Eds.), *Ecotoxicology of Amphibians and Reptiles,* p. 727–791. SETAC Press, Pensacola, FL.

Bligh, E.G. 1972. Mercury in Canadian fish, *J. Inst. Canad. Sci. Tech. Alim.*, 5(1), A6–A14.

Boney, A.D. 1971. Sub-lethal effects of mercury on marine algae, *Mar. Pollut. Bull.*, 2, 69–71.

Boney, A.D. and E.D.S. Corner. 1959. Application of toxic agents in the study of ecological resistance of intertidal red algae, *J. Mar. Biol. Assoc. U.K.*, 38, 267–275.

Boney, A.D., E.D.S. Corner, and B.W.P. Sparrow. 1959. The effects of various poisons on the growth and viability of sporelings of the red algae *Plumaria elegans* (Bonnem.), *Biochem. Pharmacol*, 2, 37–49.

Bowles, K.C., S.C. Apte, W.A. Maher, M. Kawei, and R. Smith. 2001. Bioaccumulation and biomagnification of mercury in Lake Murray, Papua New Guinea, *Canad. J. Fish. Aquat. Sci.*, 58, 888–897.

Calabrese, A., F.P. Thurberg, M.A. Dawson, and D.R. Wenzloff. 1975. Sublethal physiological stress induced by cadmium and mercury in winter flounder, *Pseudopleuronectes americanus*. In J.H. Koeman and J.H.T.W.A. Strik (Eds.), *Sublethal Effects of Toxic Chemicals on Aquatic Animals,* p. 15–21. Elsevier, Amsterdam.

Chang, L.W. 1979. Pathological effects of mercury poisoning. In J.O. Nriagu (Ed.), *The Biogeochemistry of Mercury in the Environment,* p. 519–580. Elsevier/North-Holland Biomedical Press, New York.

Choi, M.H. and J.J. Cech, Jr. 1998. Unexpectedly high mercury level in pelleted commercial fish feed, *Environ. Toxicol. Chem.*, 17, 1979–1981.

Clarkson, T.W., R. Hamada, and L. Amin-Zaki. 1984. Mercury. In J.O. Nriagu (Ed.), *Changing Metal Cycles and Human Health,* p. 285–309. Springer-Verlag, Berlin.

Clarkson, T.W. and D.O. Marsh. 1982. Mercury toxicity in man. In A.S. Prasad (Ed.), *Clinical, Biochemical, and Nutritional Aspects of Trace Elements*, Vol. 6, p. 549–568. Alan R. Liss, Inc., New York.

Clendenning, K.A. and W.J. North. 1959. Effects of wastes on the giant kelp, *Macrocystis pyrifera*, In E.A. Pearson (Ed.), *Proceedings 1st international Conference on Waste Disposal in the Marine Environment, p. 82–91.* Berkeley, CA.

Colwell, R.R. and J.D. Nelson, Jr. 1975. Metabolism of mercury compounds in microorganisms, *U.S. Environ. Protect. Agen Rept.*, 600/3-75-007, 1–84.

Colwell, R.R., G.S. Sayler, J.D. Nelson, Jr., and A. Justice. 1976. Microbial mobilization of mercury in the aquatic environment. In J.O. Nriagu (Ed.). *Environmental Biogeochemistry, Vol. 2., Metals Transfer and Ecological Mass Balances*, p. 437–487. Ann Arbor Sci. Publ., Ann Arbor, MI.

Cope, W.G., J.G. Wiener, and R.G. Rada. 1990. Mercury accumulation in yellow perch in Wisconsin seepage lakes: relation to lake characteristics, *Environ. Toxicol. Chem.*, 9, 931–940.

Corner, E.D.S. 1959. The poisoning of *Maia squinado* (Herbst) by certain compounds of mercury, *Biochem. Pharmacol.*, 2, 121–132.

Corner, E.D.S. and F.H. Rigler. 1958. The modes of action of toxic agents. III. Mercuric chloride and *n*-amylmercuric chloride on crustaceans, *J. Mar. Biol. Assoc. U.K.*, 37, 85–96.

Cunningham, P.A. and M.R. Tripp. 1973. Accumulation and depuration of mercury in the American oyster *Crassostrea virginica*, *Mar. Biol.*, 20, 14–19.

Cunningham, P.A. and M.R. Tripp. 1975a. Factors affecting the accumulation and removal of mercury from tissues of the American oyster *Crassostrea virginica*, *Mar. Biol.*, 31, 311–319.

Cunningham, P.A. and M.R. Tripp. 1975b. Accumulation, tissue distribution and elimination of $^{203}HgCl_2$ and CH_3HgCl in the tissues of the American oyster *Crassostrea virginica*, *Mar. Biol.*, 31, 321–334.

D'Agostino, A. and C. Finney. 1974. The effect of copper and cadmium on the development of *Tigriopus japonicus*. In F.J. Vernberg and W.B. Vernberg (Eds.). *Pollution and Physiology of Marine Organisms*. p. 445–463. Academic Press, NY.

Dave, G. and R. Xiu. 1991. Toxicity of mercury, copper, nickel, lead, and cobalt to embryos and larvae of zebrafish, *Brachydanio rerio*, *Arch. Environ. Contam. Toxicol.*, 21, 126–134.

Davies, A.G. 1974. The growth kinetics of *Isochrysis galbana* in cultures containing sublethal concentrations of mercuric chloride, *J. Mar. Biol. Assoc. U.K.*, 54, 167–169.

Davies, A.G. 1976. An assessment of the basis of mercury tolerance in *Dunaliella tertiolecta*, *J. Mar. Biol. Assoc. U.K.*, 56, 39–57

Dawson, M.A. 1982. Effects of long-term mercury exposure on hematology of striped bass, *Morone saxatilis*, *U.S. Natl. Mar. Fish. Serv. Fish. Bull.*, 80, 389–392.

De Flora, S., C. Bennicelli, and M. Bagnasco. 1994. Genotoxicity of mercury compounds. A review, *Mutat. Res.*, 317, 57–79.

Delbekke, K., C. Joiris, and G. Decadt. 1984. Mercury contamination of the Belgian avifauna 1970–1981, *Environ. Pollut.*, 7B, 205–221.

Delcourt, A. and J.C. Mestre. 1978. The effects of phenylmercuric acetate on the growth of *Chlamydomonas variabilis* Dang, *Bull. Environ. Contam. Toxicol.*, 20, 145–148.

De Wolf, P. 1975. Mercury content of mussels from West European coasts, *Mar. Pollut. Bull.*, 6, 61–63.

Dillon, T.M. 1977. Mercury and the estuarine marsh clam, *Rangia cuneata* Gray. I. Toxicity, *Arch. Environ. Contam. Toxicol.*, 6, 249–255.

Dillon, T.M. and J.M. Neff. 1978. Mercury and the estuarine marsh clam *Rangia cuneata* Gray. II. Uptake, tissue distribution and depuration, *Mar. Environ. Res.*, 1, 67–77.

D'Itri, P. and F.M. D'Itri. 1977. *Mercury Contamination: A Human Tragedy.* John Wiley, New York. 311 pp.

Eaton, R.D.P., D.C. Secord, and P. Hewitt. 1980. An experimental assessment of the toxic potential of mercury in ringed-seal liver for adult laboratory cats, *Toxicol. Appl. Pharmacol.*, 55, 514–521.

Eganhouse, R.P. and D.R. Young. 1978. Total and organic mercury in benthic organisms near a major submarine wastewater outfall system, *Bull. Environ. Contam. Toxicol.*, 19, 758–766.

Eisler, R. 1978. Mercury contamination standards for marine environments. In J.H. Thorp and J.W. Gibbons (Eds.), *Energy and Environmental Stress in Aquatic Systems,* p. 241–272. U.S. Dept. Energy Symp. Ser. 48, CONF-771114. Available from Natl. Tech. Infor. Serv., U.S. Dept. Commerce, Springfield, VA 22161.

Eisler, R. 1981. *Trace Metal Concentrations in Marine Organisms.* Pergamon, Elmsford, NY. 687 pp.

Eisler, R. 1984. Trace metal changes associated with age of marine vertebrates, *Biol. Trace Elem. Res.*, 6, 165–180.

Eisler, R. 1985. Selenium hazards to fish, wildlife, and invertebrates: a synoptic review, *U.S. Fish Wildl. Serv. Biol. Rep.*, 85(1.5), 1–57.

Eisler, R. 1987. Mercury hazards to fish, wildlife, and invertebrates: a synoptic review, *U.S. Fish Wildl. Serv. Biol. Rep.*, 85(1.10), 1–90.

Eisler, R. 2000. Mercury. In *Handbook of Chemical Risk Assessment: Health Hazards to Humans, Plants, and Animals. Vol. 1, Metals,* p. 313–409. Lewis Publishers, Boca Raton, FL.

Elhassani, S.B. 1983. The many faces of methylmercury poisoning, *J. Toxicol.,* 19, 875–906.

Establier, R., M. Gutierrez, and A. Aarias. 1978. Accumulation and histopathological effects of organic and inorganic mercury to the lisa (*Mugi auratus* Risso), *Invest. Pesquera*, 42, 65–80.

Evans, H.L., R.H. Garman, and V.G. Laties. 1982. Neurotoxicity of methylmercury in the pigeon, *Neuro-Toxicology*, 3(3), 21–36.

Fang, S.C. 1973. Uptake and biotransformation of phenylmercuric acetate by aquatic organisms, *Arch. Environ. Contam. Toxicol.*, 1, 18–26.

Fimreite, N. 1979. Accumulation and effects of mercury on birds. In J.O. Nriagu (Ed.). *The Biogeochemistry of Mercury in the Environment*, p. 601–627. Elsevier/North-Holland Biomedical Press, New York.

Finley, M.T. and R.C. Stendell. 1978. Survival and reproductive success of black ducks fed methylmercury, *Environ. Pollut.*, 16, 51–64.

Finley, M.T., W.H. Stickel, and R.E. Christensen. 1979. Mercury residues in tissues of dead and surviving birds fed methylmercury, *Bull. Environ. Contam. Toxicol.*, 21, 105–110.

Fjeld, E., T.O. Haugen, and L.A. Vollestad. 1998. Permanent impairment in the feeding behavior of grayling (*Thymallus thymallus*) exposed to methylmercury during embryogenesis, *Sci. Total Environ.*, 213, 247–254.

Fowler, B.A., D.A. Wolfe, and W.F. Hettler 1975. Mercury and iron uptake by cytosomes in mantle epithelial cells of quahog clams (*Mercenaria mercenaria*) exposed to mercury, *J. Fish. Res. Bd. Canada*, 32, 1767–1775.

Fowler, S.W. and G. Benayoun. 1976. Selenium kinetics in marine zooplankton, *Mar. Sci. Commun.*, 2, 43–67.

Fowler, S.W., M. Heyraud, and J. La Rosa. 1978. Factors affecting methyl and inorganic mercury dynamics in mussels and shrimp, *Mar. Biol.*, 46, 267–276.

Friedmann, A.S., H. Chen, L.D. Rabuck, and B.R. Zirkin. 1998. Accumulation of dietary methylmercury in the testes of the adult brown Norway rat: impaired testicular and epididymal function, *Environ. Toxicol. Chem.*, 17, 867–871.

Friedmann, A.S., M.C. Watzin, T. Brinch–Johnsen, and J.C. Leiter. 1996. Low levels of dietary methylmercury inhibit growth and gonadal development in juvenile walleye (*Stizostedion vitreum*), *Aquat. Toxicol.*, 35, 265–278.

Fujiki, Motto, M. Fujiki, S. Yamaguchi, R. Hirota, S. Tajima, N. Shimojo, and K. Sano. 1978. Accumulation of methyl mercury in red sea bream (*Chrysophrys major*) via the food chain. In S.A. Peterson and K.K. Randolph (Eds.), *Management of Bottom Sediments Containing Toxic Substances, Proc. 3rd U.S-Japan Experts' Meeting*, p. 87–94, November 1977. Easton, Maryland, U.S. Environ. Protect. Agen. Rep. 600/3-78-084.

Fujiyama, J., K. Hirayama, and A. Yasutake. 1994. Mechanism of methylmercury efflux from cultured astrocytes, *Biochem. Pharmacol.*, 47, 1525–1530.

Gale, T.F. and D.P. Hanlon. 1976. The permeability of the Syrian hamster placenta to mercury, *Environ. Res.*, 12, 26–31.

Galster, W. A. 1976. Mercury in Alaskan Eskimo mothers and infants, *Environ. Health Perspec.*, 15, 135–140.

Gambrell, R.P.,V.R. Collard, C.N. Reddy, and W.H. Patrick. 1977. Trace and toxic metal uptake by marsh plants as affected by Eh, pH, and salinity, *U.S. Army Eng. Water. Exp. Sta, Vicksburg, MS, Dredged Mat. Res. Prog. Tech. Rept. D77-40.* 124 pp.

Gardner, W.S., H.L. Windom, J.A. Stephens, F.E. Taylor, and R.R. Stickney. 1975. Concentrations of total mercury in fish and other coastal organisms: implications to mercury cycling. In F.G. Howell, J.B. Gentry, and M.H. Smith (Eds.). *Mineral Cycling in Southeastern Ecosystems.* p. 268–278. U.S. Energy Res. Dev. Admin., Avail. as CONF- 740513 from NTIS, U.S. Dept. Commerce, Springfield, VA.

Gauthier, J.M., H. Dubeau, and E. Rassart. 1998. Mercury-induced micronuclei in skin fibroblasts of beluga whales, *Environ. Toxicol. Chem.*, 17, 2487–2493.

Geffen, A.J., N.J.G. Pearce, and W.T. Perkins. 1998. Metal concentrations in fish otoliths in relation to body composition after laboratory exposure to mercury and lead, *Mar. Ecol. Prog. Ser.*, 165, 235–245.

Gentile, J., S.M. Gentile, G. Hoffman, J.F. Heltshe, and N. Hairston, Jr. 1983. The effects of a chronic mercury exposure on survival, reproduction and population dynamics of *Mysidopsis bahia*, *Environ. Toxicol. Chem.*, 2, 61–68.

Ghosh, S.K., S. Ghosh, J. Chaudhuri, R. Gachhui, and A. Mandal. 2004. Studies on mercury resistance in yeasts isolated from natural sources, *Bull. Environ. Contam. Toxicol.*, 72, 21–28.

Giblin, F.J. and E.J. Massaro. 1973. Pharmacodynamics of methylmercury in rainbow trout (*Salmo gairdneri*): tissue uptake, distribution, and excretion, *Toxicol. Appl. Pharmacol.*, 24, 81–91.

Gilbert, S.G., C.D. Munkers, T.M. Burbacher, and D.C. Rice. 1994. Effects of *in utero* methylmercury exposure on schedule controlled behavior in adult monkeys, *Am. J. Primatol.*, 33, 211.

Glickstein, N. 1978. Acute toxicity of mercury and selenium to *Crassostrea gigas* embryos and *Cancer magister* larvae, *Mar. Biol.*, 49, 113–117.

Glooschenko, W.A. 1969. Accumulation of ^{203}Hg by the marine diatom *Chaetoceros costatum*, *J. Phycol.*, 5, 224–226.

Gochfeld. M. 1980. Tissue distribution of mercury in normal and abnormal young common terns, *Mar. Pollut. Bull.*, 11, 362–366.

Godbold, D.L. 1991. Mercury-induced root damage in spruce seedlings, *Water Air Soil Pollut.*, 56, 823–831.

Goyer, R.A. 1986. Toxic effects of metals. In C.D. Klaassen, M.O. Amdur, and J. Doull (Eds.), *Casarett and Doull's Toxicology,* 3rd ed., p. 582–635. Macmillan, New York.

Green, F.A., Jr., J.W. Anderson, S.R. Petrocelli, B.J. Presley, and R. Sims. 1976. Effect of mercury on the survival, respiration, and growth of postlarval white shrimp, *Penaeus setiferus*, *Mar. Biol.*, 37, 75–81.

Greener, Y. and J.A. Kochen. 1983. Methyl mercury toxicity in the chick embryo, *Teratology*, 28, 23–28.

Grissom, R.E., Jr. and J.P. Thaxton. 1985. Onset of mercury toxicity in young chickens, *Arch. Environ. Contam. Toxicol.*, 14, 193–196.

Hammerschmidt, C.R., M.B. Sandheinrich, J.G. Wiener, and R.G. Rada. 2002. Effects of dietary methylmercury on reproduction in fathead minnows, *Environ. Sci. Technol.*, 36, 877–883.

Hammerschmidt, C.R., J.G. Wiener, B.E. Frazier, and R.G. Rada. 1999. Methylmercury content of eggs in yellow perch related to maternal exposure in four Wisconsin lakes, *Environ. Sci. Technol.*, 33, 999–1003.

Handy, R.D. and W.S. Penrice. 1993. The influence of high oral doses of mercuric chloride on organ toxicant concentrations and histopathology in rainbow trout, *Oncorhynchus mykiss*, *Comp. Biochem. Physiol.*, 106C, 717–724.

Hannan, P.J., P.E. Wilkniss, C. Patouillet, and R.A. Carr. 1973. Measurements of mercury sorption by algae, *U.S. Naval Res. Lab, Wash. D.C.*, Rept. 7628, 1–26.

Hannerz, L. 1968. Experimental investigations on the accumulation of mercury in water organisms, *Rep. Inst. Freshwater Res. Drotting.*, 48, 120–176.

Harada, M. 1978. Congenital Minamata disease: intrauterine methylmercury poisoning, *Teratology*, 18, 285–288.

Harada, Y. 1977. Congenital Minamata disease. In T. Tsubaki, and K. Irukayama (Eds.), *Minamata Disease,* p. 209–218. Elsevier, Amsterdam.

Harmon, S.M., J.K. King, J.B. Gladden, G.T. Chandler, and L.A. Newman. 2005. Mercury body burdens in *Gambusia holbrooki* and *Erimyzon sucetta* in a wetland mesocosm amended with sulfate, *Chemosphere*, 59, 227–233.

Harris, S.B., J.B. Wilson, and R.H. Printz. 1972. Embryotoxicity of methylmercuric chloride in golden hamster, *Teratology*, 6, 139–142.

Harrison, S.E., J.F. Klaverkamp, and R.H. Hesslein. 1990. Fates of metal radiotracers added to a whole lake: accumulation in fathead minnow (*Pimephales promelas*) and lake trout (*Salvelinus namaycush*), *Water Air Soil Pollut.*, 52, 277–293.

Harrison, W.G., E.H. Renger, and R.W. Eppley. 1978. Controlled ecosystem pollution experiment: effect of mercury on enclosed water columns. VII. Inhibition of nitrogen assimilation and ammonia regeneration by plankton in seawater samples, *Mar. Sci. Commun.*, 4, 13–22.

Hasselrot, T.B. 1968. Report on current field investigations concerning the mercury content in fish, bottom sediment, and water, *Rep. Inst. Freshwater Res. Drottingholm*, 48, 102–111.

Heagler, M.G., M.C. Newman, M. Mulvey, and P.M. Dixon. 1993. Allozyme genotype in mosquitofish, *Gambusia holbrooki*, during mercury exposure: temporal stability, concentration effects and field verification, *Environ. Toxicol. Chem.*, 12, 385–395.

Heinz, G.H. 1979. Methylmercury: reproductive and behavioral effects on three generations of mallard ducks, *J. Wildl. Manage.*, 43, 394–401.

Heinz, G.H. 1980a. Comparison of game-farm and wild-strain mallard ducks in accumulation of methylmercury, *J. Environ. Pathol. Toxicol.*, 3, 379–386.

Heinz, G.H. 1980b. Eggshell thickness in mallards fed methylmercury, *Bull. Environ. Contam. Toxicol.*, 25, 498–502.

Heinz, G.H. 1987. Mercury accumulations in mallards fed methylmercury with or without added DDE, *Environ. Res.*, 42, 372–376.

Heinz, G.H. and D.J. Hoffman. 1998. Methylmercury chloride and selenomethionine interactions on health and reproduction in mallards, *Environ. Toxicol. Chem.*, 17, 139–145.

Heinz, G.H. and L.N. Locke. 1976. Brain lesions in mallard ducklings from parents fed methylmercury, *Avian Dis.*, 20, 9–17.

Hibaya, T. and M. Oguri. 1961. Gill absorption and tissue distribution of some radionuclides (Cr-51, Hg-203, Zn-65, and Ag-110 m) in fish, *Bull. Japan. Soc. Sci. Fish.*, 27, 996–1000.

Hill, E.F. and J.H. Soares, Jr. 1984. Subchronic mercury exposure in *Coturnix* and a method of hazard evaluation, *Environ. Toxicol. Chem.*, 3, 489–502.

Hirayama, K. 1975. Transport mechanism of methylmercury: intestinal absorption, biliary excretion and distribution of methylmercury, *Kumamoto Med. J.*, 28, 151–163.

Hirayama, K. 1980. Effect of amino acids in brain uptake of methylmercury, *Toxicol. Appl. Pharmacol.*, 55, 318–323.

Hirayama, K. 1985. Effects of combined administration of thiol compounds and methylmercury chloride on mercury distribution in rats, *Biochem. Pharmacol.*, 34, 2030–2032.

Hirayama, K., M. Inouye, and T. Fujisaki. 1985. Alteration of putative amino acid levels and morphological findings in neural tissues of methylmercury-intoxicated mice, *Arch. Toxicol.*, 57, 35–40.

Hirayama, K., A. Uasutake, and T. Adachi. 1991. Mechanism of renal handling of methylmercury. In T. Suzuki, N. Imura, and T.W. Clarkson (Eds.), *Advances in Mercury Toxicology,* p. 121–134. Plenum Press, New York.

Hirayama, K., A. Yasutake, and M. Inoue. 1987. Effect of sex hormones on the fate of methylmercury and glutathione metabolism in mice, *Biochem. Pharmacol.*, 36, 1919–1924.

Hirayama, K., A. Yasutake, and M. Inoue. 1990. Role of glutathione metabolism in heavy metal poisoning. In G. Lubec and G.A. Rosenthal (Eds.), *Amino Acids: Chemistry, Biology and Medicine,* p. 970–975. ESCOM Science Publ., Amsterdam.

Hirota, R., J. Asada, S. Tajima, and M. Fujiki. 1983. Accumulation of mercury by the marine copepod *Acartia clausi*, *Bull. Japan. Soc. Sci. Fish.*, 49, 1249–1251.

Hoffman, D.J. and G.H. Heinz. 1998. Effects of mercury and selenium on glutathione metabolism and oxidative stress in mallard ducks, *Environ. Toxicol. Chem.*, 17, 161–166.

Hoffman, D.J. and J.M. Moore. 1979. Teratogenic effects of external egg applications of methyl mercury in the mallard, *Anas platyrhynchos*, *Teratology*, 20, 453–462.

Holt, D. and M. Webb. 1986. The toxicity and teratogenicity of mercuric mercury in the pregnant rat, *Arch. Toxicol.*, 58, 243–248.

Hook, J.B. and W.R. Hewitt. 1986. Toxic responses of the kidney. In C.D. Klaassen, M.O. Amdur, and J. Doull (Eds.), *Casarett and Doull's Toxicology*, third edition, p. 310–329. Macmillan, New York.

Hopkins, R. and J.M. Kain. 1971. The effect of marine pollutants on *Laminaria hyperboria*, *Mar. Pollut. Bull.*, 2, 75–77.

Huckabee, J.W. and B.G. Blaylock. 1972. Transfer of mercury and cadmium from terrestrial to aquatic ecosystems. In S.K. Dhar (Ed.). *Metal Ions in Biological Systems*, p. 125–160. Plenum, New York.

Huckabee, J.W., J.W. Elwood, and S.G. Hildebrand. 1979. Accumulation of mercury in freshwater biota. In J.O. Nriagu (Ed.). *The Biogeochemistry of Mercury in the Environment*, p. 277–302. Elsevier/North-Holland Biomedical Press, New York.

Huckabee, J.W., D.M. Lucas, and J.M. Baird. 1981. Occurrence of methylated mercury in a terrestrial food chain, *Environ. Res.*, 26, 174–181.

Inouye, M. 1989. Teratology of heavy metals: mercury and other contaminants, *Cong. Anom.*, 29, 333–344.

Inouye, M. and Y. Kajiwara. 1988a. Teratogenic interactions between methylmercury and mitomycin-C in mice, *Arch. Toxicol.*, 61, 192–195.

Inouye, M. and Y. Kajiwara. 1988b. Developmental disturbances of the fetal brain in guinea-pigs caused by methylmercury, *Arch. Toxicol.*, 62, 15–21.

Inouye, M. and Y. Kajiwara. 1988c. An attempt to assess the inheritable effect of methylmercury toxicity subsequent to prenatal exposure of mice, *Bull. Environ. Contam. Toxicol.*, 41, 508–514.

Inoue, M. and Y. Kajiwara. 1990a. Strain difference of the mouse in manifestation of hydrocephalus following prenatal methylmercury exposure, *Teratology*, 41, 205–210.

Inouye, M. and Y. Kajiwara. 1990b. Placental transfer of methylmercury and mercuric mercury in mice, *Environ. Med.*, 34, 169–172.

Inouye, M., Y. Kajiwara, and K. Hirayama. 1986. Dose- and sex-dependent alterations in mercury distribution in fetal mice following methylmercury exposure, *J. Toxicol. Environ. Health*, 19, 425–435.

Inouye, M., Y. Kajiwara, and K. Hirayama. 1991. Combined effects of low–level methylmercury and X-radiation on the developing mouse cerebellum, *J. Toxicol. Environ. Health*, 33, 47-56.

Inouye, M. and U. Murakami. 1975. Teratogenic effect of orally administered methylmercuric chloride in rats and mice, *Cong. Anom.*, 15, 1–9.

Inouye, M., K. Murao, and Y. Kajiwara. 1985. Behavioral and neuropathological effects of prenatal methylmercury exposure in mice, *Neurobehav. Toxicol. Teratol.*, 7, 227–232.

Irukayama, K., M. Fujiki, F. Kai, and T. Kondo. 1962a. Studies on the causative agent of Minamata disease. II. Comparison of the mercury compound in the shellfish from Minamata Bay with mercury compounds experimentally accumulated in the control shellfish, *Kumamoto Medic. J.*, 15, 1–12.

Irukayama, K., F. Kai, M. Fujiki, and T. Kondo. 1962b. Studies on the causative agent of Minamata disease. III. Industrial wastes containing mercury compounds from Minamata factory, *Kumamoto Medic. J.*, 15, 57–68.

Itow, T., R.E. Loveland, and M.L. Botton. 1998. Developmental abnormalities in horseshoe crab embryos caused by exposure to heavy metals, *Arch. Environ. Contam. Toxicol.*, 35, 33–40.

Jarvenpaa, T., M. Tillander, and J.K. Miettinen. 1970. Methyl-mercury: half-time of elimination in flounder, pike, and eel, *Suomen Kemist. B*, 43, 439–442.

Johnels, A.G., T. Westermark, W. Berg, P.I. Persson, and B. Sjostrand. 1967. Pike (*Esox lucius* L.) and some other aquatic organisms in Sweden as indicators of mercury contamination of the environment, *Oikos*, 18, 323–333.

Kanamadi, R.D. and S.K. Saidapur. 1991. Effect of sublethal concentration of mercuric chloride on the ovary of the frog *Rana cyanophlyctis*, *J. Herpetol.*, 25, 494–497.

Kari, T. and P. Kauranen. 1978. Mercury and selenium contents of seals from fresh and brackish water in Finland, *Bull. Environ. Contam. Toxicol.*, 19, 273–280.

Kasschau, M.R., M.M. Skaggs, and E.C.M. Chen. 1980. Accumulation of glutamate in sea anemones exposed to heavy metals and organic amines, *Bull. Environ. Contam. Toxicol.*, 25, 873–878.

Kayser, H. 1976. Waste-water assay with continuous algal cultures: the effect of mercuric acetate on the growth of some marine dinoflagellates, *Mar. Biol.*, 36, 61–72.

Kennedy, W.A. and M.S. Smith. 1972. Sablefish culture — progress in 1971, *Fish. Res. Bd. Canada Tech. Rep.*, 309, 1–15.

Khan, A.T. and J. S. Weis. 1993. Differential effects of organic and inorganic mercury on the micropyle of the eggs of *Fundulus heteroclitus*, *Environ. Biol. Fish*, 37, 323–327.

Khera, K.S. 1979. Teratogenic and genetic effects of mercury toxicity. In J.O. Nriagu (Ed.). *The Biogeochemistry of Mercury in the Environment*, p. 501–518. Elsevier/North-Holland Biomedical Press, New York.

Kikuchi, T., H. Honda, M. Ishikawa, H. Yamanaka, and K. Amano. 1978. Excretion of mercury from fish, *Bull. Japan. Soc. Sci. Fish.*, 44, 217–222.

Kim, J.P. and S. Burggraaf. 1999. Mercury bioaccumulation in rainbow trout (*Oncorhynchus mykiss*) and the trout food web in lakes Okareka, Okaro, Tarawera, Rotomahana and Rotorua, New Zealand, *Water Air Soil Pollut.*, 115, 535–546.

Kirubagaran, R. and K.P. Joy. 1988. Toxic effects of three mercurial compounds on survival, and histology of the kidney of the catfish *Clarias batrachus* (L.), *Ecotoxicol. Environ. Safety*, 15, 171–179.

Kirubagaran, R. and K.P. Joy. 1989. Toxic effects of mercurials on thyroid function of the catfish, *Clarias batrachus* (L.), *Ecotoxicol. Environ. Safety*, 17, 265–271.

Kirubagaran, R. and K.P. Joy. 1992. Toxic effects of mercury on testicular activity in the freshwater teleost, *Clarias batrachus* (L.), *J. Fish Biol.*, 41, 305–315.

Kobayashi, N. 1971. Fertilized sea urchin eggs as an indicatory material for marine pollution bioassay, preliminary experiments, *Publ. Seto Mar. Biol. Lab.*, 18, 379–406.

Koeller, P.A and G.T. Wallace. 1977. Controlled ecosystem pollution experiment: effect of mercury on enclosed water columns. V. Growth of juvenile chum salmon (*Oncorhynchus keta*), *Mar. Sci. Commun.*, 3, 395–406.

Koeman, J.H., W.H.M. Peeters, C.H.M. Koudstaal-Hol, P.S. Tijoe, and J.J.M. de Goeij. 1973. Mercury-selenium correlations in marine mammals, *Nature*, 245, 385–386.

Koeman, J.H., W.S.M. van de Ven, J.J.M. de Goeij, P.S. Tijoe, and J.L. van Haaften. 1975. Mercury and selenium in marine mammals and birds, *Sci. Total Environ.*, 3, 279–287.

Kopfler, F.C. 1974. The accumulation of organic and inorganic mercury compounds by the eastern oyster (*Crassostrea virginica*), *Bull. Environ. Contam. Toxicol.*, 11, 275–280.

Kramer, H.J. and B. Neidhart. 1975. The behavior of mercury in the system water-fish, *Bull. Environ. Contam. Toxicol.*, 14, 699–701.

Kramer, V.J. and M.C. Newman. 1994. Inhibition of glucosephosphate isomerase allozymes of the mosquitofish, *Gambusia holbrooki*, by mercury, *Environ. Toxicol. Chem.*, 13, 9–14.

Kramer, V.J., M.C. Newman, M. Mulvey, and G.R. Ultsch. 1992. Glycolysis and Krebs cycle metabolites in mosquitofish, *Gambusia holbrooki*, Girard 1859, exposed to mercuric chloride: allozyme genotype effects, *Environ. Toxicol. Chem.*, 11, 357–364.

Kucera, E. 1983. Mink and otter as indicators of mercury in Manitoba waters, *Can. J. Zool.*, 61, 2250–2256.

Latif, M.A., R.A. Bodaly, T.A. Johnstom, and R.J.P. Fudge. 2001. Effects of environmental and maternally derived methylmercury on the embryonic and larval stages of walleye (*Stizostedium vitreum*), *Environ. Pollut.*, 111, 139–148.

Laumond, F., M. Neuburger, B. Donnier, A. Fourcy, R. Bittel, and M. Aubert. 1973. Experimental investigations, at laboratory, on the transfer of mercury in marine trophic chains, In *6th Inter. Sympos. Medic. Ocean.*, p. 47–53 Portoroz, Yugoslovia, Sept. 26–30, 1973.

Lawrence, A.L. and R.P. Mason. 2001. Factors controlling the bioaccumulation of mercury and methylmercury by the estuarine amphipod *Leptocheirus plumulosus*, *Environ. Pollut.*, 111, 217–231.

Leander, J.D., D.E. McMillan, and T.S. Barlow. 1977. Chronic mercuric chloride: behavioral effects in pigeons, *Environ. Res.*, 14, 424–435.

Lee, C.J., M.C. Newman, and M. Mulvey. 1992. Time to death of mosquitofish (*Gambusia holbrooki*) during acute inorganic mercury exposure: population structure effects, *Arch. Environ. Contam. Toxicol.*, 22, 284–287.

Lee, E.H., B.H. Ryu, and S.T. Yang. 1975. Suitability of shellfish for processing, *Bull. Korean Fish. Soc.*, 8, 85–89.

Linder, G. and B. Grillitsch. 2000. Ecotoxicology of metals. In D.W. Sparling, G. Linder, and C.A. Bishop (Eds.), *Ecotoxicology of Amphibians and Reptiles*, p. 325–459. SETAC Press, Pensacola, FL.

Loftenius, A., G. Sandborgh-Englund, and J. Ekstrand. 1998. Acute exposure to mercury from amalgam: no short-time effect on the peripheral blood lymphocytes in healthy individuals, *J. Toxicol. Environ. Health*, 54A, 547–560.

Luoma, S.N. 1976. The uptake and interorgan distribution of mercury in a carnivorous crab, *Bull. Environ. Contam. Toxicol.*, 16, 719–723.

Luoma, S.N. 1977. The dynamics of biologically available mercury in a small estuary, *Estuar. Coast. Mar. Sci.*, 5, 643–652.

Maas, R.P., S.C. Patch, and K.R. Sergent. 2004. *A Statistical Analysis of Factors Associated with Elevated Hair Mercury Levels in the U.S. Population: An Interim Progress Report*. Tech. Rep. 04-136, Environ. Qual. Inst., Univ. North Carolina, Ashville. 13 pp.

MacLeod, J.C. and E. Pessah. 1973. Temperature effects on mercury accumulation, toxicity and metabolic rate in rainbow trout (*Salmo gairdneri*), *J. Fish. Res. Bd. Canada*, 30, 485–492.

March, B.E., R. Poon, and S. Chu. 1983. The dynamics of ingested methyl mercury in growing and laying chickens, *Poult. Sci.*, 62, 1000–1009.

Matsumura, A., K. Eto, N. Furuyoshi, and R. Okamura. 1993. Disturbances of accommodation in Minamata disease. A neuropathological study of methylmercury toxicity in common marmoset monkeys, *Neuro-ophthalmology*, 13, 331-339.

McCloskey, J.T., I.R. Schultz, and M.C. Newman. 1998. Estimating the oral bioavailability of methylmercury to channel catfish (*Ictalurus punctatus*), *Environ. Toxicol. Chem.*, 17, 1524–1529.

McClurg, T.P. 1984. Effects of fluoride, cadmium and mercury on the estuarine prawn *Penaeus indicus*, *Water SA*, 10, 40–45.

McEwen, L.C., R.K. Tucker, J.O. Ells, and M.A. Haegele. 1973. Mercury-wildlife studies by the Denver Wildlife Research Center. In D.R. Buhler (Ed.). *Mercury in the Western Environment*, p. 146–156. Oregon State University, Corvallis.

McKim, J.M., G.F. Olson, G.W. Holcombe, and E.P. Hunt. 1976. Long-term effects of methylmercuric chloride on three generations of brook trout (*Salvelinus fontinalis*): toxicity, accumulation, distribution, and elimination, *J. Fish. Res. Bd. Canada*, 33, 2726–2739.

Middaugh, D.P. and C.L. Rose. 1974. Retention of two mercurials by striped mullet, *Mugil cephalus*, *Water Res.*, 8, 173–177.

Miettinen, J.K, M. Heyraud, and S. Keckes. 1972. Mercury as a hydrospheric pollutant. II. Biological half-time of methyl-mercury in four Mediterranean species: a fish, a crab, and two molluscs. In M. Ruivo (Ed.), *Marine Pollution and Sea Life*, p. 295–298. Fishing Trading News (books), London.

Miettinen, J.K., M. Tillander, K. Rissanen, V. Miettinen, and Y. Ohmono. 1969. Distribution and excretion rate of phenyl- and methylmercury nitrate in fish muscles, molluscs, and crayfish, *Proc. 9th Japan. Conf. Radioisotopes, Tokyo*, 474–478.

Minigawa, K., T. Sasaki, Y. Takizawa, R. Tamura, and T. Oshina. 1980. Accumulation route and chemical form of mercury in mushroom species. *Bull. Environ. Contam. Toxicol.*, 25, 382–388.

Miura, K., T.W. Clarkson, K, Ikeda, A. Naganuma, and N. Imura. 1994. Factors determining the sensitivity to methylmercury in methylmercury-resistant PC12 cell lines. In *Proceedings of the International Symposium on "Assessment of Environmental Pollution and Health Effects from Methylmercury,"* p. 232–243. October 8–9, 1993, Kumamoto, Japan. National Institute for Minamata Disease, Minamata City, Kumamoto 867, Japan.

Monteiro, L.R., J.P. Granadeiro, and R.W. Furness. 1998. Relationship between mercury levels and diet in Azores seabirds, *Mar. Ecol. Prog. Ser.*, 166, 259–265.

Moore, D.R.J., B.E. Sample, G.W. Suter, B.R. Parkhurst, and R.S. Teed. 1999. A probabilistic risk assessment of the effects of methylmercury and PCBs on mink and kingfishers along East Fork Poplar Creek, Oak Ridge, Tennessee, USA, *Environ. Toxicol. Chem.*, 18, 2941–2953.

Moore, M.N. 1977. Lysosomal responses to environmental chemicals in some marine invertebrates. In C.S. Giam (Ed.). *Pollutant Effects on Marine Organisms*, p. 143–154. D.C. Heath, Lexington, MA.

Moore, M.N. and A.R.D. Stebbing. 1976. The quantitative cytochemical effects of three metal ions on a lysosomal hydrolase of a hydroid, *J. Mar. Biol. Assoc. U.K.*, 56, 995–1005.

Mullins, W.H., E.G. Bizeau, and W.W. Benson. 1977. Effects of phenyl mercury on captive game farm pheasants, *J. Wildl. Manage.*, 41, 302–308.

Mulvey, M., M.C. Newman, A. Chazal, M.M. Keklak, M.G. Haegler, and L.S. Hales, Jr. 1995. Genetic and demographic responses of mosquitofish (*Gambusia holbrooki* Girard 1859) populations stressed by mercury, *Environ. Toxicol. Chem.*, 14, 1411–1418.

Muto, H., M. Shinada, K. Tokuta, and Y. Takizawa. 1991. Rapid changes in concentrations of essential elements in organs of rats exposed to methylmercury chloride and mercuric chloride as shown by simultaneous multielemental analysis, *Br. J. Indust. Med.*, 48, 382–388.

Naganuma, A., N. Oda-Urano, T. Tanaka, and N. Imura. 1988. Possible role of hepatic glutathione in transport of methylmercury into mouse kidney, *Biochem. Pharmacol.*, 37, 291–296.

Nanda, D.R. and B.B. Mishra. 1997. Effect of solid waste from a chlor-alkali factory on rice plants: mercury accumulation and changes in biochemical variables. In P.N. Cheremisinoff (Ed.), *Ecological Issues and Environmental Impact Assessment*, p. 601–612. Gulf Publishing Co., Houston, TX.

Naruse, I., H. Arakawa, and Y. Fukui. 1993. Effects of methylmercury on the brain of rats reared artificially, *Tokushima J. Exp. Med.*, 40, 69–74.

Naruse, I., N. Matsumoto, and Y. Kajiwara. 1991. Toxicokinetics of methylmercury and mercuric chloride in mouse embryos *in vitro*, *Bull. Environ. Contam. Toxicol.*, 47, 689–695.

Nelson, D.A., A. Calabrese, B.A. Nelson, J.R. MacInnes, and D. R. Wenzloff. 1976. Biological effects of heavy metals on juvenile bay scallops, *Argopecten irradians*, in short-term exposures, *Bull. Environ. Contam. Toxicol.*, 16, 275–282.

Nelson, D.A., A. Calabrese, and J.R. MacInnes. 1977. Mercury stress on juvenile bay scallops, *Argopecten irradians*, under various salinity-temperature regimes, *Mar. Biol.*, 43, 293–297.

Nelson, J.D., W. Blair, F.E. Brinckman, R.R. Colwell, and W.P. Iverson. 1973. Biodegradation of phenylmercuric acetate by mercury-resistant bacteria, *Appl. Microbiol.*, 26, 321–326.

Nelson, J.D., Jr. and R.R. Colwell. 1975. The ecology of mercury-resistant bacteria in Chesapeake Bay, *Microbiol. Ecol.*, 1, 191–218.

Newman, M.C. and D.K. Doubet. 1989. Size-dependence of mercury (II) accumulation kinetics in the mosquitofish, *Gambusia affinis* (Baird and Girard), *Arch. Environ. Contam. Toxicol.*, 18, 819–825.

Nicholls, D.M., K. Teichert-Kuliszewska, and G.R. Girgis. 1989. Effect of chronic mercuric chloride exposure on liver and muscle enzymes in fish, *Comp. Biochem. Physiol.*, 94C, 265–270.

Nicholson, J.K. and D. Osborn. 1984. Kidney lesions in juvenile starlings *Sturnus vulgaris* fed on a mercury-contaminated synthetic diet, *Environ. Pollut.*, 33A, 195–206.

Niimi, A.J. and G.P. Kissoon. 1994. Evaluation of the critical body burden concept based on inorganic and organic mercury toxicity to rainbow trout (*Oncorhynchus mykiss*), *Arch. Environ. Contam. Toxicol.*, 26, 169–178.

Niimi, A.J. and L. Lowe-Jinde. 1984. Differential blood cell ratios of rainbow trout (*Salmo gairdneri*) exposed to methylmercury and chlorobenzenes, *Arch. Environ. Contam. Toxicol.*, 13, 303–311.

Odin, M., A. Fuertet-Mazel, F. Ribeyre, and A. Boudou. 1995. Temperature, pH and photoperiod effects on mercury bioaccumulation by nymphs of the burrowing mayfly *Hexagenia rigida*, *Water Air Soil Pollut.*, 80, 1003–1006.

Ohi, G., S. Nishigaki, H. Seki, Y. Tamura, T. Maki, K. Minowa, Y. Shimamura, and I. Mizoguchi. 1980. The protective potency of marine animal meat against the neurotoxicity of methylmercury: its relationship with the organ distribution of mercury and selenium in the rat, *Food Cosmet. Toxicol.*, 18, 139–145.

Oikawa, K., H. Saito, I. Kifune, T. Ohshina, M. Fujii, and Y. Takizawa. 1982. Respiratory tract retention of inhaled air pollutants. Report 1: mercury absorption by inhaling through the nose and expiring through the mouth at various concentrations, *Chemosphere*, 11, 949–951.

Okubo, K. and T. Okubo. 1962. Study on the bioassay method for the evaluation of water pollution. –II. Use of the fertilized eggs of sea urchins and bivalves, *Bull. Tokai Reg. Fish. Res. Lab.*, 32, 131–140.

Olson, K.R., H.L. Bergman, and P.O. Fromm. 1973. Uptake of methyl mercuric chloride and mercuric chloride by trout: a study of uptake pathways into the whole animal and uptake by erythrocytes *in vitro*, *J. Fish. Res. Bd. Canada*, 30, 1293–1299.

Olson, K.R. and P.O. Fromm. 1973. Mercury uptake and ion distribution in gills of rainbow trout (*Salmo gairdneri*): tissue scans with an electron microprobe, *J. Fish. Res. Bd. Canada*, 30, 1575–1578.

Olson, K.R. and R.C. Harrel. 1973. Effect of salinity on acute toxicity of mercury, copper, and chromium for *Rangia cuneata* (Pelecypoda, Mactridae), *Contrib. Mar. Sci.*, 17, 9–13.

Osowski, S.L., L.W. Brewer, O.E. Baker, and G.P. Cobb. 1995. The decline of mink in Georgia, North Carolina, and South Carolina: the role of contaminants, *Arch. Environ. Contam. Toxicol.*, 29, 418–423.

Pan, J.F. and W.X. Wang. 2004. Uptake of Hg(II) and methylmercury by the green mussel *Perna viridis* under different organic carbon conditions, *Mar. Ecol. Prog. Ser.*, 276, 125–136.

Parrish, K.M. and R.A. Carr. 1976. Transport of mercury through a laboratory two-level marine food chain, *Mar. Pollut. Bull.*, 7, 90–91.

Pass, D.A., P.B. Little, and L.A. Karstad. 1975. The pathology of subacute and chronic methyl mercury poisoning of the mallard duck (*Anas platyrhynchos*), *J. Comp. Pathol.*, 85, 7–21.

Pelletier, E. and C. Audet. 1995. Tissue distribution and histopathological effects of dietary methylmercury in benthic grubby *Myoxocephalus aenaeus*, *Bull. Environ. Contam. Toxicol.*, 54, 724–730.

Pentreath, R.J. 1976a. The accumulation of mercury from food by the plaice, *Pleuronectes platessa*, *J. Exp. Mar. Biol. Ecol.*, 24, 121–132.

Pentreath, R.J. 1976b. The accumulation of mercury by the thornback ray, *Raja clavata*, *J. Exp. Mar. Biol. Ecol.*, 25, 131–140

Perry, D.M., J.S. Weis, and P. Weis. 1988. Cytogenetic effects of methylmercury in embryos of the killifish, *Fundulus heteroclitus*, *Arch. Environ. Contam. Toxicol.*, 17, 569–574.

Petruccioli, L. and P. Turillazzi. 1991. Effect of methylmercury on acetylcholinesterase and serum cholinesterase activity in monkeys, *Macaca fascicularis*, *Bull. Environ. Contam. Toxicol.*, 46, 769–773.

Ponce, R.A. and N.S. Bloom. 1991. Effect of pH on the bioaccumulation of low level, dissolved methylmercury by rainbow trout (*Oncorhynchus mykiss*), *Water Air Soil Pollut.*, 56, 631–640.

Punzo, F. 1993. Ovarian effects of a sublethal concentration of mercuric chloride in the river frog, *Rana heckscheri* (Anura:Ranidae), *Bull. Environ. Contam. Toxicol.*, 50, 385–391.

Radoux, D. and J.M. Bouquegneau. 1979. Uptake of mercuric chloride from sea water by *Serranus cabrilla*, *Bull. Environ. Contam. Toxicol.*, 22, 771–778.

Ramamoorthy, S. and K. Blumhagen. 1984. Uptake of Zn, Cd, and Hg by fish in the presence of competing compartments, *Can. J. Fish. Aquat. Sci.*, 41, 750–756.

Ramprashad, F. and K. Ronald. 1977. A surface preparation study on the effect of methylmercury on the sensory hair cell population in the cochlea of the harp seal (*Pagophilus groenlandicus* Erxleben, 1977), *Can. J. Zool.*, 55, 223–230.

Rao, R.V.V.P., S.A. Jordan, and M.K. Bhatnagar. 1989. Ultrastructure of kidney of ducks exposed to methylmercury, lead and cadmium in combination, *J.Environ. Pathol. Toxicol. Oncol.*, 9, 19–44.

Renfro, J.L., B. Schmidt-Neilson, D. Miller, D. Benos, and J. Allen. 1974. Methylmercury and inorganic mercury: uptake, distribution, and effect on osmoregulatory mechanisms in fishes. In F.J. Vernberg and W.B. Vernberg (Eds.), *Pollution and Physiology of Marine Organisms*, p. 101–122. Academic Press, New York.

Ribeiro, C.A., C. Rouleau, E. Pelletier, C. Audet, and H. Tjalve. 1999. Distribution kinetics of dietary methylmercury in the Arctic charr (*Salvelinus alpinus*), *Environ. Sci. Technol.*, 33, 902–907.

Ribeyre, F. and A. Boudou. 1984. Bioaccumulation et repartition tissulaire du mercure — $HgCl_2$ et CH_3HgCl_2 — chez *Salmo gairdneri* apres contamination par voie directe, *Water Air Soil Pollut.*, 23, 169–186.

Rodgers, D.W. and F.W.H. Beamish. 1982. Dynamics of dietary methylmercury in rainbow trout, *Salmo gairdneri*, *Aquat. Toxicol.*, 2, 271–290.

Roesijadi, G., S.R. Petrocelli, J.W. Anderson, B.J. Presley, and R. Sims. 1974. Survival and chloride ion regulation of the porcelain crab *Petrolisthes armatus* exposed to mercury, *Mar. Biol.*, 17, 213–217.

Ronald, K., S.V. Tessaro, J.F. Uthe, H.C. Freeman, and R. Frank. 1977. Methylmercury poisoning in the harp seal (*Pagophilus groenlandicus*), *Sci. Total Environ.*, 8, 1–11.

Rucker, R.R. and D.F. Amend. 1969. Absorption and retention of organic mercurials by rainbow trout and chinook and sockeye salmon, *Prog. Fish-Cult.*, 31, 197–201.

Saboski, E.M. 1977. Effects of mercury and tin on frustular ultrastructure of the marine diatom, *Nitzschia liebethrutti*, *Water Air Soil Pollut.*, 8, 461–466.

Sample, B.E. and G.W. Suter. 1999. Ecological risk assessment in a large river-reservoir. 4. Piscivorous wildlife, *Environ. Toxicol. Chem.*, 18, 610–620.

Samson, J.C., R. Goodridge, F. Olobatuyi, and J.S. Weis. 2001. Delayed effects of embryonic exposure of zebrafish (*Danio rerio*) to methylmercury (MeHg), *Aquat. Toxicol.*, 51, 369–376.

Sastry, K.V., D.R. Rao, and S.K. Singh. 1982. Mercury induced alterations in the intestinal absorption of nutrients in the fresh water murrel, *Channa punctatus*, *Chemosphere*, 11, 613–619.

Sato, M., H. Sugano, and Y. Takizawa. 1981. Effects of methylmercury on zinc-thionein levels of rat liver, *Arch. Toxicol.*, 47, 125–133.

Sato, M. and Y. Takizawa. 1983. The effects of CCl_4 on the accumulation of mercury in rat tissues after methylmercury injection, *Toxicol. Lett.*, 15, 245–249.

Sayler, G.S., J.D. Nelson, Jr., and R.R. Colwell. 1976. Role of bacteria in bioaccumulation of mercury in the oyster *Crassostrea virginica*, *Appl. Microbiol.*, 30, 91–96.

Scheuhammer, A.M. 1988. Chronic dietary toxicity of methylmercury in the zebra finch, *Poephila guttata*, *Bull. Environ. Contam. Toxicol.*, 40, 123–130.

Schlenk, D., Y.S. Zhang, and J. Nix. 1995. Expression of hepatic metallothionein messenger RNA in feral and caged fish species correlates with muscle mercury levels, *Ecotoxicol. Environ. Safety*, 31, 282–286.

Schmidt-Nielsen, B., J. Sheline, D.S. Miller, and M. Deldonno. 1977. Effect of methylmercury upon osmoregulation, cellular volume, and ion regulation in winter flounder, *Pseudopleuronectes americanus*. In F.J. Vernberg, A. Calabrese, F.P. Thurberg, and W.B. Vernberg (Eds.), *Physiological Responses of Marine Biota to Pollutants*, p. 105–117. Academic Press, New York.

Shah, S.L. and A. Altindag. 2004. Hematological parameters of tench (*Tinca tinca* L.) after acute and chronic exposure to lethal and sublethal mercury treatments, *Bull. Environ. Contam. Toxicol.*, 73, 911–918.

Sheffy, T.B. and J.R. St. Amant. 1982. Mercury burdens in furbearers in Wisconsin, *J. Wildl. Manage.*, 46, 1117–1120.

Sick, L.V. and H.L. Windom. 1975. Effects of environmental levels of mercury and cadmium on rates of metal uptake and growth physiology of selected genera of marine phytoplankton. In F.G. Howell, J.B. Gentry, and M.H. Smith (Eds.), *Mineral Cycling in Southeastern Ecosystems*, p. 239–249. U.S. Energy Res. Dev. Admin. Available as CONF-740153 from NTIS, U.S. Dept. Commerce, Springfield VA.

Silva, P., F.H. Epstein, and R.J. Solomon. 1992. The effect of mercury on chloride secretion in the shark (*Squalus acanthias*) rectal gland, *Comp. Biochem. Physiol.*, 103C, 569–575.

Simon, O. and A. Boudou. 2001. Simultaneous experimental study of direct and direct plus trophic contamination of the crayfish *Astacus astacus* by inorganic mercury and methylmercury, *Environ. Toxicol. Chem.*, 20, 1206–1215.

Solonen, T. and M. Lodenius. 1984. Mercury in Finnish sparrowhawks *Accipiter nisus*, *Ornis Fennica*, 61, 58–63.

Spann, J.W., R.G. Heath, J.F. Kreitzer, and L.N. Locke. 1972. Ethyl mercury p-toluene sulfonanilide: lethal and reproductive effects on pheasants, *Science*, 175, 328–331.

Spinelli, J. and C. Mahnken. 1976. Effect of diets containing dogfish (*Squalus acanthias*) meal on the mercury content and growth of pen-reared coho salmon (*Oncorhynchus kisutch*), *J. Fish. Res. Bd. Canada*, 33, 1771–1778.

Spyker, J.M. and M. Smithberg. 1972. Effects of methylmercury on prenatal development in mice, *Teratology* 5, 181–190.

Stebbing, A.R.D. 1976. The effects of low metal levels on a clonal hydroid, *J. Mar. Biol. Assoc. U.K.*, 56, 977–994.

Stickel, L.F., W.H. Stickel, M.A.R. McLane, and M. Bruns. 1977. Prolonged retention of methyl mercury by mallard drakes, *Bull. Environ. Contam. Toxicol.*, 18, 393–400.

Stokes, P.M., S.I. Dreier, N.V. Farkas, and R.A.N. McLean. 1981. Bioaccumulation of mercury by attached algae in acid stressed lakes, *Can. Tech. Rep. Fish. Aquat. Sci.*, 1151, 136–148.

Stromgren, T. 1980. The effect of lead, cadmium, and mercury on the increase in length of five intertidal fucales, *J. Exp. Mar. Biol. Ecol.*, 43, 107–119.

Stutzenberger, F.J. and E.O Bennett. 1965. Sensitivity of mixed populations of *Staphylococcus aureus* and *Escherichia coli* to mercurials, *Appl. Microbiol.*, 13, 570–574.

Suda, I., K. Eto, H. Tokunaga, R. Furusawa, K. Suetomi, and H. Takahashi. 1989. Different histochemical findings in the brain produced by mercuric chloride and methyl mercury chloride in rats, *NeuroToxicology*, 10, 113–126.

Summers, A.O. and S. Silver. 1972. Mercury resistance in a plasmid-bearing strain of *Escherichia coli*, *J. Bacteriol.*, 112, 1228–1236.

Suzuki, T. 1979. Dose-effect and dose-response relationships of mercury and its derivatives. In J.O. Nriagu (Ed.). *The Biogeochemistry of Mercury in the Environment*, p. 399–431. Elsevier/North-Holland Biomedical Press, New York.

Suzuki, T., N. Matsumoto, T. Miyama, and H. Katsunuma. 1967. Placental transfer of mercuric chloride, phenyl mercury acetate and methyl mercury acetate in mice, *Ind. Health*, 5, 149–155.

Takizawa, Y. 1979. Minamata disease and evaluation of medical risk from methylmercury, *Akita J. Med.*, 5, 183–213.

Thain, J.E. 1984. Effects of mercury on the prosobranch mollusc *Crepidula fornicata*: acute lethal toxicity and effects on growth and reproduction of chronic exposure, *Mar. Environ. Res.*, 12, 285–309.

Thomas, W.H., J.T. Hollibaugh, D.L.R. Seibert, and G.T. Wallace, Jr. 1980. Toxicity of a mixture of ten metals to phytoplankton, *Mar. Ecol. Prog. Ser.*, 2, 213–220.

Thurberg, F.P., A. Calabrese, E. Gould, R.A. Greig, M.A. Dawson, and R.K. Tucker. 1977. Response of the lobster, *Homarus americanus*, to sublethal levels of cadmium and mercury. In F.J. Vernberg, A. Calabrese, F.P. Thurberg, and W.B. Vernberg (Eds.), *Physiological Responses of Marine Biota to Pollutants*, p. 185–197. Academic Press, New York.

Tillander, M., J.K. Miettinen, and I. Koivisto. 1972. Excretion rate of methyl mercury in the seal (*Pusa hispida*). In M. Ruivo (Ed.). *Marine Pollution and Sea Life*. p. 303–305. Fishing Trading News (books), London.

Tremblay, A., M. Lucotte, and I. Rheault. 1996 Methylmercury in a benthic food web of two hydroelectric reservoirs and a natural lake of northern Quebec (Canada), *Water Air Soil Pollut.*, 91, 255–269.

Tsui, M.T.K. and W.X. Wang. 2005. Influences of maternal exposure on the tolerance and physiological performance of *Daphnia magna* under mercury stress, *Environ. Toxicol. Chem.*, 24, 1228–1234.

Tsuruga, H. 1963. Tissue distribution of mercury orally given to fish, *Bull. Japan. Soc. Sci. Fish*, 29(5), 403–406 (in Japanese, English summary).

Uchiyama, M., S. Akiba, Y. Ohmomo, G. Tanaka, T. Watabe, T. Ishihara, and Y. Takizawa. 1976. ^{203}Hg-labelled methyl mercury chloride retention in man after inhalation, *Health Physics*, 31, 335–342.

Ukeles, R. 1962. Growth of pure cultures of marine phytoplankton in the presence of toxicants, *Appl. Microbiol.*, 10, 532–537.

U.S. Environmental Protection Agency (USEPA). 1980. Ambient water quality criteria for mercury, *U.S. Environ. Protection Agen. Rep. 440/5-80-058.* Available from Natl. Tech. Infor. Serv., 5285 Port Royal Road, Springfield, VA 22161.

U.S. Environmental Protection Agency (USEPA). 1985. Ambient water quality criteria for mercury — 1984. *U.S. Environ. Protection Agen. Rep. 440/5-84-026.* 136 pp. Available from Natl. Tech. Infor. Serv., 5285 Port Royal Road, Springfield, VA 22161.

U.S. National Academy of Sciences (USNAS). 1978. *An Assessment of Mercury in the Environment.* Natl. Acad. Sci., Washington, D.C., 185 pp.

U.S. Public Health Service (USPHS). 1994. *Toxicological Profile for Mercury (Update), TP-93/10.* U.S. PHS, Agen. Toxic Substances Dis. Registry, Atlanta, GA. 366 pp.

Unlu, M.Y., M. Heyraud, and S. Keckes. 1972. Mercury as a hydrospheric pollutant. I. Accumulation and excretion of $^{203}HgCl_2$ in *Tapes decussata* L. In M. Ruivo (Ed.), *Marine Pollution and Sea Life*, p. 292–295. Fishing Trading News (books), London.

Unrine, J.M. and C.H. Jagoe. 2004. Dietary mercury exposure and bioaccumulation in southern leopard frog (*Rana sphenocephala*) larvae, *Environ. Toxicol. Chem.*, 23, 2956–2963.

Unrine, J.M., C.H. Jagoe, W.A. Hopkins, and H.A. Brant. 2004. Adverse effects of ecologically relevant dietary mercury exposure in southern leopard frog (*Rana sphenocephala*) larvae, *Environ. Toxicol. Chem.*, 23, 2964–2970.

Valenti, T.W., D.S. Cherry, R.J. Neves, and J. Schmerfeld. 2005. Acute and chronic toxicity of mercury to early life stages of the rainbow mussel, *Villosa iris* (Bivalvia: Unionidae), *Environ. Toxicol. Chem.*, 24, 1242–1246.

Van de Ven, W.S. M., J.H. Koeman, and A. Svenson. 1979. Mercury and selenium in wild and experimental seals, *Chemosphere*, 8, 539–555.

Verma, S.R., and I.P. Tonk. 1983. Effect of sublethal concentrations of mercury on the composition of liver, muscles and ovary of *Notopterus notopterus*, *Water Air Soil Pollut.*, 20, 287–292.

Vernberg, W.B. and P.J. DeCoursey. 1977. Effect of sublethal metal pollutants on the fiddler crab *Uca pugilator*, *U.S. Environ. Protection Agen.*, Rep. 600/3–77-024, 1–59.

Vernberg, W.B., P.J. DeCoursey, and J. O'Hara. 1974. Multiple environmental factor effects on physiology and behavior of the fiddler crab, *Uca pugilator*. In F.J. Vernberg and W. B. Vernberg (Eds.), *Pollution and Physiology of Marine Organisms*, p. 381–425. Academic Press, New York.

Vernberg, W.B. and J. O'Hara. 1972. Temperature-salinity stress and mercury uptake in the fiddler crab, *Uca pugilator*, *J. Fish. Res. Bd. Canada*, 29, 1491–1494.

Vernberg, W.B. and J. Vernberg. 1972. The synergistic effects metabolism of the adult fiddler crab, *Uca pugilator*, *U.S. Dept. Comm. Fish Bull.*, 70 (2), 415–420.

Vidal, D.E. and A.J. Horne. 2003. Inheritance of mercury tolerance in the aquatic oligochaete *Tubifex tubifex*, *Environ. Toxicol. Chem.*, 22, 2130–2135.

Voccia, I., K. Krzystniak, M. Dunier, D. Flipo, and M. Fournier. 1994. *In vitro* mercury-related cytotoxicity and functional impairment of the immune cells of rainbow trout (*Oncorhynchus mykiss*), *Aquat. Toxicol.*, 29, 37–48.

Vosjan, J.H. and G.J. van der Hoek. 1972. A continuous culture copper ions, *Nether. J. Sea Res.*, 5, 440–444.

Walker, J.D. and R.R. Colwell. 1974. Mercury-resistant bacteria and petroleum degradation, *Appl. Microbiol.*, 27, 285–287.

Watras, C.J., R.C. Black, S. Halvorsen, R.J.M. Hudson, K.A. Morrison, and S.P. Wente. 1998. Bioaccumulation of mercury in pelagic freshwater food webs, *Sci. Total Environ.*, 219, 183–208.

Weis, J.S. 1976. Effects of mercury, cadmium, and lead salts on regeneration and ecdysis in the fiddler crab, *Uca pugilator*, *U.S. Natl. Mar. Fish. Serv. Fish. Bull.*, 74, 464–467.

Weis, J.S. 1978. Interactions of methylmercury, cadmium and salinity on regeneration in the fiddler crabs *Uca pugilator*, *U. pugnax*, and *U. minax*, *Mar. Biol.*, 49, 119–124.

Weis, J.S. and P. Weis. 1995a. Effects of embryonic exposure to methylmercury on larval prey-capture ability in the mummichog, *Fundulus heteroclitus*, *Environ. Toxicol. Chem.*, 14, 153–156.

Weis, J.S. and P. Weis. 1995b. Swimming performance and predator avoidance by mummichog (*Fundulus heteroclitus*) larvae after embryonic or larval exposure to methylmercury, *Can. J. Fish. Aquat. Sci.*, 52, 2168–2173.

Weis, P. 1984. Metallothionein and mercury tolerance in the killifish, *Fundulus heteroclitus*, *Mar. Environ. Res.*, 14, 153–166.

Weis, P. and J.S. Weis. 1978. Methylmercury inhibition of fin regeneration in fishes and its interaction with salinity and cadmium, *Estuar. Coast. Mar. Sci.*, 6, 327–334.

Weisbart, M. 1973. The distribution and tissue retention of mercury-203 in the goldfish (*Carassius auratus*), *Can. J. Zool.*, 51, 143–150.

Wiemeyer, S.N., T.G. Lamont, C.M. Bunck, C.R. Sindelar, F.J. Gramlich, J.D. Fraser, and M.A. Byrd. 1984. Organochlorine pesticide, polychlorobiphenyl, and mercury residues in bald eagle eggs — 1969–79 — and their relationships to shell thinning and reproduction, *Arch. Environ. Contam. Toxicol.*, 13, 529–549.

Wiener, J.G., and D.J. Spry. 1996. Toxicological significance of mercury in freshwater fish. In W.N. Beyer, G.H. Heinz, and A.W. Redmon-Norwood (Eds.), *Environmental Contaminants in Wildlife: Interpreting Tissue Concentrations*, p. 297–339. CRC Press, Boca Raton, FL.

Windom, H., W. Gardner, J. Stephens, and F. Taylor. 1976. The role of methylmercury production in the transfer of mercury in a salt marsh ecosystem, *Estuar. Coastal Mar. Sci.*, 4, 579–583.

Wobeser, G.A., N.O. Nielsen, and B. Scheifer. 1976. Mercury and mink. II. Experimental methyl mercury intoxication, *Can. J. Compar. Med.*, 40, 34–45.

Wood, J.M. 1973. Metabolic cycles of toxic elements in aqueous systems. In *6th Int. Symp. Medicale Ocean.*, p. 7–16. Portoroz, Yugoslavia, Sept. 26–30, 1973.

Wood, J.M., F.S. Kennedy, and C.G. Rosen. 1968. Synthesis of methylmercury compounds by extracts of a methanogenic bacterium, *Nature*, 220, 173–174.

World Health Organization (WHO). 1976. *Mercury. Environmental Health Criteria 1.* WHO, Geneva, 131 pp.

Wren, C.D. 1986. A review of metal accumulation and toxicity in wild mammals. I. Mercury, *Environ. Res.*, 40, 210–244.

Wren, C.D., D.B. Hunter, J.F. Leatherland, and P.M. Stokes. 1987a. The effects of polychlorinated biphenyls and methylmercury, singly and in combination, on mink. I. Uptake and toxic responses, *Arch. Environ. Contam. Toxicol.*, 16, 441–447.

Wren, C.D., D.B. Hunter, J.F. Leatherland, and P.M. Stokes. 1987b. The effects of polychlorinated biphenyls and methylmercury, singly and in combination on mink. II. Reproduction and kit development. *Arch. Environ. Contam. Toxicol.*, 16, 449–454.

Wrench, J.J. 1978. Biochemical correlates of dissolved mercury uptake by the oyster *Ostrea edulis*, *Mar. Biol.*, 47, 79–86.

Wright, D.A., P.M. Welbourn, and A.V.M. Martin. 1991. Inorganic and organic mercury uptake and loss by the crayfish *Orconectes propinquus*, *Water Air Soil Pollut.*, 56, 697–707.

Wu, M.-F., J.R. Ison, J.R. Wecker, and L.W. Lapham. 1985. Cutaneous and auditory function in rats following methyl mercury poisoning, *Toxicol. Appl. Pharmacol.*, 79, 377–388.

Yamaguchi, M., A. Yasutake, M. Nagano, and Y. Yasuda. 2004. Accumulation and distribution of methylmercury in freshwater- and seawater-adapted eels, *Bull. Environ. Contam. Toxicol.*, 73, 257–263.

Yasutake, A. and K. Hirayama. 1994. Acute effects of methylmercury on hepatic and renal glutathione metabolism in mice, *Arch. Toxicol.*, 68, 512–516.

Young, L.G. and L. Nelson 1974. The effects of heavy metal ions on the motility of sea urchin spermatozoa, *Biol. Bull.*, 147, 236–246.

PART 4

Case Histories

Chapter 10

Case Histories: Mercury Poisoning in Japan and Other Locations

This chapter extensively reviews the well-documented mercury poisoning episode at Minamata, Japan, and briefly reviews other mercury intoxication incidents in Niigita Prefecture (Japan), Tokuyama Bay (Japan), Guizhou (China), the Faroe Islands, the Republic of Seychelles, Ontario (Canada), and New Zealand.

10.1 MINAMATA, JAPAN

One of the earliest and most extensively documented cases of mercury poisoning occurred in the 1950s at Minamata Bay, in southwestern Kyushu, Japan — especially among fishermen and their families (Irukayama et al., 1961, 1962a, 1962b; Fujiki, 1963, 1980; Irukayama, 1967; Tsubaki et al., 1967; Kitamura, 1968; Matida and Kumada, 1969; Matida et al., 1972; Takeuchi, 1972; Kojima and Fujita, 1973; D'Itri and D'Itri, 1977; Smith and Smith, 1975; USNAS, 1978; Takizawa, 1979a, 1979b, 1993, 1994; Tamashiro, et al., 1984, 1985; Elhassani, 1983; Nishimura and Kumagai, 1983; Doi et al., 1984; Futatsuka and Eto, 1989; Takeuchi et al., 1989; Davies, 1991; Sakamoto et al., 1991; Eto et al., 1992; Matsumura et al., 1993; Silver et al., 1994; Eisler, 2000). Deaths and congenital birth defects in humans were attributed to long-term ingestion of marine fish and shellfish highly contaminated with methylmercury compounds. An abnormal mercury content of greater than 30.0 mg/kg FW was measured in fish and shellfish from the Bay (Table 10.1). The source of the mercury was in waste discharged from an acetaldehyde plant that used inorganic mercury as a catalyst; between 1932 and 1968, Minamata Bay received at least 260 tons of mercury, and perhaps as much as 600 tons. A severe neurological disorder was recognized in late 1953 and had reached epidemic proportions by 1956. At that time, the mercury level in sediments near the plant outfall was about 2010.0 mg/kg FW; this decreased sharply with increasing distance from the plant, and sediments in the Bay contained between 0.4 and 3.4 mg Hg/kg FW. Concentrations of mercury in fish, shellfish, and other organisms consumed by the Japanese decreased with increasing distance from the point of effluence and appeared to reflect sediment mercury levels.

The first recognition of Minamata disease as a new syndrome occurred in 1956, but it was not until 1959 that mercury poisoning was proposed as the cause of the disease and when fishing within 1 km of the shore was prohibited (Silver et al., 1994). Mercury catalysts were first used beginning in 1932 in a factory producing acetaldehyde, acetic acid, and vinyl chloride; about 82 tons of mercury were discharged into Minamata Bay from this factory alone — one of several using mercury catalysts — between 1932 and 1971 through the Hyakkon Drainage Outlet. It is noteworthy that

Table 10.1 Mercury Concentrations in Selected Biological Tissues and Abiotic Materials Collected from Minamata Bay, Japan, and Environs

Sample, Year of Collection, and Other Variables	Concentration (mg/kg)	Ref.[a]
Vegetation		
Seaweeds, 1961	1.0 FW	17
Phytoplankton, 1974	Max. 0.32 DW	1
Invertebrates[b]		
1961:		
Coelenterates	9.6 DW	2
Tunicates	35.0–56.0 DW	2
Molluscs:		
Pacific scallop, *Chlamys ferrei nipponensis*; soft parts	48.0 DW	2
Pacific oyster, *Crassostrea gigas*; soft parts	10.0 DW; 5.6 FW	2, 17
Bivalve, *Pinna attenuata*; soft parts	5.2–32.0 DW	2
Bivalve *Pinctada martensi*; soft parts	11.0–25.0 DW	2
Clam, *Hormomya mutabilis:*		
Ganglion	181.0 DW	3, 10, 11
Gills	87.0 DW	3, 10, 11
Ligament	62.0 DW	3, 10, 11
Mantle	63.0 DW	3, 10, 11
Muscle	25.0 (18.0–48.0) DW	3, 10, 11
Byssus	20.0 DW	3, 10, 11
Digestive gland	73.0 DW	3. 10. 11
Genital gland	64.0 DW	3, 10, 11
Octopus, *Octopus vulgaris:*		
Abdomen	62.0 DW	2
Tentacles	39.0 DW	2
Clam, *Hormomya mutabilis*; soft parts:		
December 1959	100.0 DW	12
January 1960	85.0 DW	13
April 1960	50.0 DW	13
August 1960	31.0 DW	13
January 1961	56.0 DW	13
April 1961	30.0 DW	13
December 1961	9.0 DW	13
October 1963	12.0 DW	13
Filter-feeding molluscs; soft parts:		
1962	Max. 43.0 DW	4
1963	Max. 40.0 DW	4
1965	Max. 35.0 DW	4
1967	Max. 60.0 DW	4
1969	Max. 16.0 DW	4
1971	Max. 16.0 DW	4
1972	Max. 4.0 DW	4
Zooplankton		
1974	Max. 1.1 DW	1
Crustaceans		
Crab, *Neptunus pelagicus*; muscle; 1961	39.0 DW	3
Fish[c]		
1961:		
Japanese anchovy, *Engraulis japonicus*; whole	0.27 FW	17
Largescale blackfish, *Girella punctata:*		
Viscera	18.0–27.0 DW	2
Muscle	12.0–20.0 DW	2
Scarbreast tuskfish, *Choerodon azurio:*		
Muscle	309.1 DW	3
Liver	85.0 DW	3

Table 10.1 (continued) Mercury Concentrations in Selected Biological Tissues and Abiotic Materials Collected from Minamata Bay, Japan, and Environs

Sample, Year of Collection, and Other Variables	Concentration (mg/kg)	Ref.[a]
Heart	36.4 DW	3
Gill	9.1 DW	3
Digestive gland	1.3 DW	3
Dotted gizzard shad, *Clupanodon* (*Konosirus*) *punctatus*; muscle	1.6 FW	17
Striped mullet, *Mugil cephalus*; muscle[f]:	16.6 FW	17
Black porgy, *Sparus macrocephalus:*		
Muscle	16.5 DW	3
Liver	32.2 DW	3
Heart	18.3 DW	3
Gill	9.1 DW	3
Digestive gland	4.0 DW	3
Muscle[f]	24.1 FW	17
Viscera[f]	23.3 FW	17
Japanese Spanish mackerel, *Scomberomorus niphonicus*[f]:		
Muscle	8.7 FW	17
Liver	15.3 FW	17
Muscle:		
1960	10.0–30.0 FW	14
1961	23.0 DW	4
1963	3.5 DW	4
1965	11.5 DW	4
1968	0.3 FW	14
1969	Max. 50.0 DW	9
1966–1972	< 0.6 DW	4
1974	Max. 0.6 DW	1
Birds		
1955–1980; feather:		
Fish-eating seabirds	7.1 DW	5
Omnivorous waterfowl	5.5 DW	5
Raptors	3.6 DW	5
Omnivorous terrestrial birds	1.5 DW	5
Herbivorous waterfowl	0.9 DW	5
1965–1966, found dead; feather	4.6–13.4 FW	6
Mammals		
Cat, *Felis domesticus*, 1961; hair:		
Naturally poisoned	40.0–52.0 DW	7
Experimentally poisoned	22.0–70.0 DW	7
Humans, *Homo sapiens*:		
Dying of Minamata disease; 1957–1989; autopsy results:		
Brain	0.1–21.3 FW	9
Kidney	3.1–106.0	9
Liver	2.1–70.5	9
Hair:		
Adults; 1960 vs. 1969	41.2 FW vs. 5.5 FW	14
1960–1961; children age 1.1–6.1 years	5.2–100.0 FW	19
Minamata disease victim autopsied 26 years postexposure vs. control:		
Liver:		
Total mercury	0.54 FW vs. 0.64 FW	15, 16
Methylmercury	0.04 FW vs. 0.07 FW	15, 16

(continued)

Table 10.1 (continued) Mercury Concentrations in Selected Biological Tissues and Abiotic Materials Collected from Minamata Bay, Japan, and Environs

Sample, Year of Collection, and Other Variables	Concentration (mg/kg)	Ref.[a]
Kidney:		
Total mercury	2.0 FW vs. 1.0 FW	15, 16
Methylmercury	0.02 FW vs. 0.015 FW	15, 16
Brain:		
Total mercury	0.53 FW vs. 0.076 FW	15, 16
Methylmercury	max. 0.007 FW vs. 0.009 FW	15, 16
Dead from Minamata disease:		
Acute poisoning; total Hg vs. methylmercury:		
Liver	115.8 FW vs. 6.8 FW	18
Kidney	147.9 FW vs. 8.6 FW	18
Cerebral cortex	12.2 FW vs. 2.0 FW	18
Chronic poisoning; total Hg vs. methylmercury:		
Liver	4.3 FW vs. 0.9 FW	18
Kidney	10.0 FW vs. 0.5 FW	18
Cerebral cortex	14.3 FW vs. 0.2 FW	18
Umbilical cord[d]; from fetal Minamata disease victims:		
1955–1960	Max. 3.1 DW (as methylmercury)	19
1960–1970	Max. 1.1 DW (as total mercury)	19
1970–1980	Max. 0.045 DW (as total mercury)	19
1980–1988	Max. 0.02 FW (as total mercury)	19
	Seawater	
1961; unfiltered	0.0016–0.0036 FW	8
1974:		
Filtered	0.001 FW	1
Suspended particulates	0.000075 FW	1
	Mud	
1963	28.0–713.0 DW	4
1969	19.0–908.0 DW	4
1970	8.0–253.0 DW	4
1971	14.0–586.0 DW	4
	Sediments	
1961	2010.0 FW (vs. 0.4–3.1 DW at reference sites)	17
1969	> 900.0 FW	14
1973	Max. 262.0 DW	14, 20
1973	> 15.0–600.0 DW	1
1982	46.0 DW (vs. < 5.0 FW 1 km offshore)	14
1988[e]	3.6 DW	20

Note: Concentrations are in mg Hg/kg (ppm) fresh weight (FW) or dry weight (DW).

[a] Reference: 1, Nishimura and Kumagai, 1983; 2, Matida and Kumada, 1969; 3, Fujiki, 1963; 4, Fujiki, 1980; 5, Doi et al., 1984; 6, Kojima and Fujita, 1973; 7, Jenkins, 1980; 8, USEPA, 1980; 9, Davies, 1991; 10, Irukayama et al., 1962a; 11, Irukayama et al., 1962b; 12, Irukayama et al., 1961; 13, Iruykayama, 1967; 14, Silver et al., 1994; 15, Takeuchi et al., 1989; 16, Eto et al., 1992; 17, Kitamura, 1968; 18, Takizawa, 1975; 19, Moriyama et al., 1994; 20, Nakamura, 1994.

[b] Maximum values in tissues of uncontaminated invertebrates from reference sites, in mg total Hg/kg FW, were 0.1 for clams, and 0.3 in seacucumbers (echinoderm) (Takizawa, 1979).

[c] Maximum values in tissues of fish collected from nearby reference sites, in mg total Hg/kg FW, were 0.35 in muscle, 1.9 in liver, and 0.2 in viscera (Kitamura, 1968; Takizawa, 1979a).

[d] As a traditional Japanese custom the dried umbilical cord is preserved to commemorate childbirth, especially in rural areas (Moriyama et al., 1994)

[e] By 1990, all sediments containing more than 25.0 mg total Hg/kg DW had been removed (Nakamura, 1994).

[f] Moribund, floating near surface.

production of vinyl chloride using mercury catalysts at this factory continued until 1971 (Silver et al., 1994).

10.1.1 Minamata Disease

Minamata disease is defined as neuropathy arising from intake of fish and shellfish containing high concentrations of methylmercury (Takizawa, 1979). An outbreak depends on factors that include mercury concentrations in water, bioconcentration and biomagnification of mercuric compounds by aquatic plants and animals, and continuous daily intake of mercury-contaminated fish in large quantities (Takizawa, 1979).

Minamata disease patients have neurological symptoms that include paresthesia, visual field constriction, impaired handwriting, unsteady gait, tremors, impaired hearing and speech, mental disturbances, and excessive salivation and perspiration (Araki et al., 1994). Severe fetal Minamata disease may result in cerebral palsy, blindness, deafness, microcephaly, and loss of speech and motor coordination (WHO, 1990). Pathological lesions in Minamata disease include the visual, auditory and post- and precentral cortices of the cerebrum, and cerebellar atrophy; in mild cases, the lesions tend to localize in the area striata of the occipital lobe and post central gyrus (Takeuchi et al., 1962; Igata, 1993). These findings suggest that exposure to methylmercury leads to visual, auditory, somatosensory, and autonomic system dysfunction (Araki et al., 1994).

10.1.1.1 Human Health

The first recorded case of Minamata disease was that of a child in 1953. In 1954, a total of 12 victims were documented: 7 adults and 5 children (Takizawa, 1979). In 1954, total mercury concentrations in brain from Minamata disease victims ranged between 0.35 and 5.3 mg/kg FW (Takizawa, 1994). In 1955, the total dead was 15; and in 1956, it had risen to 50: 22 adults, 21 children, and 7 fetal cases. The death rate of victims was 36.9%, being higher in summer and lower in winter, and was correlated with fish landings (Takizawa, 1979, 1993).

The mortality rate for acutely affected patients of the disease in 1957 was 32.8%. Infants born between 1955 and 1958 and diagnosed with mercury poisoning had a mental retardation rate of 29.1%, excluding congenital cases. By the end of 1960, 111 cases of poisoning were reported. By August 1965, 41 of the 111 had died. All congenital cases showed mental disturbance, lack of coordination, speech difficulty, and impaired chewing and swallowing (Harada, 1968). Among afflicted children, all had mental disturbance, disturbed coordination, and impaired walking; the most frequent symptoms among adults were visual field constriction (100.0%), difficulty in chewing and swallowing (94.0%), speech difficulty (88.0%), and impaired coordination (85.0%). A variety of ocular symptoms occur in Minamata disease, most typically concentric constriction of the visual field and disturbances of eye movements. Based on studies with marmoset monkeys, it was concluded that impaired visual disturbances in patients with methylmercury intoxication are attributed mainly to lesions in the visceral nuclei of the forebrain (Matsumura et al., 1993).

Between 1973 and 1981, total mercury content in kidney, liver, cerebrum, and cerebellum was significantly higher in Minamata victims when compared to similar data from control populations in western Japan, but no statistically significant difference in methylmercury values were found between the two groups (Takizawa, 1994). There was some overlap in total mercury concentrations in Minamata victims and other residents in the Minamata Bay area not diagnosed with Minamata disease. By March 1978, the total number of Minamata disease patients had risen to 1303, including 155 deaths, being highest among residents nearest Minamata Bay who had consumed fish and shellfish from the Bay more frequently than nonvictims (Takizawa, 1979, 1993). By 1980, there were 378 deaths among 1422 Minamata disease patients in Kumamoto Prefecture (Tamashiro et al., 1984). Of these 378, the first death occurred in 1954, with a peak incidence in 1956. The number of deaths increased rapidly after 1972, with a second peak in 1976. The mean age at death was

67.2 years. Most deaths after 1969 were from a combination of causes, including Minamata disease, noninflammatory diseases of the central nervous system, pneumonia, cardiovascular and cerebrovascular diseases, and malignant neoplasms (Tamashiro et al., 1984).

Concentrations of mercury in blood, urine, and especially hair are generally recognized as the best indicators of methylmercury exposure (Takizawa, 1993). Concentrations of total mercury in the hair of persons with known occupational exposure to mercury and with a low consumption of fish are usually less than 5.0 mg/kg FW (Takizawa, 1993). However, persons in Sweden and Finland with high consumption of mercury-contaminated fish and without symptoms of mercury intoxication often contain hair mercury levels greater than 30.0 mg/kg, and in one case 180.0 mg/kg (Berglund et al., 1971).

At the end of 1981, mortality analysis of 439 victims among 1483 patients with Minamata disease in Kumamoto Prefecture showed that the mortality rate for all causes of death was significantly higher in both sexes when compared to the general population and that older patients had significantly lower survival. Male patients dying of Minamata disease had significantly higher frequencies of liver and kidney diseases (Tamashiro et al., 1985). Obesity and alcohol consumption significantly influenced the frequency of liver dysfunction in Minamata disease victims, being higher in obese females and alcoholic males; moreover, there was no obvious relation between methylmercury exposure and liver disease (Futatsuka et al., 1994).

Analysis of age-specific mortality rates shows a significant increase in Minamata disease patients under age 30 years, owing to high mortality from *in utero* Minamata disease or childhood Minamata disease (Kinjo, 1993). Over age 30 years, mortality tended to be slightly higher in Minamata disease patients than a reference population, although death rates were statistically the same. The death rate from renal and liver diseases, as mentioned previously, was significantly higher in Minamata disease patients under age 30 than in controls. Most of the deaths from liver dysfunction were due to cirrhosis and chronic hepatitis, and from renal diseases it was from nephritis and nephrosis. Damage to human kidney and liver function is probably associated with the tendency of these organs to accumulate mercury (Kinjo, 1993).

A 71-year-old male severely afflicted with Minamata disease in 1956 died in 1982 (Takeuchi et al., 1989). On autopsy, methylmercury concentration in the brain was within the normal range; however, the total mercury remained high in the brain and mercury was clearly demonstrated in macrophages over wide areas of the brain and in neurons of specific brain areas. The half-time persistence of methylmercury in the brain was estimated at 240 to 245 days (Takeuchi et al., 1989). Total mercury and methylmercury levels in brain cerebrum tissue of patients severely afflicted with Minamata disease who died between 19 days and 26 years are shown in Table 10.2.

By 1982, there were 1800 verified human victims of mercury poisoning in a total regional population of 200,000. Symptoms evidenced by human victims included sensory impairment, constriction of visual fields, hearing loss, ataxia, and speech disturbances. Congenital cases were accompanied by disturbance of physical and mental development; about 6.0% of babies born in Minamata had cerebral palsy (vs. 0.5% elsewhere). Some recovery was evident in 1986, as judged by the finding that mercury concentrations in erythrocytes of Minamata disease victims were not significantly different from those of nearby inhabitants (Sakamoto et al., 1991).

In 1987, afflicted humans displayed symptoms of peripheral neuropathy (70.1%), ataxia (22.9%), constriction of the visual field (17.4%), tremor of the digits (10.2%), and dysarthria (9.2%); another 12.1% had no symptoms (Davies, 1991). Nearly all patients complained of fatigue, numbness of parts of the body, and muscle cramps. In 1987, a female Minamata disease victim born in 1957 presented with seizures in 1959, and died at age 30 of cerebral palsy (Eto et al., 1992). Four of her eight siblings and her parents were diagnosed with Minamata disease. The total mercury in her mother's head hair was 101.0 mg/kg FW in 1959. The mother died of rectal cancer in 1972 at age 55 years. The victim was presumed to have been exposed to methylmercury *in utero*. On autopsy, the victim's brain showed marked cerebral atrophy and severe atrophy of nerve cells in the cerebellum; mercury granules, mostly inorganic, were present in the brain, kidney, and liver,

Table 10.2 Total Mercury and Methylmercury Concentrations in Human Brain Cerebrum of Patients Dying from Diagnosed Minamata Disease after Various Intervals vs. Controls

Clinical Course	Number of Cases	Total Mercury (mg/kg FW)	Methylmercury (mg/kg FW)
Acute			
Death within 19–100 days of consumption of methylmercury-contaminated fish and shellfish from Minamata Bay	12	15.8 (8.8–21.4)	5.0 (2.5–8.4)
Long-term			
1.3–2.6 years	4	(2.1–4.9)	(0.45–0.69)
7 years	1	4.6	0.78
14–18 years	3	(4.07–4.23)	(0.70–1.01)
26 years	1	0.53	0.007
Controls	16	0.076	0.009

Source: Modified from Takeuchi, T., K. Eto, and H. Tokunaga. 1989. Mercury level and histochemical distribution in a human brain with Minamata disease following a long-term-clinical course of 26 years, *NeuroToxicology,* 10, 651–658.

suggesting that biotransformation of methylmercury to inorganic mercury had occurred. The total mercury content in the victim's hair was 62.0 mg/kg FW in 1959 at age 2 years and 5.4 mg/kg in 1974 at age 17 years. At death, she weighed 23 kg (50.6 pounds). Total mercury concentrations (methylmercury) measured at death in this case, in mg/kg FW, were 0.3 (0.01) in cerebral cortex, 0.55 (0.007) in thalamus, 0.6 (0.025) in cerebellum, 2.8 (0.025) in kidney, and 0.72 (0.01) in liver. For comparison, healthy adults contained 0.64 mg total Hg/kg FW (0.07 mg methylmercury/kg FW) in liver, 1.0 (0.015) in kidney, and 0.078 (0.009) in cerebellum (Eto et al., 1992).

Mean mercury concentrations in organs of Minamata disease victims dying between 1973 and 1985 remained elevated over those of residents not afflicted with Minamata disease and dying between 1973 and 1991 (Table 10.3). Minamata disease victims always had mean total mercury (methylmercury) concentrations, in mg/kg FW, of > 1.51 (0.05) in kidney, > 0.48 (0.035) in liver, > 0.10 (0.016) in brain cerebrum, and 0.05 (0.026) in brain cerebellum (Takizawa, 1994; Table 10.3).

As of 1988, 12,336 persons had applied for compensation as being victims of Minamata disease. Of these, only 1750 cases were approved, 6653 cases were rejected, and 3993 cases were still under investigation (Silver et al., 1994). By 1989, over 20,000 people were thought to have been affected; symptoms were mainly neurological and resulted in death, chronic disability, and congenital abnormalities (Davies, 1991). By June 1989, 1757 patients were officially diagnosed with Minamata disease, of which 765 had died and their families awarded compensation. Another 7621 people were disapproved for compensation. Another group of 918 (94 dead) were under investigation. A group of 2347 (320 dead) were awaiting official examination, and a final group of 1876 patients received health costs compensation only (Davies, 1991). It is alleged that a large proportion of the population residing near Minamata Bay, especially the older population, is still incapacitated to varying degrees by the disease and that the more chronic effects of the disease are still becoming apparent (Davies, 1991).

By 1989, about 2000 individuals in the Minamata area were officially certified to have Minamata disease and eligible for financial compensation (Futatsuka and Eto, 1989). Histopathological changes in the brain were clearly linked to organomercury insult in Minamata disease victims, and the distribution of lesions in the nervous system was characteristic, especially in the cerebral cortices and the cerebellum (Eto, 1995). Pathological studies of 112 Minamata disease victims also showed elevated frequencies of sepsis and malignant neoplasms of the thyroid gland when compared to 112 sex-age matched pair control deaths between 1970 and 1983 from various causes (senile

Table 10.3 Mercury and Methylmercury Concentrations in Organs of Minamata Disease Victims Dying between 1973 and 1985 and in Residents from Minamata Bay Not Afflicted with Minamata Disease and Dying between 1973 and 1991

Year of Death and Tissue	Concentration (mg/kg FW)
Total Mercury vs. Methylmercury	
1973:	
Kidney	1.85 vs. 0.065
Liver	0.67 vs. 0.122
Cerebellum	0.10 vs. 0.056
Cerebrum	0.14 vs. 0.062
1975:	
Kidney	2.43 vs. 0.109
Liver	0.48 vs. 0.067
Cerebellum	0.05 vs. 0.026
Cerebrum	0.08 vs. 0.036
1977:	
Kidney	1.51 vs. 0.041
Liver	0.57 vs. 0.056
Cerebellum	0.14 vs. 0.034
Cerebrum	0.12 vs. 0.037
1980:	
Kidney	3.89 vs. 0.02
Liver	0.66 vs. 0.061
Cerebellum	0.17 vs. 0.037
Cerebrum	0.15 vs. 0.02
1983:	
Kidney	2.54 vs. 0.17
Liver	0.52 vs. 0.035
Cerebellum	0.27 vs. 0.022
Cerebrum	0.23 vs. 0.016
1985:	
Kidney	1.6 vs. 0.15
Liver	0.65 vs. 0.133
Cerebellum	0.11 vs. 0.063
Cerebrum	0.10 vs. 0.060
Residents from Minamata Bay Not Afflicted with Minamata Disease; Autopsied 1973–1991	
Kidney	1.35 (0.24–2.74)
Liver	0.47 (0.24–0.86)
Cerebrum	0.09 (0.06–0.43)
Cerebellum	0.09 (0.05–0.15)

Source: Modified from Takizawa, Y. 1994. Mercury levels in several organs of residents exposed to methylmercury from Minamata Bay in the last 20 years. In *Environmental and Occupational Chemical Hazards (2)*, p. 39–45. ICMR Kobe University School of Medicine, COFM National University of Singapore.

dementia of Alzheimer's type, Parkinson's disease of idiopathic type, amyotrophic lateral sclerosis) in western parts of Japan (Futatsuka and Eto, 1989).

By 1991, after extensive reexamination of Minamata disease patients in Kumamoto and Kagoshima Prefectures, only 1385 were officially certified as victims and received compensation; however, an additional 6000 applications were still pending at the time (Takizawa, 1993).

In 1993, more than 2000 patients have been officially designated with Minamata disease, and 59 had congenital Minamata disease in Kumamoto Prefecture (Moriyama et al., 1994). Pathological findings of fetal Minamata disease indicated that the central nervous system was affected by methylmercury during gestation. Cases diagnosed before 1970 were similar to adult cases except for the higher frequency of speech disturbances, primitive reflex, salivation, and cerebellar abnormalities. Cases designated after 1970 had no specific clinical symptoms; however, symptoms were especially severe for cerebral palsy and mental deficiency, and the death rate was higher in this group than in reference populations (Moriyama et al., 1994). In cases of congenital Minamata disease, the incidence rate of cerebral palsy was comparatively elevated in fishing villages around Minamata Bay: 1.0 to 2.0% vs. 0.06 to 0.6% in the general Japanese population. And hair mercury levels were elevated in children, with a maximum recorded of 100.0 mg total mercury/kg FW. Concentrations of methylmercury in dried umbilical cords of infants born with congenital Minamata disease between 1955 and 1960 were significantly higher than in the general Japanese population: 3.1 mg methylmercury/kg DW vs. less than 0.02 mg total mercury/kg DW. Total mercury concentrations in umbilical cords decreased from the maxima of 1.1 mg/kg DW in the 1960 to 1970 decade, to 0.045 in the 1970 to 1980 decade, to less than 0.02 from 1981 to 1988, at which point these levels were the same or lower than that of the general Japanese population (Moriyama et al., 1994).

10.1.1.2 Natural Resources

Minamata disease resulted from the discharge of methylmercury from chemical factories into Minamata Bay. It is emphasized that Minamata disease is from direct methylmercury contamination rather than methylation of environmental sources of inorganic mercury (Menzer and Nelson, 1986). Once diluted and diffused in the water, it was concentrated to a high level in fish and filter-feeding shellfish by several routes, including bioconcentration and food chain biomagnification (Doi et al., 1984). When these fish and shellfish were consumed by humans, methylmercury gradually accumulated to exceed a threshold value, causing intoxication. Spontaneously poisoned cats, dogs, rats, waterfowl, and pigs behaved erratically and died; flying crows and grebes suddenly fell into the sea and drowned; and large numbers of dead fish were seen floating on the sea surface (Doi et al., 1984). In laboratory studies, cats and rats fed shellfish from the Bay developed the same signs as those seen in animals affected spontaneously. Abnormal mercury content — that is, more than 30.0 mg/kg fresh weight — was measured in fish, shellfish, and muds from the Bay, and in organs of necropsied humans and cats that had succumbed to the disease. Total mercury concentrations in tissues of fish from Minamata Bay with signs of methylmercury poisoning were about 15.0 mg/kg FW in liver and 8.0 to 24.0 mg/kg FW in muscle (Wiener and Spry, 1996). Mercury contamination of fish and sediments was still evident in 1981, although discharges from the acetaldehyde plant ceased in 1971 (Doi et al., 1984). Plans are underway to reopen Minamata Bay for fishing in the future; however, certain species of fish are still likely to contain unsafe concentrations of mercury for human consumption for some years (Davies, 1991).

Most mercury found in fish occurred as methylmercury, even when — as was the case at the Minamata Bay site — the initial release of mercury is inorganic. Inorganic mercury at Minamata was methylated chemically with methylcobalamin as the carbon methyl group donor (Baldi et al., 1993; Choi and Bartha, 1993). The methylcobalamin is synthesized by a range of bacterial species, especially *Desulfovibrio desulfuricans*, which is considered a major methylating source in anaerobic sediments (Choi and Bartha, 1993). At low (0.1 mg Hg^{2+}/kg) sediment mercury concentrations, up to 37.0% of the added mercury was methylated during fermentative growth of *D. desulfuricans*. But under conditions of sulfate reduction, mercury methylation was less efficient. The fermentative mercury methylating cultures were comparatively sensitive to Hg^{2+}, whereas the sulfate reducing cultures that did not methylate at a high rate were more resistant to Hg^{2+} (Choi and Bartha, 1993). Some strains of *D. desulfuricans* — in addition to producing methylmercury — can also convert methylmercury to methane and Hg^o, as was true for aerobic species of bacteria isolated from Minamata Bay sediments (Baldi et al., 1993).

There is a strong relation between the food of birds from Minamata and the mercury content in feathers; the content is highest in fish-eating seabirds and lowest in herbivorous waterfowl (Doi et al., 1984). This same relation held in birds collected from China and Korea, although concentrations were significantly lower in those nations (Doi et al., 1984). There are close correlations between mercury contents of zooplankton and suspended particulate matter, and of sediments and fish muscle, suggesting a pathway from sediment to fish by way of suspended matter and zooplankton. The conversion from inorganic mercury to methylmercury is believed to have occurred primarily in zooplankton (Nishimura and Kumagai, 1983).

10.1.2 Mitigation

In aquatic environments where point sources of industrial contamination have been identified, the elimination of mercury discharges has usually improved environmental quality. Such improvement has been reported for Minamata Bay; for sediments in Saguenay Fjord, Quebec, when chloralkali wastes were limited; for fish residues in Lake St. Clair, Canada, after two chloralkali plants were closed; and in various sections of Europe and North America when industrial discharges were eliminated (Barber et al., 1984).

In Minamata Bay, 72 strains of mercury-resistant *Pseudomonas* spp. bacteria were isolated from sediments near the drainage outlet to the Bay (Nakamura et al., 1986). *Pseudomonas* spp. dominated the bacteria with the highest resistance to mercury, although *Bacillus* spp. strains were the most numerous among all bacterial species isolated from Minamata Bay sediments. Total bacterial concentrations in Minamata Bay were the same as a nearby reference site, but mercury-resistant bacteria were found in higher numbers in Minamata Bay. In 1984, additional bacteria were isolated. *Bacillus* strains dominated in sediments containing up to 23.0 mg total Hg/kg and *Pseudomonas* strains in sediments with higher (> 52.0 mg total Hg/kg) mercury concentrations. The mercury-resistant *Pseudomonas* strains, when compared to *Bacillus* strains, were more resistant to inorganic mercury, methylmercury, and phenylmercury (Nakamura et al., 1986).

The gradual decrease in mercury content of Minamata Bay sediments from 1959 (when organomercury was first suggested as the cause of Minamata disease) and the late 1980s (when dredging removed most remaining mercury) suggests that natural processes — mainly microbial — had removed 75.0 to 90.0% of the mercury in Minamata Bay sediments (Tsubaki and Irukayama, 1977; Silver et al., 1994). However, the mechanisms of action were not known with certainty at that time (Nakamura et al., 1990). Multiple organomercurial-volatilizing bacteria that can volatilize Hg^o from methylmercury chloride, ethylmercury chloride, phenylmercury acetate, fluorescein mercuric acetate, and *p*-chloromercuric benzoate were found in the sediments of Minamata Bay (Nakamura, 1994; Nakamura and Silver, 1994). The bacteria detoxify mercury compounds by two separate enzymes — organomercurial lyase and mercuric reductase — acting sequentially. Organomercurial lyase cleaves the C–Hg bond of certain mercurials, and mercuric reductase then reduces Hg^{2+} to volatile Hg^o; bacteria remove mercury from the environment as mercury vapor (Nakamura, 1994).

In 1988, a total of 4604 bacterial strains were isolated from Minamata Bay (1428 strains) and from noncontaminated environs (3176 strains) and screened for their ability to volatilize mercuric chloride (Nakamura, 1994). Up to 38.0% of all bacterial strains isolated from Minamata Bay could grow on agar media containing 40.0 mg mercuric chloride/L vs. only 0.8% of bacteria from reference locations. A total of 67 *Pseudomonas* strains and 91 *Bacillus* strains from Minamata Bay were significantly more resistant to mercuric chloride than were 64 strains of *Pseudomonas* and 100 strains of *Bacillus* from reference locations. *Pseudomonas* spp. were more resistant to mercuric chloride than were *Bacillus* spp., but both detoxify the mercury compounds in Minamata Bay by organomercurial lyase and mercuric reductase enzymes. The organomercurials-volatilizing bacterial strains were found only in sediments of Minamata Bay. A total of 78 strains of *Bacillus* that can degrade both inorganic and organic mercurials were collected from Minamata Bay and were identified as *Bacillus subtilis*, *B. firmus*, *B. lentus*, *B. badius*, and two unidentified strains. The ability to detoxify mercury compounds is chromosomally encoded and similar to that of *Bacillus* spp. isolated from Boston Harbor, Massachusetts, USA (Nakamura, 1994).

Bacterial detoxification and volatilization of mercury may accelerate natural processes in contaminated environments (Barkay et al., 1991; Horn et al., 1994); however, social concerns were associated with bioremediation of mercury from sewage, with subsequent release of mercury into the atmosphere. Mercury releases from these limited sources do not appear to be a public health problem, although major concerns center on systems that retain the detoxified mercury (Horn et al., 1994). Retention of the reduced elemental mercury in bioreactors may allay these concerns (Frischmuth et al., 1991; Brunke et al., 1993). Genetically engineered bacteria that degrade organomercurials with subsequent retention on alginate beads is also suggested (Kono et al., 1985).

Enzymatic detoxification was determined as the major resistance mechanism in all species of mercury-resistant bacteria (Furukawa and Tonomura, 1971; Komura and Izaki, 1971; Summers and Silver, 1972). For example, mercuric reductase was essential for volatilization of Hg^o from Hg^{2+}, and various organomercurial hydrolases were responsible for volatilization of methane (CH_4) from methylmercury, for ethane (C_2H_4) from ethylmercury, and for benzene from phenylmercury (Silver et al., 1994). Minamata Bay bacterial isolates can also volatilize Hg^o from added inorganic and organic mercurials (Nakamura et al., 1988; Nakamura, 1989). Genes that govern the chemistry of mercury detoxification were abundant in bacteria found in Minamata Bay and other mercury-polluted sites; these genetic strains of mercury-resistant bacteria show promise for bioremediation of mercury pollution (Misra, 1992; Silver and Walderhaug, 1992; Nakamura, 1994; Silver et al., 1994).

In 1984, the clearance rate for mercury in Minamata Bay sediments was estimated at 18.2 years, with 90.0% clearance via natural processes estimated by the year 2000 (Silver et al., 1994). This natural clean-up rate was judged unacceptably low, and in 1984 dredging was initiated to remove all sediments containing more than 25.0 mg total mercury/kg. By 1987, 1.5 million m^3 of contaminated sediments had been removed from 2.09 km^2 of Bay areas and used as landfill at an isolated 58-ha site. This site, with an estimated 7.5 tons of mercury, is now the site of Minamata Disease Park, replete with playing fields and a museum. The landfill was capped with a layer of vinyl plastic sheet, then by volcanic ash, and topped with soil. The mercury at the site will be exposed to physical and microbial activities and subsequently volatilized to the atmosphere or leached into Minamata Bay (Silver et al., 1994). Near this site, a hospital and rehabilitation area was established in 1965 for victims of Minamata disease, and in 1978 the National Institute for Minamata Disease was established to research this disease (Silver et al., 1994).

By 1989, it was shown that mercury-volatilizing bacterial strains comprised 5.3% of all bacterial strains isolated from Minamata Bay, or 3 times more abundant than control isolates. Moreover, the number of bacterial isolates from Minamata Bay able to volatilize Hg^o from phenylmercury was 20 times greater than reference isolates (Nakamura, 1989). The development of mutant bacterial strains with the ability to detoxify inorganic and organic mercurials is continuing (Horn et al., 1994).

Mercury bioremediation research is now mainly limited to aqueous sources because mercury bound to soils and sediments is largely unavailable to bacterial cells over short exposures of hours or days (Silver et al., 1994). Techniques to measure bioavailable mercury in water using mercury-resistant bacteria (Selifonova et al., 1993) show promise for use in the decontamination of soils and sediments. Also, more research is needed on isolation of anaerobic strains of bacteria from mercury-contaminated sediments and their ability to methylate and demethylate mercury under field conditions (Silver et al., 1994).

10.2 NIIGATA PREFECTURE, JAPAN

A second outbreak of mercury intoxication occurred in Niigata Prefecture, Japan, in 1964 to 1965, in the area along the Agano River, as a result of methylmercury wastes discharged from an acetaldehyde factory (Takizawa, 1993). Afflicted individuals presented with numbness, hearing difficulties, cerebellar ataxia, speech disturbance, visual field constriction, and difficulty in walking (Tsubaki et al., 1967). Mercury concentrations were greatly elevated in hair, blood, and urine of the earliest victims diagnosed in 1965 (Table 10.4). In 1965, there were 26 diagnosed cases of Minamata disease in Niigata Prefecture. This increased to 30 in 1968, to 121 in 1972, and to 552 certified cases in 1991, with another 25 cases pending certification (Takizawa, 1993). At least 690 residents along the Agano River, including those from Niigata Prefecture, have been officially designated as having Minamata disease (Kinjo et al., 1995). At least 55 deaths have been documented (Takizawa, 1979a).

Alkylated mercury compounds were detected in the hair of afflicted individuals, with a maximum concentration of 659.0 mg/kg FW vs. nondetectable levels in controls (Takizawa, 1970). Fish captured from the Agano River in 1965 of the type eaten by all victims contained up to 41.0 mg total mercury/kg FW (Table 10.4). It was concluded at that time that organomercury poisoning was caused by eating fish contaminated with methylmercury. No further cases of mercury poisoning were documented after prohibiting fishing in the Agano River (Takizawa, 1970). The source of the methylmercury was untreated mercury-containing wastes generated by an acetaldehyde factory and discharged directly into the Agano River (Takizawa et al., 1972; Futatsuka et al., 1994). Within the factory drain, sludge contained up to 461.0 mg Hg/kg FW. Aquatic mosses collected upstream from the acetaldehyde plant had no detectable concentrations of methylmercury; however, aquatic mosses collected at the waste outlet in September 1966 contained 131.0 mg total Hg/kg FW. This decreased rapidly with increasing distance from the outlet, with 0.16 mg/kg measured in mosses 300 m downstream. Water from the Agano River contained up to 1.0 mg total mercury/L and species of

Table 10.4 Total Mercury Concentrations in Tissues of First 23 Reported Cases of Minamata Disease in Niigata Prefecture, Japan, 1965, and in Fish and Shellfish from Agano River, Niigata Prefecture, 1965

Organism and Tissue	Total Mercury Concentration
Humans[a]	
Hair	56.8–570.0 mg/kg FW
Blood	64.0–908.0 μg/L
Urine	92.0–915.0 μg excreted daily
Fish and Shellfish; Edible Tissues	
Fish "Nigoi"	6.6 (0.17–24.0) mg/kg FW
Fish "Ugui"	2.6 (0.0–8.4) mg/kg FW
Fish, various	1.8 (0.0–41.0) mg/kg FW
Shellfish	2.1 (0.0–16.0) FW

[a] Data are presented for the first 23 of 26 cases diagnosed and does not include 3 deaths. All 23 had consumed river fish abundantly.

Source: Modified from Takizawa, Y. 1993. Overview on the outbreak of Minamata Disease — epidemiological aspects. In *Proceedings of the International Symposium on Epidemiological Studies on Environmental Pollution and Health Effects of Methylmercury,* p. 3–26,. October 2, 1992, Kumamoto, Japan. Published by National Institute for Minamata Disease, Kumamoto 867, Japan.

commonly eaten fish by local residents contained up to 10.0 mg total mercury/kg FW in 1966. By 1966, 47 victims were documented, including 7 deaths. Victims successfully sued the owners of the acetaldehyde plant, and were awarded damages by the court in September 1971 (Takizawa et al., 1972). A domestic cat thought to have died from eating mercury-contaminated fish was exhumed; its skin had elevated mercury concentrations (79.5 mg total Hg/kg DW vs. 2.0 to 7.1 in controls), as did humerus (3.1 mg total Hg/kg DW vs. 0.6) and femur (2.4 mg total Hg/kg DW vs. 0.4) (Takizawa et al., 1972).

Fish collected in the Agano River in 1965 contained up to 41.0 mg total Hg/kg FW muscle, with means of various species ranging between 1.8 and 6.6 mg/kg FW (Takizawa, 1979a). Clams and oysters contained a maximum of 16.0 mg total Hg/kg FW soft parts. During the period from 1968 to 1972, total mercury concentrations in fish muscle declined from 0.67 to 0.21 mg/kg FW; for methylmercury, these values were 0.51 mg/kg FW in 1968 and 0.21 mg/kg FW in 1972 (Takizawa, 1979a).

Mercury concentrations in tissues of 26 victims in 1965 ranged between 57.0 and 570.0 mg/kg DW in hair (May to July), between 53.0 and 696.0 μg excreted in urine daily (June to August), and between 64.0 and 908.0 μg/L blood (September to November; Takizawa, 1979a). The lowest level of blood mercury associated with clinical effects was 200.0 μg/L (Takizawa, 1993). The relation between hair mercury concentrations and occurrence of Minamata disease in Niigata — based on analysis of hair from 205 male and 968 female residents of mercury-contaminated areas in Niigata — showed that hair concentrations of 40.0 to 70.0 mg Hg/kg DW was the threshold dose above which mercury effects were expected (Futatsuka et al., 1994; Kinjo et al., 1995). The biological half-time persistence in individual hair samples ranged from 30 to 120 days, with a mean of 70 days (Kinjo et al., 1995).

Minamata disease patients in Niigata Prefecture frequently contained mercury concentrations in hair greater than 50.0 mg/kg FW (127 persons), greater than 100.0 mg/kg FW (36 persons), and greater than 200.0 mg/kg FW (6 persons; Takizawa, 1993). The Swedish Expert Group (1971), using data from the Niigata epidemic, stated that symptoms onset may occur at hair mercury levels as low as 50.0 mg/kg FW; for blood, this was 200.0 μg/L (Skerfving, 1972). Blood and hair concentrations correspond to an average daily intake of 0.3 mg mercury, as methylmercury, for a typical male (OECD, 1974; WHO, 1976). However, Marsh et al. (1975), on reexamination of Minamata Bay data, concluded that onset of Minamata disease symptoms occurred at higher levels; specifically, blood methylmercury concentrations greater than 340.0 μg/L and hair concentrations greater than 200.0 mg/kg FW. The communication of Marsh et al. (1975) was never published in the peer-reviewed technical literature; accordingly, their conclusions need verification.

10.3 TOKUYAMA BAY, JAPAN

Mercury contamination is also documented in Tokuyama Bay, Japan, which received 6.6 metric tons of mercury wastes between 1952 and 1975 in wastewater from two chloralkali plants, although sediment analysis suggests that as much as 380.0 tons of mercury were released (Nakanishi et al., 1989). Unlike Minamata Bay, however, there were no human sicknesses reported and the hair of residents contained 0.0-5.0 mg Hg/kg FW vs. 15.0 to 100.0 mg Hg/kg FW in Minamata residents. In 1970, a maximum concentration of 3.3 mg total Hg/kg FW was reported in tissues of *Squilla*, a crustacean, from Tokuyama Bay. In 1973, a health safety limit was set of 0.4 mg total Hg/kg FW in edible fish and shellfish tissues, with a maximum of 0.3 mg methylmercury/kg FW permitted; at least five species of fish had more than 0.4 mg total Hg/kg FW and fishing was prohibited. Contaminated sediments (> 15.0 mg total Hg/kg) were removed by dredging and reclamation between 1974 and 1977. In 1979, the mercury content of all fish species except for one species, was less than 0.4 mg total Hg/kg FW and fishing remained prohibited. By 1983, all fish and shellfish contained less than 0.4 mg Hg/kg FW and fishing was allowed (Nakanishi et al., 1989).

10.4 GUIZHOU, CHINA

Contamination of soils and soil leachates with mercury-containing wastes is documented for the Guizhou Organic Chemical Factory, owned by the People's Republic of China. Allegedly, this factory used outdated technology to manufacture acetic acid (Yasuda et al., 2004). In the 30-year period from 1971 to 2000, this plant produced acetic acid from acetaldehyde synthesized by the addition reaction of acetylene and water using mercury as a catalyst — a process essentially the same as that used in Japan causing Minamata disease. The estimated total loss of mercury to the environment over the working life of this plant was about 135 metric tons. The drainage ditch from the acetaldehyde plant runs through paddy fields into the Zhujia River; the river branches into an irrigation canal 3.5 km downstream. The irrigation canal provides water to 120 ha of farmland. Total mercury levels in soils at the contaminated site were elevated (15.7 mg/kg DW, range 0.06 to 321.4 mg/kg DW) when compared to a reference site (0.1 mg/kg DW, range 0.05 to 0.22 mg/kg DW). Soils took up about 24.0% of the total mercury waste discharged. For soil leachates, these values were 0.4 µg/L (max. 8.3 µg/L) at the contaminated site and 0.1 µg/L (max. 0.5 µg/L) at the reference site (Yasuda et al., 2004). It is presumed that this site will be investigated further for evidence of mercury damage to biota and human health.

10.5 FAROE ISLANDS

The Faroe Islands are located approximately midway between Scotland and Iceland in the northeast Atlantic Ocean (Weihe and Grandjean, 1994). The population of about 45,000 lives on 17 small islands and are descendants of Norwegian settlers who came to the islands about 1000 years ago. A typical Faroese adult consumes each day 72.0 g of fish, 12.0 g of whale muscle, and 7.0 g of whale blubber; fish comprised 44.0% of dinner meals and whale 9.5%. Most of the fish consumed in the Faroe Islands is Atlantic cod, *Gadus morhua*; cod muscle contains, on average, 0.07 mg total Hg/kg FW muscle, almost all of which is methylated. Muscle tissue of the long-finned pilot whale, *Globicephala melas* — the whale consumed by the Faroese — contained 3.3 mg total Hg/kg FW, about half of which is methylmercury; higher concentrations occurred in whale liver, mainly as inorganic mercury. In both muscle and liver, selenium content varied but tended to approach equimolar concentrations of total mercury (Weihe and Grandjean, 1994). Based on data from a questionnaire study and the mercury concentration in whale and cod, it was calculated that average total mercury intake daily per person over age 14 years was about 36.0 µg, equivalent to 0.51 µg/kg body weight (BW) for a 70-kg person. If 67.0% of the mercury in fish and whale is methylmercury, an average blood level of 25.0 µg/L would be expected, and this is in agreement with the actual findings in the Faroe Islands (Weihe and Grandjean, 1994; Table 10.5). Total mercury concentrations in blood from umbilical cord (µg/L FW) and in maternal hair were directly related to maternal consumption of pilot whale during pregnancy; it was not related to fish consumption (Weihe and Grandjean, 1994).

Prenatal exposure to methylmercury varies considerably in the Faroe Islands. Exposure levels in 15.0 to 25.0% of all births were elevated above recommended limits for mercury, although no adverse pregnancy effects were related to mercury exposure levels (Weihe and Grandjean, 1994). To assess developmental neurobehavioral toxicity associated with intrauterine exposure to mercury, follow-up of children should be maintained until the children reach school age (Weihe and Grandjean, 1994). Accordingly, in the period April to June 1993, 423 Faroese children born between March 1986 and February 1987 were examined and tested by pediatricians, psychologists, ophthalmologists, physiologists, and hearing specialists for evidence of mercury intoxication (Araki et al., 1994; Grandjean and Weihe, 1994; Weihe and Grandjean, 1994; White et al., 1994). Results of these tests and examinations showed the following: marked differences between sexes and some evidence of prior exposure to low levels of methylmercury based on electroneurophysiological tests

Table 10.5 Mercury Concentrations in Whole Blood and Serum of 53 Adult Women from the Faroe Islands

Tissue	Concentration (µg/L FW)
Whole blood:	
Total mercury	12.1 (2.6–50.1)
Inorganic mercury	2.1 (< 0.1–5.0)
Serum:	
Total mercury	2.1 (< 0.1–8.3)
Inorganic mercury	1.7 (< 0.1–3.0)

Source: Modified from Weihe, P. and P. Grandjean. 1994. Sources and magnitude of mercury exposure in the Faroe Islands; overall design of the cohort study. In *Proceedings of the International Symposium on "Assessment of Environmental Pollution and Health Effects from Methylmercury,"* p. 112–126. October 8–9, 1993, Kumamoto, Japan. National Institute for Minamata Disease, Minamata City, Kumamoto 867, Japan.

(Araki et al., 1994); children below age 6 years had slower performance and therefore some neuropsychological tests could not be completed (White et al., 1994).

Significant dose-related adverse associations between prenatal mercury exposure and performance were demonstrated in various memory, attention, language, and spatial perception tests (Grandjean et al., 1997, 1999, 2004). Effects on delayed brainstem auditory responses occurred at concentrations greater than 5.8 µg methylmercury/L in umbilical cord blood, a concentration at which developmental effects became apparent (Murata et al., 2004); however, fetal blood mercury concentrations were only 30.0% higher than maternal blood mercury concentrations (Budtz-Jorgensen et al., 2004).

Neurobehavioral effects of intrauterine methylmercury exposure in the Faroe Islands are modified by alcohol consumption, mercury concentrations in umbilical cord blood, and diet (Grandjean and Weihe, 1994). Alcohol consumption and number of dental amalgams both affect blood mercury levels, but these did not appear to affect Faroese women as they were mainly alcohol abstainers (Weihe and Grandjean, 1994). Nevertheless, the toxicokinetic interaction between ethanol and inorganic mercury is well known (Dunn et al., 1981) and should be evaluated in future investigations. Maternal alcohol consumption caused a decrease in mercury concentrations in umbilical cord blood, probably owing to interaction between ethanol and mercury. Any alcohol-related effect on neurobehavioral development would then also be associated with lower levels of mercury exposure.

High mercury concentrations in cord blood was associated with increased birth weight (Grandjean and Weihe, 1994; Table 10.6). A low birth weight of less than 2.5 kg is often associated with delayed or deficient neurobehavioral development; however, the higher birth weights among Faroese mothers may be related to prolonged ingestion of mercury-contaminated fish. High birth weight is presumably caused by n-3 polyunsaturated fatty acids in marine fish, which may lead to a prolongation of the gestation period. Thus, children with high mercury exposures were somewhat protected against low birth weight and its associated neurobehavioral risks (Grandjean and Weihe, 1994). Continued postnatal intake of these fatty acids and other components of human milk may confer a developmental advantage. But mercury is also transferred with the milk and results in a higher hair mercury concentration in 1-year-old children who were nursed for a long period vs. those who were nursed only briefly. Authors aver that toxic effects of postnatal mercury exposure could be counterbalanced by the increased intake of n-3 fatty acids (Grandjean and Weihe, 1994).

Table 10.6 Relation between Number of Fish Dinners per Week, Birth Weight, and Mercury Concentrations in Umbilical Cord during Pregnancy for Nonsmoking Faroese Mothers

Number of Fish Meals per Week	Mean Birth Weight (kg [25th and 75th percentiles])	Mercury in Cord Blood (µg/L)
0	3.4 (3.25–3.95)	4 (1–20)
1	3.6 (3.4–4.0)	21 (8–32)
2	3.85 (3.5–4.1)	21 (13–35)
3	3.8 (3.5–4.15)	27 (13–43)
4 or more	3.75 (3.45–4.1)	27 (18–37)

Source: Modified from Olsen et al., 1986, 1992; Grandjean and Weihe, 1994.

In addition to methylmercury, seafood also contains various polychlorinated biphenyls (PCBs), which may have neurotoxic effects (Eisler, 2000). If PCB exposure is considerable, either prenatally or through human milk, then an association between methylmercury and PCBs could result in a serious confounding bias (Grandjean and Weihe, 1994). All of these factors must be considered in the design, analysis, and interpretation of epidemiological data from intrauterine methylmercury exposure studies (Grandjean and Weihe, 1994).

10.6 REPUBLIC OF SEYCHELLES

The Republic of Seychelles is located in the center of the Indian Ocean. In 1989, a joint project was initiated between the Seychelles government and the University of Rochester located in New York state (U.S.) (Davidson et al., 1994). A study cohort of 779 mother–infant pairs were examined to provide data on normal growth and development of Seychelles children and to relate neurodevelopmental outcomes in children based on fetal exposure to methylmercury (Davidson et al., 1994). Accurate measurement of neurodevelopmental outcomes in this study were complicated by several factors. These factors included the subtle effects of low doses of methylmercury, which were not detectable using standard neurodevelopmental tests. Moreover, the specific developmental processes that may be affected by low doses of methylmercury were not quantifiable because standard measures were unable to measure specific processes. Other problems, which were resolved, included cultural and language differences between test formulators and the Seychelles test subjects, and a lack of trained personnel to administer the tests. By age 29 months, each child had been tested three times. Preliminary results suggest that the mental developmental index was normal but the psychomotor developmental index was 20.0% faster than U.S. norms. Accelerated psychomotor infant development was also found in previous studies of other African cultures and is believed to reflect child rearing in naturally enriching environments (Davidson et al., 1994).

Elevated maternal hair mercury concentration was associated with only one of the 48 neurodevelopmental endpoints examined, namely, prolonged time to complete a pegboard test with the nonpreferred hand (Myers et al., 2003). However, this was one of the few neurobehavioral tests not subject to translation difficulties that can degrade the validity of culture-bound tests of higher cognitive function (Landrigan and Goldman, 2003).

10.7 NEW ZEALAND

The mercury poisoning outbreak in Minamata, Japan, showed that fetuses exposed to methylmercury *in utero* could produce congenital cerebral palsy in children at a rate 50 times higher than that of the general Japanese population. In New Zealand, some decades later, the finding of high methylmercury intake from consumption of fish and sharks led to a study to test the hypothesis that comparatively low methylmercury exposure levels might cause early changes to brain function in children exposed *in utero* (Kjellstrom, 1993).

In the New Zealand study, the target group was 11,000 women giving birth to children in the northern part of North Island in 1978. Of these mothers, a questionnaire survey showed that 935 of them were frequent consumers of fish during pregnancy (more than three fish meals weekly) and that 73 of these mothers — who produced 74 children, including one set of twins — had hair mercury concentrations that ranged between 6.0 and 86.0 mg/kg. Analysis of fish consumed by these mothers showed that the main form of mercury in these fish was organomercurials, probably methylmercury (Kjellstrom, 1993).

At age 4 years, all children were tested for social interactions, general intelligence, and various psychological functions. Children from the 73 mothers with elevated hair mercury levels performed worse than matched reference children; however, differences could not be related to mercury exposure except when initial maternal hair mercury levels were greater than 15.0 mg/kg. In 1985, when the children were 6 years old and attending school, they were retested. Results showed that children from mothers with initial hair mercury concentrations greater than 6.0 mg/kg performed significantly worse on psychological tests than the reference groups; significant confounding variables included ethnic group and maternal age.

Kjellstrom (1993) recommends that groups of new mothers in potential high-risk communities (i.e., economically deprived fishing communities in developing countries) should be studied for methylmercury exposure by hair analysis and that corrective measures should be taken to reduce exposures.

10.8 ONTARIO, CANADA

The English-Wabigoon River System in northwestern Ontario, Canada, was contaminated in the 1960s by mercury-containing wastes from a chloralkali plant (Harada et al., 1976, 1977). The first indication of a problem was discovered in 1970 when fish collected from the system contained up to 27.8 mg total Hg/kg FW muscle (Fimreite and Reynolds, 1973), at which time all fishing was prohibited in the System. Fish-eating mammals, including river otters (*Lutra canadensis*) and mink (*Mustela vison*), became scarce (Fimreite and Reynolds, 1973). Turkey vultures, *Cathartes aura*, had difficulty flying and one had a liver mercury burden of 96.0 mg/kg FW (Fimreite, 1972). A feral cat (*Felis domesticus*) captured nearby showed neurological and pathological signs of Minamata disease (Harada et al., 1976); mercury concentrations were high in cat brain (16.0 mg/kg FW), liver (67.0), and fur (392.0 mg/kg FW), which fits the Minamata disease syndrome (Takeuchi et al., 1977). Cats fed local fish for 90 days developed signs of Minamata disease (Chabonneau et al., 1974). Two reservations — home for indigenous tribes — were located within the confines of the mercury-contaminated area. Hair mercury levels in these people in 1970 were as high as 96.0 mg/kg in Grassy Narrows township and 198.0 mg/kg in the nearby town of White Dog (Ontario Ministry of Health, 1974).

By the mid-1970s, hair mercury levels were in decline (Ontario Ministry of Health, 1974); however, results of other studies conducted in 1975 indicated that hair mercury levels remained elevated (20.0 to 30.0 mg/kg), with maximum values recorded of 80.3 and 105.0 mg Hg/kg in Grassy Narrow and White Dog, respectively (Clarkson, 1976; Harada et al., 1976). The Japanese investigators (Harada et al., 1976, 1977) concluded that Minamata disease was present in these populations as early as 1975, based on neurological symptoms alone. Compensation to tribal members with elevated hair mercury levels was provided through the Mercury Disability Board established in 1986 and funded by owners of the chloralkali plant and the provincial and federal governments (Harada et al., 2005).

In 2002, the population (N = 57) was reexamined for neurological symptoms and hair mercury concentrations (Harada et al., 2005). Four groups were evident: (1) a light Minamata disease group with hair mercury concentrations of 1.4 (0.1 to 5.1) mg/kg, and a low percentage of subjective and objective clinical symptoms; (2) a Minamata disease group with hair mercury concentrations of 2.2 (0.8 to 4.5) mg/kg and a higher percentage of symptoms that included sensory impairment of extremities, motor ataxia, tunnel vision, impaired hearing, tremors, impaired speech, difficulty in standing, and impaired motor coordination; (3) a Minamata disease group with complications, having hair mercury concentrations of 3.3 (0.2 to 18.1) mg/kg and the highest number of neurological symptoms per individual; and (4) others, with hair mercury of 2.7 (0.1 to 6.6) mg/kg, but no significant neurological symptoms (Harada et al., 2005). The Mercury Disability Board did not accept the finding of Minamata disease based on impairment of the sensory extremities alone, and disallowed compensation to individuals afflicted with only light symptoms and psychogenic reactions; however, compensation was provided to 21 of the 57 individuals examined in the Minamata disease group and the Minamata disease group with complications (Harada et al., 2005). Future surveys are planned.

10.9 SUMMARY

Up to 600 tons of methylmercury were discharged into Minamata Bay, Japan, between 1932 and 1971 from acetaldehyde manufacturing plants. Human fatalities were documented beginning in 1953 from consumption of methylmercury-contaminated fish and shellfish from the Bay. By 1993, about 2000 victims of Minamata disease were identified, including more than 100 deaths and 59 congenital birth defects from a total regional population of about 200,000; however, at least 10,000 additional cases are pending. Mercury levels in the Minamata Bay ecosystem are now near normal as a result of dredging and natural processes. A second outbreak of mercury intoxication occurred in Niigata Prefecture, Japan in 1964 to 1965, from methylmercury wastes discharged from an acetaldehyde factory. By 1991, at least 552 cases of Minamata disease were diagnosed in Niigata Prefecture, including 55 deaths. Mercury contamination is also documented in Tokuyama Bay, Japan, which received at least 6.6 tons of mercury wastes from two chloralkali plants between 1952 and 1975, contaminating resident fish and shellfish; fishing was prohibited in Tokuyama Bay in 1973. There were no human sicknesses reported in Tokuyama Bay environs; by 1983, fishing had resumed because resident fish and shellfish had less than 0.4 mg total mercury/kg FW.

In Guizhou, China, about 135 tons of mercury-containing wastes from an acetic acid manufacturer were discharged into a nearby river between 1971 and 2000, heavily contaminating farmland soils and soil leachates. Mercury damage to biota and human health is unknown at this time.

In Ontario, Canada, mercury contamination from a chloralkali plant in the 1960s contaminated fisheries and native indigenous peoples. Mercury levels have declined but neurological symptoms persist. Canadian authorities — unlike Japanese investigators — maintain that there is mercury contamination but not Minamata disease.

In the Faroe Islands, the Republic of Seychelles, and New Zealand, ongoing investigations focus on quantification of mercury in the diets of expectant mothers; neurobehavioral toxicity assessment of fetuses during intrauterine exposure to mercury; development of sophisticated psychological and other tests to periodically measure imbalances and deficiencies in children exposed to mercury *in utero*; and development of techniques to quantitate the subtle effects of mercury intoxication.

REFERENCES

Araki, S., K. Murata, K. Yokoyama, F. Okajima, P. Grandjean, and P. Weihe. 1994. Neuroelectrophysiological study of children in low-level methylmercury exposure in Faroe Islands: methodology and preliminary findings. In *Proceedings of the International Symposium on Assessment of Environmental Pollution and Health Effects from Methylmercury,* p. 141–151. October 8–9, 1993, Kumamoto, Japan. National Institute for Minamata Disease, Minamata City, Kumamoto 867, Japan.

Baldi, F., M. Pepi, and M. Filipelli. 1993. Methylmercury resistance in *Desulfovibrio desulfuricans* in relation to methylmercury degradation, *Appl. Environ. Microbiol.*, 59, 2479–2485.

Barber, R.T., P.J. Whaling, and D.M. Cohen. 1984. Mercury in recent and century-old deep-sea fish, *Environ. Sci. Technol.*, 18, 552–555.

Barkay, T., R. Turner, A. VandenBrook, and C. Liebert. 1991. The relationships of Hg (II) volatilization from a freshwater pond to the abundance of *mer* genes in the gene pool of the indigenous microbial community, *Microbial Ecol.*, 21, 151–161.

Berglund, F., M. Berlin, G. Birke, U. von Euler, L. Friberg, B. Halmsteds, E. Jonsson, S. Ramel, S. Skerfving, A. Swensson, and S. Tejning. 1971. Methylmercury in fish: a toxicological and epidemiological evaluation of risks, *Nord. Hyg. Tidskr.*, 4 (Suppl.), 226–285.

Brunke, M., W.D. Deckwer, A. Frischmuth, J.M. Horn, H. Lunsdorf, M. Rhode, M. Rohricht, K.N. Timmis, and P. Weppen. 1993. Microbial retention of mercury from waste streams in a laboratory column containing *merA* gene bacteria, *FEMS Microbiol. Rev.,* 11, 145–152.

Budtz-Jorgensen, E., P. Grandjean, P.J. Jorgensen, P. Weihe, and N. Keiding. 2004. Association between mercury concentrations in blood and hair in methylmercury-exposed subjects at different ages, *Environ. Res.*, 95, 385–393.

Chabonneau, S.M., L.C. Munro, E.A. Nera, R.F. Willes, T. Kuiper-Goodman, F. Iverson, C.A. Moodie, D.R. Stoltz, F.A.J. Armstrong, J.F. Uthe, and H.C. Grice. 1974. Subacute toxicity of methylmercury in the adult cat, *Toxicol. Appl. Pharmacol.*, 27, 569–581.

Clarkson, T.W. 1976. Exposure to methyl mercury in Grassy Narrows and White Dog Reserves, *Report to Medical School, Univ. Rochester*, Rochester, NY.

Choi, S.C. and R. Bartha. 1993. Cobalamin-mediated mercury methylation by *Desulfovibrio desulfuricans* LS, *Appl. Environ. Microbiol.*, 59, 290–295.

Davidson, P.W., G.J. Myers, C. Cox, C. Shamlaye, D.O. Marsh, T.W. Clarkson, and M.A. Tanner. 1994. Measuring neurological outcomes of young children following prenatal dietary methylmercury exposures. In *Proceedings of the International Symposium on Assessment of Environmental Pollution and Health Effects from Methylmercury,* p. 96–111. October 8–9, 1993, Kumamoto, Japan. National Institute for Minamata Disease, Minamata City, Kumamoto 867, Japan.

Davies, F.C.W. 1991. Minamata disease: a 1989 update on the mercury poisoning epidemic in Japan, *Environ. Geochem. Health*, 13, 35–38.

D'Itri, P. and F.M. D'Itri. 1977. *Mercury Contamination: A Human Tragedy.* John Wiley, New York. 311 pp.

Doi, R., H. Ohno, and M. Harada. 1984. Mercury in feathers of wild birds from the mercury-polluted area along the shore of the Shiranui Sea, *Sci. Total Environ.*, 40, 155–167.

Dunn, J.D., T.W. Clarkson, and L. Magos. 1981. Interaction of ethanol and inorganic mercury: generation of mercury vapor in vivo, *J. Pharmacol. Exp. Ther.*, 216, 19–23.

Eisler, R. 2000. Mercury. In *Handbook of Chemical Risk Assessment: Health Hazards to Humans, Plants, and Animals, Vol. 1, Metals,* p. 313–409. Lewis Publishers, Boca Raton, FL.

Elhassani, S.B. 1983. The many faces of methylmercury poisoning, *J. Toxicol.,* 19, 875–906.

Eto, K. 1995. Effect assessment methodology--pathological view points. In *Proceedings of the International Workshop on Environmental Mercury Pollution and its Health Effects in Amazon River Basin,* p. 142–153. Rio de Janeiro, 30 November–2 December, 1994. Published by National Institute for Minamata Disease, Minamata City, Kumamoto 867, Japan.

Eto, K., S. Oyanagi, Y. Itai, H. Tokunaga, Y. Takizawa, and I. Suda. 1992. A fetal type of Minamata Disease. An autopsy case report with special reference to the nervous system, *Molec. Chem. Neuropathol.*, 16, 171–186.

Fimreite, N. 1972. Mercury Contamination of Aquatic birds in Northwestern Ontario, *Report of Univ. Tromso*, Tromso, Norway.

Fimreite, N. and L.M. Reynolds. 1973. Mercury contamination in fish of northwestern Ontario, *J. Wildl. Manage.*, 37, 62–68.

Frischmuth, A., P. Weppen, and W.P. Deckwer. 1991. Quecksilberentfernung und wassrigen Medien durch aktive mikrobielle Prozess, *Bioengineering,* 3, 38–48.

Fujiki, M. 1963. Studies on the course that the causative agent of Minamata disease was formed, especially on the accumulation of the mercury compound in the fish and shellfish of Minamata Bay, *J. Kumamoto Med. Soc.*, 37, 494–521.

Fujiki, M. 1980. The pollution of Minamata Bay by mercury and Minamata disease. In R.A. Baker (Ed.), *Contaminants and Sediments,* Vol. 2, p. 493–500. Ann Arbor Science Publ., Ann Arbor, MI.

Furukawa, A. and K. Tonomura. 1971. Enzyme system involved in the decomposition of phenyl mercuric acetate by mercury-resistant *Pseudomonas*, *Agricul. Biol. Chem.*, 35, 604–610.

Futatsuka, M. and K. Eto. 1989. A case-control study of mortality in Minamata disease based on pathological findings, *Kumamoto Med. J.*, 41(3), 73–79.

Futatsuka, M., T. Kitano, W. Junji, Y. Kinjo, H. Kato, and Y. Takizawa. 1994. Recent findings in the epidemiological studies on Minamata Disease. In *Proceedings of the International Symposium on Assessment of Environmental Pollution and Health Effects from Methylmercury,* p. 1–18, October 8–9, 1993, Kumamoto, Japan. National Institute for Minamata Disease, Minamata City, Kumamoto 867, Japan.

Grandjean, P., E. Budtz-Jorgensen, R.F. White, P.J. Jorgensen, P. Weihe, and F. Debes. 1999. Methylmercury exposure biomarkers as indicators of neurotoxicity in children age 7 years, *Am. J. Epidemiol.*, 150, 301–305.

Grandjean, P., K. Murata, E. Budtz-Jorgensen, and P. Weihe. 2004. Cardiac autonomic activity in methylmercury neurotoxicity: 14-year follow-up of a Faroese birth cohort, *J. Pediatr.*, 144, 169–176.

Grandjean, P. and P. Weihe. 1994. Neurobehavioral effects of intrauterine methylmercury exposure: bias problems in epidemiological studies. In *Proceedings of the International Symposium on Assessment of Environmental Pollution and Health Effects from Methylmercury,* p. 152–162, October 8–9, 1993, Kumamoto, Japan. National Institute for Minamata Disease, Minamata City, Kumamoto 867, Japan.

Grandjean, P., P. Weihe, R.F. White, F. Debes, S. Araki, and K. Yokoyama. 1997. Cognitive deficit in 7-year-old children with prenatal exposure to methylmercury, *Neurotoxicol. Teratol.*, 19, 417–428.

Harada, M., T. Fujino, T. Akagi, and S. Nishigaki. 1976. Epidemiological and clinical study and historical background of mercury pollution on Indian reservations in northwestern Ontario, Canada, *Bull. Inst. Constit. Med. Kumamoto Univ.*, 26, 169–184.

Harada, M., T. Fujino, T. Akagi, and S. Nishigaki. 1977. Mercury contamination in human hair at Indian reservation in Canada, *Kumamoto Med. J.*, 30, 57–64.

Harada, M., T. Fujino, T. Oorui, S. Nakachi, T. Kizaki, Y. Hitomi, N. Nakano, and H. Ohno. 2005. Followup study of mercury pollution in indigenous tribe reservations in the province of Ontario, Canada, 1975–2002. *Bull. Environ. Contam. Toxicol.*, 74, 689–697.

Harada, Y. 1968. Infantile Minamata diseases. In M. Katsuna (Ed.), *Minamata Disease*, p. 73–177. Shuhan, Kumamoto.

Horn, J.M., M. Brunke, W.D. Deckwer, and K.N. Timmis. 1994. *Pseudomonas putida* strains which constitutively overexpress mercury resistance for biodetoxification of organomercurial pollutants, *Appl. Environ. Microbiol.*, 60, 357–362.

Igata, A. 1993. Epidemiological and clinical features of Minamata disease, *Environ. Res.*, 63, 157–169.

Irukayama, K. 1967. The pollution of Minamata Bay and Minamata disease. In J.P. Maroto and F. Josa (Eds.), *Advances in Water Pollution Research,* Vol. 3, p. 153–180. Proc. Third Int. Conf., Munich.

Irukayama, K., M. Fujiki, F. Kai, and T. Kondo. 1962a. Studies on the causative agent of Minamata disease. II. Comparison of the mercury compound in the shellfish from Minamata Bay with mercury compounds experimentally accumulated in the control shellfish, *Kumamoto Med. J.*, 15, 1–12.

Irukayama, K., F. Kai, M. Fujiki, and T. Kondo. 1962b. Studies on the causative agent of Minamata disease. III. Industrial wastes containing mercury compounds from Minamata factory, *Kumamoto Med. J.*, 15, 57–68.

Irukayama, K., T. Kondo, F. Kai, and M. Fujiki. 1961. Studies on the origin of the causative agent of Minamata disease. I. Organic mercury compound in the fish and shellfish from Minamata Bay, *Kumamoto Med. J.*, 14, 158–169.

Jenkins, D.W. 1980. In *Biological Monitoring of Toxic Trace Metals, Volume 2, Toxic Trace Metals in Plants and Animals of the World, Part II, Mercury*, p. 779–982. U.S. Environmental Protection Agency, Rep. 600/3–80–091.

Kinjo, Y. 1993. Recent epidemiological findings on Minamata Disease. In *Proceedings of the International Symposium on Epidemiological Studies on Environmental Pollution and Health Effects of Methylmercury*, p. 31–36. October 2, 1992, Kumamoto, Japan. Published by National Institute for Minamata Disease, Kumamoto 867, Japan.

Kinjo, Y., Y. Takizawa, Y. Shibata, M. Watanabe, and H. Kato. 1995. Threshold dose for adults exposed to methylmercury in Niigata Minamata disease outbreak, *Environ. Sci.*, 3, 91–101.

Kitamura, S. 1968. Determination of mercury content of inhabitants, cats, fishes, and shells in Minamata District and the mud of Minamata Bay. In M. Katsuna (Ed.), *Minamata Disease,* p. 257–266. Shuhan, Kumamoto.

Kjellstrom, T. 1993. Effects of methylmercury exposure *in utero*; studies in New Zealand and proposals for future studies. In *Proceedings of the International Symposium on Epidemiological Studies on Environmental Pollution and Health Effects of Methylmercury,* p. 67–78. October 2, 1992, Kumamoto, Japan. Published by National Institute for Minamata Disease, Kumamoto, 867, Japan.

Kojima, K. and M. Fujita. 1973. Summary of recent studies in Japan on methyl mercury poisoning, *Toxicology*, 1, 43–62.

Komura, I. and K. Izaki. 1971. Mechanism of mercuric chloride resistance in microorganisms. I. Vaporization of a mercury compound from mercuric chloride by multiple drug resistant strains of *Escherichia coli*, *J. Biochem. (Tokyo)*, 70, 885–893.

Kono, M., K. O'Hara, Y. Arai, H. Fukuda, M. Asakawa, and H. Nakahara. 1985. Inactivation and sedimentation of organomercurials by *Klebsiella pneumoniae*, *FEMS Microbiol. Lett.*, 28, 213–217.

Landrigan, P.J. and L. Goldman. 2003. Prenatal methylmercury exposure in the Seychelles, *Lancet*, 362, 666.

Marsh, D.O., M.D. Turner, and J.C. Smith. 1975. Loaves and fishes — some aspects of methylmercury in foodstuffs, unpublished report submitted to U.S. Food and Drug Administration.

Matida, Y. and H. Kumada. 1969. Distribution of mercury in water, bottom mud and aquatic organisms of Minamata Bay, the River Agano and other water bodies in Japan, *Bull. Freshwater Fish. Res. Lab. (Tokyo)*, 19(2), 73–93.

Matida, Y., H. Kumada, S. Kimura, Y. Saiga, T. Nose, M. Yokote, and H. Kawatsu. 1972. Toxicity of mercury compounds to aquatic organisms and accumulation of the compounds by the organisms, *Bull. Freshwat. Fish. Res. Lab.*, 21, 197–227.

Matsumura, A., K. Eto, N. Furuyoshi, and R. Okamura. 1993. Disturbances of accommodation in Minamata disease. A neuropathological study of methylmercury toxicity in common marmoset monkeys, *Neuroophthalmology*, 13, 331–339.

Menzer, R.E. and J.O. Nelson. 1986. Water and soil pollutants. In C.D. Klaassen, M.O. Amdur, and J. Doull (Eds.), *Casarett and Doull's Toxicology, third edition,* p. 825–853. Macmillan, New York.

Misra, T.K. 1992. Bacterial resistance to inorganic mercury salts and organomercurials, *Plasmid*, 27, 4–16.

Moriyama, H., M. Futatsuka, and Y. Kinjo. 1994. Fetal Minamata Disease — A Review. In *Proceedings of the International Symposium on "Assessment of Environmental Pollution and Health Effects from Methylmercury,"* p. 64–72. October 8–9, 1993, Kumamoto, Japan. National Institute for Minamata Disease, Minamata City, Kumamoto 867, Japan.

Murata, K., P. Weihe, E. Budtz-Jorgensen, P.J. Jorgensen, and P. Grandjean. 2004. Delayed brainstem auditory response evoked potential in 14-year-old children exposed to methylmercury, *J. Pediatr.*, 144, 177–183.

Myers, G.J., P.W. Davidson, C. Cox, C.F. Shamlaye, D. Palumbo, and E. Cernichiari. 2003. Prenatal methylmercury exposure from the ocean fish consumption in the Seychelles child development study, *Lancet*, 361, 1686–1692.

Nakamura, K. 1989. Volatilization of fluorescein mercuric acetate by marine bacteria from Minamata Bay, *Bull. Environ. Contam. Toxicol.*, 42, 785–790.

Nakamura, K. 1994. Mercury compounds-decomposing bacteria in Minamata Bay. In *Proceedings of the International Symposium on "Assessment of Environmental Pollution and Health Effects from Methylmercury,"* p. 198–209. October 8–9, 1993, Kumamoto, Japan. National Institute for Minamata Disease, Minamata City, Kumamoto 867, Japan.

Nakamura, K., T. Fujisaki, and Y. Shibata. 1988. Mercury-resistant bacteria in the sediment of Minamata Bay, *Nippon Suisan Gakkashi*, 54, 1359–1363.

Nakamura, K., T. Fujisaki, and H. Tamashiro, 1986. Characteristics of Hg-resistant bacteria isolated from Minamata Bay sediment, *Environ. Res.*, 40, 58–67.

Nakamura, K., M. Sakamoto, H. Uchiyama, and O. Yagi. 1990. Organomercurial-volatilizing bacteria in the mercury-polluted sediment of Minamata Bay, Japan, *Appl. Environ. Microbiol.*, 56, 304–305.

Nakamura, K. and S. Silver. 1994. Molecular analysis of mercury-resistant *Bacillus* isolates from sediment of Minamata Bay, Japan, *Appl. Environ. Microbiol.*, 60, 4596–4599.

Nakanishi, H., M. Ukita, M. Sekine, and S. Murakami. 1989. Mercury pollution in Tokuyama Bay, *Hydrobiologia*, 176/177, 197–211.

Nishimura, H. and M. Kumagai. 1983. Mercury pollution of fishes in Minamata Bay and surrounding water: analysis of pathway of mercury, *Water Air Soil Pollut.*, 20, 401–411.

OECD. 1974. In *Mercury and the Environment,* p. 129–170. Organization for Economic Cooperation and Development, Paris.

Olsen, S., J.D. Sorensen, N.J. Secher, M. Hedegaard, T.B. Henriksen, H.S. Hansen, and A. Grant. 1992. Randomised controlled trial of effect of fish-oil supplementation on pregnancy duration, *Lancet*, 339, 1003–1007.

Olsen, S.F., H.S. Hansen, T.I.A. Sorensen, B. Jensen, N.J. Secher, S. Sommer, and L.B. Knudsen. 1986. Intake of marine fat, rich in (n-3) polyunsaturated fatty acids, may increase birth weight by prolonged gestation, *Lancet*, 327, 367–369.

Ontario Ministry of Health. 1974. *Methyl Mercury in Northwestern Ontario,* Ontario Ministry of Health, Toronto, Ontario, Canada.

Sakamoto, M., A. Nakano, Y. Kinjo, H. Higashi, and M. Futasuka. 1991. Present mercury levels in red blood cells of nearby inhabitants about 30 years after the outbreak of Minamata disease, *Ecotoxicol. Environ. Safety*, 22, 58–66.

Selifonova, O., R. Burlage, and T. Barkay. 1993. Bioluminescent sensors for detection of bioavailable Hg (II) in the environment, *Appl. Environ. Microbiol.*, 59, 3083–3090.

Silver, S., G. Endo, and K. Nakamura. 1994. Mercury in the environment and the laboratory, *J. Japan. Soc. Water Environ.*, 17, 235–243.

Silver, S. and M. Walderhaug. 1992. Gene regulation of plasmid- and chromosome-determined inorganic ion transport in bacteria, *Microbiol. Rev.*, 56, 195–228.

Skerfving, S. 1972. *Toxicity of Methylmercury with Special Reference to Exposure via Fish*. Available from Karolinska Inst., Stockholm.

Smith, W.E. and A.M. Smith. 1975. *Minamata*. Holt, Rinehart, and Winston, New York. 192 pp.

Summers, A.O. and S. Silver. 1972. Mercury resistance in a plasmid-bearing strain of *Escherichia coli*, *J. Bacteriol.,* 112, 1228–1236.

Takeuchi, T. 1972. Distribution of mercury in the environment of Minamata Bay and inland Ariake Sea. In Hartung, R. and B.D. Dinman (Eds.), *Environmental Mercury Contamination,* p. 79–81. Ann Arbor Science Publ., Ann Arbor, MI.

Takeuchi, T., F.M. D'Itri, P.V. Fischer, C.S. Annett, and M. Okabe. 1977. The outbreak of Minamata Disease (methyl mercury poisoning) in cats on northwestern Ontario reserves, *Environ. Res.*, 13, 215–228.

Takeuchi, T., K. Eto, and H. Tokunaga. 1989. Mercury level and histochemical distribution in a human brain with Minamata disease following a long-term-clinical course of 26 years, *NeuroToxicology*, 10, 651–658.

Takeuchi, T., N. Morikawa, H. Matsumoto, and Y. Shiraishi. 1962. A pathological study of Minamata disease in Japan, *Acta Neuropathol.*, 2, 40–57.

Takizawa, Y. 1970. Studies on the Niigata episode of Minamata disease outbreak — investigation of causative agents of organic mercury poisoning in the district along the River Agano, *Acta Med. Biol.*, 17, 293–297.

Takizawa, Y. 1975. Studies on the distribution of mercury in several body organs — application of activography to examination of mercury distribution of Minamata disease patients. In *Studies on the Health Effects of Alkylmercury in Japan,* p. 5–9. Environment Agency Japan, Tokyo.

Takizawa, Y. 1979a. Minamata disease and evaluation of medical risk from methylmercury, *Akita J. Med.*, 5, 183–213.

Takizawa, Y. 1979b. Epidemiology of mercury poisoning. In J.O. Nriagu (Ed.), *The Biogeochemistry of Mercury in the Environment,* p. 325–365. Elsevier/North Holland Biomedical Press, Amsterdam.

Takizawa, Y. 1993. Overview on the outbreak of Minamata Disease — epidemiological aspects. Pages 3–26. In *Proceedings of the International Symposium on Epidemiological Studies on Environmental Pollution and Health Effects of Methylmercury,* p. 3–26. October 2, 1992, Kumamoto, Japan. Published by National Institute for Minamata Disease, Kumamoto 867, Japan.

Takizawa, Y. 1994. Mercury levels in several organs of residents exposed to methylmercury from Minamata Bay in the last twenty years. In *Environmental and Occupational Chemical Hazards (2)*, p. 39–45. ICMR Kobe University School of Medicine, COFM National University of Singapore.

Takizawa, Y., T. Kosaka, R. Sugai, I. Sasagawa, C. Sekiguchi, and K. Minagawa. 1972. Studies on the cause of the Niigata episode of Minamata disease outbreak, *Acta Med. Biol.*, 19, 193–206.

Tamashiro, H., H. Akagi, M. Arakaki, M. Futatsuka, and L. H. Roht. 1984. Causes of death in Minamata disease: analysis of death certificates, *Int. Arch. Occupat. Environ. Health*, 54, 135–146.

Tamashiro, H., M. Arakaki, H. Akagi, M. Futatsuka, and L.H. Roht. 1985. Mortality and survival for Minamata disease, *Int. J. Epidemiol.*, 14, 582–588.

Tsubaki, T. and K. Irukayama (Eds.). 1977. *Minamata Disease: Methylmercury Poisoning in Minamata and Niigata, Japan*. Kodansha Ltd., Tokyo.

Tsubaki, T., T. Sato, K. Kondo, K. Shirakawa, K. Kanbayashi, K. Hirota, K. Yamada, and I. Murone. 1967. Outbreak of intoxication by organic mercury compound in Niigata Prefecture. An epidemiological and clinical study, *Jpn. J. Med.*, 6(3), 132–133.

U.S. Environmental Protection Agency (USEPA). 1980. *Ambient Water Quality Criteria for Mercury,* U.S. Environ. Protection Agen. Rep. 440/5-80-058. Available from Natl. Tech. Infor. Serv., 5285 Port Royal Road, Springfield, VA 22161.

U.S. National Academy of Sciences (USNAS). 1978. *An Assessment of Mercury in the Environment*. Natl. Acad. Sci., Washington, D.C., 185 pp.

Weihe, P. and P. Grandjean. 1994. Sources and magnitude of mercury exposure in the Faroe Islands; overall design of the cohort study. In *Proceedings of the International Symposium on "Assessment of Environmental Pollution and Health Effects from Methylmercury,"* p. 112–126. October 8–9, 1993, Kumamoto, Japan. National Institute for Minamata Disease, Minamata City, Kumamoto 867, Japan.

White, R.F., F. Debes, R. Dahl, and P. Grandjean. 1994. Development and field testing of a neuropsychological test battery to assess the effects of methylmercury exposure in the Faroe Islands. In *Proceedings of the International Symposium on "Assessment of Environmental Pollution and Health Effects from Methylmercury,"* p.127–140. October 8–9, 1993, Kumamoto, Japan. National Institute for Minamata Disease, Minamata City, Kumamoto 867, Japan.

Wiener, J.G. and D.J. Spry. 1996. Toxicological significance of mercury in freshwater fish. In W.N. Beyer, G.H. Heinz, and A.W. Redmon-Norwood (Eds.), *Environmental Contaminants in Wildlife: Interpreting Tissue Concentrations*, p. 297–339. CRC Press, Boca Raton, FL.

World Health Organization (WHO). 1976. *Mercury. Environmental Health Criteria 1*. WHO, Geneva, 131 pp.

World Health Organization (WHO). 1990. *Methylmercury. Environmental Health Criteria 101*. WHO, Geneva. 144 pp.

Yasuda, Y., A. Matsuyama, A. Yasutake, M. Yamaguchi, R. Aramake, L. Xiaojie, J. Pin, A. Yumin, L. Li, L. Mei, C. Wei, and Q. Liya. 2004. Mercury distribution in farmlands downstream from an acetaldehyde producing chemical company in Qingzhen City, Guizhou, People's Republic of China, *Bull. Environ. Contam. Toxicol.*, 72, 445–451.

CHAPTER 11

Case Histories: Mercury Hazards from Gold Mining

The use of liquid mercury (Hg^o) to separate microgold (Au^o) particles from sediments through the formation of amalgam (Au-Hg) with subsequent recovery and reuse of mercury is a technique that has been in force for at least 4700 years (Lacerda, 1997a). However, this process is usually accompanied by massive mercury contamination of the biosphere (Petralia, 1996). It is estimated that gold mining currently accounts for about 10.0% of the global mercury emissions from human activities (Lacerda, 1997a). This chapter documents the history of mercury in gold production and ecotoxicological aspects of the amalgamation process in various geographic regions, with emphasis on Brazil and North America.

Useful general reviews on mercury and mercury amalgamation of gold include those by Montague and Montague (1971), D'Itri and D'Itri (1977), U.S. National Academy of Sciences (1978), Nriagu (1979), Porcella et al. (1995), Da Rosa and Lyon (1997), Nriagu and Wong 1997), De Lacerda and Salomons (1998), Eisler (2000, 2003, 2004a, 2004b), and Fields (2001).

11.1 HISTORY

The use of mercury in the mining industry to amalgamate and concentrate precious metals dates from about 2700 BCE when the Phoenicians and Carthaginians used it in Spain. The technology became widespread by the Romans in 50 CE and is similar to that employed today (Lacerda, 1997a; Rojas et al., 2001). In 177 CE, the Romans banned elemental mercury use for gold recovery in mainland Italy, possibly in response to health problems caused by this activity (De Lacerda and Salomons, 1998). Gold extraction using mercury was widespread until the end of the first millennium (Meech et al., 1998). In the Americas, mercury was introduced in the 16th century to amalgamate Mexican gold and silver. In 1849, during the California gold rush, mercury was widely used, and mercury poisoning was allegedly common among miners (Meech et al., 1998). In the 30-year period between 1854 and 1884, gold mines in California's Sierra Nevada range released between 1400 and 3600 tons of mercury to the environment (Fields, 2001); dredge tailings from this period still cover more than 73 km^2 in the Folsom-Natomas region of California, and represent a threat to current residents (De Lacerda and Salomons, 1998). In South America, mercury was used extensively by the Spanish colonizers to extract gold, releasing nearly 200,000 metric tons of mercury into the environment between 1550 and 1880 as a direct result of this process (Malm, 1998). At the height of the Brazilian gold rush in the 1880s, more than 6 million people were prospecting for gold in the Amazon region alone (Frery et al., 2001).

It is doubtful whether there would have been gold rushes without mercury (Nriagu and Wong, 1997). Supplies that entered the early mining camps included hundreds of flasks of mercury weighing 34.5 kg each, consigned to the placer diggings and recovery mills. Mercury amalgamation provided an inexpensive and efficient process for the extraction of gold, and itinerant gold diggers could rapidly learn the process. The mercury amalgamation process absolved the miners from any capital investment on equipment, and this was important where riches were obtained instantaneously and ores contained only a few ounces of gold per ton and could not be economically transported elsewhere for processing (Nriagu and Wong, 1997). Mercury released to the biosphere between 1550 and 1930 due to gold mining activities, mainly in Spanish colonial America, but also in Australia, southeast Asia, and England, may have exceeded 260,000 metric tons (Lacerda, 1997a). Exceptional increases in gold prices in the 1970s, concomitant with worsening socioeconomic conditions in developing regions of the world, resulted in a new gold rush in the southern hemisphere involving more than 10 million people on all continents. At present, mercury amalgamation is used as the major technique for gold production in South America, China, Southeast Asia, and some African countries. Most of the mercury released to the biosphere through gold mining can still participate in the global mercury cycle through remobilization from abandoned tailings and other contaminated areas (Lacerda, 1997a).

From 1860 to 1925, amalgamation was the main technique for gold recovery worldwide, and was common in the United States until the early 1940s (Greer, 1993). The various procedures in current use can be grouped into two categories (De Lacerda and Salomons, 1998; Korte and Coulston, 1998):

1. *Recovery of gold from soils and rocks containing 4.0 to 20.0 grams of gold per ton.* The metal-rich material is passed through grinding mills to produce a metal-rich concentrate. In Colonial America, mules and slaves were used instead of electric mills. This practice is associated with pronounced deforestation, soil erosion, and river siltation. The concentrate is moved to small amalgamation ponds or drums, mixed with liquid mercury, squeezed to remove excess mercury, and taken to a retort for roasting. Any residue in the concentrate is returned to the amalgamation pond and reworked until the gold is extracted.
2. *Gold extracted from dredged bottom sediments.* Stones are removed by iron meshes. The material is then passed through carpeted riffles for 20 to 30 h, which retains the heavier gold particles. The particles are collected in barrels, amalgamated, and treated as in (1). However, residues of the procedure are released into the rivers. Vaporization of mercury and losses due to human error also occur (De Lacerda and Salomons, 1998).

The organized mining sector abandoned amalgamation because of economic and environmental considerations. But small-scale mine operators in South America, Asia, and Africa, often driven by unemployment, poverty, and landlessness, have resorted to amalgamation because they lack affordable alternative technologies. Typically, these operators pour liquid mercury over crushed ore in a pan or sluice. The amalgam, a mixture of gold and mercury (Au-Hg), is separated by hand, passed through a chamois cloth to expel the excess mercury — which is reused — then heated with a blowtorch to volatilize the mercury. About 70.0% of the mercury lost to the environment occurs during the blowtorching. Most of these atmospheric emissions quickly return to the river ecosystem in rainfall and concentrate in bottom sediments (Greer, 1993).

Residues from mercury amalgamation remain at many stream sites around the globe. Amalgamation should not be applied because of health hazards and is, in fact, forbidden almost everywhere; however, it remains in use today, especially in the Amazon section of Brazil. In Latin America, more than a million gold miners collect between 115 and 190 tons of gold annually, emitting more than 200 tons of mercury in the process (Korte and Coulston, 1998). The world production of gold is about 225 tons annually, with 65 tons of the total produced in Africa. It is alleged that only 20.0% of the mined gold is recorded officially. About a million people are employed globally on nonmining aspects of artisanal gold, 40.0% of them female with an average yearly income of U.S.$600 (Korte and Coulston, 1998). The total number of gold miners in the world using mercury amalgamation to produce gold ranges from 3 to 5 million, including 650,000 from Brazil; 250,000 from Tanzania; 250,000 from Indonesia; and 150,000 from Vietnam (Jernelov and Ramel, 1994). To provide a living — marginal at best — for this large number of miners, gold production and mercury use would come to thousands of tons annually; however, official figures account for only 10.0% of the production level (Jernelov and Ramel, 1994). At least 90.0% of the gold extracted by individual miners in Brazil is not registered with authorities for a variety of reasons, some financial. Accordingly, official gold production figures reported in Brazil and probably most other areas of the world are grossly under-reported (Porvari, 1995). Cases of human mercury contamination have been reported from various sites around the world ever since mercury was introduced as the major mining technique to produce gold and other precious metals in South America hundreds of years ago (De Lacerda and Salomons, 1998). Contamination in humans is reflected by elevated mercury concentrations in air, water, diet, and in hair, urine, blood, and other tissues. However, only a few studies actually detected symptoms or clinical evidence of mercury poisoning in gold mining communities (Eisler, 2003).

After the development of the cyanide leaching process for gold extraction, mercury amalgamation disappeared as a significant mining technology (De Lacerda and Salomons, 1998). But when the price of gold soared from U.S.$58/troy ounce in 1972 to $430 in 1985, a second gold rush was triggered, particularly in Latin America, and later in the Philippines, Thailand, and Tanzania (De Lacerda and Salomons, 1998). In modern Brazil, where there has been a gold rush since 1980, at least 2000 tons of mercury were released, with subsequent mercury contamination of sediments, soils, air, fish, and human tissues; a similar situation exists in Colombia, Venezuela, Peru, and Bolivia (Malm, 1998). Estimates of global anthropogenic total mercury emissions range from 2000 to 4000 metric tons per year, of which 460 tons are from small-scale gold mining (Porcella et al., 1995, 1997). Major contributors of mercury to the environment from recent gold mining activities include Brazil (3000 tons since 1979), China (596 tons since 1938), Venezuela (360 tons since 1989), Bolivia (300 tons since 1979), the Philippines (260 tons since 1986), Colombia (248 tons since 1987), the United States (150 tons since 1969), and Indonesia (120 tons since 1988) (Lacerda, 1997a).

The most mercury-contaminated site in North America is the Lahontan Reservoir and environs in Nevada (Henny et al., 2002). Millions of kilograms of liquid mercury used to process gold and silver ore mined from Virginia City, Nevada, and vicinity between 1859 and 1890, along with waste rock, were released into the Carson River watershed. The inorganic elemental mercury was readily methylated to water-soluble methylmercury. Over time, much of this mercury was transported downstream into the lower reaches of the Carson River, especially the Lahontan Reservoir and Lahontan wetlands near the terminus of the system, with significant damage to wildlife (Henny et al., 2002).

11.2 ECOTOXICOLOGICAL ASPECTS OF AMALGAMATION

Mercury emissions from historic gold mining activities and from present gold production operations in developing countries represent a significant source of local pollution. Poor amalgamation distillation practices account for a significant part of the mercury contamination, followed by inefficient amalgam concentrate separation and gold melting operations (Meech et al., 1998). Ecotoxicological aspects of mercury amalgamation of gold are presented below for selected geographic regions, with special emphasis on Brazil and North America.

11.2.1 Brazil

High mercury levels found in the Brazilian Amazon environment are attributed mainly to gold mining practices, although elevated mercury concentrations are reported in fish and human tissues in regions far from any anthropogenic mercury source (Fostier et al., 2000). Since the late 1970s, many rivers and waterways in the Amazon have been exploited for gold using mercury in the mining process as an amalgamate to separate the fine gold particles from other components in the bottom gravel (Malm et al., 1990). Between 1979 and 1985, at least 100 tons of mercury were discharged into the Madeira River basin, with 45.0% reaching the river and 55.0% passing into the atmosphere. As a result of gold mining activities using mercury, elevated concentrations of mercury were measured in bottom sediments from small forest streams (up to 157.0 mg Hg/kg DW), in stream water (up to 10.0 μg/L), in fish (up to 2.7 mg/kg FW muscle), and in human hair (up to 26.7 mg/kg DW) (Malm et al., 1990). Mercury transport to pristine areas by rainwater, water currents, and other vectors could be increased with increasing deforestation, degradation of soil cover from gold mining activities, and increased volatilization of mercury from gold mining practices (Davies, 1997; Fostier et al., 2000). Population shifts due to gold mining are common in Brazil. For example, from 1970 to 1985, the population of Rondonia, Brazil, increased from about 111,000 to 904,000, mainly due to gold mining and agriculture. One result was a major increase in deforested areas and in gold production from 4 kg Au/year to 3600 kg/year (Martinelli et al., 1988). Mercury is lost during two distinct phases of the gold mining process. In the first phase, sediments are aspirated from the river bottom and passed through a series of seines. Metallic mercury is added to the seines to separate and amalgamate the gold. Part of this mercury escapes into the river, with risk to fish and livestock that drink river water, and to humans from occupational exposure and from ingestion of mercury-contaminated fish, meat, and water. In the second phase, the gold is purified by heating the amalgam — usually in the open air — with mercury vapor lost to the atmosphere. Few precautions are taken to avoid inhalation of the mercury vapor by the workers (Martinelli et al., 1988; Palheta and Taylor, 1995).

In Brazil, four stages of mercury poisoning were documented leading to possible occurrence of Minamata disease (Harada, 1994). The first route involves inorganic mercury poisoning among miners and gold shop workers directly exposed to elemental mercury used for gold extraction. Inhalation of mercuric vapor via the respiratory tract and absorption through the skin are considered the major pathways. Miners and gold shop workers who have been exposed directly to mercury vapors show clinical symptoms of inorganic mercury poisoning, including dizziness, headache, palpitations, tremors, numbness, insomnia, abdominal pain, dyspnea, and memory loss. Serious cases also show hearing difficulty, speech disorders, gingivitis, impotence, impaired eyesight, polyneuropathy, and disturbances in taste and smell. In a second stage, inorganic mercury discharged into the biosphere is converted to organomercurials via bacterial and other processes with resultant contamination of air, soil, and water. In the third stage, the organomercurials are bioaccumulated and biomagnified by fish and filter-feeding bivalve molluscs. Finally, humans who consume mercury-contaminated fish and shellfish evidence increased concentrations of mercury in blood, urine, and hair, which, if sufficiently high, are associated with the onset of Minamata disease (Harada, 1994).

11.2.1.1 Mercury Sources and Release Rates

All mercury used in Brazil is imported, mostly from the Netherlands, Germany, and England, reaching 340 tons in 1989 (Lacerda, 1997b). For amalgamation purposes, mercury in Brazil is sold in small quantities (200.0 grams) to a great number (about 600,000) of individual miners. Serious ecotoxicological damage is likely because much — if not most — of the human population in these regions depend on local natural resources for food (Lacerda, 1997b). In 1972, the amount of gold produced in Brazil was 9.6 tons, and in 1988 it was 218.6 tons; an equal amount of mercury is estimated to have been discharged into the environment (Camara et al., 1997). In Brazil, industry was responsible for almost 100.0% of total mercury emissions to the environment until the early 1970s, at which time existing mercury control policies were enforced with subsequent declines in mercury releases (Lacerda, 1997b). Mercury emissions from gold mining were insignificant up to the late 1970s, but by the mid-1990s it accounted for 80.0% of total mercury emissions. About 210 tons of mercury are now released to the biosphere each year in Brazil: 170 tons from gold mining, 17 tons from the chloralkali industry, and the rest from other industrial sources. Emission to the atmosphere is the major pathway of mercury release to the environment, with the gold mining industry accounting for 136 tons annually in Brazil (Lacerda, 1997b). During an 8-year period in the 1980s, about 2000 metric tons of mercury were used to extract gold in Brazil (Harada, 1994).

About 55.0% of the mercury used in gold mining operations is lost to the atmosphere during the burning of amalgam (Forsberg et al., 1995). The resulting mercury vapor (Hg^o) can be transported over considerable distances. Atmospheric transport of mercury from gold mining activities, coupled with high natural background concentrations of mercury, may produce mercury contamination in pristine areas of the Amazon (Forsberg et al., 1995).

At least 400,000 — and perhaps as many as a million — small-scale gold miners, known as *garimpos,* are active in the Brazilian Amazon region on more than 2000 sites (Pessoa et al., 1995; Veiga et al., 1995). It is estimated that each *garimpo* is indirectly responsible for another four to five people, including builders and operators of production equipment, dredges, aircraft (at least 1000), small boats or engine-driven canoes (at least 10,000), and about 1100 pieces of digging and excavation equipment. It is conservatively estimated that this group discharges 100 tons of mercury into the environment each year. There are five main mining and concentration methods used in the Amazon region to extract gold from rocks and soils containing 0.6 to 20.0 grams of gold per ton (Pessoa et al., 1995):

1. *Manual.* This involves the use of primitive equipment, such as shovels and hoes. About 15.0% of the *garimpos* use this method, usually in pairs. Gold is recovered in small concentration boxes with crossed riffles. Very few tailings are discharged into the river.
2. *Floating dredges with suction pumps.* This is considered inefficient, with large loss of mercury and low recovery of gold.
3. *Rafts with underwater divers directing the suction process.* This is considered a hazardous occupation, with many fatalities. Incidentally, there is a comparatively large mercury loss using this procedure.
4. *Hydraulic disintegration.* This involves breaking down steep banks using a high-pressure water jet pump.
5. *Concentration mills.* Gold recovered from underground veins is pulverized and extracted, sometimes by cyanide heap leaching.

The production of gold by *garimpos* (small-and medium-scale, often clandestine and transitory, mineral extraction operations) is from three sources: (1) extraction of auriferous materials from river sediments; (2) from veins where gold is found in the rocks; and (3) alluvial, where gold is found on the banks of small rivers (Camara et al., 1997). The alluvial method is most common and includes installation of equipment and housing, hydraulic pumping (high-pressure water to bring down the pebble embankment), concentration of gold by mercury, and burning the gold to remove the mercury. The latter step is responsible for about 70.0% of the mercury entering the environment. The gold is sold at specialized stores where it is again fired. Metallic mercury can also undergo methylation in the river sediments and enter the food chain (Camara et al., 1997).

Elemental mercury discharged into the Amazon River basin due to gold mining activities is estimated at 130 tons annually (Pfeiffer and Lacerda, 1988). Between 1987 and 1994 alone, more than 3000 metric tons of mercury were released into the biosphere of the Brazilian Amazon region from gold mining activities, especially into the Tapajos River basin (Boas, 1995; Castilhos et al., 1998). Local ecosystems receive about 100 tons of metallic mercury yearly, of which 45.0% enters river systems and 55.0% the atmosphere (Akagi et al., 1995). Mercury lost to rivers and soils as Hg^o is comparatively unreactive and contributes little to mercury burdens in fish and other biota (De Lacerda, 1997). Mercury entering the atmosphere is redeposited with rainfall at 90.0 to 120.0 $\mu g/m^2$ annually, mostly as Hg^{2+} and particulate mercury; these forms are readily methylated in floodplains, rivers, lakes, and reservoirs (De Lacerda, 1997). Health hazards to humans include direct inhalation of mercury vapor during the processes of burning the Hg-Au amalgam and consuming mercury-contaminated fish. Methylmercury, the most toxic form of mercury, is readily formed (Akagi et al., 1995). High levels of methylmercury in fish collected near gold mining areas and in the hair of humans living in fishing villages downstream of these areas (Martinelli et al., 1988; Malm et al., 1990; Eisler, 2004a) suggest that the reaction that converts discharged Hg^o to Hg^{2+} is present in nature before Hg^{2+} is methylated to CH_3Hg^+. Yamamoto et al. (1995a) indicate that oxidation of Hg^o to Hg^{2+} occurs in the presence of sulfhydryl compounds, including L-cysteine and glutathione. Because sulfhydryl compounds are known to have a high affinity for Hg^{2+}, the conversion of Hg^o to Hg^{2+} may be due to an equilibrium shift between Hg^o and Hg^{2+} induced by the added sulfhydryl compounds (Yamamoto et al., 1995b).

About 130 tons of Hg^o are released annually by alluvial gold mining to the Amazonian environment, either directly to rivers or into the atmosphere, after reconcentration, amalgamation, and burning (Reuther, 1994). In the early 1980s, the Amazon region in northern Brazil was the scene of the most intense gold rush in the history of Brazil (Hacon et al., 1995). Metallic mercury was used to amalgamate particulate metallic gold. Refining of gold to remove the mercury is considered the source of environmental mercury contamination; however, other sources of mercury emissions in Amazonia include tailings deposits and burning of tropical forests and savannahs (Hacon et al., 1995). In 1989 alone, gold mining in Brazil contributed 168 metric tons of mercury to the environment (Aula et al., 1995).

Lechler et al. (2000) assert that natural sources of mercury and natural biogeochemical processes contribute heavily to reported elevated mercury concentrations in fish and water samples collected up to 900 km downstream from local gold mining activities. Based on analysis of water, sediments, and fish samples systematically collected along a 900-km stretch of the Madeira River in 1997, they concluded that the elevated mercury concentrations in samples were due mainly to natural sources and that the effects of mercury released from gold mining sites were localized (Lechler et al., 2000). This must be verified.

11.2.1.2 Mercury Concentrations in Abiotic Materials and Biota

Since 1980, during the present gold rush in Brazil, at least 2000 tons of mercury have been released into the environment (Malm, 1998). Elevated mercury concentrations are reported in virtually all abiotic materials, plants, and animals collected near mercury-amalgamation gold mining sites (Table 11.1). Mercury concentrations in samples show high variability, and this may be related to seasonal differences, geochemical composition of the samples, and species differences (Malm, 1998). In 1992, more than 200 tons of mercury were used in the gold mining regions of Brazil (Von Tumpling et al., 1995). One area, near Pocone, has been mined for more than 200 years. In the 1980s, about 5000 miners were working 130 gold mines in this region. Mercury was used to amalgamate the preconcentrated gold particles for the separation of the gold from the slag. Mercury-contaminated wastes from the separation process were combined with the slag from the reconcentration process and collected as tailings. The total mercury content in tailings piles in this geographic locale was estimated at about 1600 kg, or about 12.0% of all mercury used in the past 10 years. Surface runoff from tropical rains caused extensive erosion of tailings piles — some 4.5 m high — with contaminated material reaching nearby streams and rivers. In the region of Pocone, mercury concentrations in waste tailings material ranged from 2.0 to 495.0 μg/kg, occupied 4.9 km^2, and degraded an estimated 12.3 km^2 (Von Tumpling et al., 1995).

Tropical ecosystems in Brazil are under increasing threat of development and habitat degradation from population growth and urbanization, agricultural expansion, deforestation, and mining (Lacher and Goldstein, 1997). Where mercury has been released into the aquatic system as a result of unregulated gold mining, subsequent contamination of invertebrates, fish, and birds was measured and biomagnification of mercury was documented from gastropod molluscs (*Ampullaria* spp.) to

Table 11.1 Total Mercury Concentrations in Abiotic Materials, Plants, and Animals near Active Brazilian Gold Mining and Refining Sites

Location, Sample, and Other Variables	Concentration[a]	Ref.[b]
Amazon Region		
Livestock; gold field vs. reference site:		
Hair:		
Cattle, *Bos* sp.	0.2 mg/kg dry weight (DW) vs. 0.1 mg/kg DW	1
Pigs, *Sus* sp.	0.9 mg/kg DW vs. 0.2 mg/kg DW	1
Sheep, *Ovis aires*	0.2 mg/kg DW vs. 0.1 mg/kg DW	1
Blood:		
Cattle	12.0 μg/L vs. 5.0 μg/L	1
Pigs	18.0 μg/L vs. 13.0 μg/L	1
Sheep	3.0 μg/L vs. 1.0 μg/L	1
Humans:		
Blood:		
Miners	(2.0–29.0) μg/L	1
Villagers	(3.0–10.0) μg/L	1
River dwellers	(1.0–65.0) μg/L	1
Reference site	(2.0–10.0) μg/L	1
Urine:		
From people in gold processing shops vs. maximum allowable level vs. reference site	269.0 (10.0–1,168.0) μg/L vs. < 50.0 μg/L vs. 12.0 (1.5–74.3) μg/L	16
Miners	(1.0–155.0) μg/L	1
Villagers	(1.0–3.0) μg/L	1
Reference site	(0.1–7.0) μg/L	1
Hair:		
Pregnant women vs. maximum allowable for this cohort	3.6 (1.4–8.0) mg/kg FW vs. < 10.0 mg/kg FW	16
Miners	(0.4–32.0) mg/kg DW	1
Villagers	(0.8–4.6) mg/kg DW	1
River dwellers	(0.2–15.0) mg/kg DW	1
Reference site	< 2.0 mg/kg DW	1
Soils:		
Forest soils; 20–100 m from amalgam refining area vs. reference site	2.0 (0.4–10.0) mg/kg DW vs. 0.2 (max. 0.3) mg/kg DW	16
Urban soils; 5–350 m from amalgam refining area vs. reference site	7.5 (0.5–64.0) mg/kg DW vs. 0.4 (0.03–1.3) mg/kg DW	16
Alta Floresta and Vicinity		
Air; near mercury emission areas from gold purification vs. indoor gold shop	(0.02–5.8) μg/m³ vs. (0.25–40.6) μg/m³	2, 3
Fish muscle, carnivorous species	0.3–3.6 mg/kg fresh weight (FW)	3
Soil	(0.05–4.1) mg/kg DW	2, 4
Madeira River and Vicinity		
Air	(10.0–296.0) μg/m³	4
Aquatic macrophytes:		
Leaves; floating vs. submerged	0.9–1.0 mg/kg DW vs. 0.001 mg/kg DW	4
Victoria amazonica	0.9 mg/kg DW	5
Eichornia crassipes	(0.04–1.01) mg/kg DW	5
Echinocloa polystacha	< 0.008 DW	5
Fish eggs, detritovores	(0.05–3.8) mg/kg FW	5
Fish muscle:		
Carnivores vs. omnivores	0.5–2.2 mg/kg FW vs. 0.04–1.0 mg/kg FW	5
Carnivorous species vs. noncarnivorous species	max. 2.9 mg/kg FW vs. max. 0.65 mg/kg FW	4

(continued)

Table 11.1 (continued) Total Mercury Concentrations in Abiotic Materials, Plants, and Animals near Active Brazilian Gold Mining and Refining Sites

Location, Sample, and Other Variables	Concentration[a]	Ref.[b]
7 species:		
Herbivores	0.08 mg/kg FW; max. 0.2 mg/kg FW	6
Omnivores	0.8 mg/kg FW; max. 1.7 mg/kg FW	6
Piscivores	0.9 mg/kg FW; max. 2.2 mg/kg FW	6
Maximum	2.7 mg/kg FW	7, 15
Water	Max. 8.6 to 10.0 μg/L	7, 15
Sediments	19.8 mg/kg DW; max. 157.0 mg/kg DW	7, 15
Mato Grosso		
Freshwater molluscs (*Ampullaria* spp., *Marisa planogyra*); soft parts	Max. 1.2 mg/kg FW	8
Sediments	Max. 0.25 mg/kg FW	8
Negro River		
Fish muscle; fish-eating species vs. herbivores	Max. 4.2 mg/kg FW vs. max. 0.35 mg/kg FW	4
Pantanal		
Clam, *Anodontitis trapesialis*; soft parts	0.35 mg/kg FW	9
Clam, *Castalia* sp.; soft parts	0.64 mg/kg FW	9
Parana River; water; dry season vs. rainy season	0.41 μg/L vs. 2.95 μg/L	4
Pocone and Vicinity		
Air	< 0.14–1.68 μg/m³	4
Fish muscle; carnivores vs. noncarnivores	Max. 0.68 mg/kg FW vs. max. 0.16 mg/kg FW	4
Surface sediments	0.06–0.08 mg/kg DW	4
Porto Velho		
Air	0.1–7.5 μg/m³	4
Soils; near gold dealer shops vs. reference site	(0.4–64.0) mg/kg DW vs. (0.03–1.3) mg/kg DW	4
Rio Negro Basin; March 1993		
Fish muscle:		
Detritovores	0.1 mg/kg FW	19
Omnivores	0.35 mg/kg FW	19
Carnivores	0.73 mg/kg FW; max. 2.6 mg/kg FW	19
Human hair:		
Rio Negro area	75.5 (5.8–171.2) mg/kg FW	19
Reference area 400 km upwind	23.1 (6.1–39.4) mg/kg FW	19
Recommended maximum	< 50.0 mg/kg FW	19
Tapajos River Basin		
Air in goldshops	Max. 292.0 μg/m³	18
Tucunare, *Cichla monoculus*; 1992–2001; mercury-contaminated gold mining area vs. reference site:		
Muscle	0.71 mg total Hg/kg FW vs. 0.23 mg total Hg/kg FW	17
Erythrocyte number, in millions	2.00 vs. 2.56	17
Hematocrit	40.0 vs. 44.8	17
Leukocyte count	36,224 vs. 53,161	17
Fish muscle:		
Frequently	> 2.0 mg/kg FW	18
Max.	5.9 mg/kg FW	18
Safe	< 0.5 mg/kg FW	18

Table 11.1 (continued) Total Mercury Concentrations in Abiotic Materials, Plants, and Animals near Active Brazilian Gold Mining and Refining Sites

Location, Sample, and Other Variables	Concentration[a]	Ref.[b]
Fish muscle; carnivores vs. noncarnivores	Max. 2.6 mg/kg FW vs. max. 0.31 mg/kg FW	4
Fish muscle; contaminated site vs. reference site 250 km downstream:		
Carnivorous fishes	0.42 mg/kg FW vs. 0.23 mg/kg FW	10
Noncarnivorous fishes	0.06 mg/kg FW vs. 0.04 mg/kg FW	10
Fishermen; blood	31.0–46.9 μg/L (vs. 12.6 μg/L 800 km downstream)	18
Gold miners; hair	22.2 mg/kg FW; max. 113.2 mg/kg FW	18
Gold brokers; blood	Max. 0.29 mg/L	18
Gold miners; exposed 16.3 years; symptoms evident after 4.4 years:		
Blood	22.0 μg/L	20
Urine	35.4 μg/L	
Gold miners and gold shop workers; 1986—1992:		
Blood	30.5 (4.0–130.0) μg/L	20
Urine	32.7 μg/L; max. 151.0 μg/L	20
Gold shop workers; exposed about 5.3 years; mercury intoxication symptoms evident after 2.5 years:		
Blood	51.0 μg/L	20
Urine	61.0 μg/L	20
House dust	150.0 mg/kg FW	18
Mud in river beds	2.8–143.5 mg/kg FW	18
Nonoccupational exposure:		
Blood; males vs. females	32.0 μg/L vs. 19.0 μg/L	20
Hair; total mercury vs. methylmercury; 1992:		
Males	58.5 (12.0–151.2) mg/kg FW vs. 49.8 (11.1–132.6) mg/kg FW	20
Females	15.7 (7.2–29.5) mg/kg FW vs. 13.2 (6.1–26.3) mg/kg FW	20
Residents consuming local fish; hair; 1992	1.5–151.2 mg/kg FW (90.0% methylHg)	20
Sediments; mining area vs. reference site:		
Total mercury	0.14 mg/kg FW vs. (0.003–0.009) mg/kg FW	4
Methylmercury	0.8 μg/kg FW vs. 0.07–0.19 μg/kg FW	4
Soil	0.7–1,370.0 mg/kg FW	18
Water; unfiltered	Max. 6.7 mg/L	18
Teles River Mining Site		
Air	(0.01–3.05) μg/m^3	4
Fish muscle	Max. 3.8 mg/kg FW	4
Tucurui Reservoir and Environs		
Aquatic macrophytes:		
Floating vs. submerged	0.12 mg/kg DW vs. 0.03 mg/kg DW	4
Floating plants; roots vs. shoots	Max. 0.098 mg/kg DW vs. max. 0.046 mg/kg DW	11
Fish muscle, 7 species	0.06–2.6 mg/kg FW; max. 4.5 mg/kg FW	12
Fish muscle; carnivores vs. noncarnivores	Max. 2.9 mg/kg FW vs. max. 0.16 mg/kg FW	4
Gastropods; soft parts vs. eggs	0.06 (0.01–0.17) mg/kg FW vs. ND	12
Turtle, *Podocnemis unifilis*; egg	0.01 (0.007–0.02) mg/kg FW	12, 21
Caiman (crocodile), *Paleosuchus* sp.; muscle vs. liver	1.9 (1.2–3.6) mg/kg FW vs. 19.0 (11.0–30.0) mg/kg FW	12, 21
Human hair; November 1990–March 1991:		
Fishermen (13.8 fish meals per week)	47.0 (4.0–240.0) DW	21
Power company employees (1.1 fish meals/week)	11.0 (0.9–37.0) DW	21
Parakana Indians (2.0 fish meals/week)	8.5 (3.3–12.0) DW	21

(continued)

Table 11.1 (continued) Total Mercury Concentrations in Abiotic Materials, Plants, and Animals near Active Brazilian Gold Mining and Refining Sites

Location, Sample, and Other Variables	Concentration[a]	Ref.[b]
Capybara (mammal), *Hydrochoerus hydrochaeris*:		
Hair	0.16 (0.12–0.19) mg/kg DW	12, 21
Liver	0.01 (0.006–0.01) mg/kg FW	12, 21
Muscle	0.015 (0.007–0.026) mg/kg FW	12, 21
Sediments (up to 240.0 μg Hg/m² deposited monthly, 1990–1991)	0.13 (0.07–0.22) mg/kg DW	11, 21
Various Locations, Brazil		
Air:		
Mining areas	Max. 296.0 μg/m³	4
Rio de Janeiro	(0.02–0.007) μg/m³	4
Rural areas vs. urban areas	0.001–0.015 μg/m³ vs. 0.005–0.05 μg/m³	4
Bromeliad epiphyte (plant), *Tillandsia usenoides*; exposure for 45 days; dry season vs. rainy season:		
Near mercury emission sources	12.2 (1.9–22.5) mg/kg FW vs. 5.2 (2.5–9.5) mg/kg FW	14
Inside gold shop	4.3 (0.6–26.8) mg/kg FW vs. 1.7 (0.2–5.3) mg/kg FW	14
Local controls	0.2 (< 0.08–0.4) mg/kg FW vs. 0.09 (< 0.08–0.12) mg/kg FW	14
Rio de Janeiro controls	0.2 (< 0.08–0.4) mg/kg FW vs. < 0.08 mg/kg FW	14
Fish muscle:		
Near gold mining areas	0.21–2.9 mg/kg FW	13
Global, mercury contaminated	1.3–24.8 mg/kg FW	13
Reference sites; carnivorous species vs. noncarnivorous species	Max. 0.17 mg/kg FW vs. max. < 0.10 mg/kg FW	4
Lake water; gold mining areas vs. reference sites	0.04–8.6 μg/L vs. < 0.03 μg/L	1
River water; mining areas vs. reference sites	0.8 μg/L vs. < 0.2 μg/L	1
Sediments; gold mining areas vs. reference sites	0.05–19.8 mg/kg DW vs. < 0.04 mg/kg DW	13
Soils (forest); gold mining areas vs. reference sites	0.4–10.0 mg/kg DW vs. 0.03–0.34 mg/kg DW	4

[a] Concentrations are shown as means, range (in parentheses), maximum (max.), and nondetectable (ND).
[b] Reference: 1, Palheta and Taylor, 1995; 2, Hacon et al., 1995; 3, Hacon et al., 1997; 4, De Lacerda and Salomons, 1998; 5, Martinelli et al., 1988; 6, Dorea et al., 1998; 7, Pfeiffer et al., 1989; 8, Vieira et al., 1995; 9, Callil and Junk, 1999; 10, Castilhos et al., 1998; 11, Aula et al., 1995; 12, Aula et al., 1994; 13, Pessoa et al., 1995; 14, Malm et al., 1995a; 15, Malm et al., 1990; 16, Malm et al., 1995b; 17, Castilhos et al., 2004; 18, Harada, 1994; 19, Forsberg et al., 1995; 20, Branches et al., 1994; 21, Lodenius, 1993.

birds (snail kite, *Rostrhamus sociabilis*) and from invertebrates and fish to waterbirds and humans (Lacher and Goldstein, 1997). Indigenous peoples of the Amazon living near gold mining activities have elevated levels of mercury in hair and blood. Other indigenous groups are also at risk from mercury contamination as well as from malaria and tuberculosis (Greer, 1993). The miners, mostly former farmers, are also victims of hard times and limited opportunities. Small-scale gold mining offers an income and an opportunity for upward mobility (Greer, 1993). Throughout the Brazilian Amazon, about 650,000 small-scale miners are responsible for about 90.0% of Brazil's gold production and for the discharge of 90 to 120 tons of mercury to the environment every year. About 33.0% of the miners had elevated concentrations in tissues over the tolerable limit set by the World Health Organization (WHO) (Greer, 1993). In Brazil, it is alleged that health authorities are unable to detect conclusive evidence of mercury intoxication due to difficult logistics and the poor health conditions of the mining population, which may mask evidence of mercury poisoning. There is a strong belief that a silent outbreak of mercury poisoning has the potential for regional disaster (De Lacerda and Salomons, 1998).

In the Madeira River Basin, mercury levels in certain sediments were 1500 times higher than similar sediments from nonmining areas, and dissolved mercury concentrations in the water column were 17 times higher than average for rivers throughout the world (Greer, 1993). High concentrations of mercury were measured in fish and sediments from a tributary of the Madeira River affected by alluvial small-scale mining (Reuther, 1994). The local safety limit of 0.1 mg Hg/kg DW sediment was exceeded by a factor of 25, and the safety level for fish muscle of 0.5 mg Hg/kg FW muscle was exceeded by a factor of 4. Both sediments and fish act as potential sinks for mercury because existing physicochemical conditions in these tropical waters (low pH, high organic load, high microbial activity, elevated temperatures) favor mercury mobilization, methylation, and availability (Reuther, 1994). In Amazonian river sediments, mercury methylation accounts for less than 2.2% of the total mercury in sediments (De Lacerda and Salomons, 1998). In soils, mercury mobility is low, in general (De Lacerda and Salomons, 1998). There is an association between the distribution of mercury-resistant bacteria in sediments and the presence of mercury compounds (Cursino et al., 1999). Between 1995 and 1997, mercury concentrations were measured in sediment along the Carmo stream, Minas Gerais, located in gold prospecting areas. Most sediments contained more than the Brazilian allowable limit of 0.1 mg Hg/kg DW. Mercury-resistant bacteria were present in sediments at all sites and ranged from 27.0 to 77.0% of all bacterial species, with a greater percentage of species showing resistance at higher mercury concentrations (Cursino et al., 1999).

The Pantanal is one of the largest wetlands in the world and extends over 300,000 km^2 along the border area of Brazil, Bolivia, Argentina, and Paraguay (Guimaraes et al., 1995). Half this surface is flooded annually. Since the 18th century, gold has been extracted from quartz veins in Brazil using amalgamation as a concentration process, resulting in metallic mercury releases to the atmosphere, soils, and sediments. The availability to aquatic biota of Hg^o released by gold mining activities is limited to its oxidation rate to Hg^{2+} and then by conversion to methylmercury (CH_3Hg^+), which is readily soluble in water (Guimaraes et al., 1998). The Pantanal in Brazil, at 140,000 km^2, is an important breeding ground for storks, herons, egrets, and other birds, as well as a refuge for threatened or endangered mammals, including jaguars (*Panthera onca*), giant anteaters (*Myrmecophaga tridactyla*), and swamp deer (*Cervus duvauceli*) (Alho and Viera, 1997). Gold mining is common in the northern Pantanal. There are approximately 700 operating gold-mining dredges along the Cuiba River. Unregulated gold mines have contaminated the area with mercury, and 35.0 to 50.0% of all fishes collected from this area contain more than 0.5 mg Hg/kg FW muscle, the current Brazilian and international (World Health Organization) standard for fish consumed by humans (Alho and Viera, 1997). Gastropod molluscs that are commonly eaten by birds contained 0.02 to 1.6 mg Hg/kg FW soft parts. Mercury concentrations in various tissues of birds that ate these molluscs were highest in the anhinga (*Anhinga anhinga*) at 0.4 to 1.4 mg/kg FW and the snail kite at 0.3 to 0.6 mg/kg FW, and were lower in the great egret (*Casmerodius* [formerly *Ardea*] *albus*) at 0.02 to 0.04 mg/kg FW and limpkin (*Aramus guarauna*) at 0.1 to 0.5 mg/kg FW. The high mercury levels detected, mainly in fishes, show that the mercury used in gold mining and released into the environment has reached the Pantanal and spread throughout the ecosystem with potential biomagnification (Alho and Viera, 1997).

Floating plants accumulate small amounts of mercury (Table 11.1) but their sheer abundance makes them likely candidates for mercury phytoremediation. For example, in the Tucurui Reservoir in the state of Para, it is estimated that 32 tons of mercury are stored in floating plants, mostly *Scurpus cubensis* (Aula et al., 1995). Mercury methylation rates in sediments and floating plants were evaluated in Fazenda Ipiranga Lake, 30 km downstream from gold mining fields near Pantanal during the dry season of 1995 (Guimaraes et al., 1998). Sediments and roots of dominant floating macrophytes (*Eichornia azurea*, *Salvina* sp.) were incubated *in situ* for 3 days with about 43.0 μg Hg^{2+}/kg DW added as $^{203}HgCl_2$. Net methylation was about 1.0% in sediments under floating macrophytes, being highest at temperatures in the 33 to 45°C range and high concentrations of sulfate-reducing bacteria. Methylation was inhibited above 55°C, under saline conditions, and under conditions of low sulfate. Radiomercury-203 was detectable to a depth of 16 cm in the sediments, coinciding with the depth reached by chironomid larvae. Methylation was up to 9 times greater in the roots of floating macrophytes than in the underlying surface sediments: an average of 10.4%

added Hg^{2+} was methylated in *Salvina* roots in 3 days and 6.5% in *Eichornia* roots (Guimaraes et al., 1998). Using radiomercury-203 tracers, no methylation was observed under anoxic conditions in organic-rich, flocculent surface sediments due to the formation of HgS — a compound that is much less available for methylation than is Hg^{2+} (Guimaraes et al., 1995). Authors conclude that floating macrophytes should be considered when evaluating mercury methylation rates in tropical ecosystems (Guimaraes et al., 1998).

Clams collected near gold mining operations had elevated concentrations of mercury (up to 0.64 mg Hg/kg FW) in soft tissues (Table 11.1). Laboratory studies suggest that mercury adsorbed to suspended materials in the water column is the most likely route for mercury uptake by filter-feeding bivalve molluscs (Callil and Junk, 1999).

Mercury concentrations in fish collected near gold mining activities in Brazil were elevated, and decreased with increasing distance from mining sites (Table 11.1). In general, muscle is the major tissue of mercury localization in fishes, and concentrations are higher in older, larger, predatory species (Aula et al., 1994; Eisler, 2000; Lima et al., 2000). In the Tapajos River region, which receives between 70 and 130 tons of mercury annually from gold mining activities, mercury concentrations in fish muscle were highest in carnivorous species, lowest in herbivores, and intermediate in omnivores (Lima et al., 2000). However, only 2.0% of fish collected in 1988 (vs. 1.0% in 1991) from the Tapajos region exceeded the Brazilian standard of 0.5 mg total mercury/kg FW muscle, and all violations were from a single species of cichlid (tucunare/speckled pavon, *Cichla temensis*) (Lima et al., 2000). Tucunare from the contaminated Tapajos region, when compared to a reference site, accumulated mercury 3.5 to 4 times more rapidly (0.8 to 1.4 μg daily vs. 0.2 to 0.38), and had significantly lower erythrocyte counts, hematocrits, and leukocyte counts (Castilhos et al., 2004; Table 11.1). A 1991 survey of 11 species of fishes collected from a gold mining area in Cachoeira de Teotonio revealed that almost all predatory species had greater than 0.5 mg Hg/kg FW muscle vs. less than 0.5 mg/kg FW in conspecifics collected from Guajara, a distant reference site (Padovani et al., 1995). Limits on human food consumption were set for individual species on the basis of mercury concentrations in muscle — specifically, no restrictions on some species, mostly herbivores and omnivores; some restrictions on some omnivores and small predators; and severe restrictions on larger predators (Padovani et al., 1995).

Fish that live near gold mining areas have elevated concentrations of mercury in their flesh and are at high risk of reproductive failure (Oryu et al., 2001). Mercury concentrations of 10.0 to 20.0 mg/kg FW in fish muscle are considered lethal to the fish, and 1.0 to 5.0 mg/kg FW sublethal; predatory fish frequently contain 2.0 to 6.0 mg Hg/kg FW muscle. Mercury-contaminated fish pose a hazard to humans and other fish consumers, including the endangered giant otter (*Pteronura brasiliensis*) and the jaguar. Giant otters eat mainly fish and are at risk from mercury intoxication: 1.0 to 2.0 mg Hg/kg FW diet is considered lethal. Jaguars consume fish and giant otters; however, no data are available on the sensitivity of this top predator to mercury (Oryu et al., 2001). Caiman crocodiles are also threatened by gold mining and related mercury contamination of habitat, increased predation by humans, extensive agriculture, and deforestation (Brazaitis et al., 1996). Caiman crocodiles, *Paleosuchus* spp., from the Tucurui Reservoir area had up to 3.6 mg Hg/kg FW muscle and 30.0 mg Hg/kg FW liver (Aula et al., 1994). Extensive habitat destruction and mercury pollution attributed to mining activity was observed at 19 localities in Mato Grosso and environs. Crocodiles (*Caiman* spp., *Melanosuchus niger*, *Crocodilus crocodilus*) captured from these areas were emaciated, algae covered, in poor body condition, and heavily infested with leeches (Brazaitis et al., 1996).

High levels of mercury in urine (81.0 to 102.0 μg/L) and blood (25.0 to 39.0 μg/L) were positively associated with the amount of amalgam burnt each week. For residents of a fishing village, mercury urine concentrations were highest among residents who refined amalgam by burning and consumed fish frequently, with maximum levels recorded of 108.0 μg/L in urine and 254.0 μg/L in blood (Cleary, 1995). The hair mercury concentration in gold miners in the Tapajos River basin averaged 22.2 mg/kg FW, with a maximum of 113.3 mg/kg FW (Harada, 1994). Also, blood mercury levels were high (max. 291.0 μg/L FW), with organomercurials accounting for about 5.3%; however, the inorganic mercurials are quickly excreted (Harada, 1994). Gold miners, gold shop workers, and neighbors of gold shops had abnormally high mercury levels in urine (up to 151.0 μg/L) and blood (up to 130.0 μg/L), and some had symptoms indicative of mercury intoxication (Table 11.1; Branches et al., 1994). However, residents of the same area who had no previous contact with metallic mercury and its compounds have shown elevated concentrations of total mercury in hair (up to 151.0 mg/kg FW, about 90.0% methylmercury). Motor difficulties were seen in some individuals with greater than 50.0 mg total Hg/kg FW hair (Branches et al., 1994). In the Tucurui Reservoir area, where the main source of mercury is from gold mining activities upstream, Lodenius (1993) avers that a human male needs 0.3 mg of mercury daily to reach a hair mercury concentration of 50.0 mg/kg DW. To receive this amount of mercury from fish containing 1.0 mg total Hg/kg FW muscle, a daily ingestion of 330.0 grams of fish muscle was calculated (Lodenius, 1993), although this needs verification.

11.2.1.3 Mitigation

There are two populations at significant risk from mercury intoxication in Brazilian gold mining communities: (1) riverine populations — with high levels of mercury in hair — that routinely eat mercury-contaminated fish, and (2) gold dealers in indoor shops exposed to Hg^{o} vapors (De Lacerda and Salomons, 1998). These two critical groups should receive special attention regarding exposure risks. Riverine populations, especially children and women of child-bearing age, should avoid consumption of carnivorous fishes (Harada, 1994; De Lacerda and Salomons, 1998). And in gold dealer shops, adequate ventilation and treatment systems for mercury vapor retention must be installed (De Lacerda and Salomons, 1998).

Confounding variables in evaluating mercury risk to these groups include the amount of mercury spilled, transportation and methylation rates, mercury uptake among fish species, date on which mining began in a given area, and patterns of fish consumption in rural Amazonia (Cleary, 1995). Analysis of mercury samples alone is not sufficient in evaluating risk and should be complemented by a questionnaire survey, as well as various clinical, neurological, and psychological tests. Risks are greatest from Hg^{o} vapor to gold traders when the Hg-Au amalgam is refined through burning indoors and from organomercurials to children from riverine populations consuming carnivorous fish contaminated by mercury spillage in mining areas. Hair sampling from pregnant women is the most appropriate initial indicator of organomercury contamination. Children between the ages of 7 and 12 years and pregnant women are the most important target groups. Children, rather than infants or toddlers, are recommended because tests of neurological function require children to understand simple instructions and to cooperate with researchers (Cleary, 1995).

Tests measuring memory and coordination are also desirable, provided that they are easy to administer, require only basic equipment, and there is some training of a nurse or paramedic who will actually administer the test (Cleary, 1995). Cleary recommends that the full battery of tests need only be applied to children, and that a hair sample and questionnaires covering dietary habits and clinical history should be sufficient for women at risk. In isolated communities, the research team is also expected by the population to provide general health care, and research teams should be accompanied by health professionals working primarily to provide general health care. The minimum team necessary for conducting mercury-related epidemiological work over a 7- to 10-day period, in a community of up to 1500 people should consist of eight individuals: one physician to perform general clinical examinations; one physician to perform neurological tests; one nurse/paramedic to administer memory and physical coordination tests; one nurse to administer clinical history and dietary questionnaire; one physician and two nurses to provide general health care to the community; and one coordinator to control patient access (Cleary, 1995).

An occupational health and safety program was launched to educate Brazilian adolescents about industrial hazards, including health risks associated with mercury amalgamation of gold (Camara et al., 1997). Adolescents, together with young adults, constitute a large portion of the *garimpos* and are in a critical physical and psychological growth phase. The number of adolescents participating in gold extraction increases with decreasing family income and with the structure of the labor market as approved by public policy and the courts. The program was designed to transfer information on risks associated with mining and other dangerous occupations, to apprise them of their choices, and to promote training. The method had been successfully tested earlier in selected urban and rural schools of different economic strata. The community selected was Minas Gerais, where gold was discovered more than 300 years ago and is the main source of community income. The municipality does not have a sewer system and the water supply is not treated. The program was initiated in April 1994 over a 5-week period. The 70 students were from the 5th to 8th grade. Almost all worked, most in mining — where exposure to mercury was illegal. Students successfully improved safety habits and recognition of potential accident sites. A generalized occupational safety program, such as this one, is recommended for other school districts (Camara et al., 1997). Continuous and systematic monitoring of mercury levels in fish and fishermen is also recommended (Harada, 1994).

11.2.2 Remainder of South America

A major portion of the mercury contamination noted in Central and South America was from amalgamation of silver by the Spanish for at least 300 years, starting in 1554 (Nriagu, 1993). Until the middle of the 18th century, about 1.5 kg of mercury was lost to the environment for every kilogram of silver produced. Between 1570 and 1820, mercury loss in this geographic region averaged 527 tons annually, or a total of about 126,000 tons during this time period. Between 1820 and 1900, another 70,000 tons of mercury were lost to the environment through silver production, for a total of 196,000 tons of mercury during the period between 1570 and 1900. By comparison, the input of mercury into the Brazilian Amazon associated with gold mining is 90 to 120 tons of mercury annually. Under the hot tropical conditions typical of Mexico and parts of South America, mercury in abandoned mine wastes or deposited in aquatic sediments is likely to be methylated and released to the atmosphere where it cycles for considerable periods. It now seems reasonable to conclude that the Spanish American silver mines were responsible, in part, for the high background concentrations of mercury in the global environment now being reported (Nriagu, 1993).

11.2.2.1 Colombia

In Colombia, mercury intoxication was reported among fishermen and miners living in the Mina Santa Cruz marsh, possibly from ingestion of mercury-contaminated fish. This marsh received an unknown amount of mercury-contaminated gold mine wastes (Olivero and Solano, 1998). However, mercury concentrations in marsh sediments, fish muscle, and macrophytes were low. Examination of abiotic and biotic materials from this location in 1996 showed maximum mercury concentrations of 0.4 mg/kg FW in sediment (range 0.14 to 0.4); 0.4 mg/kg DW in roots of *Eichornia crassipes*, an aquatic macrophyte (range 0.1 to 0.4); and 1.1 mg/kg FW in fish muscle. Among eight species of fish examined, the mean mercury concentrations in muscle ranged between 0.03 and 0.38 mg/kg FW, and the range for all observations extended from 0.01 to 1.1 mg/kg (Olivero and Solano, 1998).

11.2.2.2 Ecuador and Peru

Mercury concentrations in water (up to 1.1 μg/L) and sediments (up to 5.8 mg/kg DW) downstream of the Portovela-Zaruma cyanide-gold mining area in Ecuador during the 1988 dry season exceeded that country's recommended safe value for aquatic life protection of less than 0.1 μg/L and the no probable effect level of less than 0.45 mg/kg DW sediments (Tarras-Wahlberg et al., 2000).

Mercury concentrations in hair and urine from children of Andean gold miners in the Nambija, Ecuador region — where mercury is used extensively in gold recovery and is the probable major source of elemental and methylmercury exposure — were measured in 80 children with no known dental amalgams (Counter et al., 2005). Urine samples were indicative of the inorganic mercury burden; the mean mercury concentration in urine was 10.9 μg/L and ranged between 1.0 and 166.0 μg/L. Hair samples were used as an indicator of dietary methylmercury; mercury concentrations ranged between 1.0 and 135.0 mg/kg DW, with a mean of 6.0 mg/kg DW (Counter et al., 2005). Previous studies (Counter et al., 1998, 2002) of the same population showed blood mercury concentrations ranging between 2.0 and 89.0 μg/L with a mean value of 18.2 μg/L; the mean exceeded the World Health Organization reference level of 5.0 μg Hg/L (WHO, 1991). Counter et al. (2005) concluded that the mercury concentrations in blood, urine, and hair of these children were indicative of significant chronic mercury exposure and that adverse neurodevelopmental outcomes were possible.

In Peru, mercury in scat of the giant otter, and in the otter's fish diet, was measured in samples collected between 1990 and 1993 near a gold mine (Gutleb et al., 1997). Total mercury in fish muscle ranged between 0.05 and 1.54 mg/kg FW. In 68.0% of fish muscle samples analyzed, the total mercury levels exceeded the proposed (by Gutleb et al., 1997) maximum tolerated level of 0.1 mg/kg FW in fish for the European otter, *Lutra lutra*, and 17.6% exceeded 0.5 mg/kg FW, the recommended maximum level for human consumption. Most (61.0 to 97.0%) of the mercury in fish muscle was in the form of methylmercury. In otter scat, no methylmercury and a maximum of 0.12 mg total Hg/kg DW was measured. Gutleb et al. (1997) concluded that the concentrations of mercury in fish flesh may pose a threat to humans and wildlife feeding on the fish, and that all mercury discharges into tropical rainforests should cease immediately in order to protect human health and endangered wildlife.

11.2.2.3 Suriname

In Suriname (population 400,000), gold mining has existed on a small scale since 1876, producing about 200 kg of gold yearly through the 1970s and 1209 kg in 1988 using amalgamation as the extraction procedure of choice (De Kom et al., 1998). In the 1990s, gold mining activities increased dramatically due to hyperinflation and poverty; up to 15,000 workers produced 10,000 kg of crude gold in 1995 using mercury methodology unchanged for at least a century (De Kom et al., 1998). Riverine food fishes from the vicinity of small-scale gold mining activities in Suriname were contaminated with mercury, and concentrations in muscle of fish-eating fishes collected there exceeded the maximum permissible concentration of 0.5 mg total Hg/kg FW in 57.0% of the samples collected (Mol et al., 2001). Mercury pollution is now recognized as one of the main environmental problems in tropical South America, as judged by increasingly elevated concentrations of mercury in water, sediments, aquatic macrophytes, freshwater snails, fish, and fish-eating birds. Human health is compromised through occupational mercury exposures from mining, and from ingestion of methylmercury-contaminated fish (Mol et al., 2001).

11.2.2.4 Venezuela

In Venezuela, miners exploring the region for gold over a 10-year period, cut and burned virgin forests, excavated large pits through the floodplain with pumps and high-pressure hoses, hydraulically dredged stream channels, and used mercury (called exigi by miners) to isolate gold from sediments (Nico and Taphorn, 1994). The use of mercury in Venezuelan gold mining is common and widespread, with annual use estimated at 40 to 50 tons. Between 1979 and 1985, at least 87 tons of mercury were discharged into river systems (Nico and Taphorn, 1994). Forest soils in Venezuela near active gold mines contained as much as 129.3 mg Hg/kg DW vs. 0.15 to 0.28 mg/kg from reference sites (Davies, 1997; De Lacerda and Salomons, 1998). Maximum mercury concentrations measured in nine species of fish collected in 1992 from gold mining regions in the upper Cuyuni River system, Venezuela (in mg total Hg/kg FW) were 2.6 in liver, 0.9 in muscle, 0.8 in stomach contents, and 0.1 in ovary (Nico and Taphorn, 1994). The highest value of 8.9 mg/kg FW muscle was from a 63-cm long aimara, *Hoplias macrophthalmus*, a locally important food fish (Nico and Taphorn, 1994).

In 1994, Venezuela officially produced 8.7 tons of gold worth U.S.$ 100 million, with potential reserves of thousands of tons (LesEnfants, 1995). Venezuela is the fourth-largest producer of gold in Latin America after Brazil, Colombia, and Chile. Most of the unreported gold produced — estimated at 14 tons annually — is from the activities of 35,000 informal miners who use at least 40 tons of mercury each year in the mining zone of the Guayana Region of Venezuela. The Caroni River, the main tributary of the Orinoco River, annually receives 1.2 tons of mercury as residuals from mining activities (LesEnfants, 1995). There is chronic mercury contamination among miners and residents of the mining zones, with about 72.0% of all people sampled showing symptoms of mercury intoxication. These symptoms include blindness, hypertension, memory loss, and central nervous system disorders. In the lower Caroni Basin, 64.0% of the miners have concentrations of mercury in urine above the recommended tolerable limits. Sediments from the Caroni River contaminated by mercury from gold mining contained up to 3.7 mg Hg/kg FW vs. 0.02 from a reference site; and fish liver from the contaminated site contained up to 3.4 mg Hg/kg FW. It is noteworthy that the use of mercury, cyanide, and other toxic substances is prohibited in water bodies and exploitation sites according to Resolution 81, dated April 6, 1990 (LesEnfants, 1995).

11.2.3 Africa

Gold mining in Tanzania dates back to 1884 during Germany's colonial rule, with greater than 100,000 kg of gold produced officially since 1935. Between 150,000 and 250,000 Tanzanians are now involved in small-scale gold mining, with extensive use of mercury in gold recovery; about 6 tons of mercury are lost to the environment annually from gold mining activities. The current gold rush, which began in the 1980s and reached a peak in 1983, was stimulated, in part, by the relaxation of mining regulations by the state-controlled economy. Gold ore processing involves crushing and grinding the ore, gravity separation involving simple panning or washing, amalgamation with mercury of the gold-rich concentrate, and firing in the open air. Additional firing is conducted at the gold ore processing sites, in residential compounds, and inside residences. Further purification is done in the national commercial bank branches before the bank purchases the gold from the miners (Ikingura and Akagi, 1996; Ikingura et al., 1997). Mercury concentrations recorded in water at the Lake Victoria goldfields in Tanzania averaged 0.68 (0.01 to 6.8) μg/L vs. 0.02 to 0.33 μg/L at an inland water reference site and 0.02 to 0.35 μg/L at a coastal water reference site (Ikingura et al., 1997). Mercury concentrations in sediments, forest soils, and tailings at the Lake Victoria goldfields were 0.02 to 136.0 mg/kg in sediments, 3.4 (0.05 to 28.0) mg/kg in soils vs. 0.06 mg/kg at a reference site, and 16.2 (0.3 to 31.2) mg/kg in tailings (Ikingura and Akagi, 1996; Ikingura et al., 1997; De Lacerda and Salomons, 1998). The highest mercury concentrations recorded in any fish species from Lake Victoria or in rivers draining from gold processing sites were in the marbled lungfish, *Protopterus aethiopicus*, with 0.24 mg Hg/kg FW in muscle and 0.56 mg Hg/kg FW in liver; however, a catfish, *Clarias* sp., from a tailings pond contained 2.6 mg Hg/kg FW muscle and 4.5 in liver (Van Straaten, 2000).

In Obuasi, Ghana, elevated mercury concentrations were recorded in various samples collected from 14 sites near active gold mining towns and environs during 1992 to 1993 (Amonoo-Neizer et al., 1996). Mercury concentrations (in mg/kg DW) were 0.9 (0.3 to 2.5) in soil, 2.0 max. in whole mudfish (*Heterobranchus bidorsalis*), 2.3 max. in edible portions of plantain (*Musa paradisiaca*), 3.3 max. in edible portions of cassava (*Manihot esculenta*), 9.1 max. in whole elephant grass (*Pennisetum purpureum*), and 12.4 max. in whole water fern (*Ceratopterus cornuta*).

In Kenya, Ogola et al. (2002) report elevated mercury concentrations near gold mines in tailings (max. 1920 mg/kg DW) and surficial stream sediments (max. 348.0 mg/kg DW). Tailings also contained elevated concentrations of arsenic (76.0 mg/kg DW) and lead (510.0 mg/kg DW); stream sediments contained up to 1.9 mg As/kg DW, and up to 11,075.0 mg Pb/kg DW.

11.2.4 The People's Republic of China

Dexing County in Jiangxi Province was the site of about 200 small-scale gold mines using mercury amalgamation to extract gold between 1990 and 1995 (Lin et al., 1997). Gold firing was usually conducted in private residences. One result of this activity was mercury contamination of air (1.0 mg/m^3 in workshop, 2.6 mg/m^3 in workroom vs. 10 to 20 ng/m^3 at reference sites), waste water (0.71 mg/L), and solid tailings (max. 189.0 mg/kg) — with serious implications for human health. Since September 1996, however, most small-scale mining activities are prohibited through passage of national environmental legislation (Lin et al., 1997).

11.2.5 The Philippines

The gold rush began in eastern Mindanao in the 1980s, resulting in the development of several mining communities with more than 100,000 residents. Between 1986 and 1988, about 140 tons of mercury were released into the environment from 53 mining communities (Appleton et al., 1999). In the early 1990s, about 200,000 small-scale gold miners produced approximately 15 tons of gold each year and in the process released 25 tons of mercury annually, with concomitant contamination of aquatic ecosystems (Greer, 1993). In 1993, 15,000 gold panners/miners were engaged in gold production using mercury for gold extraction (Torres, 1995). Much of the discharged mercury is in the river sediments, where it can be recycled through flash flooding, consumption by bottom-feeding fish, or microbial digestion and methylation (Greer, 1993). In mercury-contaminated areas of Davao del Norte in Mindanao, mercury concentrations were as high as 2.6 mg/kg FW in muscle of carnivorous fishes, and 136.4 μg/m^3 in air (De Lacerda and Salomons, 1998). In 1995, drainage downstream of gold mines in Mindanao was characterized by extremely high levels of mercury in solution (2.9 mg/L) and in bottom sediments (greater than 20.0 mg Hg/kg) (Appleton et al., 1999). The Environment Canada sediment quality mercury toxic effect threshold for the protection of aquatic life (less than 1.0 mg Hg/kg DW) was exceeded for a downstream distance of 20 km. In sections of the stream used for fishing and potable water supply, the surface water mercury concentrations — for at least 14 km downstream — exceeded the World Health Organization's international standard for drinking water (set at less than 1.0 μg Hg/L) and the U.S. Environmental Protection Agency's water quality criteria for the protection of aquatic life (less than 2.1 μg/L) (Appleton et al., 1999). At present, about 1.6 tons of mercury are released for every ton of gold produced in the Philippines, with releases usually highest to the atmosphere and lower to rivers and soils. However, mercury releases can be reduced by at least an order of magnitude through the use of closed amalgamation systems and the use of retorts in the roasting process (De Lacerda and Salomons, 1998).

In May 1987, the first diagnosed mercury vapor poisonings were reported in the Philippines (Torres, 1995). One death and eleven intoxication cases resulted when a gold amalgam yielded 2 kg of gold after 8 hours of blowtorching. Workplace mercury levels in air exceeded the NIOSH (U.S. National Institute of Occupational Safety and Health) 8-h weighted average at various locations by 25.0 to 93.0%, and 50.0% exceeded the air mercury levels set by the World Health Organization. In 1988 to 1989, 6.0% of the workers in small-scale gold processing industries in the Philippines had elevated blood mercury levels. In 1991, 590 children, age 3 to 6 years, from gold processing workers, showed a significant association between gross motor and personal social development delay with exposure to inorganic mercury vapors.

11.2.6 Siberia

Measurements of air, soil, surface water, and groundwater near gold mines in eastern Transbaikalia, Siberia, where mercury amalgamation was used to extract gold from ores, showed no evidence of significant mercury contamination in any compartment measured (Tupyakov et al., 1995). Mean mercury concentrations (in ng/ m^3) in air near the gold mines vs. a reference site were 2.0 vs. 2.2. For soils it was 0.03 mg/kg vs. 0.02; and for surface water and ground water, these values were 0.01 μg/L vs. 0.006, and 0.008 μg/L vs. 0.005, respectively.

11.2.7 Canada

Gold was discovered in Nova Scotia in 1861 (Nriagu and Wong, 1997). Since then, more than 1.2 million troy ounces of gold were produced from about 65 mines in the province, with about 20.0% from mines located in Goldenville. Processing of the 3.3 million tons of ores from Goldenville mills between 1862 and 1935 released an estimated 63 tons of mercury to the biosphere, mostly prior to 1910. Mine tailings at one Goldenville site still contain large quantities of mercury (up to 2600.0 mg Hg/kg) as unreacted elemental mercury, unrecovered Au-Hg amalgams, and mercury compounds formed by side reactions during the amalgamation process. Current mercury levels in muds and sediments at the mine site range from 100.0 to 250.0 mg/kg, well above the 0.4 to 2.0 μg/kg typically found in uncontaminated stream sediments (Nriagu and Wong, 1997). Concentrations of other elements in Goldenville mine tailings were also high: up to 2000.0 mg As/kg, 1400.0 mg Pb/kg, 500.0 mg Cu/kg, and 80.0 mg Se/kg. Except for some mosses, no vegetation grows on the tailings-covered areas. Stream sediments located 200 m downstream of the mine were 10 to 1000 times higher than upstream concentrations for Hg, As, Cd, Cu, Pb, and Se. Little is known about the forms of mercury in contaminated mine sites that can affect the loss rate of mercury to the environment from the waste materials (Nriagu and Wong, 1997). Aquatic macrophytes from gold mining areas in Nova Scotia, where mercury was used extensively, accumulated significant quantities of mercury (De Lacerda and Salomons, 1998). Mercury concentrations were higher in emergent species (16.3 mg/kg DW) when compared to floating and submerged species (0.54 to 0.56 mg/kg DW). And aquatic macrophytes growing on tailings had higher concentrations of mercury in roots than in shoots (De Lacerda and Salomons, 1998).

In Quebec, gold mines at Val d'Or used mercury amalgamation techniques throughout most of the 20th century to produce gold (Meech et al., 1998). Most abandoned sites and environs now show high mercury concentrations in sediments (up to 6.0 mg/kg) and in fish muscle (up to 2.6 mg/kg FW). As alluvial gold deposits have become exhausted in North America, cyanidation has become the primary method for primary gold production, and amalgamation is now limited to individual prospectors (Meech et al., 1998).

In Yellowknife, NWT, about 2.5 tons of mercury were discharged between the mid-1940s and 1968, together with tailings, into Giauque Lake. This lake is now listed as a contaminated site under the Environment Canada National Contaminated Site Program (Meech et al., 1998). In 1977 to 1978, about 70.0% of the bottom sediments contained more than 0.5 mg Hg/kg DW (Moore and Sutherland, 1980). Also in 1977 to 1978, lake trout (*Salvelinus namaycush*) were found to contain 3.8 mg Hg/kg FW muscle; for northern pike (*Esox lucius*) and round whitefish (*Prosopium cylindraceum*), these values were 1.8 and 1.2 mg/kg FW, respectively. Authors concluded that contamination of only a small part of the lake results in high levels in fish throughout the lake, and this is probably due to fish movement from mercury-contaminated to uncontaminated areas (Moore and Sutherland, 1980). In 1995, sediments from Giauque Lake were dated using radiolead-210 and radiocesium-137, and analyzed for mercury (Lockhart et al., 2000). The peak mercury concentrations in sediment cores ranged between 2.0 and 3.0 mg Hg/kg DW. The history of mercury deposition derived from the dated sediment cores agreed well with the known history of input from gold mining (Lockhart et al., 2000).

11.2.8 The United States

Gold mining in the United States is ubiquitous; however, persistent mercury hazards to the environment were considered most severe from activities conducted during the latter portion of the 19th century, especially in Nevada. In the 50-year period from 1850 to 1900, gold mining in the United States consumed 268 to 2820 tons of mercury yearly, or about 70,000 tons during that period (De Lacerda and Salomons, 1998). Mercury contamination from gold mining in the Sierra Nevada region of California during the late 1800s and early 1900s was extensive in watersheds where placer gold was recovered and processed by amalgamation (Hunerlach et al., 1999). Elemental mercury from these operations continues to enter local and downstream water bodies via transport of contaminated sediments by river flooding (Hunerlach et al., 1999).

A 1998 study that measured total mercury accumulations in predatory fishes collected nationwide showed that mercury levels in muscle were significantly correlated with methylmercury concentrations in water, pH of the water, percent wetlands in the basin, and the acid volatile content of the sediment (Brumbaugh et al., 2001). These four variables — especially methylmercury levels in water — accounted for 45.0% of the variability in mercury concentrations of fish, normalized by total length. A methylmercury water concentration of 0.12 ng/L was, on average, associated with a fish fillet concentration of 0.3 mg Hg/kg FW for an age 3 (years) fish when all species were considered. Sampling sites with the highest overall mercury concentrations in water, sediment, and fish were highest in the Nevada Basins and environs (from historic gold mining activities), followed, in order, by the South Florida Basin, Sacramento River Basin in California, Santee River Basin and Drainages in South Carolina, and the Long Island and New Jersey Coastal Drainages (Table 11.2). Elevated mercury concentrations in fish, except for Nevada, are not necessarily a result of gold mining activities. The mercury criterion for human health protection set by the U.S. Environmental Protection Agency in 2001 is now 0.3 mg Hg/kg FW diet, down from 0.5 mg/kg FW previously (Brumbaugh et al., 2001).

Beginning in 1859, gold from the Comstock Lode near Virginia City, Nevada, was processed at 30 sites using a crude mercury amalgamation process, discharging about 6,750,000 kg mercury to the environment during the first 30 years of mine operation (Miller et al., 1996). Over time, mercury-contaminated sediments were eroded and transported downstream by fluvial processes. The most heavily contaminated wastes — with a total estimated volume of 710,700 m^3 — contained 31,500 kg mercury, 248 kg gold, and 37,000 kg silver. If site remediation is conducted, extraction of the gold and silver — worth about U.S.$12 million — would defray a significant portion of the cleanup costs (Miller et al., 1996).

In the Carson River-Lahontan Reservoir (Nevada) watershed, approximately 7100 tons of metallic mercury were released into the watershed between 1859 and 1890 as a by-product of silver and gold ore refining (Wayne et al., 1996). Between 1859 and 1920, about 5 million troy ounces of gold (180 kg) were produced from this area, as well as 2500 kg of silver, mostly from using amalgamation (Lawren, 2003). In 1901, cyanide leaching was introduced and eventually replaced amalgamation as the dominant gold recovery process (Lawren, 2003; Eisler, 2004b). During the past 130 years, mercury has been redistributed throughout 500 km^2 of the basin, and mercury concentrations at this site rank among the highest reported in North America (Gustin et al., 1995). Mercury contamination was still severe in this region in 1993 (Table 11.2). Nevada authorities have issued health advisories against fish consumption from the Carson River; an 80-km stretch of the river has been declared a Superfund site (Da Rosa and Lyon, 1997). Mercury-contaminated tailings were dispersed throughout the lower Carson River, Lahontan Reservoir, and the Carson Sink by floods that occurred 19 times between 1861 and 1997 (Hunerlach et al., 1999; Lawren, 2003). Low levels of methylmercury in surface waters of the Carson River-Lahontan Reservoir are attributed to increasing pH and increasing concentrations of anions (of selenium [SeO_4^{2-}], molybdenum [MoO_4^{2-}], and tungsten [WO_4^{2-}]), both of which are inhibitory to sulfate-reducing bacteria known to play a key role in methylmercury production in anoxic sediments (Bonzongo et al., 1996a).

Table 11.2 Total Mercury Concentrations in Abiotic Materials, Plants, and Animals near Historic Gold Mining and Refining Sites in the United States

Location, Sample, and Other Variables	Concentration[a]	Ref.[c]
Alaska; Sediments		
Northeastern Bering Sea; 80 m below sea floor; typical vs. anomalies	0.03 mg/kg DW vs. 0.2–1.3 mg/kg DW	1
Modern beach; mean vs. max.	0.22 mg/kg DW vs. 1.3 mg/kg DW	1
Nearshore subsurface gravels; mean vs. max.	0.06 mg/kg DW vs. 0.6 mg/kg DW	1
Seward Peninsula: mean vs. max.	0.1 mg/kg DW vs. 0.16 mg/kg DW	1
Georgia; Former Gold Mining Areas		
Historical values (1829–1940):		
Sediments	0.12–12.0 mg/kg DW	14
Soils	0.2–0.6 mg/kg DW	14
Samples collected in 1990s:		
Surface waters	Max. 0.013 μg/L	12
Sediments	Max. 4.0 mg/kg DW near source mines; < 0.1 mg/kg DW 10–15 km downstream	13
Clams and mussels, soft parts	0.7 mg/kg DW	13
Nationwide; 1998; 106 Sites; Freshwater Fish; Muscle		
Lahontan Reservoir, Nevada (near historic gold mining sites)	3.34 mg/kg FW	2
South Florida Basin	0.95 mg/kg FW	2
Santee River, South Carolina	0.70 mg/kg FW	2
Sacramento River Basin, California	0.46 mg/kg FW	2
Yellowstone River Basin	0.44 mg/kg FW	2
Acadian-Ponchartrain Basin	0.39 mg/kg FW	2
Others	< 0.3 mg/kg FW[b]	2
Nevada; Migratory Waterbirds; 1997–1998; Nesting on Lower Carson River vs. Reference Site (Humboldt River)		
Double-crested cormorant, *Phalacrocorax auritus*; adults:		
Blood	17.1 (11.6–22.0) μg/L vs. 3.1 μg/L	3
Brain	11.3 (8.9–14.9) mg/kg FW vs. 1.2 mg/kg FW	3
Kidney	69.4 (36.2–172.9) mg/kg FW vs. 9.0 mg/kg FW	3
Liver	134.0 (82.4–222.2) mg/kg FW vs. 18.0 mg/kg FW	3
Snowy egret, *Egretta thula*; adults:		
Blood	5.9 (2.8–10.3) μg/L vs. 3.1 μg/L FW	3
Brain	2.3 (2.0–3.8) mg/kg FW vs. 0.9 mg/kg FW	3
Kidney	11.1 (5.7–21.7) mg/kg FW vs. 3.1 mg/kg FW	3
Liver	43.7 (16.0–109.9) mg/kg FW vs. 7.9 mg/kg FW	3
Black-crowned night-heron, *Nycticorax nycticorax*; adults:		
Blood	6.6 (3.2–14.2) μg/L vs. 2.5 μg/L FW	3
Brain	1.7 (0.8–5.5) mg/kg FW vs. 0.5 mg/kg FW	3
Kidney	6.1 (3.4–15.1) mg/kg FW vs/ 1.8 mg/kg FW	3
Liver	13.5 (5.0–49.6) mg/kg FW vs. 2.1 mg/kg FW	3
Nevada; Migratory Waterbirds; 1997–1998; Lahontan Reservoir		
Stomach contents; adults; total mercury vs. methylmercury:		
Black-crowned night-heron	0.5 mg/kg FW vs. 0.48 mg/kg FW	3
Snowy egret	1.0 mg/kg FW vs. 0.0 mg/kg FW	3
Double-crested cormorant	1.4 (0.8–2.2) mg/kg FW vs. 1.2 (0.8–1.6) mg/kg FW	3

Table 11.2 (continued) Total Mercury Concentrations in Abiotic Materials, Plants, and Animals near Historic Gold Mining and Refining Sites in the United States

Location, Sample, and Other Variables	Concentration[a]	Ref.[c]
Nestlings/fledglings; Lahontan Reservoir vs. reference site:		
Black-crowned night-heron:		
Blood	3.2 (1.2–5.7) µg/L vs. 0.6 µg/L FW	3
Brain	1.1 (0.6–1.7) mg/kg FW vs. 0.1 mg/kg FW	3
Feathers	32.3 (21.9–67.1) mg/kg FW vs. 5.5 (3.4–16.6) mg/kg FW	3
Kidney	3.4 (1.8–6.3) mg/kg FW vs. 0.5 mg/kg FW	3
Liver	4.0 (1.1–7.8) mg/kg FW vs. 0.7 mg/kg FW	3
Snowy egret		
Blood	2.7 µg/L vs. 0.4 µg/L FW	3
Brain	0.7 mg/kg FW vs. 0.1 mg/kg FW	3
Feathers	30.6 (14.7–59.8) mg/kg FW vs. 7.8 mg/kg FW	3
Kidney	2.2 mg/kg FW vs. 0.3 mg/kg FW	3
Liver	2.7 mg/kg FW vs. 0.4 mg/kg FW	3
Double-crested cormorant:		
Blood	5.4 (3.4–7.2) µg/L vs. 0.8 (0.6–1.2) µg/L	3
Brain	2.7 (2.1–3.3) mg/kg FW vs. 0.5 (0.4–0.5) mg/kg FW	3
Feathers	66.3 (54.0–87.3) mg/kg FW vs. 11.8 (10.9–13.5) mg/kg FW	3
Kidney	6.2 (2.1–9.8) mg/kg FW vs. 1.2 (1.1–1.3) mg/kg FW	3
Liver	10.9 (8.7–14.9) mg/kg FW vs. 1.8 (1.4–2.6) mg/kg FW	3
Nevada; Abiotic Materials		
Air; over Comstock Lode tailings vs. reference site	0.23 µg/m³ vs. 0.01–0.04 µg/m³	4
Carson River-Lahontan Reservoir; Sampled in Early 1990s		
Surface waters: total mercury vs. methylmercury	Max. 7.6 µg/L vs. max. 0.007 µg/L	5
Surface waters: upstream vs. downstream	0.004 µg/L vs. 1.5–2.1 µg/L	6
Water; filtered vs. unfiltered	0.113 µg/L vs. max. 0.977 µg/L	8
Unfiltered reservoir water vs. reference site	0.053–0.391 µg/L vs. 0.001–0.0033 µg/L	7
Mill tailings	3.0–1,610.0 mg/kg FW	7
Atmospheric vapor vs. reference site	0.002–0.294 µg/m³ vs. 0.001–0.004 µg/m³	7
Sediments vs. reference site	27.0 (2.0–156.0) mg/kg FW vs. 0.01–0.05 mg/kg FW	7
Sediments; Carson River; below mines vs. above mines	Max. 881.0 mg/kg DW vs. 0.03–6.1 mg/kg DW	8
Sediments; Lahontan Reservoir; before dam vs. after dam	0.011 mg/kg DW vs. 1.0–60.0 mg/kg DW	8
Sediments; 1990s; 60 km downstream of Comstock Lode discharges	> 100.0 mg/kg FW	11
Alluvial fan soils; Six Mile Canyon:		
Premining (1859)	0.5 (0.07–1.9) mg/kg DW	11
Post mining:		
Fan deposits	103.7 (1.3–368.9) mg/kg DW	11
Modern channel	4.8 (1.1–9.3) mg/kg DW	11
Nevada; Carson River vs. Reference Site; Fish and Ducks; Sampled in Early 1990s		
Fish muscle	Max. 5.5 mg/kg FW vs. max. < 0.5 mg/kg FW	7
Duck muscle	> 0.5 mg/kg FW vs. < 0.5 mg/kg FW	7
North Carolina; near Marion (gold mining activity using mercury in the 1800s to early 1900s); 1991; Maximum Concentrations		
Heavy metal concentrates	784.0 mg/kg	9
Gold grains	448.0 mg/kg	9

(continued)

Table 11.2 (continued) Total Mercury Concentrations in Abiotic Materials, Plants, and Animals near Historic Gold Mining and Refining Sites in the United States

Location, Sample, and Other Variables	Concentration[a]	Ref.[c]
Sediments	7.4 mg/kg	9
Moss (max. gold content of 1.85 mg/kg)	4.9 mg/kg	9
Fish muscle	< 0.2 mg/kg FW	9
Stream and well waters	< 2.0 μg/L (limit of detection)	9
South Dakota; Lake Oahe; Gold Mining Discontinued in 1970; Sampled 1970–1971		
Water	< 0.2 μg/L	10
Sediments	0.03–0.62 mg/kg DW; < 0.03–0.33 mg/kg FW	10
Fish muscle	0.02–1.05 mg/kg FW; 13.0% of samples had > 0.5 mg/kg FW, especially older, larger northern pike, *Esox lucius*	10

[a] Concentrations are shown as means, range (in parentheses), maximum (max.), and nondetectable (ND).
[b] Current mercury criterion to protect human health is < 0.3 mg/kg FW, down from < 0.5 mg/kg FW (Brumbaugh et al., 2001).
[c] Reference: 1, Nelson et al., 1975; 2, Brumbaugh et al., 2001; 3, Henny et al., 2002; 4, De Lacerda and Salomons, 1998; 5, Bonzongo et al., 1996a; 6, Bonzongo et al., 1996b; 7, Gustin et al., 1995; 8, Wayne et al., 1996; 9, Callahan et al., 1994; 10, Walter et al., 1973; 11, Miller et al., 1996; 12, Mastrine et al., 1999; 13, Leigh, 1994; 14, Leigh, 1997.

Methylmercury concentrations ranged from 0.1 to 0.7 ng/L in this ecosystem and were positively associated with total suspended solids (Bonzongo et al., 1996b; Lawren, 2003). Removal of mercury from the water column was attributed to binding to particles, sedimentation, and volatilization of dissolved gaseous mercury.

The Lahontan Reservoir supports game and commercial fisheries, and the Lahontan Valley wetlands — home to many species of birds — is considered the most mercury-contaminated natural system in the United States (Henny et al., 2002). Mercury concentrations in sediments from this site and environs ranged from 10.0 to 30.0 mg/kg, or about 80 times higher than uncontaminated sediments; elevated concentrations of mercury were documented from the site in annelid worms, aquatic invertebrates, fishes, frogs, toads, and birds (Da Rosa and Lyon, 1997; Lawren, 2003). In 1997 to 1998, three species of fish-eating birds nesting along the lower Carson River were examined for mercury contamination: double-crested cormorants (*Phalacrocorax auritus*), snowy egrets (*Egretta thula*), and black-crowned night-herons (*Nycticorax nycticorax*) (Henny et al., 2002). The high concentrations of total mercury observed in livers (means, in mg/kg FW, of 13.5 in herons, 43.7 in egrets, and 69.4 in cormorants) and kidneys (6.1 in herons, 11.1 in egrets, 69.4 in cormorants) of adult birds (Table 11.2) were possible due to a threshold-dependent demethylation coupled with sequestration of the resultant inorganic mercury (Henny et al., 2002). Demethylation and sequestration processes also appeared to have reduced the amount of methylmercury distributed to eggs — although the short time spent by adults in the contaminated area was a factor in the lower-than-expected mercury concentrations in eggs (Henny et al., 2002; Table 11.2). Most eggs had mercury concentrations, as methylmercury, below 0.8 mg/kg FW, the threshold concentration where reproductive problems might be expected (Heinz, 1979; Heinz and Hoffman, 2003). After hatching, young birds were fed diets by parent birds averaging 0.36 to 1.18 mg methylmercury/kg FW for 4 to 6 weeks through fledging (Henny et al., 2002). Eisler (2000) recommends avian dietary intakes of less than 0.1 mg Hg/kg FW ration. Mercury concentrations in organs of the fledglings were much lower than those found in adults, but evidence was detected of toxicity to immune, detoxicating, and nervous systems (Henny et al., 2002). Immune deficiencies and neurological impairment of fledglings may affect survival when burdened with stress associated with learning to forage and ability to complete the first migration. Oxidative stress was noted in young cormorants containing the highest concentrations of mercury, as evidenced by increasing thiobarbituric acid-reactive substances and altered glutathione metabolism (Henny et al., 2002). Henny et al. (2002), based on their studies with fish-eating birds, conclude that:

1. Adults tolerate relatively high levels of mercury in critical tissues through demethylation processes that occur above threshold concentrations.
2. Adults demethylate methylmercury to inorganic mercury, which is excreted or complexed with selenium and stored in liver and kidney. This change in form and sequestering process reduces the amount of methylmercury circulating in tissues and in the amount available for deposition in eggs.
3. The low concentrations of methylmercury in eggs are attributed to the short duration of time spent in the area prior to egg laying, and to the demethylation and sequestering processes within the birds.
4. Young of these migratory fish-eating birds experience neurological and histological damage associated with exposure to dietary mercury.

In the southeastern United States, gold mining was especially common between 1830 and 1849, and again during the 1880s. Historical gold mining activities contributed significantly to mercury problems in this region, as evidenced by elevated mercury concentrations in surface waters (Mastrine et al., 1999). Mercury in surface waters was positively correlated with total suspended solids and with bioavailable iron. Vegetation in the southeast is comparatively heavy and by controlling erosion may reduce the total amount of mercury released from contaminated mining sites to the rivers. Mercury concentrations in surface waters of southeastern states were significantly lower — by orders of magnitude — than those from western states where amalgamation extraction techniques were also practiced, and this may reflect the higher concentrations of mercury used and the sparser vegetation of the western areas (Mastrine et al., 1999).

In northern Georgia, gold was discovered in 1829 and mined until about 1940 (Leigh, 1994, 1997). Extensive use of mercury probably began in 1838 when stamp mills (ore crushers) were introduced to help recover gold from vein ore, with about 38.0% of all mercury used in gold mining escaping into nearby streams. Mercury concentrations in historical floodplain sediments near the core of the mining district were as high as 4.0 mg/kg DW, but decreased with increasing distance downstream to less than 0.1 mg/kg at 10 to 15 km from the source mines. Near Dahlonega, Georgia, historical sediments contained as much as 12.0 mg Hg/kg, and mean values in streambanks near the mining district of 0.2 to 0.6 mg Hg/kg DW. Mercury-contaminated floodplain sediments pose a potential hazard to wildlife. Clams and mussels, for example, collected from these areas more than 50 years after mining had ceased contained elevated (0.7 mg/kg DW) mercury concentrations in soft parts. Erosion of channel banks and croplands, with subsequent transport of mercury-contaminated sediments from the mined watersheds, is likely to occur for hundreds or thousands of years. Mercury was the only significant trace metal contaminant resulting from former gold mining activities in Georgia, exceeding the U.S. Environmental Protection Agency's "heavily polluted" guideline for sediments of greater than 1.0 mg Hg/kg. Other metals examined did not exceed the "heavily polluted" sediment guidelines of greater than 50.0 mg/kg for copper, greater than 200.0 mg/kg for lead, and greater than 200.0 mg/kg for zinc (Leigh, 1994, 1997).

In South Dakota, most of the fish collected from the Cheyenne River arm of Lake Oahe in 1970 contained elevated (greater than 0.5 mg/kg FW muscle) concentrations of mercury (Walter et al., 1973). Elemental mercury was used extensively in this region between 1880 and 1970 to extract gold from ores and is considered to be the source of the contamination. In 1970, when the use of mercury in the gold recovery process was discontinued at this site, liquid wastes containing 5.5 to 18.0 kg of mercury were being discharged daily into the Cheyenne River arm (Walter et al. ,1973).

In Nome, Alaska, gold mining is responsible, in part, for the elevated mercury levels (max. 0.45 mg/kg DW) measured in modern beach sediments (Nelson et al., 1975). However, higher concentrations (max. 0.6 mg/kg) routinely occur in buried Pleistocene sediments immediately offshore and in modern, nearby unpolluted beach sediments (1.3 mg Hg/kg). This suggests that the effects of mercury contamination from mining are less than natural concentration processes in the Seward Peninsula region of Alaska (Nelson et al., 1975).

11.3 SUMMARY

Mercury contamination of the environment from historical and ongoing mining practices that rely on mercury amalgamation for gold extraction is widespread. Contamination was particularly severe in the immediate vicinity of gold extraction and refining operations. However, mercury — especially in the form of water-soluble methylmercury — can be transported to pristine areas by rainwater, water currents, deforestation, volatilization, and other vectors. Examples of gold mining-associated mercury pollution are shown for Canada, the United States, Africa, China, the Philippines, Siberia, and South America. In parts of Brazil, for example, mercury concentrations in all abiotic materials, plants, and animals — including endangered species of mammals and reptiles — collected near ongoing mercury-amalgamation gold mining sites were far in excess of allowable mercury levels promulgated by regulatory agencies for the protection of human health and natural resources. Although health authorities in Brazil are unable to detect conclusive evidence of human mercury intoxication, the potential exists in the absence of mitigation for epidemic mercury poisoning of the mining population and environs. In the United States, environmental mercury contamination is mostly from historical gold mining practices; however, portions of Nevada now remain sufficiently mercury-contaminated to pose a hazard to reproduction of carnivorous fishes and fish-eating birds.

REFERENCES

Akagi, H., O. Malm, F.J.P. Branches, Y. Kinjo, Y. Kashima, J.R.D. Guimaraes, R.B. Oliveira, K. Haraguchi, W.C. Pfeiffer, Y. Takizawa, and H. Kato. 1995. Human exposure to mercury due to goldmining in the Tapajos River Basin, Amazon, Brazil: speciation of mercury in human hair, blood and urine, *Water Air Soil Pollut.*, 80, 85–94.

Alho, C.J.R. and L.M. Viera. 1997. Fish and wildlife resources in the Pantanal wetlands of Brazil and potential disturbances from the release of environmental contaminants, *Environ. Toxicol. Chem.*, 16, 71–74.

Amonoo-Neizer, E.H., D. Nyamah, and S.B. Bakiamoh. 1996. Mercury and arsenic pollution in soil and biological samples around the mining town of Obuasi, Ghana, *Water Air Soil Pollut.*, 91, 363–373.

Appleton J.D., T.M. Williams, N. Breward, A. Apostol, J. Miguel, and C. Miranda. 1999. Mercury contamination associated with artisanal gold mining on the island of Mindanao, the Philippines, *Sci. Total Environ.*, 228, 95–109.

Aula, I., H. Braunschweiler, T. Leino, I. Malin, P. Porvari, T. Hatanaka, M. Lodenius, and A. Juras. 1994. Levels of mercury in the Tucurui reservoir and its surrounding area in Para, Brazil. In C.J. Watras and J.W. Huckabee, (Eds.), *Mercury Pollution: Integration and Synthesis,* p. 21–40. Lewis Publishers, Boca Raton, FL.

Aula, I., H. Braunschweiler, and I. Malin. 1995. The watershed flux of mercury examined with indicators in the Tucurui reservoir in Para, Brazil, *Sci. Total Environ.*, 175, 97–107.

Boas, R.C.V. 1995. Mineral extraction in Amazon and the environment. The mercury problem. In *Chemistry of the Amazon: Biodiversity, Natural Products, and Environmental Issues,* p. 295–303. American Chemical Society, Washington, D.C., ACS Symposium Series 588.

Bonzongo, J.C., K.J. Heim, Y. Chen, W.B. Lyons, J.J. Warwick, G.C. Miller, and P.J. Lechler. 1996a. Mercury pathways in the Carson River-Lahontan Reservoir system, Nevada, USA, *Environ. Toxicol. Chem.*, 15, 677–683.

Bonzongo, J.C., K.J. Heim, J.J. Warwick, and W.B. Lyons. 1996b. Mercury levels in surface waters of the Carson River-Lahontan Reservoir system, Nevada: influence of historic mining activities, *Environ. Pollut.*, 92, 193–201.

Branches, F.J.P., M. Harada, H. Akagi, O. Malm, H. Kato, and W.C. Pfeiffer. 1994. Human mercury contamination as a consequence of goldmining activity in the Tapajos River Basin, Amazon, Brazil. In *Proceedings of the International Symposium on Assessment of Environmental Pollution and Health Effects from Methylmercury*, p. 19–32. October 8–9, 1993, Kumamoto, Japan. National Institute for Minamata Disease, Minamata City, Kumamoto 867, Japan.

Brazaitis, P., G.H. Rebelo, C. Yamashita, E.A. Odierna, and M.E. Watanabe. 1996. Threats to Brazilian crocodile populations, *Oryx*, 30, 275–284.

Brumbaugh, W.G., D.P. Krabbenhoft, D.R. Helsel, J.G. Wiener, and K.R. Echols. 2001. A National Pilot Study of Mercury Contamination of Aquatic Ecosystems along Multiple Gradients: Bioaccumulation in Fish, *U.S. Geol. Surv., Biol. Sci. Rep. USGS/BRD/BSR 2001-0009,* 25 pp.

Callahan, J.E., J.W. Miller, and J.R. Craig. 1994. Mercury pollution as a result of gold extraction in North Carolina, U.S., *Appl. Geochem.*, 9, 235–241.

Callil, C.T. and W.J. Junk. 1999. Concentration and incorporation of mercury by mollusc bivalves *Anodontites trapesialis* (Lamarck, 1819) and *Castalia ambigua* (Lamarck, 1819) in Pantanal of Pocone-MT, Brasil, *Biociecias*, 7, 3–28. In Portuguese with English abstract.

Camara, V.M., M.I.D.F. Filhote, M.I.M. Lima, F.V. Aleira, M.S. Martins, T.O. Dantes, and R.R. Luiz. 1997. Strategies for preventing adolescent mercury exposure in Brazilian gold mining areas, *Toxicol. Indus. Health*, 13, 285–297.

Castilhos, Z.C., N. Almonsy, P.S. Souto, L.C.C. Pereira da Silva, A.R. Linde, and E.D. Bidone. 2004. Bioassessment of ecological risk of Amazonian ichthyofauna to mercury, *Bull. Environ. Contam. Toxicol.*, 72, 671–679.

Castilhos, Z.C., E.D. Bidone, and L.D. Lacerda. 1998. Increase of the background human exposure to mercury through fish consumption due to gold mining at the Tapajos River region, Para State, Amazon, *Bull. Environ. Contam. Toxicol.*, 61, 202–209.

Cleary, D. 1995. Mercury contamination in the Brazilian Amazon: an overview of epidemiological issues. In *Proceedings of the International Workshop on Environmental Mercury Pollution and its Health Effects in Amazon River Basin*, p. 61–72. Rio de Janeiro, 30 November–2 December 1994. Published by National Institute for Minamata Disease, Minamata City, Kumamoto 867, Japan.

Counter, S.A., L.H. Buchanan, G. Laurell, and F. Ortega. 1998. Blood mercury and auditory neuro-sensory responses in children and adults in the Nambija gold mining area of Ecuador, *Neurotoxicology*, 19, 185–196.

Counter, S.A., L.H. Buchanan, and F. Ortega. 2005. Mercury levels in urine and hair of children in an Andean gold-mining settlement, *Int. J. Occup. Environ. Health*, 11, 132–137.

Counter, S.A., L.H. Buchanan, F. Ortega, and G. Laurell. 2002. Elevated blood mercury and neuro-otological observations in children of the Ecuadorian gold mines, *J. Toxicol. Environ.Health*, 65 (Part A), 149–163.

Cursino, L., S.M Olberda, R.V. Cecilio, R.M. Moreira, E. Chartone-Souza, and A.M.A. Nascimento. 1999. Mercury concentrations in the sediment at different gold prospecting sites along the Carmo stream, Minas Gerais, Brazil, and frequency of resistant bacteria in the respective aquatic communities, *Hydrobiologia*, 394, 5–12.

Da Rosa, C.D. and J.S. Lyon (Eds.), 1997. *Golden Dreams, Poisoned Streams.* Mineral Policy Center, Washington, D.C., 20006, 269 pp.

Davies, B.E. 1997. Deficiencies and toxicities of trace elements and micronutrients in tropical soils: limitations of knowledge and future research needs, *Environ. Toxicol. Chem.*, 16, 75–83.

De Kom, J.F.M., G.B. van der Voet, and F.A. de Wolff. 1998. Mercury exposure of maroon workers in the small scale gold mining in Suriname, *Environ. Res.*, 77A, 91–97.

De Lacerda, L.D. 1997. Atmospheric mercury and fish contamination in the Amazon, *Cien. Cult. J. Brazil. Assoc. Adv. Sci.*, 49, 54–57.

De Lacerda, L.D. and W. Salomons. 1998. *Mercury from Gold and Silver Mining: A Chemical Time Bomb?.* Springer, Berlin, 146 pp.

D'Itri, P. and F.M. D'Itri. 1977. *Mercury Contamination: A Human Tragedy.* John Wiley, New York. 311 pp.

Dorea, J.G., M.B. Moreira, G. East, and A.C. Barbosa. 1998. Selenium and mercury concentrations in some fish species of the Madeira River, Amazon Basin, Brazil, *Biol. Trace Elem. Res.*, 65, 211–220.

Eisler, R. 2000. *Mercury.* In *Handbook of Chemical Risk Assessment: Health Hazards to Humans, Plants, and Animals. Vol. 1, Metals,* p. 313-409. Lewis Publishers, Boca Raton, FL.

Eisler, R. 2003. Health risks of gold miners: a synoptic review, *Environ. Geochem. Health*, 25, 325–345.
Eisler, R. 2004a. Mercury hazards from gold mining to humans, plants, and animals, *Rev. Environ. Contam. Toxicol.*, 181, 139–198.
Eisler, R. 2004b. *Biogeochemical, Health, and Ecotoxicological Perspectives on Gold and Gold Mining*. CRC Press, Boca Raton, FL. 355 pp.
Fields, S. 2001. Tarnishing the earth: gold mining's dirty secret. *Environ. Health Perspect.*, 109, A474–A482.
Forsberg, B.R., M.C.S. Forsbeerg, C.R. Padovani, E. Sargentini, and O. Malm. 1995. High levels of mercury in fish and human hair from the Rio Negro Basin (Brazilian Amazon): natural background or anthropogenic contamination?. In *Proceedings of the International Workshop on Environmental Mercury Pollution and Its Health Effects in Amazon River Basin,* p. 33–38. Rio de Janeiro, 30 November–2 December 1994. Published by National Institute for Minamata Disease, Minamata City, Kumamoto 867, Japan.
Fostier, A.H., M.C. Forti, J.R.D. Guimareaes, A.J. Melfi, R. Boulet, C.M.E. Santo, and F.J. Krug. 2000. Mercury fluxes in a natural forested Amazonian catchment (Serra do Navio, Amapa State, Brazil), *Sci. Total Environ.*, 260, 201–211.
Frery, N., R. Maury-Brachet, E. Maillot, M. Deheeger, B. de Merona, and A. Boudou. 2001. Gold-mining activities and mercury contamination of native Amerindian communities in French Guiana: key role of fish in dietary uptake, *Environ. Health Perspec.*, 109, 449–456.
Greer, J. 1993. The price of gold: environmental costs of the new gold rush, *The Ecologist*, 23(3). 91–96.
Guimaraes, J.R.D., O. Malm, and W.C. Pfeiffer. 1995. A simplified technique for measurements of net mercury methylation rates in aquatic systems near goldmining areas, Amazon, Brazil, *Sci. Total Environ.*, 175, 151–162.
Guimaraes, J.R.D., M. Meili, O. Malm, and E.M.S. Brito. 1998. Hg methylation in sediments and floating meadows of a tropical lake in the Pantanal floodplain, Brazil, *Sci. Total Environ.*, 213, 165–175.
Gustin, M.S., G.E. Taylor, Jr., and T.L. Leonard. 1995. contaminated mill tailings in the Carson River drainage basin, NV, *Water Air Soil Pollut.*, 80, 217–220.
Gutleb, A.C., C. Schenck, and E. Staib. 1997. Giant otter (*Pteronura brasiliensis*) at risk? Total mercury and methylmercury levels in fish and otter scats, Peru, *Ambio*, 26, 511–514.
Hacon, S., P. Artaxo, F. Gerab, M.A. Yamasoe, R.C. Campos, L.F. Conti, and L.D. De Lacerda. 1995. Atmospheric mercury and trace elements in the region of Alta Floresta in the Amazon basin, *Water Air Soil Pollut.*, 80, 273–283.
Hacon, S., E.R. Rochedo, R. Campos, G. Rosales, and L.D. Lacerda. 1997. Risk assessment of mercury in Alta Floresta, Amazon Basin — Brazil, *Water Air Soil Pollut.*, 97, 91–105.
Harada, M. 1994. Preliminary field study in Tapajos River Basin, Amazon. In *Proceedings of the International Symposium on Assessment of Environmental Pollution and Health Effects from Methylmercury*, p. 33–40. October 8–9, 1993, Kumamoto, Japan. National Institute for Minamata Disease, Minamata City, Kumamoto 867, Japan.
Heinz, G.H. 1979. Methylmercury: reproductive and behavioral effects on three generations of mallard ducks, *J. Wildl. Manage.*, 43, 394–401.
Heinz, G.H. and D.J. Hoffman. 2003. Embryotoxic thresholds of mercury: estimates from individual mallard eggs, *Arch. Environ. Toxicol. Chem.*, 44, 257–264.
Henny, C.J., E.F. Hill, D.J. Hoffman, M.G. Spalding, and R.A. Grove. 2002. Nineteenth century mercury: hazard to wading birds and cormorants of the Carson River, Nevada, *Ecotoxicology*, 11, 213–231.
Hunerlach, M.P., J.J. Rytuba, amd C.N. Alpers. 1999. Mercury contamination from hydraulic placer-gold mining in the Dutch Flat mining district, California, *U.S. Geol. Surv. Water-Resour. Invest.*, Rept. 99-4018B, 179–189.
Ikingura, J.R., M.K.D. Mutakyahwa, J.M.J. Kahatano. 1997. Mercury and mining in Africa with special reference to Tanzania, *Water Air Soil Pollut.*, 97, 223–232.
Ikingura, J.R. and H. Akagi. 1996. Monitoring of fish and human exposure to mercury due to gold mining in the Lake Victoria goldfields, Tanzania, *Sci. Total Environ.*, 191, 59–68.
Jernelov, A. and C. Ramel. 1994. Mercury in the environment, *Ambio*, 23, 166.
Korte F. and F. Coulston. 1998. Some considerations on the impact of ecological chemical principles in practice with emphasis on gold mining and cyanide, *Ecotoxicol. Environ. Safety*, 41, 119–129.
Lacerda, L.D. 1997a. Global mercury emissions from gold and silver mining, *Water Air Soil Pollut.*, 97, 209–221.

Lacerda, L.D. 1997b. Evolution of mercury contamination in Brazil, *Water Air Soil Pollut.*, 97, 247–255.

Lacher, Jr., T.E. and M.I. Goldstein. 1997. Tropical ecotoxicology: status and needs, *Environ. Toxicol. Chem.*, 16, 100–111.

Lawren, S.J. 2003. Mercury in the Carson River Basin, Nevada. In J.E Gray (Ed.), *Geologic Studies of Mercury by the U.S. Geological Survey,* p. 29–34. USGS Circular 1248. For sale by USGS Information Services, Box 25286, Denver Federal Center, Denver, CO 80225.

Lechler, P.J., J.R. Miller, L.D. Lacerda, D. Vinson, J.C. Bonzongo, W.B. Lyons, and J.J. Warwick. 2000. Elevated mercury concentrations in soils, sediments, water, and fish of the Madeira River basin, Brazilian Amazon: a function of natural enrichments?, *Sci. Total Environ.*, 260, 87–96.

Leigh, D.S. 1994. Mercury contamination and floodplain sedimentation from former gold mines in north Georgia, *Water Resour. Bull.*, 30, 739–748.

Leigh, D.S. 1997. Mercury-tainted overbank sediment from past gold mining in north Georgia, USA, *Environ. Geol.*, 30, 244–251.

LesEnfants, Y. 1995. Gold mining and mercury contamination in Venezuela. In *Proceedings of the International Workshop on "Environmental Mercury Pollution and Its Health Effects in Amazon River Basin,"* p. 17–20. Rio de Janeiro, 30 November–2 December 1994. Published by National Institute for Minamata Disease, Minamata City, Kumamoto 867, Japan.

Lima, A.P.S., R.C.S. Muller, J.E.S. Sarkis, C.N. Alves, M.H.S. Bentes, E. Brabo, and E.O. Santos. 2000. Mercury contamination in fish from Santarem, Para, Brazil, *Environ. Res.*, 83A, 117–122.

Lin, Y., M. Guo, and W. Gan. 1997. Mercury pollution from small gold mines in China, *Water Air Soil Pollut.*, 97, 233–239.

Lockhart, W.L., R.W. Macdonald, P.M. Outridge PM, P. Wilkinson, J.B. DeLaronde, and J.W.M. Rudd. 2000. Tests of the fidelity of lake sediment core records of mercury deposition to known histories of mercury contamination, *Sci. Total Environ.*, 260, 171–180.

Lodenius, M. 1993. Mercury contamination in the Tucurui Reservoir, Brazil. In *Proceedings of the International Symposium on Epidemiological Studies on Environmental Pollution and Health Effects of Methylmercury,* p. 44–49. October 2, 1992, Kumamoto, Japan. Published by National Institute for Minamata Disease, Kumamoto 867, Japan.

Malm, O. 1998. Gold mining as a source of mercury exposure in the Brazilian Amazon, *Environ. Res.*, 77A, 73–78.

Malm, O., F.J.P. Branches, H. Akagi, M.B. Castro, W.C. Pfeiffer, M. Harada, W.R. Bastos, and H. Kato. 1995a. Mercury and methylmercury in fish and human hair from the Tapajos river basin, Brazil, *Sci. Total Environ.*, 175, 141–150.

Malm, O., M.B. Castro, W.R. Bastos, F.J.P. Branches, J.R.D. Guimaraes, C.E. Zuffo, and W.C. Pfeiffer. 1995b. An assessment of Hg pollution in different goldmining areas, Amazon Brazil, *Sci. Total Environ.*, 175, 127–140.

Malm, O., W.C. Pfeiffer, C.M.M. Souza, and R. Reuther. 1990. Mercury pollution due to gold mining in the Madeira River Basin, Brazil, *Ambio*, 19, 11–15.

Martinelli, L.A., J.R. Ferreira, B.R. Forsberg, and R.L. Victoria. 1988. Mercury contamination in the Amazon: a gold rush consequence, *Ambio*, 17, 252–254.

Mastrine, J.A., J.C.J. Bonzongo, and W.B. Lyons. 1999. Mercury concentrations in surface waters from fluvial systems draining historical precious metals mining areas in southeastern U.S.A., *Appl. Geochem.*, 14, 147–158.

Meech, J.A., M.M Veiga, and D. Tromans. 1998. Reactivity of mercury from gold mining activities in darkwater ecosystems, *Ambio*, 27, 92–98.

Miller, J.R., J. Rowland, P.J. Lechler, M. Desilets, and L.C. Hsu. 1996. Dispersal of mercury-contaminated sediments by geomorphic processes, Sixmile Canyon, Nevada, USA: implications to site characterization and remediation of fluvial environments, *Water Air Soil Pollut.*, 86, 373–388.

Mol, J.H., J.S. Ramlal, C. Lietar, and M. Verloo. 2001. Mercury contamination in freshwater, estuarine, and marine fishes in relation to small-scale gold mining in Suriname, South America, *Environ Res.*, 86A, 183–197.

Montague, K. and P. Montague. 1971. *Mercury.* Sierra Club, New York. 158 pp.

Moore, J.W. and D.J. Sutherland. 1980. Mercury concentrations in fish inhabiting two polluted lakes in northern Canada, *Water Res.*, 14, 903–907.

Nelson, C.H., D.E. Pierce, K.W. Leong. and F.F.H. Wang. 1975. Mercury distribution in ancient and modern sediment of northeastern Bering Sea, *Mar. Geol.*, 18, 91–104.

Nico, L.G. and D.C. Taphorn. 1994. Mercury in fish from gold-mining regions in the upper Cuyuni River system, Venezuela, *Fresenius Environ. Bull.*, 3, 287–292.

Nriagu, J.O. (Ed.), 1979. *The Biogeochemistry of Mercury in the Environment.* Elsevier/North-Holland Biomedical Press, New York. 696 pp.

Nriagu, J.O. 1993. Legacy of mercury pollution, *Nature*, 363, 589.

Nriagu, J. and H.K.T. Wong. 1997. Gold rushes and mercury pollution. In A. Sigal and H. Sigal (Eds.), *Mercury and Its Effects on Environment and Biology,* p. 131–160. Marcel Dekker, New York.

Ogola, J.S., W.V. Mitullah, and M.A. Omulo. 2002. Impact of gold mining on the environment and human health: a case study in the Migori gold belt, Kenya, *Environ. Geochem. Health*, 24, 141–158.

Olivero, J. and B. Solano. 1998. Mercury in environmental samples from a waterbody contaminated by gold mining in Columbia, South America, *Sci. Total Environ.,* 217, 83–89.

Oryu, Y., O. Malm, I. Thornton, I. Payne, and D. Cleary. 2001. Mercury contamination of fish and its implications for other wildlife of the Tapajos Basin, Brazilian Amazon, *Conserv. Biol.*, 15, 438–446.

Padovani, C.R., B.R. Forsberg, and T.P. Pimental. 1995. Contaminacao mercurial em peixes do Rio Madeira: resultados e recomendacoes para consumo humano, *Acta Amazonica*, 25, 127–136 (in Portuguese, English summary).

Palheta, D. and A. Taylor. 1995. Mercury in environmental and biological samples from a gold mining area in the Amazon region of Brazil, *Sci. Total Environ.*, 168, 63–69.

Pessoa, A., G.S. Albuquerque, and M.L. Barreto. 1995. The "garimpo" problem in the Amazon region. In *Chemistry of the Amazon: Biodiversity, Natural Products, and Environmental Issues,* p. 281–294. American Chemical Society, Washington, D.C., ACS Symposium Series 588.

Petralia, J.F. 1996. *Gold! Gold! A Beginner's Handbook and Recreational Guide: How & Where to Prospect for Gold!.* Sierra Outdoor Products Co., P.O. Box 2497, San Francisco, CA 94126–2497. 143 pp.

Pfeiffer, W.C., L.D. de Lacerda, O. Malm, C.M.M. Souza, E.G. da Silveira, and B.R. Bastos. 1989. Mercury concentrations in inland waters of gold-mining areas in Rondonia, Brazil, *Sci. Total Environ.*, 87/88, 233–240.

Pfeiffer, W.C. and L.D. Lacerda. 1988. Mercury input into the Amazon region, Brazil, *Environ. Tech. Lett.*, 9, 325–330.

Porcella, D., J. Huckabee, and B. Wheatley. 1995. Mercury as a global pollutant, *Water Air Soil Pollut.,* 80, 1–1336.

Porcella, D.B., C. Ramel, and A. Jernelov. 1997. Global mercury pollution and the role of gold mining: an overview, *Water Air Soil Pollut.*, 97, 205–207.

Porvari, P. 1995. Mercury levels of fish in Tucurui hydroelectric reservoir and in River Moju in Amazonia, in the state of Para, Brazil, *Sci. Total Environ.*, 175, 109–117.

Reuther, R. 1994. Mercury accumulation in sediment and fish from rivers affected by alluvial gold mining in the Brazilian Madeira River basin, Amazon, *Environ. Monitor. Assess.*, 32, 239–258.

Rojas, M., P.L. Drake, and S.M. Roberts. 2001. Assessing mercury health effects in gold workers near El Callao, Venezuela, *J. Occup. Environ. Med.*, 43, 158–165.

Tarras-Wahlberg, N.H., A. Flachier, G. Fredriksson, S. Lane, B. Lundberg, and O. Sangfors. 2000. Environmental impact of small scale and artisanal gold mining in southern Ecuador, *Ambio*, 29, 484–491.

Torres, E.B. 1995. Epidemiological investigation of mercury exposure and health effects in the Philippines. In *Proceedings of the International Workshop on Environmental Mercury Pollution and Its Health Effects in Amazon River Basin*, p. 3–8. Rio de Janeiro, 30 November–2 December 1994. Published by National Institute for Minamata Disease, Minamata City, Kumamoto 867, Japan.

Tupyakov, A.V., T.G. Laperdina, A.I. Egerov, V.A. Banshchikov, M.V. Mel'nikova, and O.B. Askarova. 1995. Mercury concentration in the abiogenic environmental components near gold and tungsten mining and concentration complexes in Eastern Transbaikalia, *Water Resour.*, 22, 163–169.

U.S. National Academy of Sciences (USNAS). 1978. *An Assessment of Mercury in the Environment*. National Academy of Sciences, Washington, D.C., 185 pp.

Van Straaten, P. 2000. Human exposure to mercury due to small scale gold mining in northern Tanzania, *Sci. Total Environ.*, 259, 45–53.

Veiga, M.M., J.A. Meech, and R. Hypolito. 1995. Educational measures to address mercury pollution from gold-mining activities in the Amazon, *Ambio*, 24, 216–220.

Vieira, L.M., C.J.R. Alho, and G.A.L. Rerreira. 1995. Mercury contamination in sediment and in molluscs of Pantanel, Mato Grosso, Brazil, *Rev. Bras. Zool.*, 12, 663–670.

Von Tumpling Jr., W., P. Zeilhofer, U. Ammer, J. Einax, and R.D. Wilken. 1995. Estimation of mercury content in tailings of the gold mine area of Pocone, Mato Grosso, Brazil, *Environ. Sci. Pollut. Res. Int.*, 2, 225–228.

Walter, C.M., F.C. June, and H.G. Brown. 1973. Mercury in fish, sediments, and water in Lake Oahe, South Dakota, *J. Water Pollut. Control Feder.*, 45, 2203–2210.

Wayne, D.M., J.J. Warwick, P.J. Lechler, G.A. Gill, and W.B. Lyons. 1996. Mercury contamination in the Carson River, Nevada: a preliminary study of the impact of mining wastes, *Water Air Soil Pollut.*, 92, 391–408.

World Health Organization (WHO). 1991. *Inorganic Mercury. Environmental Health Criteria 118.* International Programme on Chemical Safety, WHO, Geneva.

Yamamoto, M., H. Hou, K. Nakamura, A. Yasutake, and T. Fujisaki. 1995a. Stimulation of elemental mercury oxidation by SH compounds, *Bull. Environ. Contam. Toxicol.*, 54, 409–413.

Yamamoto, M., H. Hou, K. Nakamura, A. Nakano, T. Ando, and S. Akiba. 1995b. Stimulation of elemental mercury oxidation by SH compounds in aquatic environments, *Jpn. J. Toxicol. Environ. Health*, 41, 3.

Part 5

Proposed Mercury Criteria, Concluding Remarks

CHAPTER 12

Proposed Mercury Criteria for Protection of Natural Resources and Human Health

Proposed mercury criteria for the protection of representative crops, aquatic organisms, birds, and mammals are shown in Table 12.1, and for human health in Table 12.2. These criteria vary widely between nations and even between localities in the same nation. In almost every instance, these criteria are listed as concentrations of total mercury, with most, if not all, of the mercury present as an organomercury species. In some cases, recommended mercury criteria are routinely exceeded, as is the case for brown bears (*Ursus arctos*) in the Slovak Republic (Zilincar et al., 1992), and in Italian seafood products recommended for human consumption (Barghigiani and De Ranieri, 1992).

12.1 AGRICULTURAL CROPS

Proposed mercury criteria for crop protection (Table 12.1) include < 0.2 μg/L in irrigation water, and < 0.2 to < 0.5 mg/kg DW in soils of several countries, although higher levels are allowable in cropland soils of New Jersey (< 1.0 mg/kg DW) and the Former Soviet Union (< 2.1 mg/kg DW). Sludge and other wastes applied to European soils should contain < 1.5 mg Hg/kg, but higher levels of < 10.0 to 25.0 mg Hg/kg are permissible in solid wastes applied to agricultural soils of Iowa, Maine, Vermont, and California (Table 12.1).

12.2 AQUATIC LIFE

In 1980, the U.S. Environmental Protection Agency's proposed mercury criteria for freshwater aquatic life protection were 0.00057 μg/L (24–h average), not to exceed 0.0017 μg/L at any time; these criteria seemed to afford a high degree of protection to freshwater biota, as judged by survival, bioconcentration, and biomagnification. Literature documented in USEPA (1980) showed that mercury concentrations in water of 0.1 to 2.0 μg/L were fatal to sensitive aquatic species and that concentrations of 0.03 to 0.1 μg/L were associated with significant sublethal effects. The 1980 proposed freshwater criteria provided safety factors for acute toxicities of 175 to 3508 based on the 24-h average, and 58 to 1176 based on the maximum permissible concentration (Table 12.1). For protection against sublethal effects, these values were 53 to 175 based on the 24-h mean, and 18 to 59 based on the maximum permissible concentration (Table 12.1). However, more recent freshwater criteria of 0.012 μg/L, not to exceed 2.4 μg/L (Table 12.1; USEPA, 1985), dramatically reduce the level of protection afforded aquatic biota: safety factors for acute toxicities are now 8 to 167 based on the 96-h average, and only 0.04 to 0.8 based on the maximum permissible concentration. For

Table 12.1 Proposed Mercury Criteria for the Protection of Selected Natural Resources

Resource and Other Variables	Criterion or Effective Mercury Concentration	Ref.[a]
	Crops	
Irrigation water, Brazil	< 0.2 µg/L	1
Land application of sludge and solid waste; maximum permissible concentration:		
Europe; soils with pH 6.0–7.0	< 1.0–< 1.5 mg/kg sludge	39
Iowa, Maine, Vermont	10.0 mg/kg waste	2
California	20.0 mg/kg waste	2
Vermont, sewage sludge:		
Loamy sand and sandy loam soils	< 6.0 kg Hg/ha	39
Fine sandy loam, loam, and silt loam soils	< 11.0 kg Hg/ha	39
Clay loam, clay, silty clay	< 22.0 kg Hg/ha	39
Soils, Australia; urban areas	< 1.0 mg/kg FW	45
Soils, Germany	< 2.0 mg/kg dry weight (DW)	3
Soils, Canada, agricultural lands	< 0.5 mg/kg DW	39
Soils, Finland:		
Recommended	< 0.2 mg/kg DW	37
Maximum allowable	< 5.0 mg/kg DW	37
Soils, Former Soviet Union	< 2.1 mg/kg DW	39
Soils, Japan, contaminated	> 3.0 mg/kg DW	39
Soils, Netherlands:		
Target	< 0.3 mg/kg DW	45
Normal	< 0.5 mg/kg DW	39
Moderate contamination: requires additional study	> 2.0 mg/kg DW	39
Contaminated; immediate cleanup required	> 10.0 mg/kg DW	39
Soils, New Jersey	< 1.0 mg/kg DW	39
	Aquatic Life	
Diet, piscivorous fishes	< 100.0 µg total mercury/kg fresh weight (FW) whole prey fish[f]	44
Freshwater:		
Total mercury	< 0.00057 µg/L, 24 h average; not to exceed 0.0017 µg/L at any time	4
Total mercury	< 0.1 µg/L	5, 38
Total mercury; adverse effects expected with chronic exposure	> 0.012 µg/L	47
Inorganic mercury	Adverse effects at > 0.23 µg Hg^{2+}/L	7
Methylmercury	< 0.01 µg/L	6
Total mercury	< 0.012 µg/L, 4-day average (not to be exceeded more than once every 3 years); < 2.4 µg/L, 1-hour average (not to be exceeded more than once every 3 years)[b]	7
Inland surface waters, India	< 10.0 µg/L from point source discharge	8
Public water supply, Wisconsin	< 0.079 µg/L	2
Saltwater	Total recoverable mercury < 0.025 µg/L, 24-h average; not to exceed 3.7 µg/L at any time[g]	4, 7
Saltwater	< 0.025 µg/L, 4-day average; < 2.1 µg/L, 1-hour average	7
Sediments:		
California:		
Low toxic effect	> 0.15 mg/kg DW	9
Acceptable	< 0.51 mg/kg DW	9
Bivalve molluscs, abnormal larvae	> 0.51 mg/kg DW	9
Hazardous	> 1.2–1.3 mg/kg DW	9

Table 12.1 (continued) Proposed Mercury Criteria for the Protection of Selected Natural Resources

Resource and Other Variables	Criterion or Effective Mercury Concentration	Ref.[a]
Canada, marine and freshwater:		
Safe	< 0.14 mg/kg DW	6
Adverse effects expected	> 2.0 mg/kg DW	9
Ecuador	< 0.45 mg/kg DW	38
Great Lakes:		
Nonpolluted	< 1.0 mg/kg DW	39
Heavily polluted	> 1.0 mg/kg DW	39
Wisconsin sediment disposal into Great Lakes	< 0.1 mg/kg DW	39
Ontario guideline for disposal of dredged sediments into lake	< 0.3 mg/kg DW	39
Washington state, safe	< 0.41 mg/kg DW	9
Tissue residues:		
Goldfish, *Carassius auratus*; impaired egg production and spawning expected; gonad	> 0.76 mg/kg FW	43
Rainbow trout, *Oncorhynchus mykiss*		
Lethal:		
Eggs	> 70.0 μg/kg FW	10
Muscle, adults	> 10.0 mg/kg FW	10
Whole body	10.0–20.0 mg/kg FW	11
Adverse effects probable, whole body	1.0–5.0 mg/kg FW	11
Impaired spawning and reduced survival of early life stages expected; gonad	0.49 mg/kg FW	43
Brook trout, *Salvelinus fontinalis*; whole body, nonlethal	< 5.0 mg/kg FW	4,7
Various species of freshwater adult fishes; adverse effects expected:		
Brain:		
Toxic	> 3.0 mg/kg FW	10
Potentially lethal	> 7.0 mg/kg FW	10
Muscle:		
Toxic	> 5.0–8.0 mg/kg FW	10
Lethal	10.0–20.0 mg/kg FW	10
Whole body:		
Adverse effects	> 1.0 mg/kg FW	7
No observed effect	< 3.0 mg/kg FW	10
Amphibians		
South African clawed frog, *Xenopus laevis*; impaired gamete function and reduced early life survival expected; gonad	> 0.48 mg/kg FW	43
Birds		
Tissue residues:		
Safe:		
Brain, muscle	< 15.0 mg/kg FW	12
Feather	< 5.0 mg/kg FW	12, 13
Kidney, seabirds	< 30.0 mg/kg FW	14
Kidney, not seabirds	< 20.0 mg/kg FW	12,1 4
Liver:		
Normal	1.0–10.0 mg/kg FW	15
Toxic to sensitive species	> 5.0–6.0 mg/kg FW	15, 16
Hazardous, possibly fatal	> 20.0 mg/kg FW	15, 17
Egg:		
Mallard, *Anas platyrhynchos*; safe	< 0.8–< 1.0 mg/kg FW	18, 19
Ring-necked pheasant, *Phasianus colchicus*; safe	0.5–< 0.9 mg/kg FW	20,21

(continued)

Table 12.1 (continued) Proposed Mercury Criteria for the Protection of Selected Natural Resources

Resource and Other Variables	Criterion or Effective Mercury Concentration	Ref.[a]
Common tern, *Sterna hirundo*; normal reproduction vs. reduced hatching and fledging success	< 1.0 mg/kg FW vs. 2.0–4.7 mg/kg FW	21
Various species; safe	< 0.5 to < 2.0 mg/kg FW	14, 22
Waterbirds; adverse effects	1.0 to 3.6 mg/kg FW	16
Feather, acceptable	< 9.0 mg/kg FW	23
Methylmercury — poisoned:		
Brain	15.0–20.0 mg/kg FW	12
Liver	20.0–60.0 mg/kg FW	12
Kidney	20.0–60.0 mg/kg FW	12
Muscle	15.0–30.0 mg/kg FW	12
Diet, fish-eating birds	< 20.0 μg Hg/kg FW ration, as methylmercury	24
Diet, fish-eating birds	< 100.0 μg total Hg/kg FW in prey fish[f]	44
Diet, non fish-eating birds	50.0–< 100.0 μg Hg/kg FW ration, as methylmercury	18, 25
Diet, loon	< 300.0 μg/kg FW ration[c]	26, 27
Diet	< 1.0–< 3.0 mg/kg DW	14, 28
Daily intake	< 640.0 μg/kg body weight (BW)	20,29–31
Daily intake	< 32.0 μg/kg BW[d]	31
Mammals		
Daily intake	< 250.0 μg/kg BW	32
Diet; fish-eating mammals	< 100.0 μg Hg/kg FW ration, as methylmercury	24
Diet; fish-eating mammals	< 100.0 μg total Hg/kg FW whole prey fish[f]	44,50
Diet	< 1.1 mg/kg FW ration	14
Diet; methylmercury poisoning of minks and otters	> 1.0 mg/kg FW ration	40–42, 46
Diet; minks and otters; adverse effects	> 0.12 to 1.4 mg total mercury/kg FW ration	51
Drinking water:		
Feral and domestic animal water supply, Wisconsin	< 0.002 μg/L	2
Terrestrial vertebrate wildlife	< 0.0013 μg/L[e]	31
Soils; terrestrial ecosystem protection; agricultural and residential land use vs. commercial and industrial use	< 2.0 mg/kg DW vs. < 30.0 mg/kg DW	6
Tissue residues:		
Acceptable, most species:		
Kidney	< 1.1 mg/kg FW	23
Liver, kidney	< 30.0 mg/kg FW	14
Blood	< 1.2 mg/kg FW	33
Brain	< 1.5 mg/kg FW	33
Hair	< 2.0 mg/kg FW	33
Acceptable; otters, *Lutra* spp.:		
Hair	< 1.0 –< 5.0 mg/kg DW	48
Liver	< 4.0 mg/kg FW	49
Mercury-poisoned minks and otters:		
Brain	> 10.0 mg/kg FW	40–42, 46
Liver	> 20.0–> 100.0 mg/kg FW	40–42, 46
Florida panther, *Felis concolor coryi*; blood:		
Reproduction normal (1.46 kittens per female annually)	< 250.0 μg/kg FW	34
Reproduction inhibited (0.167 kittens per female annually)	> 500.0 μg/kg FW	34
European otter, *Lutra lutra*; liver		
Normal	< 4.0 mg/kg FW	35
Adverse sublethal effects possible	> 10.0 mg/kg FW	35

Table 12.1 (continued) Proposed Mercury Criteria for the Protection of Selected Natural Resources

Resource and Other Variables	Criterion or Effective Mercury Concentration	Ref.[a]
Wildlife protection, Slovak Republic:		
Fat	< 1.0 μg/kg FW	36
Muscle	< 50.0 μg/kg FW	36
Liver, kidney	< 100.0 μg/kg FW	36

[a] Reference: 1, Palheta and Taylor, 1995; 2, U.S. Public Health Service (USPHS), 1994; 3, Zumbroich, 1997; 4, U.S. Environmental Protection Agency (USEPA), 1980; 5, Dave and Xiu, 1991; 6, Gaudet et al., 1995; 7, USEPA, 1985; 8, Abbasi and Soni, 1983; 9, Gillis et al., 1993; 10, Wiener and Spry, 1996; 11, Niimi and Kissoon, 1994; 12, Heinz, 1996; 13, U.S. National Academy of Sciences (USNAS), 1978; 14, Thompson, 1996; 15, Wood et al., 1996; 16, Zillioux et al., 1993; 17, Littrell, 1991; 18, Heinz, 1979; 19, Heinz and Hoffman, 2003; 20, Spann et al., 1972; 21, Mora, 1996; 22, Fimreite, 1979; 23, Beyer et al., 1997; 24, Yeardley et al., 1998; 25, March et al., 1983; 26, Scheuhammer et al., 1998; 27, Gariboldi et al., 1998; 28, Scheuhammer, 1988; 29, Mullins et al., 1977; 30, McEwen et al., 1973; 31, Wolfe and Norman, 1998; 32, Ramprashad and Ronald, 1977; 33, Suzuki, 1979; 34, Roelke et al., 1991; 35, Mason and Madsen, 1992; 36, Zilincar et al., 1992; 37, Peltola and Astrom, 2003; 38, Tarras–Wahlberg et al., 2000; 39, Beyer, 1990; 40, Aulerich et al., 1974; 41, Dansereau et al., 1999; 42, Wren et al., 1987a; 43, Birge et al., 2000; 44, Atkeson et al., 2003; 45, Zarcinas et al., 2004a; 46, Wren et al., 1987b; 47, USEPA, 1992; 48, Sheffy and St. Amant, 1982; 49, Wren, 1986; 50, Fonseca et al., 2005; 51, Halbrook et al., 1994.

[b] All mercury that passes through a 0.45-micromillimeter membrane filter after the sample is acidified to pH 1.5 to 2.0 with nitric acid. Derived from bioconcentration factor of 81,700 for methylmercury and the fathead minnow, *Pimephales promelas* (USEPA, 1985).

[c] Reproduction declined in loons, *Gavia immer*, when mercury in prey exceeded 300.0 μg total mercury/kg FW.

[d] No observed adverse effect level with uncertainty factor of 20.

[e] Based on food chain biomagnification in aquatic webs.

[f] Provisional U.S. Fish and Wildlife Service standard for protection of fish-eating wildlife and animals that feed on them. Also considered safe for propagation and maintenance of healthy well-balanced populations of fish and other wildlife (Atkeson et al., 2003).

[g] Based on bioconcentration factor of 40,000 for methylmercury and the American oyster, *Crassostrea virginica* (USEPA, 1985).

protection against sublethal effects, these values were 2 to 8 based on the 4-day average, and only 0.01 to 0.04 based on the maximum permissible concentration, or essentially no significant protection. The proposed saltwater criteria of USEPA (1980) for mercury and marine life were unsatisfactory. Proposed saltwater values of 0.025 μg/L (24-h average), not to exceed 3.7 μg/L at any time (Table 12.1), provided safety factors of 4 to 8O against acute toxicity (based on 24-h average), but less than 0.5 based on the maximum permissible level. For protection against sublethal damage effects, the safety factors computed were 1.2 to 4 (based on 24-h average) and less than 0.03 (based on maximum allowable concentration). The more recent saltwater criteria of 0.025 μg/L, not to exceed 2.1 μg/L (Table 12.1; USEPA, 1985), does not appear to offer a substantive increase in protection to marine life, when compared to criteria proposed 5 years earlier (USEPA, 1980). It seems that some downward modification is needed in the proposed mercury saltwater criteria if marine and estuarine biota are to be provided even minimal protection.

The significance of elevated mercury residues in tissues of aquatic organisms is not fully understood. Induction of liver metallothioneins and increased translatability of MRNA are biochemical indicators of the response of fish to mercury exposure (Angelow and Nicholls, 1991; Schlenk et al., 1995), and more research is recommended on this and other indicators of mercury stress. Concentrations exceeding 1.0 mg Hg/kg fresh weight can occur in various tissues of selected species of fish and aquatic mammals eaten by humans. But it would be incorrect to assume that aquatic food chains — especially marine food chains — incorporate mercury exclusively from anthropogenic activities (Barber et al., 1984). Some organisms, however, do contain mercury tissue residues associated with known adverse effects to the organism and its predators. Thus, whole body residues of 5.0 to 7.0 mg Hg/kg fresh weight in brook trout eventually proved fatal to that species (USEPA, 1980). To protect sensitive species of mammals and birds that regularly consume fish and other aquatic organisms, total mercury concentrations in these food items should probably not exceed 100.0 μg/kg for avian protection, or 1100.0 μg/kg for small mammals (Table 12.1).

Since long-lived, slow-growing, high-trophic-position aquatic organisms usually contain the highest tissue mercury residues (Eisler, 1981, 2000), some fisheries managers have proposed a legal maximum limit based on fish length or body weight (Lyle, 1984; Chvojka, 1988), or alternatively, constraining the mean mercury concentration of the entire catch to a nominated level. In the Australian shark fishery, for example, implementation of a maximum length restriction (to a designated level of 500.0 μg Hg/kg), would result in retention of less than half of the present catch of seven species (Lyle, 1984). Also in Australia, a maximum total length of 92 cm is proposed for the taking of yellowtail kingfish (*Seriola grandis*) and would effectively remove 23.0% of the total weight of the catch and 9.0% of the numbers (Chvojka 1988). If the total length of the yellowtail kingfish is reduced to 73 cm, a length that ensures that almost all fish contained < 500.0 μg Hg/kg FW muscle, this would preclude 59.0% by weight and 30.0% by numbers (Chvojka, 1988). Other strategies to control mercury burdens in predatory fishes include control of forage fish, overfishing, and various chemical treatments. In lakes with pelagic forage fish, there is less than a 5.0% probability of finding elevated mercury levels in muscle of lake trout less than 30 cm in total length vs. 45 cm in lakes where pelagic forage fish were absent. In the case of lake trout lakes with no pelagic forage fish, every effort should be made to avoid their introduction (Futter, 1994). Overfishing of top-level predators is recommended as a means of lowering methylmercury levels in certain types of lakes, and is attributed to the more rapid growth of the predators and by changes in the dietary intake of methylmercury (Verta, 1990). Treatment of lakes with selenium compounds is one of the few known methods of lowering the mercury content of fish muscle to < 1.0 mg Hg/kg FW (Paulsson and Lundbergh, 1989). Treatments that have achieved partial success in reducing mercury content in fish tissues include liming of lakes, wetlands, and drainage areas (Lindqvist et al., 1991). More research is needed on mercury protectants because several are known to cause substantial reductions in tissue mercury concentrations in fishes and plants (Siegel et al., 1991). Thiamine and various group VI derivatives, including sulfur, selenium, and tellurium compounds, protect against organomercury poisoning by their antagonistic effects; thiamine was the most effective of the derivatives against the widest spectrum of organisms and test systems (Siegel et al., 1991).

More research is also needed on mercury removal technology. In the Florida Everglades, for example, using prototype wetlands of 1545 ha, removal of agricultural nutrients from stormwater reduced total mercury and methylmercury concentrations in water by as much as 70.0% in the first 2 years of operation. Moreover, total mercury concentrations in largemouth bass were about 0.1 mg Hg/kg FW muscle throughout the project site vs. 0.5 mg Hg/kg FW in adjacent areas (Miles and Fink, 1998).

12.3 BIRDS

Tissue residues of mercury, as methylmercury, considered harmful to adult birds ranged from 8.0 mg/kg FW in brain to 15.0 in muscle to 20.0 mg/kg FW in liver and kidney (Heinz, 1996). Among sensitive avian species, adverse effects, mainly on reproduction, have been reported at total mercury concentrations (in mg/kg fresh weight) of 5.0 in feather, 0.9 in egg, 0.05 to 0.1 in diet, and daily administered doses of 0.64 mg/kg body weight (Table 12.1; Eisler, 2000). The proposed mercury concentration to protect sensitive species of birds that regularly consume fish and aquatic invertebrates is < 0.1 mg/kg FW in these food items (Eisler, 2000; Table 12.1).

Although low mercury concentrations (e.g., 0.05 mg/kg in the diets of domestic chickens) sometimes produced no adverse effects on chickens, the tissue residues of mercury were sufficiently elevated to pose a hazard to human consumers (March et al., 1983). In contrast, with eggs of the bald eagle containing 0.15 mg Hg/kg FW and low hatch, it is probable that other contaminants present — especially organochlorine compounds — had a greater effect on hatchability than did mercury (Wiemeyer et al., 1984).

12.4 MAMMALS

Mammals, such as the domestic cat and the harp seal, showed birth defects, histopathology, and elevated tissue residues at doses of 250.0 μg Hg/kg body weight daily (Table 12.1). Tissue residues of mercury, as methylmercury, considered harmful to adult inland mammals ranged between 8.0 mg/kg FW in brain, 15.0 in muscle, and 20.0 mg/kg FW in liver and kidney (Heinz, 1996). Mink fed dietary levels of 1.1 mg Hg/kg had signs of mercury poisoning; mercury residues in mink brain at this dietary level ranged from 7.1 to 9.3 mg/kg (Kucera, 1983). Tissue residues in kidney, blood, brain, and hair in excess of 1.1 mg Hg/kg in nonhuman mammals are usually considered presumptive evidence of significant mercury contamination (Table 12.1). To protect sensitive species of small mammals that regularly consume fish and other aquatic organisms, total mercury concentrations in these food items should probably not exceed 100.0 μg total mercury/kg FW (Eisler, 2000; Table 12.1).

For most species of mammals, recommended mercury criteria include daily intake of less than 250.0 μg total mercury/kg body weight; diets that contain less than 1.1 mg total mercury/kg FW; and for livestock, less than 0.002 μg total mercury/L in the drinking water supply (Table 12.1). Tissue mercury concentrations in sensitive mammals (in mg total mercury/kg FW) should probably not exceed 10.0 in liver, 2.0 in hair, 1.5 in brain, and 0.5 in blood (Table 12.1).

12.5 HUMAN HEALTH

Proposed mercury criteria for the protection of human health are numerous and disparate (Table 12.2). Proposed mercury air criteria, for example, in the workplace range from < 10.0 μg/m^3 for organomercury compounds to < 50.0 μg/m^3 for elemental mercury vapor; however, much lower criteria are proposed by Texas (< 0.05 μg/m^3 for 1 year), New York (< 0.167 μg/m^3 per year), and other jurisdictions (Table 12.2). Drinking water criteria for total mercury range between < 1.0 and < 2.0 μg/L (USPHS, 1994; Gemici, 2004), except for Brazil with < 0.2 μg/L (Palheta and Taylor, 1995). Current fish consumption recommendations are based on risk assessments for children and women of child-bearing age (Weil et al., 2005). Dietary criteria for mercury range between 10.0 and 1500.0 μg/kg FW, with lower values associated with seafoods, organomercurials, and pregnancy. Accordingly, proposed mercury levels in fish and seafood should not exceed 250.0 μg/kg FW for expectant mothers, and 400.0 to 1000.0 μg/kg FW for adults worldwide (Table 12.2). It is again emphasized that total mercury concentrations exceeding 1.0 mg/kg fresh weight naturally occur in edible tissues of some species of fish and aquatic mammals regularly eaten by humans (Barber et al., 1984). Proposed tolerable weekly intakes range between < 3.3 and 5.0 μg/kg body weight (Table 12.2). Daily mercury intake for pregnant women should not exceed < 0.6 to 1.1 μg total mercury/kg body weight vs. 4.3 μg/kg BW for others (Clarkson, 1990). It has been suggested that humans can safely ingest up to 8.4 mg Hg/kg body weight daily (Birke et al., 1972; USPHS, 1994), but this requires verification. The minimum toxic intake for humans is estimated to range between 0.6 and 1.1 μg methylmercury/kg body weight daily (Clarkson, 1990). Recommended total mercury concentrations in human tissues include < 5.8 μg/L in blood, < 0.7 mg/kg whole body, and < 6.0 to 50.0 mg/kg in hair (Table 12.2).

Methylmercury concentration in scalp hair during pregnancy is considered the most reliable indicator for predicting the probability of psychomotor retardation in the child. The U.S. Environmental Protection Agency has established a Reference Dose of 1.0 mg total Hg/kg DW in hair as indicative of mercury exposure. At this level, women of child-bearing age are advised to stop

Table 12.2 Proposed Mercury Criteria for the Protection of Human Health

Variables	Criterion or Effective Mercury Concentration	Ref.[a]
	Air; Safe	
California	< 0.00 µg/m^3	1
North Dakota	< 0.0005 µg/m^3 for 8 h	1
Kansas	0.0024 µg/m^3 annually	1
Montana	< 0.008 µg/m^3 for 24 h	1
Texas	< 0.05 µg/m^3 for 1 year	1
New York	< 0.167 µg/m^3 per year	1
Connecticut	1.0 µg/m^3 for 1 h; < 1.0 µg/m^3 for 8 h	1
Arizona	< 1.5 µg/m^3 for 1 h	1
Virginia	< 1.7 µg/m^3 for 24 h	1
General population; metallic mercury vapor	< 15.0 µg/m^3 for 24 h	29
Workplace:		
Organic mercury	< 10.0 µg/m^3	1, 30
Metallic mercury vapor	< 50.0 µg/m^3	1, 29, 35
	Air; Adverse Effects Possible	
Skin	> 30.0 µg/m^3	1
Emissions from individual industrial sites	> 2300.0–3200.0 g mercury daily	1
French municipal waste incinerator	> 0.3 µg/m^3 for > 24 h	54
	Chloralkali Plants, Canada	
Wastewater effluents	< 2.5 g mercury daily per ton of chlorine produced	36
Air emissions	< 2.0 µg/m^3 daily	36
	Drinking Water	
Brazil	< 0.2 µg/L	2
Turkey	< 1.0 µg/L	27
International	< 1.0 µg/L	1
United States, most states	< 2.0 µg/L	1, 45
Bottled water	< 2.0 µg/L	1
	Effluent Limitations from Wastewater Treatment Plants	
Delaware, Oklahoma, Texas	< 5.0 µg/L	1
Illinois, Wisconsin	< 0.5 µg/L	1
New Jersey	< 2.0 µg/L	1
Tennessee	< 50.0 µg/L	1
	Diet	
Australia:		
General diet	< 10.0 to < 100.0 µg/kg fresh weight (FW) ration	3
General diet	< 20.0 µg/kg FW	43
Seafood	< 500.0 µg/kg FW	43
Benelux countries	< 30.0 µg/kg FW ration	3
Brazil	< 50.0 µg/kg FW ration	3
Canada	< 500.0 µg/kg FW ration	4
Malaysia; vegetables; fruit; vegetable and fruit juices; tomato pulp, paste, and puree; tea; coffee, cocoa	< 50.0 µg/kg "as consumed"	42
Thailand:		
General foods	< 20.0 µg/kg FW	43
Seafoods	< 500.0 µg/kg FW	43
United States	< 1000.0 µg/kg FW ration	4, 5

Table 12.2 (continued) Proposed Mercury Criteria for the Protection of Human Health

Variables	Criterion or Effective Mercury Concentration	Ref.[a]
Japan:		
Total mercury intake:		
Adverse effects expected	> 250.0 μg daily	34
Nontoxic	< 25.0 μg daily; < 0.5 μg/kg BW daily	34
Symptoms of mercury poisoning	Consumption of > 500.0 g fish muscle daily containing 10.0 mg methylmercury chloride/kg FW muscle	34
Methylmercury	< 600.0 μg/kg BW daily[c]	34
Permissible Tolerable Weekly Intake		
Total mercury	< 5.0 μg/kg BW	1
Total mercury	Maximum of 4.28 μg/kg BW	6
Methylmercury	< 3.3 μg/kg BW	1,7
Methylmercury	< 170.0 μg	34, 37
Methylmercury; Japan; estimated weekly intake	42.0–52.0 μg	37
Methylmercury	< 200.0 μg/60–kg person	34
Fish consumption advisory; Florida vs. most states	> 0.5 mg total Hg/kg FW edible aquatic product vs. > 1.0 mg total Hg/kg FW edible aquatic product	8
Fish and Seafood, Edible Parts		
United States	< 300.0 μg total Hg/kg FW	9, 23
United States	< 300.0 μg methylmercury/kg FW	46
United States; Food and Drug Administration action level	> 1.0 mg Hg/kg FW	47
Florida:		
Safe	< 500.0 μg/kg FW[d]	38
Limited consumption	> 500.0 μg/kg FW and < 1500.0 μg/kg FW[e]	38
No consumption advised	> 1500.0 μg/kg FW	38, 49
Japan:		
Total mercury	< 400.0 μg/kg FW	33, 34, 37
Methylmercury	< 300.0 μg/kg FW	33, 34, 37
Slovak Republic	< 500.0 μg/kg FW	48
Acceptable intake:		
60-kg adult	25.0 μg daily	10
70-kg adult	200.0 μg weekly	10
Adult	500.0 μg weekly	11
Pregnant Women, Diet		
All	< 250.0 μg/kg FW	10
Japan	< 400.0 μg/kg FW	3, 32
Canada, Germany, United States, Brazil	< 500.0 μg/kg FW	1, 3, 12, 13,
Italy	< 700.0 μg/kg FW	6
Finland, Israel, Sweden	< 1000.0 μg/kg FW	3, 11, 14
Florida; Consumption of Contaminated Fish Containing 2–3 mg Methylmercury/kg FW Muscle		
Nonpregnant adults	Less than 454.0 g fish muscle weekly	24
Women of child-bearing age and children less than 15 years of age	Less than 454.0 g fish muscle monthly	24
Shark Flesh		
Containing 0.5–1.5 mg Hg/kg FW	Consumption limited to once weekly by healthy nonpregnant adults	25
Containing more than 1.5 mg Hg/kg FW	Consumption prohibited	25

(continued)

Table 12.2 (continued) Proposed Mercury Criteria for the Protection of Human Health

Variables	Criterion or Effective Mercury Concentration	Ref.[a]
	Foods of Animal Origin	
Livestock tissues	< 500.0 µg/kg FW	15
Wildlife tissues	< 50.0 µg/kg FW	16
Breast muscle:		
Domestic poultry	< 500.0 µg/kg FW	3
Ducks (wildlife)	< 1000.0 µg/kg FW	17
	Foods of Vegetable Origin	
Mercury–treated grain	< 1000.0 µg/kg FW	1
Vegetables	< 50.0 µg/kg DW	26
	All Foods	
Adult Weekly Intake		
As methylmercury	< 100.0 µg	3
As total mercury	< 150.0 µg	3
As methylmercury	< 200.0 µg	18, 19
As total mercury	< 300.0 µg	6, 19
Adult daily intake; nonpregnant vs. pregnant	< 4.3 µg/kg BW vs. < 0.6–1.1 µg/kg BW	20
Oral dose; maximum tolerated concentration; administered as phenylmercuric acetate	8.4 mg Hg/kg body weight daily[b]	1
	Human Tissue Residues	
Blood:		
Recommended, children	< 5.0 µg/L	50
No adverse neurodevelopmental effects in children exposed during pregnancy	< 5.8 µg methylmercury/L umbilical cord blood	57
Acceptable	< 5.8 µg/L	60
Associated with loss of IQ in children exposed during pregnancy	> 5.8 µg methylmercury/L umbilical cord blood	58
Developmental effects observed	> 5.8 µg methylmercury/L umbilical cord blood	55, 56
Safe	< 200.0 µg/kg FW	34
No symptoms observed	< 200.0 µg/kg FW	21
Asymptomatic (infants), but serious	120.0–630.0 µg/kg FW	37
Neurological symptoms	> 200.0 µg/kg FW	34
Critical (infants)	1050.0–> 3000.0 µg/kg FW	37
Some deaths expected (adults)	> 3100.0 µg/kg FW	37
Brain, neurological symptoms	> 6.0 mg/kg FW	34
Hair:		
Developmental effects observed	> 1.0 mg/kg DW	59
Recommended, children	< 2.0 mg/kg DW	53
Safe, China	< 4.0 mg/kg FW	40
Safe	< 6.0 mg/kg FW	22
USEPA reference dose indicative of human exposure	> 1.0 mg/kg DW	31
No observable symptoms	0.0–300.0 mg/kg FW	30
3-point reduction in Wechsler Intelligence Scale IQ of New Zealand children exposed as fetus during pregnancy	> 6.0 mg/kg DW	55, 56
No observable symptoms	5.0–30.0 mg/kg FW	34
Safe	< 50.0 mg/kg FW	34, 39
Minamata disease victims	50.0–200.0 mg/kg FW	34
Mildly affected (numbness of extremities, slight tremors; mild ataxia)	120.0–600.0 mg/kg FW	30

Table 12.2 (continued) Proposed Mercury Criteria for the Protection of Human Health

Variables	Criterion or Effective Mercury Concentration	Ref.[a]
Hearing difficulties, tunnel vision, partial paralysis	200.0–600.0 mg/kg FW	30
Severe (some combination of complete paralysis, loss of vision, loss of hearing, loss of speech, coma)	400.0–1600.0 mg/kg FW	30
Urine:		
Recommended, children	< 3.0 μg/L	52
Recommended, children	< 5.0 μg/L	51
Neurological disturbances	> 0.6 mg total Hg/L	41
Whole body:		
Zero clinical effect	< 0.7 mg/kg body weight	44
Effects observed	> 100.0 mg accumulated methylmercury	34
Fatal	> 1000.0 mg accumulated methylmercury	34
	Soils	
Residential and parklands	< 1.0 mg/kg DW	28
Commercial and industrial	< 2.0 mg/kg DW	28
Agricultural; Malaysia	< 0.362 mg/kg DW	42

[a] Reference: 1, U.S. Public Health Service (USPHS), 1994; 2, Palheta and Taylor, 1995; 3, U.S. National Academy of Sciences (USNAS), 1978; 4, Bodaly et al., 1984; 5, Kannan et al., 1998; 6, Barghigiani and De Ranieri, 1992; 7, Petruccioli and Turillazzi, 1991; 8, Facemire et al., 1995; 9, Brumbaugh et al., 2001; 10, Khera, 1979; 11, Lodenius et al., 1983; 12, Lathrop et al., 1991; 13, Hylander et al., 1994; 14, Krom et al., 1990; 15, Best et al., 1981; 16, Krynski et al., 1982; 17, Lindsay and Dimmick, 1983; 18, Yeardley et al., 1998; 19, Buzina et al., 1989; 20, Clarkson, 1990; 21, Galster, 1976; 22, Lodenius et al., 1983; 23, Fields, 2001; 24, Fleming et al., 1995; 25, Roelke et al., 1991; 26, Zumbroich, 1997; 27, Gemici, 2004; 28, Beyer, 1990; 29, U.S. Department of Health, Education, and Welfare (USHEW), 1977; 30, World Health Organization (WHO), 1976; 31, Maas et al., 2004; 32, Silver et al., 1994; 33, Nakanishi et al., 1989; 34, Takizawa, 1979; 35, Oikawa et al., 1982; 36, Paine, 1994; 37, Takizawa, 1979a; 38, Atkeson et al., 2003; 39, Harada, 1994; 40, Soong, 1994; 41, Satoh, 1995; 42, Zarcinas et al., 2004a; 43, Zarcinas et al., 2004b; 44, Marsh et al., 1975; 45, Alaska Dept. Environmental Conservation, 1994; 46, USEPA, 2001; 47, Gray, 2003; 48, Andreji et al., 2005; 49, Adams and Onorato, 2005; 50, Counter et al., 2005; 51, WHO, 1990; 52, Tsuji et al., 2003; 53, Ozuah et al., 2003; 54, Glorennec et al., 2005; 55, Kjellstrom et al., 1986; 56, Kjellstrom et al., 1989; 57, NRC, 2000; 58, Trasende et al., 2005; 59, Grandjean et al., 1997; 60, Weil et al., 2005.

[b] Assuming continuous exposure until steady-state balance for methylmercury is achieved. This process will take up to 1 year in most cases (Clarkson, 1990).

[c] Based on monkey experiments that showed no symptoms at 30.0 mg methylmercury/kg BW daily for 2 years via diet, and an uncertainty factor of 50.

[d] Women of childbearing age should consume less than 227.0 g (8 oz.) of freshwater fish over 7 days; children less than 10 years old should eat less than 113.0 g during this same period (Atkeson et al., 2003).

[e] Fish consumption should be limited to once monthly by children or women of child-bearing age and not more than once weekly by other adults (Atkeson et al., 2003).

consumption of fish that may have elevated mercury levels (Maas et al., 2004); however, these results should be considered preliminary pending analysis of additional data and because current mercury criteria for human hair range between < 6.0 and 50.0 mg/kg DW (Table 12.2). Other studies suggest that methylmercury concentration in cord blood is a sensitive biomarker of exposure *in utero* and correlates well with neurobehavioral outcomes (NRC, 2000). Moreover, concentrations greater than 5.8 μg methylmercury/L cord blood are associated with loss of IQ in children exposed during pregnancy (Table 12.2).

A major source of mercury vapor in the general population is dental amalgam (WHO, 1991), and several European countries — Austria, Germany, Finland, Norway, United Kingdom, Sweden — have advised their dentists to specifically curtail installation of mercury-containing amalgam fillings in pregnant women (Anderson et al., 1998). Incidentally, Hujoel et al. (2005), in a study conducted in Washington state, U.S.), found no evidence that mercury-containing dental fillings placed during pregnancy increased the risk of low birth weight. Further evaluations of the safety profiles of dental amalgams, and other types of materials used intraorally, are needed to establish guidelines for the use of dental materials in expectant mothers (Hujoel et al., 2005).

In terms of potential health hazards, the dangerous mercury level in human blood is about 2.0 mg/kg; the normal background in about 80.0% of people without occupational exposure to mercury is less than 0.005 mg/kg (Saha, 1972). Because methylmercury has a biological half-life in humans of 70 to 200 days (Lofroth, 1970), it was not unexpected that mutagenic and teratogenic effects of mercury have been reported at levels well below those associated with poisoning (Kazantzis, 1971; Keckes and Miettinen, 1972). The findings of elevated blood mercury levels, increased chromosome breakage, and easy passage of mercury through placental membranes among women who regularly eat fish containing 1.0 to 17.0 mg Hg/kg fresh weight (FW) suggest to some health authorities that the previous mercury recommendation level in the United States of 0.5 mg Hg/kg FW (recently reduced to 0.3 mg Hg/kg FW) does not have an appreciable level of safety for pregnant women (Skerfving et al., 1970; Peakall and Lovett, 1972; Establier, 1975). In Sweden, the official limit for mercury in fish muscle is 1.0 mg/kg FW. Human risk in consumption of mercury-contaminated fish within this guideline is considered negligible in that country (Berglund and Berlin, 1969); however, some Scandinavian scientists recommend that fish consumption should be limited to one meal per week (Lofroth, 1970). It has been advocated that human use of mercury should be curtailed because the difference between tolerable background levels of mercury and harmful levels in the environment is exceptionally small (Eisler, 1978). Specifically, it has been recommended that all usage of methylmercury compounds should be prohibited and that production of other organomercury compounds be drastically reduced.

In Canadian aboriginal peoples, a 20-year follow-up study on methylmercury levels has been initiated, with emphasis on age, sex, location, relation between maternal and fetal levels, and a reassessment of potential risk in communities where the highest known methylmercury levels have been found (Wheatley and Paradis, 1995). Similar studies are recommended for avian and mammalian wildlife.

At this point, it seems that four courses of action are warranted for protection of human health and sensitive natural resources. First, toxic mercurials in agriculture and industry should be replaced by less toxic substitutes. In Sweden, for example, clinical mercury thermometers have been prohibited since January 1, 1992, for import, manufacture, and sales (Gustafsson, 1995). Since January 1993, the same prohibition was applied to other measuring instruments and electrical components containing mercury. Beginning in the year 2000, Sweden prohibited mercury in all processes and products, including thermometers and sphygmomanometers, and replaced them with available substitutes (Gustafsson, 1995). In Quebec (Canada) hospitals, medical instruments containing mercury are being replaced with mercury-free instruments because of inadequate maintenance and disposal of existing instruments (Guerrier et al., 1995). Second, controls should be applied at the point of origin to prevent the discharge of potentially harmful mercury wastes. Point sources need to be identified and regulated (Facemire et al., 1995). In Sweden, discharges from point sources in the 1950s and 1960s averaged 20 to 30 metric tons annually. Since the end of the 1960s, the annual emission of mercury in Sweden has been reduced to about 3.5 tons through better emission control legislation, improved technology, and reduction of polluting industrial production (Lindqvist et al., 1991). Third, continued periodic monitoring of fishery and wildlife resources is important, especially in areas with potential for reservoir development, in light of the hypothesis that increased flooding increases the availability of mercury to biota. The use of museum collections for mercury analysis is strongly recommended for monitoring purposes. For example, the Environmental Specimen Bank at the Swedish Museum of Natural History constitutes a base for ecotoxicological research and for spatial and trend monitoring of mercury and other contaminants in Swedish fauna (Odsjo et al., 1997). And finally, additional research is needed on mercury accumulation and detoxification in comparatively pristine ecosystems. Key uncertainties in understanding the process of mercury uptake in aquatic ecosystems, for example, include relations between water chemistry and respiratory uptake, quantitative estimates of intestinal tract methylation and depuration, and degree of seasonal variability in mercury speciation and methylation–demethylation processes (Post et al., 1996).

In the specialized case of environmental mercury contamination from historic and current gold mining activities, I recommend more research on physical and biological mercury removal technologies, development of nonmercury technologies to extract gold with minimal environmental damage, measurement of loss rates of mercury through continued periodic monitoring of fishery and wildlife resources in mercury-contaminated areas, and mercury accumulation and detoxification rates in comparatively pristine ecosystems. In view of the demonstrable adverse effects of uncontrolled mercury releases into the biosphere from gold production, it is imperative that all use of liquid mercury in gold amalgamation should cease at the earliest opportunity and that the ban be made permanent.

12.6 SUMMARY

To protect sensitive agricultural crops, mercury concentrations should not exceed 0.2 μg/L in irrigation water, 0.2 to < 0.5 mg/kg dry weight (DW) in soils, and 1.5 mg/kg DW in sludges applied to croplands.

For the protection of aquatic life, different mercury criteria have been recommended. However, many regulatory authorities argue that satisfactory protection is afforded with total mercury concentrations less than 0.012 μg/L in freshwater, < 0.025 μg/L in marine environments, < 0.15 mg/kg DW in marine sediments, and mercury tissue concentrations less than 70.0 μg/kg fresh weight (FW) in eggs, < 480.0 μg/kg FW in gonads, and < 1.0 mg/kg FW in whole organisms. Mercury concentrations considered acceptable to protect representative species of birds (in mg total mercury/kg FW) include less than 0.5–< 1.0 in eggs, < 5.0 in feather, < 5.0 in liver, and < 20.0 in kidneys. Diets of sensitive fish-eating birds should contain less than 20.0 μg Hg as methylmercury/kg FW or < 100.0 μg total mercury/kg FW; daily intake should not exceed 640.0 μg total mercury/kg body weight. For most species of non-human mammals, recommended mercury levels include daily intake of < 250.0 μg total mercury/kg FW body weight, diets that contain < 1.1 mg total mercury/kg FW, and < 0.002 μg total mercury in drinking water supply of livestock; however, diets of fish-eating mammals should contain < 100.0 μg total mercury/kg FW. Tissue mercury concentrations in sensitive mammals (in mg total mercury/kg FW) should probably not exceed 10.0 in liver, 2.0 in hair, 1.5 in brain, and 0.5 in blood. All of these proposed criteria provide, at best, minimal protection.

Proposed mercury criteria for human health protection are numerous and disparate. For example, mercury criteria for air in the workplace range from < 10.0 μg/m^3 for organomercurials to < 50.0 μg/m^3 for metallic mercury vapor; however, much lower criteria are proposed by Texas (< 0.05 μg/m^3 for 1 year), New York (0.0167 μg/m^3 per year), and others. Mercury drinking water criteria range between < 1.0 and < 2.0 μg total mercury/L, except for Brazil with < 0.2 μg/L. Proposed dietary criteria for mercury range between 10.0 and 1500.0 μg mercury/kg FW ration (exception: < 250.0 μg/kg FW diet for pregnant women), with lower values associated with seafoods and organomercurials. Permissible tolerable weekly mercury intakes proposed for the general population range between < 3.3 and 5.0 μg/kg body weight; however, daily intake for pregnant women should be restricted to < 0.6 to 1.1 μg total mercury/kg body weight. Proposed tolerable mercury concentrations include < 0.2 mg/kg FW in blood, < 0.7 mg/kg whole body, and < 6.0 to 50.0 mg/kg in hair; however, adverse neurobehavioral effects have been associated with cord blood concentrations > 5.8 μg methylmercury/L and > 1.0 mg methylmercury/kg DW hair.

REFERENCES

Abbasi, S.A. and R. Soni. 1983. Stress-induced enhancement of reproduction in earthworms *Octochaetus pattoni* exposed to chromium (VI) and mercury (II) — implications in environmental management, *Int. J. Environ. Stud.*, 22, 43–47.

Adams, D.H. and G.V. Onorato. 2005. Mercury concentrations in red drum, *Sciaenops ocellatus*, from estuarine and offshore waters of Florida, *Mar. Pollut. Bull.*, 50, 291–300.

Anderson, B.A., D. Renholt–Bindslev, and I.R. Cooper. 1998. Dental Amalgam — A Report with Reference to the Medical Devices Directive 93/42/EEC from an Ad Hoc Working Group Mandated by DGIII of the European Commission. Angelholm, Sweden: Nordiska Dental AB <http://www.nordiskadental.se/EUamalgam/eumain.htm>.

Andreji, J., I. Stranai, P. Massanyi, and M. Valent. 2005. Concentrations of selected metals in muscle of various fish species, *J. Environ. Sci. Health*, 40, 899–912.

Alaska Department of Environmental Conservation. 1994. Drinking Water Regulations. *State of Alaska DEC,* Rept. 18-ACC-80, 1–195.

Angelow, R.V. and D.M. Nicholls. 1991. The effect of mercury exposure on liver mRNA translatability and metallothionein in rainbow trout, *Comp. Biochem. Physiol.*, 100C, 439–444.

Atkeson, T., D.Axelrad, C. Pollman, and G. Keeler. 2003. Integrating Atmospheric Mercury Deposition and Aquatic Cycling in the Florida Everglades: An Approach for Conducting a Total Maximum Daily Load Analysis for an Atmospherically Derived Pollutant. Integrated Summary. Final Report. 272 pp. Available from Mercury and Applied Science MS6540, Florida Dept. Environmental Conservation, 2600 Blair Stone Road, Tallahassee, FL 32399–2400. See also <http://www.floridadep.org/labs/mercury/index.htm>

Aulerich, R.J., R.K. Ringer, and J. Iwamota. 1974. Effects of dietary mercury on mink, *Arch. Environ. Contam. Toxicol.,* 2, 43–51.

Barber, R.T., P.J. Whaling, and D.M. Cohen. 1984 — Mercury in recent and century-old deep-sea fish, *Environ. Sci. Technol.*, 18, 552–555.

Barghigiani, C. and S. De Ranieri. 1992. Mercury content in different size classes of important edible species of the northern Tyrrhenian Sea, *Mar. Pollut. Bull.*, 24, 114–116.

Berglund, F. and M. Berlin. 1969. Human risk evaluation for various populations in Sweden due to methylmercury in fish. In M.W. Miller and G.G. Berg (Eds.), *Chemical Fallout, Current Research on Persistent Pesticides,* p. 423–432. Chas. C Thomas, Springfield, IL.

Best, J.B., M. Morita, J. Ragin, and J. Best, Jr. 1981. Acute toxic responses of the freshwater planarian, *Dugesia dorotocephala*, to methylmercury, *Bull. Environ. Contam. Toxicol.*, 27, 49–54.

Beyer, W.N. 1990. Evaluating soil contamination, *U.S. Fish Wildl. Serv. Biol. Rep.*, 90(2), 1–25.

Beyer, W.N., M. Spalding, and D. Morrison. 1997. Mercury concentrations in feathers of wading birds from Florida, *Ambio*, 26, 97–100.

Birge, W.J., A.G. Westerman, and J.A. Spromberg. 2000. Comparative toxicology and risk assessment of amphibians. In D.W. Sparling, G. Linder, and C.A. Bishop (Eds.), *Ecotoxicology of Amphibians and Reptiles*, p. 727–791. SETAC Press, Pensacola, FL.

Birke, G., A.G. Johnels, L.O. Plantin, B. Sjostrand, S. Skerfring, and T. Westermark. 1972. Studies on humans exposed to methylmercury through fish consumption, *Arch. Environ. Health*, 25, 77–91.

Bodaly, R.A., R.E. Hecky, and R.J.P. Fudge. 1984. Increases in fish mercury levels in lakes flooded by the Churchill River diversion, northern Manitoba, *Can. J. Fish. Aquat. Sci.*, 41, 682–691.

Brumbaugh, W.G., D.P. Krabbenhoft, D.R. Helsel, J.G. Wiener, and K.R. Echols. 2001. A National Pilot Study of Mercury Contamination of Aquatic Ecosystems along Multiple Gradients: Bioaccumulation in Fish. *U.S. Geol. Surv., Biol. Sci. Rep. USGS/BRD/BSR 2001-0009,* 25 pp.

Buzina, R., K. Suboticanec, J. Vukusic, J. Sapunar, and M. Zorica. 1989. Effect of industrial pollution on seafood content and dietary intake of total and methylmercury, *Sci. Total Environ.*, 78, 45–57.

Chvojka, R. 1988. Mercury and selenium in axial white muscle of yellowtail kingfish from Sydney, Australia, *Mar. Pollut. Bull.*, 19, 210–213.

Clarkson, T.W. 1990. Human health risks from methylmercury in fish, *Environ. Toxicol. Chem.*, 9, 957–961.

Counter, S.A., L.H. Buchanan, and F. Ortega. 2005. Mercury levels in urine and hair of children in an Andean gold-mining settlement, *Int. J. Occup. Environ. Health*, 11, 132–137.

Danserau, M., N. Lariviere, D.d. Tremblay, and D. Belanger. 1999. Reproductive performance of two generations of female semidomesticated mink fed diets containing organic mercury contaminated freshwater fish, *Arch. Environ. Contam. Toxicol.*, 36, 221–226.

Dave, G. and R. Xiu. 1991. Toxicity of mercury, copper, nickel, lead, and cobalt to embryos and larvae of zebrafish, *Brachydanio rerio*, *Arch. Environ. Contam. Toxicol.*, 21, 126–134.

Eisler, R. 1978. Mercury contamination standards for marine environments. In J.H. Thorp and J.W. Gibbons (Eds.), *Energy and Environmental Stress in Aquatic Systems,* p. 241–272. U.S. Dept. Energy Symp. Ser. 48, CONF–771114. Available from Natl. Tech. Infor. Serv., U.S. Dept. Commerce, Springfield, VA 22161.

Eisler, R. 1981. *Trace Metal Concentrations In Marine Organisms.* Pergamon, Elmsford, New York. 687 pp.

Eisler, R. 2000. Mercury. In *Handbook of Chemical Risk Assessment: Health Hazards to Humans, Plants, and Animals,* Vol. 1, Metals, p. 313–409. Lewis Publishers, Boca Raton, FL.

Establier, R. 1975. Contenido en mercurio de las anguillas (*Anguilla anguilla*) de la desembocadura del rio Guadalquiver y esteros de les salines de la zona de Cadiz, *Invest. Pesquera*, 39(1), 249–255.

Facemire, C., T. Augspurger, D. Bateman, M. Brim, P. Conzelmann, S. Delchamps, E. Douglas, L. Inmon, K. Looney, F. Lopez, G. Masson, D. Morrison, N. Morse, and A. Robison. 1995. Impacts of mercury contamination in the southeastern United States, *Water Air Soil Pollut.*, 80, 923–926.

Fields, S. 2001. Tarnishing the earth: gold mining's dirty secret. *Environ. Health Perspect.*, 109, A474–A482.

Fimreite, N. 1979. Accumulation and effects of mercury on birds. In J.O. Nriagu (Ed.), *The Biogeochemistry of Mercury in the Environment*, p. 601–627. Elsevier/North–Holland Biomedical Press, New York.

Fleming, L.E., S. Watkins, R. Kaderman, B. Levin, D.R. Ayyar, M. Bizzio, D. Stephens, and J.A. Bean. 1995. Mercury exposure to humans through food consumption from the Everglades of Florida, *Water Air Soil Pollut.*, 80, 41–48.

Fonseca, F.R.D., O. Malm, and H.F. Waldemarin. 2005. Mercury levels in tissues of giant otters (*Pteronura brasiliensis*) from the Rio Negro, Pantanal, Brazil, *Environ. Res.*, 98, 368–371.

Futter, M.N. 1994. Pelagic food-web structure influences probability of mercury contamination in lake trout (*Salvelinus namaycush*), *Sci. Total Environ.*, 145, 7–12.

Galster, W. A. 1976. Mercury in Alaskan Eskimo mothers and infants, *Environ. Health Perspec.*, 15, 135–140.

Gariboldi, J.C., C.H. Jagoe, and A.L. Bryan Jr. 1998. Dietary exposure to mercury in nestling wood storks (*Mycteria americana*) in Georgia, *Arch. Environ. Contam. Toxicol.*, 34, 398–405.

Gaudet, C., S. Lingard, P. Cureton, K. Keenleyside, S. Smith, and G. Raju. 1995. Canadian environmental guidelines for mercury, *Water Air Soil Pollut.*, 80, 1149–1159.

Gemici, U. 2004. Impact of acid mine drainage from the abandoned Halikoy mercury mine (western Turkey) on surface and groundwater, *Bull. Environ. Contam. Toxicol.*, 72, 482–489.

Gillis, C.A., N.L. Bonnevie, and R.J. Wenning. 1993. Mercury contamination in the Newark Bay estuary, *Ecotoxicol. Environ. Safety*, 25, 214–226.

Glorennec, P., D. Zmirou, and D. Bard. 2005. Public health benefits of compliance with current E.U. emissions standards for municipal waste incinerators: a health risk assessment with the CalTox multimedia exposure model, *Environ. Int.*, 31, 693–701.

Grandjean, P., P. Weihe, R.F. White, F. Debes, S. Araki, and K. Yokoyama. 1997. Cognitive deficit in 7-year-old children with prenatal exposure to methylmercury, *Neurotoxicol. Teratol.*, 19, 417–428.

Gray, J.E. (Ed.), 2003. *Geologic Studies of Mercury by the U.S. Geological Survey.* USGS Circular 1248. Available from USGS, Box 25286, Denver, CO 80225.

Guerrier, P., J.P. Weber, R. Cote, M. Paul, and M. Rhainds. 1995. The accelerated reduction and elimination of toxics in Canada: the case of mercury-containing medical instruments in Quebec hospital centres, *Water Air Soil Pollut.*, 80, 1199–1202.

Gustafsson, E. 1995. Swedish experiences of the ban on products containing mercury, *Water Air Soil Pollut.*, 80, 99–102.

Halbrook, R.S., J.H. Jenkins, P.B. Bush, and N.D. Seabolt. 1994. Sublethal concentrations of mercury in river otters: monitoring environmental contamination, *Arch. Environ. Contamin. Toxicol.*, 27, 306–310.

Harada, M. 1994. Preliminary field study in Tapajos River Basin, Amazon. Pages 33–40 in *Proceedings of the International Symposium on Assessment of Environmental Pollution and Health Effects from Methylmercury*, p. 33–40. October 8–9, 1993, Kumamoto, Japan. National Institute for Minamata Disease, Minamata City, Kumamoto 867, Japan.

Heinz, G.H. 1979. Methylmercury: reproductive and behavioral effects on three generations of mallard ducks, *J. Wildl. Manage.*, 43, 394–401.

Heinz, G.H. 1996. Mercury poisoning in wildlife. In A. Fairbrother, L.N. Locke, and G.L. Hoff (Eds.), *Noninfectious Diseases of Wildlife, 2nd edition,* p. 118–127. Iowa State University Press, Ames.

Heinz, G.H. and D.J. Hoffman. 2003. Embryotoxic thresholds of mercury: estimates from individual mallard eggs, *Arch. Environ. Toxicol. Chem.*, 44, 257–264.

Hujoel, P.P., M. Lydon–Rochelle, A.M. Bollen, J.S. Woods, W. Guertsen, and M.A. del Aguila. 2005. Mercury exposure from dental filling placement during pregnancy and low birth weight risk, *Am. J. Epidemiol.*, 161, 734–740.

Hylander, L.D., E.C. Silva, L.J. Oliveira, S.A. Silva, E.K. Kuntze, and D.X. Silva. 1994. Mercury levels in Alto Pantantal: a screening study, *Ambio*, 23, 478–484.

Kannan, K., R.G. Smith Jr., R.F. Lee, H.L. Windom, P.T. Heitmuller, J.M. Macauley, and J.K. Summers. 1998. Distribution of total mercury and methyl mercury in water, sediment, and fish from south Florida estuaries, *Arch. Environ. Contam. Toxicol.*, 34, 109–118.

Kazantzis, G. 1971. The poison chain for mercury in the environment, *Int. J. Environ. Stud.*, 1, 301–306.

Keckes, S. and J.J. Miettinen. 1972. Mercury as a marine pollutant. In M. Ruivo (Ed.), *Marine Pollution and Sea Life*, p. 276–289. Fishing Trading News (books) Ltd., London.

Khera, K.S. 1979. Teratogenic and genetic effects of mercury toxicity. In J.O. Nriagu (Ed.), *The Biogeochemistry of Mercury in the Environment*, p. 501–518. Elsevier/North–Holland Biomedical Press, New York.

Kjellstrom, T., P. Kennedy, S. Wallis, and C. Mantell. 1986. Physical and Mental Development of Children With Prenatal Exposure to Mercury from Fish. Stage I: Preliminary Tests at Age 4. *Natl. Swedish Environ. Protect. Bd.,* Rept. 3080, Solna, Sweden.

Kjellstrom, T, P. Kennedy, S. Wallis, A. Stewart, L. Friberg, and B. Lind. 1989. Physical and Mental Development of Children with Prenatal Exposure to Mercury from Fish. Stage II: Interviews and Psychological Tests at age 6, *Natl. Swedish Environ. Protect. Bd.,* Rept. 3642, Solna, Sweden.

Krom, M.D., H. Hornung, and Y. Cohen. 1990. Determination of the environmental capacity of Haifa Bay with respect to the input of mercury, *Mar. Pollut. Bull.*, 21, 349–354.

Krynski, A., J. Kaluzinski, M. Wlazelko, and A. Adamowski. 1982. Contamination of roe deer by mercury compounds, *Acta Theriol.*, 27, 499–507.

Kucera, E. 1983. Mink and otter as indicators of mercury in Manitoba waters, *Can. J. Zool.*, 61, 2250–2256.

Lathrop, R.C., P.W. Rasmussen, and D.R. Knauer. 1991. Mercury concentrations in walleyes from Wisconsin (USA) lakes, *Water Air Soil Pollut.*, 56, 295–307.

Lindqvist, O., K. Johansson, M. Aastrup, A. Andersson, L. Bringmark, G. Hovsenius, L. Hakanson, A. Iverfeldt, M. Meili, and B. Timm. 1991. Mercury in the Swedish environment — recent research on causes, consequences and corrective methods. *Water Air Soil Pollut.*, 55, 1–261.

Lindsay, R.C. and R.W. Dimmick. 1983. Mercury residues in wood ducks and wood duck foods in eastern Tennessee, *J. Wildl. Dis.*, 19, 114–117.

Littrell, E.E. 1991. Mercury in western grebes at Lake Berryessa and Clear Lake, California, *Califor. Fish Game*, 77, 142–144.

Lodenius, M., A. Seppanen, and M. Herrnanen. 1983. Accumulation of mercury in fish and man from reservoirs in northern Finland, *Water Air Soil Pollut.*, 19, 237–246.

Lofroth, G. 1970. Methylmercury, a review of health hazards and side effects associated with the emission of mercury compounds into natural systems, *Nat. Sci. Res. Coun., Ecol. Res. Commun., Stockholm, Sweden, Bull.*, 4, 1–56.

Lyle, J.M. 1984. Mercury concentrations in four carcharhinid and three hammerhead sharks from coastal waters of the Northern Territory, *Austral. J. Mar. Freshwater Res.*, 35, 441–451.

Maas, R.P., S.C. Patch, and K.R. Sergent. 2004. *A Statistical Analysis of Factors Associated with Elevated Hair Mercury Levels in the U.S. Population: An Interim Progress Report*. Tech. Rep. 04–136, Environ. Qual. Inst., Univ. North Carolina, Ashville. 13 pp.

March, B.E., R. Poon, and S. Chu. 1983. The dynamics of ingested methyl mercury in growing and laying chickens, *Poult. Sci.*, 62, 1000–1009.

Marsh, D.O., M.D Turner, and J.C. Smith. 1975. Loaves and fishes—some aspects of methylmercury in foodstuffs. Unpublished report submitted to U.S. Food and Drug Administration.

Mason, C.F. and A.B. Madsen. 1992. Mercury in Danish otters (*Lutra lutra*), *Chemosphere*, 25, 865–867.

McEwen, L.C., R.K. Tucker, J.O. Ells, and M.A. Haegele. 1973. Mercury-wildlife studies by the Denver Wildlife Research Center. In D.R. Buhler (Ed.), *Mercury in the Western Environment,* p. 146–156. Oregon State Univ., Corvallis.

Miles, C.J. and L.E. Fink. 1998. Monitoring and mass budget for mercury in the Everglades Nutrient Removal Project, *Arch. Environ. Contam. Toxicol.*, 35, 549–557.

Mora, M.A. 1996. Organochlorines and trace elements in four colonial waterbird species nesting in the lower Laguna Madre, Texas, *Arch. Environ. Contam. Toxicol.*, 31, 533–537.

Mullins, W.H., E.G. Bizeau, and W.W. Benson. 1977. Effects of phenyl mercury on captive game farm pheasants, *J. Wildl. Manage.*, 41, 302–308.

Nakanishi, H., M. Ukita, M. Sekine, and S. Murakami. 1989. Mercury pollution in Tokuyama Bay, *Hydrobiologia*, 176/177, 197–211.

National Research Council (NRC). 2000. *Toxicological Effects of Methylmercury.* National Academy Press, Washington, D.C.

Niimi, A.J. and G.P. Kissoon. 1994. Evaluation of the critical body burden concept based on inorganic and organic mercury toxicity to rainbow trout (*Oncorhynchus mykiss*), *Arch. Environ. Contam. Toxicol.*, 26, 169–178.

Odsjo, T., A. Bignert, M. Olsson, L. Asplund, U. Eriksson, L. Haggberg, K. Litzen, C. de Wit, C. Rappe, and K. Aslund. 1997. The Swedish environmental specimen bank — application in trend monitoring of mercury and some organohalogenated compounds, *Chemosphere*, 34, 2059–2066.

Oikawa, K., H. Saito, I. Kifune, T. Ohshina, M. Fujii, and Y. Takizawa. 1982. Respiratory tract retention of inhaled air pollutants. Report 1: mercury absorption by inhaling through the nose and expiring through the mouth at various concentrations, *Chemosphere*, 11, 949–951.

Ozuah, P.O., M.S. Lesser, J.S. Woods, H. Choi, and M. Markowitz. 2003. Mercury exposure in an urban pediatric population, *Ambul. Pediatr.*, 3, 24–26.

Paine, P.J. 1994. Compliance with Chlor-alkali Mercury Regulations, 1986–1989: Status Report. *Report EPS 1/HA/2, 1–43*. Available from Environmental Protection Publications, Environment Canada, Ottawa, Ontario K1A OH3.

Palheta, D. and A. Taylor. 1995. Mercury in environmental and biological samples from a gold mining area in the Amazon region of Brazil, *Sci. Total Environ.*, 168, 63–69.

Paulsson, K. and K. Lundbergh. 1989. The selenium method for treatment of lakes for elevated levels of mercury in fish, *Sci. Total Environ.*, 87/88, 495–507.

Peakall, D.B. and R.J. Lovett. 1972. Mercury: its occurrence and effects in the ecosystem, *Bioscience*, 22(1), 20–25.

Peltola, P. and M. Astrom. 2003. Urban geochemistry: a multimedia and multielement survey of a small farm in northern Europe, *Environ. Geochem. Health*, 25, 397–419.

Petruccioli, L. and P. Turillazzi. 1991. Effect of methylmercury on acetylcholinesterase and serum cholinesterase activity in monkeys, *Macaca fascicularis*, *Bull. Environ. Contam. Toxicol.*, 46, 769–773.

Post, J.R., R. Vandenbos, and D.J. McQueen. 1996. Uptake rates of food-chain and waterborne mercury by fish: field measurements, a mechanistic model, and an assessment of uncertainties, *Can. J. Fish. Aquat. Sci.*, 53, 395–407.

Ramprashad, F. and K. Ronald. 1977. A surface preparation study on the effect of methylmercury on the sensory hair cell population in the cochlea of the harp seal (*Pagophilus groenlandicus* Erxleben, 1977), *Can. J. Zool.*, 55, 223–230.

Roelke, M.E., D.P. Schultz, C.F. Facemire, and S.F. Sundlof. 1991. Mercury contamination in the free-ranging endangered Florida panther (*Felis concolor coryi*), *Proc. Am. Assoc. Zoo Veterin.*, 1991, 277–283.

Saha, J.G. 1972. Significance of mercury in the environment: suggestions for further research. In *Radiotracer Studies of Chemical Residues in Food and Agriculture,* p. 81–86. Int. Atom. Ener. Agen., Vienna.

Satoh, H. 1995. Toxicological properties and metabolism of mercury; with an emphasis on a possible method for estimating residual amounts of mercury in the body. In *Proceedings of the International Workshop on "Environmental Mercury Pollution and its Health Effects in Amazon River Basin,"* p. 106–112. Rio de Janeiro, 30 November–2 December 1994. Published by National Institute for Minamata Disease, Minamata City, Kumamoto 867, Japan.

Scheuhammer, A.M. 1988. Chronic dietary toxicity of methylmercury in the zebra finch, *Poephila guttata*, *Bull. Environ. Contam. Toxicol.*, 40, 123–130.

Scheuhammer, A.M., C.M. Atchison, A.H.K. Wong, and D.C. Evers. 1998. Mercury exposure in breeding common loons (*Gavia immer*) in central Ontario, Canada, *Environ. Toxicol. Chem.*, 17, 191–196.

Schlenk, D., Y.S. Zhang, and J. Nix. 1995. Expression of hepatic metallothionein messenger RNA in feral and caged fish species correlates with muscle mercury levels, *Ecotoxicol. Environ. Safety*, 31, 282–286.

Sheffy, T.B. and J.R. St. Amant. 1982. Mercury burdens in furbearers in Wisconsin, *J. Wildl. Manage.*, 46, 1117–1120.

Siegel, B.Z., S.M. Siegel, T. Correa, C. Dagan, G. Galvez, L. Leeloy, A. Padua, and E. Yaeger. 1991. The protection of invertebrates, fish, and vascular plants against inorganic mercury poisoning by sulfur and selenium derivatives, *Arch. Environ. Contam. Toxicol.*, 20, 241–246.

Silver, S., G. Endo, and K. Nakamura. 1994. Mercury in the environment and the laboratory, *J. Japan. Soc. Water Environ.*, 17, 235–243.

Skerfving, S., K. Hansson, and J. Lindstem. 1970. Chromosome breakage in humans exposed to methyl mercury through fish consumption, *Arch. Environ. Health*, 21, 133–139.

Soong, T. 1994. Epidemiological research on the health effect of residents along the Sonhua River polluted by methylmercury. In *Proceedings of the International Symposium on "Assessment of Environmental Pollution and Health Effects from Methylmercury,"* p. 165–169. October 8–9, 1993, Kumamoto, Japan. National Institute for Minamata Disease, Minamata City, Kumamoto 867, Japan.

Spann, J.W., R.G. Heath, J.F. Kreitzer, and L.N. Locke. 1972. Ethyl mercury *p*–toluene sulfonanilide: lethal and reproductive effects on pheasants, *Science*, 175, 328–331.

Suzuki, T. 1979. Dose–effect and dose–response relationships of mercury and its derivatives. In J.O. Nriagu (Ed.), *The Biogeochemistry of Mercury in the Environment,* p. 399–431. Elsevier/North–Holland Biomedical Press, New York.

Takizawa, Y. 1979. Minamata disease and evaluation of medical risk from methylmercury, *Akita J. Med.*, 5, 183–213.

Takizawa, Y. 1979a. Epidemiology of mercury poisoning. In J.O. Nriagu (Ed.), *The Biogeochemistry of Mercury in the Environment,* p. 325–365. Elsevier/North Holland Biomedical Press, Amsterdam.

Tarras–Wahlberg, N.H., A. Flachier, G. Fredriksson, S. Lane, B. Lundberg, and O. Sangfors. 2000. Environmental impact of small scale and artisanal gold mining in southern Ecuador, *Ambio*, 29, 484–491.

Thompson, D.R. 1996. Mercury in birds and terrestrial mammals. In W.N. Beyer, G.H. Heinz, and A.W. Redmon–Norwood (Eds.), *Environmental Contaminants in Wildlife: Interpreting Tissue Concentrations*, p. 341–356. CRC Press, Boca Raton, FL.

Trasande, L., P.J. Landrigan, and C. Schecter. 2005. Public health and economic consequences of methyl mercury toxicity to the developing brain, *Environ. Health Perspect.*, 113, 590–596.

Tsuji, J.S., P.R.D. Williams, and M.R. Edwards. 2003. Evaluation of mercury in urine as an indicator of exposure to low levels of mercury vapor, *Environ. Health Perspect.*, 11, 623–630.

U.S. Department of Health, Education, and Welfare (USHEW). 1977. *Occupational Diseases: a Guide to Their Recognition*. USHEW Publ. 77–1811, Washington, D.C.

U.S. Environmental Protection Agency (USEPA). 1980. Ambient Water Quality Criteria for Mercury. *U.S. Environ. Protection Agen. Rep. 440/5–80–058*. Available from Natl. Tech. Infor. Serv., 5285 Port Royal Road, Springfield, VA 22161.

U.S. Environmental Protection Agency (USEPA). 1985. *Ambient Water Quality Criteria for Mercury — 1984.* U.S. Environ. Protection Agen. Rep. 440/5–84–026. 136 pp. Available from Natl. Tech. Infor. Serv., 5285 Port Royal Road, Springfield, VA 22161.

U.S. Environmental Protection Agency (USEPA). 1992. Water quality standards; establishment of numeric criteria for priority toxic pollutants; states' compliance; final rule, *Fed. Reg. 40 CFR Part 131*, 57 (246), 60847–60916.

U.S. Environmental Protection Agency (USEPA). 2001. Water Quality Criterion for the Protection of Human Health — Methylmercury. *U.S. Environmental Protection Agency Report EPA–823–R–01–001*.

U.S. National Academy of Sciences (USNAS). 1978. *An Assessment of Mercury in the Environment*. National Academy Of Science, Washington, D.C., 185 pp.

U.S. Public Health Service (USPHS). 1994. *Toxicological Profile for Mercury* (Update). TP–93/10. U.S. PHS, Agen. Toxic Substances Dis. Registry, Atlanta, GA. 366 pp.

Verta, M. 1990. Changes in fish mercury concentrations in an intensively fished lake, *Can. J. Fish. Aquat. Sci.*, 47, 1888–1897.

Weil, M., J. Bressler, P. Parsons, K. Bolla, T. Glass, and B. Schwartz. 2005. Blood mercury levels and neurobehavioral function, *J. Am. Med. Assoc.*, 293, 1875–1882.

Wheatley, B. and S. Paradis. 1995. Exposure of Canadian aboriginal peoples to methylmercury, *Water Air Soil Pollut.*, 80, 3–11.

Wiemeyer, S.N., T.G. Lamont, C.M. Bunck, C.R. Sindelar, F.J. Gramlich, J.D. Fraser, and M.A. Byrd. 1984. Organochlorine pesticide, polychlorobiphenyl, and mercury residues in bald eagle eggs — 1969–79 — and their relationships to shell thinning and reproduction, *Arch. Environ. Contam. Toxicol.*, 13, 529–549.

Wiener, J.G., and D.J. Spry. 1996. Toxicological significance of mercury in freshwater fish. In W.N. Beyer, G.H. Heinz, and A.W. Redmon–Norwood (Eds.), *Environmental Contaminants in Wildlife: Interpreting Tissue Concentrations,* p. 297–339. CRC Press, Boca Raton, FL.

Wolfe, M. and S. Norman. 1998. Effects of waterborne mercury on terrestrial wildlife at Clear Lake: evaluation and testing of a predictive model, *Environ. Toxicol. Chem.*, 17, 214–217.

, Wood, P.B., J.H. White, A. Steffer, J.M. Wood, C.F. Facemire, and H.F. Percival. 1996. Mercury concentrations in tissues of Florida bald eagles, *J. Wildl. Manage.*, 60, 178–185.

World Health Organization (WHO). 1976. *Mercury. Environmental Health Criteria 1.* WHO, Geneva, 131 pp.

World Health Organization (WHO). 1990. *Methylmercury. Environmental Health Criteria 101.* WHO, Geneva, 144 pp.

World Health Organization (WHO). 1991. *Inorganic Mercury. Environmental Health Criteria 118.* WHO, Geneva, 168 pp.

Wren, C.D. 1986. A review of metal accumulation and toxicity in wild mammals, *Environ. Res.*, 40, 210–244.

Wren, C.D., D.B. Hunter, J.F. Leatherland, and P.M. Stokes. 1987a. The effects of polychlorinated biphenyls and methylmercury, singly and in combination, on mink. I. Uptake and toxic responses, *Arch. Environ. Contam. Toxicol.*, 16, 441–447.

Wren, C.D., D.B. Hunter, J.F. Leatherland, and P.M. Stokes. 1987b. The effects of polychlorinated biphenyls and methylmercury, singly and in combination on mink. II. Reproduction and kit development. *Arch. Environ. Contam. Toxicol.*, 16, 449–454.

Yeardley, R.B., J.M. Lazorchak, and S.G. Paulsen. 1998. Elemental fish tissue contamination in northeastern U.S. lakes: evaluation of an approach to regional assessment, *Environ. Toxicol. Chem.*, 17, 1875–1884.

Zarcinas, B.A., C.F. Ishak, M.J. McLaughlin, and G. Cozens. 2004a. Heavy metals in soils and crops in southeast Asia. 1. Peninsular Malaysia, *Environ. Geochem. Health*, 26, 343–357.

Zarcinas, B.A., P. Pongsakul, M.J. McLaughlin, and G. Cozens. 2004b. Heavy metals in soils and crops in southeast Asia. 2. Thailand, *Environ. Geochem. Health*, 26, 359–371.

Zilincar, V.J., B. Bystrica, P. Zvada, D. Kubin, and P. Hell. 1992. Die Schwermeallbelastung bei den Braunbaren in den Westkarpaten, *Z. Jagdwiss.*, 38, 235–243.

Zillioux, E.J., D.B. Porcella, and J.M. Benoit. 1993. Mercury cycling and effects in freshwater wetland ecosystems, *Environ. Toxicol. Chem.*, 12, 2245–2264.

Zumbroich, T. 1997. Heavy metal pollution of vegetables in a former ore-mining region. In P.N. Cheremisinoff (Ed.), *Ecological issues and Environmental Impact Assessment,* p. 207–215. Gulf Publishing Co, Houston, TX.

CHAPTER 13

Concluding Remarks

Mercury has been used by humans for at least 2300 years, most recently as a fungicide in agriculture, in the manufacture of chlorine and sodium hydroxide, as a slime control agent in the pulp and paper industry, in the production of plastics and electrical apparatus, and in mining and smelting operations. Mercury burdens in some environmental compartments are estimated to have increased up to 5 times precultural levels, primarily as a result of human activities. The construction of artificial reservoirs, for example, which release mercury from flooded soils, has contributed to the observed elevation of mercury concentrations in fish tissues from these localities. Elevated levels of mercury in living organisms in mercury-contaminated areas may persist for as long as 100 years after the source of pollution has been discontinued. One major consequence of increased mercury use, coupled with careless waste disposal practices, has been a sharp increase in the number of epidemics of fatal mercury poisonings in humans, wildlife, and aquatic organisms. Most authorities agree on six points:

1. Mercury and its compounds have no known biological function, and the presence of the metal in the cells of living organisms is undesirable and potentially hazardous.
2. Forms of mercury with relatively low toxicity can be transformed into forms of very high toxicity, such as methylmercury, through biological and other processes.
3. Mercury can be bioconcentrated in organisms and biomagnified through food chains.
4. Mercury is a mutagen, teratogen, and carcinogen, and causes embryocidal, cytochemical, and histopathological effects.
5. Some species of fish and wildlife contain high concentrations of mercury that are not attributable to human activities.
6. Anthropogenic use of mercury should be curtailed, as the difference between tolerable natural background levels of mercury and harmful effects in the environment is exceptionally small.

Most authorities also agree on another six points regarding anthropogenic use:

1. The sale, manufacture, or use of methylmercury compounds should be prohibited through legislation and enforcement.
2. The use of all other mercurials should be sharply curtailed.
3. Alternate technologies should be developed to replace mercury in agriculture, industry, and elsewhere.
4. Mercury should be removed from contaminated waterways and other ecosystems impacted through human activities.
5. New monitoring guidelines should be developed — and existing guidelines refined — to evaluate the success of mercury mitigation programs.
6. Mercury emissions from coal-fired power plants should be reduced using best-available technologies.

Additional research is needed to more fully assess the potential hazards of mercury and its compounds to living organisms. Some of the more pressing needs are listed in alphabetical order.

AMALGAMATION CONTAMINANT PROBLEMS

The use of mercury to recover gold has resulted in extensive and persistent contamination of the biosphere, with direct — and often fatal — consequences to all members of the immediate biosphere, including humans. The use of mercury for this purpose, which accounts for about 10.0% of all anthropogenic mercury emissions, should be abandoned, and improved remediation methodologies developed for mercury-contaminated environments using physical, chemical, and biological technologies.

ANTHROPOGENIC EMISSIONS

Combustion of mercury-containing fossil fuels accounts for up to 60.0% of the global mercury burden contributed by human activities. In the United States, coal-fired power plants are now considered the only significant source that continues to be unregulated, although available technologies may reduce mercury emissions by up to 98.0%. Implementation of these technologies is recommended at installations now producing unhealthy depositions at calculated target sites. Removal of the source of anthropogenic mercury, such as closure of chloralkali plants, usually results in a slow decrease in the mercury content of sediments and biota of the mercury-receiving channel. However, the mercury loss rate depends, in part, on the initial degree of contamination, the chemical form of mercury, the physical and chemical conditions of the ecosystem, and the hydraulic turnover rate. The need to quantitate these variables is necessary for implementation of effective remediation measures.

ATMOSPHERIC TRANSPORT

Atmospheric transport of anthropogenic mercury may contaminate remote ecosystems hundreds of kilometers distant. Additional data are needed to predict deposition rates and deposition sites from distant mercury sources.

CHEMICAL ANALYSIS

There is a need for less expensive and more sensitive instrumentation to accurately measure mercury species in abiotic materials and biological tissues. The current procedure involves sample preparation using graphite furnace and measurement using atomic absorption spectrometry, which can detect methylmercury concentrations greater than 10.0 μg/kg in muscle of marine fishes.

CRITERIA

Mercury criteria for protection of living organisms are numerous, disparate, and — in many cases — unsatisfactory. For example, mercury criteria proposed by the U.S. Environmental Protection Agency for the protection of freshwater aquatic life are 0.012 μg/L medium (4-day average), not to exceed 2.4 μg/L on an hourly average; however, these criteria offer only limited protection to freshwater ecosystems. The saltwater criteria of 0.025 μg Hg/L medium (4-day average), not to exceed 2.1 μg/L hourly, are unsatisfactory for the protection of marine life. It is emphasized that the establishment of satisfactory mercury criteria over large geographic areas frequently involves a number of social, political, and economic concerns that currently override sound scientific data — to the dismay of the environmental and health sciences community. Improved communication between scientists and information transfer specialists, including the press and elected officials, is highly recommended.

DENTAL AMALGAMS

Further evaluations of the safety profiles of dental amalgams and other types of materials used intraorally are needed to establish credible guidelines for the use of mercury-containing dental materials during pregnancy. The use of mercury in preparation of dental amalgams should be reconsidered because it constitutes the main source of mercury exposure in some populations.

EMISSION CONTROLS

Although natural sources of mercury to the environment, such as volcanic eruptions, continue to outweigh anthropogenic sources such as mercury from gold mining, industrialization, arms manufacturing, and fossil fuel power plants, many anthropogenic sources can be controlled or eliminated with available technologies. For example, mercury concentrations in surface soils near coal-fired power plants in the United Kingdom can be reduced up to 40.0% using flue gas sulphurization technology. Although legally banned in most countries, the use of mercury compounds as slimicides in pulp and paper mills and as catalysts in the production of chlorine and sodium hydroxide continues; in both cases, alternatives are available and should be implemented.

ENVIRONMENTAL LEGISLATION

Reduction of mercury concentrations in mercury-contaminated invertebrates as a result of human activities may be possible through environmental legislation and subsequent enforcement. In one case, mercury concentrations in mussels from Bergen Harbor, Norway, were reduced 60.0% in 10 years following passage of environmental legislation restricting mercury discharges into the harbor accompanied by vigorous enforcement. Future environmental legislation should incorporate funds for enforcement as well as research and monitoring studies.

FETOTOXICITY

Intrauterine methylmercury exposure studies comprise one of the most important research areas now under investigation. Specifically, these investigations focus on quantification of prenatal exposure to mercurials from maternal consumption of marine fishes and mammals, and subsequent neurobehavioral risks in resultant children on reaching school age. Animal surrogate models should be established and long-term studies initiated using radiolabeled and stable inorganic and organic mercury compounds in combination with known dietary contaminants.

MECHANISMS OF LETHAL ACTION

Concentrations of total mercury lethal to sensitive, representative, nonhuman species range from 0.1 to 2.0 μg/L of medium for aquatic organisms; from 2200.0 to 31,000.0 μg/kg body weight (acute oral) and 4000.0 to 40,000.0 μg/kg (dietary) for birds; and from 100.0 to 500.0 μg/kg body weight (daily dose) and 1000.0 to 5000.0 μg/kg diet for mammals. Organomercury compounds, especially methylmercury, are always more toxic than inorganic mercury compounds. Numerous biological and abiotic factors modify the toxicity of mercury compounds — sometimes by an order of magnitude or more — but the mechanisms of action are not clear and require elucidation.

MERCURY POISONING

Because neurobehavioral disturbances were observed in Hg^o vapor poisoning cases 20 to 35 years after exposure, and because Hg^o vaporizes readily and mercury vapor is comparatively toxic via inhalation, it is recommended that the use of elemental mercury should be discontinued wherever possible. More rapid development is recommended of mercury-antagonistic drugs — such as thiols — administered in cases of acute inorganic mercury poisoning, and mercury-protectant drugs — including thiamin, selenium-, sulfur-, tellurium-, and other Group VI derivatives — to guard against organomercurial intoxication.

MITIGATION

Four courses of action now seem warranted. First, toxic mercurials in agriculture and industry should be replaced by less toxic substitutes. Second, controls should be applied at the point of origin to prevent the discharge of potentially harmful mercury wastes. Third, continued periodic monitoring of mercury in fish and wildlife is needed for identification of potential problem areas, and for the evaluation of ongoing mercury curtailment programs. And fourth, additional research is merited on mechanisms of mercury accumulation and detoxification in comparatively pristine ecosystems.

MOBILIZATION

Mobilization rates of mercury from major global reservoirs are needed to effectively predict risk, especially rates from oceanic sediments that contain 98.75% of the estimated 334 billion metric tons of mercury now in the environment, and from oceanic waters with an estimated 1.24%.

MONITORING

Continued monitoring of mercury content in abiotic materials, in crops and other vegetation, in aquatic and terrestrial invertebrates, and in human and vertebrate animal tissues is necessary to effectively determine compliance with existing mercury criteria for protection of health of humans and natural resources, establish mercury fluxes in different ecosystems, aid in prediction of future problem areas and species, and as the basis of mercury remedial actions. Monitoring emphasis should be on key geographic areas and species, short sampling periods — to be repeated in the future if results warrant — and little deviation in collection methods, sample preservation, and chemical analytical methodologies. Multidisciplinary research teams are recommended and should include ecologists, toxicologists, statisticians, and chemists, at a minimum.

PRODUCTION

Accurate data on mercury production and consumption are not known with certainty. These data would be useful in the calculation of local mercury emissions and in the prediction of environmental risk.

RESIDUES IN BIRDS

Many species of birds, including endangered species, are at risk owing to the consumption of prey containing inordinately high concentrations of mercury. Recovery of the endangered wood stork in the southeastern United States, for example, is considered jeopardized owing to the high mercury content in the diet of nestling wood storks. Management plans to reduce mercury emissions in the southeastern states should give high priority to the welfare of endangered birds and other endangered species in their jurisdiction. Mercury concentrations in muscle and liver of many species of waterfowl commonly hunted are excessive and considered harmful to human consumers; advisories should be issued until the levels in tissues fall below the limits set by regulatory agencies.

RESIDUES IN FISH AND ELASMOBRANCHS

Older adults of commercial species of sharks, swordfish, tuna, marlin, and other long-lived predators routinely contain more than 2.0 mg Hg/kg FW muscle, almost all in the form of methylmercury, placing human consumers of these products at unreasonable risk. Fishery managers are advised to impose a maximum size regulation for capture and landing of older adults of long-lived piscivores containing elevated mercury loadings, possibly through adoption of selective capture gear, thereby reducing mercury availability to humans; at the same time, this move would improve the reproductive potential of many near-depleted stocks because larger females are considered more efficient producers of eggs. All information gathered by state and federal agencies on mercury concentrations in edible fishery products of commerce should be transmitted regularly to consumers to assist them in decision making about the risks from fish consumption.

RESIDUES IN HUMANS

Recently, the U.S. Environmental Protection Agency stated that concentrations greater than 1.0 mg total Hg/kg DW hair are indicative of mercury exposure and that women of child-bearing age should restrict consumption of fish with elevated mercury content. In view of the fact that many populations exceed this hair mercury concentration, namely 45.2% of New Yorkers and 34.2% of Floridians, it seems prudent to reexamine this conclusion.

Consumption of high dietary mercury intake from seal meat and blubber daily by pregnant aboriginal women is associated with elevated mercury concentrations in maternal and fetal tissues, but with no measurable adverse effect on the fetus or resulting infant. This finding requires verification. Other investigators have stated that accumulation of mercury in tissues is associated with an excess risk of acute myocardial infarction and increased risk of death from coronary heart disease and other circulatory problems. This needs to be verified and, if true, appropriate countermeasures developed.

RESIDUES IN MAMMALS

High natural concentrations of mercury in livers of pinniped mammals, typically greater than 143.0 mg total mercury/kg FW, does not seem injurious to animal health owing, perhaps, to the biological unavailability of the metal. Mercury metabolism in marine mammals should be clarified, with emphasis on potential risk to animal health and to human consumers of pinniped tissues.

RESIDUES IN REPTILES

Mercury contamination of the Florida Everglades — primarily from anthropogenic activities — resulted in high levels of mercury, mostly methylmercury, in tissues of alligators. Alligator muscle, which is sometimes consumed locally, contains more than 6.0 mg Hg/kg FW, a level considered unsafe for human consumption over extended periods, especially to the developing human fetus. This situation must be resolved through public education, passage of appropriate legislation, and strict enforcement of existing laws.

SIGNIFICANCE OF MERCURY RESIDUES

The significance of mercury concentrations in field collections of plants and animals on organism health and on the health of consumers is not known with certainty, despite a relatively large database. The mechanisms of mercury accumulation and retention in the biosphere need clarification. Additional data are required on the chemical form of mercury in the organism and on factors that affect mercury bioavailability. For the protection of sensitive species of mammals and birds that regularly consume fish and other aquatic organisms, total mercury concentrations in these prey items should probably not exceed 0.1 mg/kg fresh weight for birds and 1.1 mg/kg for small mammals. The significance of elevated mercury levels in tissues of fish and wildlife is not fully understood; some species of marine pinnipeds, for example, normally contain high concentrations of mercury in various tissues without apparent adverse effects. Usually, however, concentrations in excess of 1.1 mg/kg fresh weight of tissue (liver, kidney, blood, brain, hair) should be considered as presumptive evidence of an environmental mercury problem, but this requires verification.

SUBLETHAL EFFECTS

Significant adverse sublethal effects were observed among selected aquatic species at water concentrations of 0.03 to 0.1 μg Hg/L. For some birds, adverse effects — predominantly on reproduction — have been associated with total mercury concentrations (in mg/kg fresh weight) of 5.0 in feather, 0.9 in egg, and 0.05 to 0.10 in diet; and with daily intakes of 0.64 mg/kg body weight. Sensitive nonhuman mammals showed significant adverse effects of mercury when daily intakes were 0.25 mg/kg body weight, when dietary levels were 1.1 mg/kg, or when tissue concentrations exceeded 1.1 mg/kg. Future research needs should emphasize sublethal interaction effects of mercury with other environmental contaminants, emphasizing effects on reproduction and metabolism.

TRANSFORMATIONS

Mathematical models are needed that can accurately predict the transformation rates of inorganic and organic mercury species in aquatic, terrestrial, and atmospheric environments; flux rates of different mercury species between compartments; and ultimately, bioavailability and accumulation of individual mercury species — particularly methylmercury species — by living organisms. Additional research is recommended on rates of inorganic mercury-ligand formation in water and runoff and its effects on methylmercury formation in soils, quantification of the sources and transport characteristics of mercury species in terrestrial environments, and identification of the mercury species in terrestrial watershed runoff in dissolved and particulate fractions.

TRANSPORT BY PLANTS

The role of terrestrial vegetation in expediting transport of elemental mercury from the geosphere to the atmosphere is considerable in areas heavily contaminated with mercury from historical gold mining activities; however, mechanisms of action affecting flux rates are poorly defined. The role of macrophytes, such as *Spartina*, in mercury cycling in saltmarsh environments is considerable. The role of higher plants in mercury cycling in other ecosystems should be evaluated.

VACCINE PRESERVATIVE

The use of ethylmercury thiosalicylate in Mexico and other countries to preserve medical vaccines must be evaluated, especially where a fetus may be subjected to this organomercurial from exposure of the expectant mother.

General Index

A

B

C

D

E

F

G

H

I

J

K

L

M

N

O

P

Q

R

S

T

U

V

W

Y

Z

Species Index

ALGAE AND HIGHER PLANTS

AMPHIBIANS

BACTERIA

BIRDS

FISHES AND ELASMOBRANCHS

FUNGI, LICHENS, MOSSES

INVERTEBRATES, AQUATIC

INVERTEBRATES, TERRESTRIAL

MAMMALS

REPTILES

YEASTS